Textile Engineering: Design and Manufacturing

Textile Engineering: Design and Manufacturing

Edited by
Dick Baia

C WILLFORD PRESS

www.willfordpress.com

Published by Willford Press,
118-35 Queens Blvd., Suite 400,
Forest Hills, NY 11375, USA

ISBN: 978-1-68285-588-1

Cataloging-in-Publication Data

Textile engineering : design and manufacturing / edited by Dick Baia.
    p. cm.
Includes bibliographical references and index.
ISBN 978-1-68285-588-1
1. Textile fabrics. 2. Textile design. 3. Textile industry--Technological innovations. I. Baia, Dick.
TS1445 .T49 2019
677--dc21

For information on all Willford Press publications
visit our website at www.willfordpress.com

WILLFORD PRESS

Contents

Preface

Textile engineering is a branch of engineering that deals with the principles, processes and scientific methodologies employed in the manufacture of textile fabrics and yarns. Research in this field delves into the advancement of textile equipment, machinery and processing techniques, design of newer textiles that suit industry requirements, environmental testing and investigating better and innovative techniques of dyeing. Scientists are investigating new methodologies to reduce the environmental footprint of engineering processes and incorporate nanotechnology and polymer science, to make textiles of superior quality. This book covers all the technological advances and studies that have broadened the horizons of textile design and manufacturing. It will be beneficial to textile engineers, chemical engineers and students interested in this discipline.

After months of intensive research and writing, this book is the end result of all who devoted their time and efforts in the initiation and progress of this book. It will surely be a source of reference in enhancing the required knowledge of the new developments in the area. During the course of developing this book, certain measures such as accuracy, authenticity and research focused analytical studies were given preference in order to produce a comprehensive book in the area of study.

This book would not have been possible without the efforts of the authors and the publisher. I extend my sincere thanks to them. Secondly, I express my gratitude to my family and well-wishers. And most importantly, I thank my students for constantly expressing their willingness and curiosity in enhancing their knowledge in the field, which encourages me to take up further research projects for the advancement of the area.

Editor

Antibacterial Activity of Cationised Cotton Dyed with Some Natural Dyes

Kamel MM, Helmy HM*, Meshaly HM and Abou- Okeil A

Textile Research Division, National Research Centre, Dokki, Cairo, Egypt

Abstract

The present study was taken up as an exploratory study to test if some natural dyes have inherent antimicrobial activity with a view to develop protective clothing using these natural dyes. For enhancing the dyeability of cotton fabric with natural dyes, three cationising agents namely; Monochloro-S-Triazine, Quat 188 and Solfix E were used. Four natural dyes namely; Madder, Logwood, Cutch and Chelidonium majus were tested against common pathogens *Escherichia Coli, Staphylococcus aureus, Aspergillus favus* and *Candida albicans. Chelidonium majus* dye was most effective and showed maximum zone of inhibition there by indicating best antimicrobial activity against all the microbes tested. The color yield of dyed samples was evaluated using K/S measurements. The total difference ΔE CIE (L*, a*, b*) was also measured. The dyed samples were tested to washing, acid and alkaline perspiration, dry and wet rubbing and light fastness. The durability of the antibacterial effect of the dyed cotton fabric was measured after 5, 10 and 20 washing cycles.

Keywords: Cotton; Natural dyes; Antimicrobial; Cationisation; Antibacterial; Macromolecules

Introduction

Increasing the global competition in textile industries created a several challenges for textile researchers. There has been upsurge interest in apparel technology all over the world for much demanding functionality of the products like wrinkle resistance, water repelling, fade resistance and resistance to microbial invasion. Among these, development of antimicrobial textile finish is highly indispensable and relevant since garments are in direct contact with human body [1]. Antimicrobial textiles with improved functionality find a variety of applications such as health and hygiene products, specially the garments worn in contact to the skin and many medical applications such as wound dressings, inflectional control and barrier materials. Cotton fabrics provide ideal environment for microbial growth [2]. Several challenges have been created for apparel researchers due to increasing global demand in textile. Therefore, textile finishes with added value particularly for medical cloths are greatly appreciated and there is an increasing demand on global scale. The consumers are aware of hygienic life style and there is a necessity of textile product with antimicrobial properties. Several antimicrobial agents viz., triclosan, quarternay ammonium compounds and recently nano silver are available for textile finishing [3-11]. However, due to their cost and synthetic in nature which creates environmental problems, natural dyes in textile coloration are gaining significant momentum [12]. This new line of interest is due to stringent environmental standards imposed by many countries due to the usage of synthetic dyes which causes allergic reaction and toxicity. The use of natural products such as chitosan [13] and natural dyes [14-16] for antimicrobial finishing of textiles has been widely reported. Other natural herbal products, such as Aloe Vera, tea, tree oil, Eucalyptus oil and leaf extracts, can also be used for this purpose. Greater interest has emerged in the field of apparel technology using natural colorants, on account of their compatibility with deodorizing properties [12]. Comprehensive literature is available on natural dyes obtained from plants [17].

The aim of the present work is to study the effect of natural dyes (Madder, Logwood, Cutch and *Chelidonium majus*) on the antibacterial properties of cotton fabrics in presence of different cationising agents.

Experimental

Materials

Natural colouring matter: Colouring substance used in this work was extracted from Madder, Logwood, Cutch and *Chelidonium* plants.

Fabrics: Cotton fabrics, mill-scoured and bleached (130 g/m²) were kindly supplied by Misr El-Beida Dyers Company, Kafr El-Dawar, Alexandria, Egypt.

Methods

Extraction of natural coloring matter: Madder (Table 1a):, Logwood (Table 1b): Cutch (Table 1c): and *Chelidonium majus* (Table 1d): were crushed to the powder form, and then the coloring matter was extracted using (20 g of the powder in 100 ml water) at the boil for one hour. At the end, the solution was filtered off and left to cool down.

Dyeing method:

Dyeing of cotton fabrics: Cotton fabric samples (8.5 g each) were dyed with the natural coloring matter extracted from Madder, Logwood, Cutch and *Chelidonium majus* at liquor ratio 1:40. Dyeing was carried out at pH 5.5 using sodium chloride 5 g/L for 60 minutes at 100°C. The fabric samples were entered to the dyeing solution in a water bath at 70°C then raised to 100°C. The fabrics were dyed for 60 minutes and the dyed samples were rinsed with cold water and washed for 30 minutes in a bath containing 3 g/L of non-ionic detergent at 45°C. Finally, the fabrics were rinsed and air dried.

Preparation of Monochloro-S-Triazine (Cationic agent) (Table 2a): Monochloro-S-triazine cationic agent were prepared using a similar method described previously [18]. N-ethyel 2-anilino

---

*Corresponding author: Helmy HM, Textile Research Division, National Research Centre, Dokki, Cairo, Egypt, E-mail: Hany.helmy70@gmail.com

(a) Madder (dye 1)	(b) Logwood (dye 2)	
	Haematein	Haematoxylin

(C) Cutch (dye 3)	(C) *Chelidonium majus* (dye 4)

Table 1: Structure of colouring groups of different plants used in this work.

(a): Monochloro-S-Triazine (Cationic Agent)	(b): 3-chloro-2-hydroxy propyl trimethyl ammonium chloride (69%) solution in water (Quat 188)

(c): Polyaminochlorohydrin quaternary ammonium (Solfix E)

Where R[1], R[2]=alkyl, and PA=polyamine

Table 2: Structure of cationising agents.

4,6dichloro-1, 3, 5-triazine (0.01 mol) was dissolved in hot acetone (80ml) and N, N dimethyl ethylendiamine (.01 mol) in acetone (30 ml) was carefully added, after the initial reaction has subsided the mixture was refluxed for 1h, cooled and intermediate filtered off and washed with acetone. The product was dissolved in hot water, the solution carefully neutralized with sodium hydroxide solution and the precipitate filtered off and dried in vacuum below 40°C, and recrystallized from acetone. The free base was dissolved in acetone and a two molar excess of the appropriate alkyl halide (CH3I) the solution was refluxed for up to 30 h, after which the colour less precipitate was filtered off from the hot solution, washed well with acetone, and dried in a vacuum below 40°C to provide the desired product [18] shown in Table 2a.

Cationisation of cotton fabric:

Cationisation of cotton using the synthesized macromolecule cationising agent (CA): Mercerized cotton fabrics were treated with (CA) at concentration of 10% (owf) in a bath with liquor ratio of 1:50 at 40°C/1h. Then Na_2CO_3, 20 g/l was added and the temperature is raised to 80°C for 30 min.

The reaction was continued for further 2 hours at 80°C. Then the treated fabrics were neutralized using 2% acetic acid at 40°C/5 min. The treated fabrics were washed using 3 g/l nonionic detergent at 60°C/30 min, followed by washing with water and finally drying at room temp.

Cationisation of cotton using (Quat 188) (Table 2b): Solution of Quat (50 g/L, 69% solution) and sodium hydroxide 15 g/L was pad applied to the cotton fabric at 100% wet pick-up. To minimize reactant hydrolysis, the alkali was added to the padding bath just prior to application. The padded samples were then dried at 80°C for 5 minutes and cured at 150°C for 5 minutes, then samples were rinsed with water and neutralized with dilute acetic acid (2 g/L), rinsed with water followed by soaping with 3 g/L nonionic detergent (Hostapal® CV, Clariant) at the boil for 30 min, rinsed with water and air dried [19].

Cationisation of cotton using Solfix E (Table 2c): Cotton fabrics were treated with Solfix E at a concentration of 30% owf at 50°C in a bath with a liquor-to-goods ratio of 30:1 for 1 hour. Then 40 ml of sodium carbonate solution (12% w/v) was added to the treatment bath portion wise at the same temperature and the reaction was continued for further 2 hours. The bath was then drained, the fabric thoroughly rinsed with water followed by soaping with 3 g/l of nonionic detergent (Hostapal® CV. Clarient) at the boil for 30 min, rinsing with water and air dried.

Measurement of the antimicrobial activity using the agar diffusion disc method: A filter paper sterilized disc saturated with measured quantity of the sample is placed on plate containing solid bacteria medium (nutrient agar broth) or fungal medium (Dox S medium) which has been heavily seeded with the spore suspension of the tested organism. After inoculation, the diameter of the clean zone of inhibition surrounding the sample is taken as a measure of the inhibitory power of the sample against the particular test organism [20-22].

Testing

Color measurements of dyed fabrics: The color yield of both dyed and mordanted samples was evaluated by light reflectance technique using the Perkin-Elmer, UV/V Spectrophotometer (Model, Lambda 3B). The color strength (K/S value) was assessed using Kubelka-Munk equation.

Colour-difference formula ΔE CIE (L*, a*, b*):

The total difference ΔE CIE (L*, a*, b*) was measured using the Hunter-Lab spectrophotometer (model: Hunter Lab DP-9000), Reston, Virginia.

The total difference ΔE CIE (L*, a*, b*) between two colors each given in terms of L*, a*, b* is calculated from:

$$\Delta E^* = [(\Delta L^*)2 + (\Delta a^*)2 + (\Delta b^*)2]1/2$$

Where:

ΔE* value: is a measure of the perceived colour size of the colour difference between standard and sample and cannot indicate the nature of that difference.

ΔL*value: indicates any difference in lightness, (+) if sample is lighter than standard, (-) if darker.

Δa* and Δb* values: indicate the relative positions in CIELAB space of the sample and the standard, from which some indication of the nature of the difference can be seen.

Fastness properties: The dyed samples were tested to washing, acid and alkaline perspiration, dry and wet rubbing and light fastness according to AATCC standard methods.

Durability of antibacterial properties: The durability of the antibacterial effect of the dyed cotton fabric was measured after 5, 10 and 20 washing cycles.

Results and Discussion

Cationisation of cotton fabric using different macromolecules cationising agents

Cationisation of cotton fabric using Quat (Table 2b): Similar to the case of reactive dyeing of cellulosic fibers, the reaction of Quat with cotton fabrics would lead to the formation of ether linkage between Quat and cellulose. However, this desirable reaction will be affected with the side reaction that takes place between Quat and water, thus lowering the efficiency of the treatment process [23,24]. Pad-dry curing technique was used in this work for the pretreatment. The concentration of Quat 50 g/l were used for cotton treatment.

Cationisation of cotton fabric using Solfix E (Table 2c): Solfix E is a polyaminochlorohydrin quaternary ammonium polymer with epoxide functionality that can react with cellulose via ether formation in the presence of alkali. Therefore the reactivity of the reagent in aqueous alkaline medium during the pretreatment process is similar to that of reactive dyes during the dyeing process of cellulosic fibers. Especially regarding the hydrolytic effect of water that competes with ether formation.

Based on the well-documented cationisation information in the literature [25,26] the best pretreatment can be obtained by avoiding as much as possible the competing hydrolytic side reaction. To achieve this, the cotton fabric was impregnated with Solfix E in the pretreatment bath at 50°C for 1 h before performing the alkali fixation process.

The effect of cationic reagent concentration on the nitrogen content of the cationised cotton fabrics

In a previous report 0.16% nitrogen content was obtained using more than 60 g cationic reagent per 100 g of cotton fabric, however, in this work 0.13% nitrogen content was achieved using less than 40 g of cationic reagent per 100 g of cotton fabric. The effect of cationic reagent concentration on the nitrogen content of the cationised cotton fabrics can be clear that as the nitrogen content of the cotton fabric increases as the concentration of cationic reagent increases. Leveling off at a reagent concentration of 30% owf. This result can be analyzed by the fact that the pretreatment process is a solid-liquid phase reaction, which proceeds by the movement of the cationic molecule from the liquid phase to the solid surface of the fiber by virtue of its ionic interaction with the surface negative charge of cotton fabric. The cationic reagent diffuses within the fabric and becomes covalently fixed to the available primary hydroxyl groups on the fibers, There for, it is expected that the process is diffusion-controlled and competes with hydrolytic effect of water. Similar behavior has been reported for cationisation using other cationic agents [27] (Scheme 1).

Effect of using different cationising agents on the color strength

Table 3 illustrated the results of using different cationising agent and its effects on the color strength and color data. For cotton samples dyed with natural coloring matter extracted from Madder, Logwood, Cutch and *Chelidonium majus*.

From Table 3, it can be observed that for dyed cotton samples, by using Monochloro-S-Triazine as cationising agent with madder and cutch plants, the highest color strength was achieved. Contrary, using of Solfix E and Quat gave the highest color strength with Logwood and *Chelidonium majus* plants, respectively.

L*, a* and b* are presented the lightness, red/green ratio and yellow/blue ratio, respectively. Table 3 showed that, L*, a* and b* data are in agreement with color strength results. Values of a* and b* located in the positive region for all dyed samples, indicating that the dyed samples had a yellowish-reddish color. Samples dyed with madder and logwood exhibited more reddish color while the others (Cutch and *Chelidonium majus*) had more yellowish color.

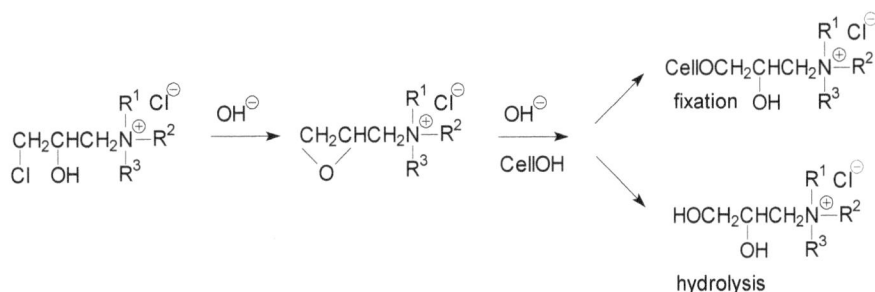

Where:

For CHTAC R^1, R^2, R^3 = CH_3

and for PAQAC R^1, R^2 = alkyl and R^3 = polyamine

Plants	Wave length	Different types of Cationising agents	(K/S) Color strength	Color data (CIE Lab)			
				L*	a*	b*	(ΔE)
Madder	500	Blank	0.76	67.69	12.45	7.04	0
		Quat*	0.75	68.01	12.97	7.14	0.62
		Solfix E**	1.85	55.18	10.19	14.89	14.94
		Cationising agent***	2.05	53.27	14.25	8.28	14.58
Logwood	400	Blank	2.93	55.62	6.28	19.53	0
		Quat*	3.22	52.72	5.92	15.96	4.61
		Solfix E**	6.77	42.12	6.72	15.81	14.01
		Cationising agent***	4.00	48.79	5.62	16.10	7.65
Cutch	400	Blank	1.75	63.66	11.59	19.18	0
		Quat*	2.17	58.41	12.12	18.19	5.37
		Solfix E**	2.68	57.28	11.46	17.60	6.57
		Cationising agent***	2.83	52.66	12.50	16.84	11.33
Chelidonium	400	Blank	3.24	58	3.25	21.45	61.94
		Quat*	3.24	57.4	4.05	19.59	60.81
		Solfix E**	1.28	69.62	2.03	17.06	71.71
		Cationising agent***	2.28	61.95	3.03	19.36	64.97

Quat*----- 3-chloro-2-hydroxy propyl trimethyl ammonium chloride (69% solution in water)

Solfix E**--Polyaminochlorohydrin quaternary ammonium

Cationising agent***------------ Monofunctional cationic agent of monochlorotriazine type

Table 3: Effect of using different Cationising agent on the color strength and color data for cotton fabric dyed with natural coloring matter extracted from Madder, Logwood, Cutch and *Chelidonium majus*.

Plants	Different types of Cationizing agents	Crocking		Acidic Perspiration			Alkaline Perspiration			Washing fastness			Light fastness	
		Dry	Wet	St.*	St.**	Alt.	St.*	St.**	Alt.	St.*	St.**	Alt.	40h	ΔE
Madder	Blank	4	3	3-4	4-5	4	4-5	4-5	4	4-5	4-5	4	5-6	2.62
	Quat*	3-4	2-3	3-4	4	4	4-5	4-5	4	4-5	4-5	4	4-5	4.96
	Sulfix E**	4	2-3	4	4	4	4-5	4-5	4	4-5	4-5	3-4	4	5.29
	Cationizing agent***	3-4	2	4	4-5	4	3-4	4	4	4-5	4-5	4	5-6	2.45
Logwood	Blank	4	2-3	3	4	3-4	3-4	4	4	4-5	4-5	4	4-5	4.68
	Quat*	3-4	2	3-4	4	4	4	4-5	4	4-5	4-5	4	5-6	2.94
	Sulfix E**	3-4	2	3	4	4	3-4	4	4	4-5	4-5	4	5	3.22
	Cationizing agent***	3-4	1-2	3	4	4	3	3-4	3	4-5	4-5	4	6	1.57
Cutch	Blank	3-4	3-4	4	4	4	4	4-5	4	3-4	4	3-4	6	1.28
	Quat*	3-4	3	4	4	4	3-4	4	4	3-4	4	3-4	6	1.79
	Sulfix E**	4-5	3-4	4	4-5	4	4	4-5	4	3	4	3-4	5	3.52
	Cationizing agent***	3-4	2	4	4-5	4	3-4	4-5	3	4	4	4	5-6	2.90
Chelidonium	Blank	4	2-3	2-3	2	3	2	2	3	2	2	3	5-6	2.19
	Quat*	3	2	2-3	2-3	3	2-3	2	3	3-4	3	3-4	6	1.43
	Sulfix E**	3-4	3-4	3	3	3	3-4	3	3-4	3-4	3	3-4	5-6	2.67
	Cationizing agent***	3-4	3	2-3	2	3	2-3	2	3	3	3	3	5-6	2.69

Table 4: Color fastness for cotton fabric dyed with natural coloring matter extracted from Madder, Logwood, Cutch and *Chelidonium*

St. *staining on cotton St. **staining on wool Alt. Alteration

From all colorimetric data, the difference in color data may be attributed to the difference in acidity for each plant used. Also, it depends on the concentration of the three cationic agents used in the Pre-treatment.

Color fastness results

From Table 4 it can be seen that, fastness properties of fabrics dyed with Madder, Logwood, Cutch and *Chelidonium* plants were assessed using blue scale. Comparing the fastness properties of the fabrics dyed with different plants, it is found that for all mordants, the dry crocking fastness is good, while the wet one is fair. Also, the acid and alkaline perspiration are good to very good. While, the washing fastness for the *Chelidonium* plant is the lowest one among the four plants. Finally, the light fastness is very good.

Antibacterial results

Table 5 shows the antibacterial properties of dyed fabrics against both *Staphylococcus aureus* (G+) and *Escherichia coli* (G-) bacteria expressed as inhibition zone. It is clear from Table 5 that all samples exhibit antibacterial properties against both types of bacteria. It is clear that the antibacterial properties against *Staphylococcus aureus* (G +) was found to be greater than that of *Escherichia coli* (G-) which

Sample	Inhibition zone diameter (mm/1 cm sample)					
	Escherichia coli (Gram -ve) bacteria			Staphylococcus aureus (Gram +ve) bacteria		
	5W	10W	20W	5W	10W	20W
Madder + quat	12	11	13	15	15	15
Madder + cationizing agent	12	12	14	17	15	15
Madder + solofix	13	12	15	17	16	17
Logwood + quat	14	12	13	16	16	14
Logwood + cationizing agent	13	12	13	16	16	15
Logwood + solofix	13	11	14	15	15	15
Cutch + quat	12	11	13	15	14	15
Cutch + cationizing agent	12	11	13	15	14	14
Cutch + solofix	11	12	13	15	15	15
Chelidonium + quat	12	12	12	16	16	14
Chelidonium + cationizing agent	13	11	13	16	16	15
Chelidonium + solofix	12	11	12	16	13	15

W: washing cycle

Table 5: Antibacterial activity of the dyed cotton with the four natural dyes after 5, 10 and 20 washing cycles.

can be attributed to the differences in the structure between the two types of bacteria. Table 5 also shows the durability of the antibacterial properties of the treated fabrics with different number of washing cycle (5, 10, and 20 washing cycles). It can be seen from Table 5 that the antibacterial properties are not affected by repeated washing up to 20 washing cycles but in some cases the antibacterial properties increased. It is obvious that the results of antibacterial properties against staph (G+) were found to be greater than that of E. coli (G-) as mentioned above due to the same reason of differences in the structure of bacteria. The antibacterial properties of the treated samples can be attributed to the presence of different types of cationising agents in all cases (Quat, Cationising agent and Solfix). The ability of these cationising agents to form true covalent bonds with cotton fabrics leads to the durability of the antibacterial properties against repeated washing.

Conclusion

Cationic groups were successfully formed on the surface of cotton using 3-chloro-2-hydroxy propyltrimethyl ammonium chloride, Polyaminochlorohydrin quaternary ammonium and Monochloro-S-triazine cationic agent, and the cationisation of fabrics strongly increases adsorption of the four natural dyes on the surface of the fibers due to the change of surface charge on the cellulose fibers. Cationised cotton dyed with these dyes exhibited stronger antibacterial activity. Cationic modification can be used for modification of cellulosic fibers to increase the antibacterial activity and increase the stability after repeated washing cycles.

References

1. Sathianarayanan MP, Bhat NV, Kokate SS, Walunj VE (2010) Antibacterial finish for cotton fabric from herbal products. Ind J Fibre Text Res 35: 50-58.

2. Hebeish A, El-Rafie MH, El-Sheikh MA, Seleem AA, El-Naggar ME (2014) Antimicrobial wound dressing and anti-inflammatory efficacy of silver nanoparticles. Int J Biol Macromol 65: 509-515

3. El-Rafie MH, El-Naggar ME, Ramadan MA, Fouda MMG, Al-Deyab SS, et al. (2011) Environmental synthesis of silver nanoparticles using hydroxypropyl starch and their characterization. Carbohydr Polym 86: 630-635.

4. Hebeish A, El-Naggar ME, Fouda MMG, Ramadan MA, Al-Deyab SS, et al. (2011) Highly effective antibacterial textiles containing green synthesized silver nanoparticles. Carbohydr Polym 86: 936-940.

5. Thilagavathi G, Kannaian T (2010) Combined Antimicrobial And Aroma Finishing Treatment for Cotton, Using Micro Encapsulate d Geranium (Pelargonium Graveolens L' Herit . Ex Ait.) Leaves Extract. Ind J Nat Pro Res 1: 348-352.

6. Emam HE, Mowafi S, Mashaly HM, Rehan M (2014) Production of antibacterial colored viscose fibers using in situ prepared spherical Ag nanoparticles. Carbohydr Polym 110: 148-155

7. Emam HE, Mowafi S, Mashaly HM, Rehan M (2014) Production of antibacterial colored viscose fibers using in situ prepared spherical Ag nanoparticles. CARBPOL 110: 148-155.

8. El-Rafie MH, Ahmed HB, Zahran MK (2014) Characterization of nanosilver coated cotton fabrics and evaluation of its antibacterial efficacy. CARBPOL 107: 174-181.

9. Emam HE, Manian AP, Široká B, Duelli H, Merschak P, et al. (2014) Copper(I) oxide surface modified cellulose fibers-Synthesis, characterization and antimicrobial properties. Surf Coat Technol 254: 344-351.

10. Zahran MK, Ahmed HB, El-Rafie MH (2014) Surface modification of cotton fabrics for antibacterial application by coating with AgNPs–alginate composite. CARBPOL 108: 145-152.

11. Emam HE, Manian AP, Široká B, Duelli H, Redl B, et al. (2013) Treatments to impart antimicrobial activity to clothing and household cellulosic-textiles – why "Nano"-silver? J Cleaner Production 39: 17-23.

12. Lee Y-H, Hwang E-K, H-D. K (2009) Colorimetric Assay and Antibacterial Activity of Cotton, Silk, and Wool Fabrics Dyed with Peony, Pomegranate, Clove, Coptis chinenis and Gallnut Extracts. Materials 2: 10-21.

13. Lee J, Han N, Lee W, Choi J, Kim J (2002) Synthesis of temporarily solubilized reactive disperse dyes and their application to the polyester/cotton blend fabric. Fibers Polym 3: 85-90

14. Gupta D, Khare SK, Laha A (2004) Antimicrobial properties of natural dyes against Gram-negative bacteria. Coloration Technology 120: 167-171.

15. Singh R, Jain A, Panwar S, Gupta D, Khare SK (2005) Antimicrobial activity of some natural dyes. Dyes and Pigments 66: 99-102.

16. Gupta D, Jain A, Panwar S (2005) Anti-UV and Antimicrobial Properties of Some Natural Dyes on ton. Ind J Fibre and textile Res 30: 190-195

17. Korac RR, Khambholja KM (2011) Potential of herbs in skin protection from ultraviolet radiation. Pharmacognosy reviews 5: 164-173.

18. Clipson JA, Roberts GAF (1989) Differential dyeing cotton. 1 – Preparation and evaluation of differential dyeing cotton yarn. J Soci of Dyers and Colourists 105: 158-162.

19. El-Shishtawy RM, Youssef YA, Ahmed NSE, Mousa AA (2004) Acid dyeing isotherms of cotton fabrics pretreated with mixtures of reactive cationic agents. Coloration Technology 120: 195-200.

20. Stokes EJ (1975) Review of medical microbiology. Ernest Jawtz, San Francisco, Joseph L. Melnick, Houston, and Edward A. Adelberg, New Haven. Eleventh edition. 260 × 180 mm. Pp. 528. Illustrated. 1974. Los Altos, California: Lange Medical Publications. £8.50. British J Surgery 62: 334-334.

21. Muanza DN, Kim BW, Euler KL, Williams L (1994) Antibacterial and anti fungal activities of nine medicinal plants from Zaire. J Pharmaco 32: 337-345.

22. Irobi O, Moo-Young M, Anderson WA (1996) Antimicrobial activity of Annatto (Bixa orellana) extract. Int J Pharmaco 34: 87-90.

23. Ankar P, MA S, Ghorpade B, Tiwari V (2000) Ultrasonic energized dyeing of cotton fabric with eucalyptus bark. Asian Text J 1: 30-32.

24. Jang J, Ko SW, Carr CM (2001) Investigation of the improved dyeability of cationised cotton via photografting with UV active cationic monomers. Coloration Technology 117: 139-146.

25. Mohamed Hashem, Smith PHaB (2003) Reaction Efficiency for Cellulose Cationization Using 3-Chloro-2- Hydroxypropyl Trimethyl Ammonium Chloride. Text Res J 73: 1017.

26. Ibrahim NA, Fahmy HM, Hassan TM, Mohamed ZE (2005) Effect of cellulase treatment on the extent of post-finishing and dyeing of cotton fabrics. J Mat Pro Tech 160: 99-106.

27. El-Shishtawy RM, Nassar SH (2002) Cationic pretreatment of cotton fabric for anionic dye and pigment printing with better fastness properties. Coloration Technology 118: 115-120.

The Treatment of Industrial Effluents for the Discharge of Textile Dyes Using by Techniques and Adsorbents

Rahman F*

Department of Petroleum and Chemical Engineering, Faculty of Engineering, Institute Technology Brunei, Brunei

Abstract

Nowadays, the extraction of textile dyes from the wastewater in industry becomes an environmental worldwide issue. Water contamination is a big threat of not only for state of the environment but human body causes some chronic diseases. Textile dyes are worn to several types of products by fabrication, for paper, leather, plastic and some products, used in human daily life. Despite of containing various hazardous chemicals into textile dyes, it is necessary to be discharged from effluents of waste-water of industry through treatment as quick as possible. A number of technologies of different processes are effectively carried out for the treatment of industrial waste water by removal of colors. A variety of textile dyes is having different chemical structure with different properties dealing with the activity of industrial reaction. By reviewing of effects of textile dyes such as toxicity and mutagenicity, bacteria and organism embedded a prologue of the expulsion of metals to the environment. Certainly, for the dominion of textile dyes removal, adsorption can be regarded as effective method used by activated carbon, bentonite clay as adsorbent for the wastewater treatment in industry. Predominantly, it is a critical review of literature conferred the removal of textile dyes for the treatment of industrial wastewater by using techniques, technologies, adsorbents thoroughly. Certainly, Adsorption is the sole and ultimate approach for removal of textile dyes through the industrial wastewater treatment. This literature shows the feasibility of minimum cost adsorbent in term of maximum outcome of industrial wastewater treatment for textile dyes removal.

Keywords: Derivatives; Techniques; Processes; Adsorbents; Textile dyes

Introduction

The textile industrial wastewater is largely yielded by the textile industry all day long to night, might be called as a manufacturer through some operations related to the industrial production. However, all the textile industries consume a large quantity of water for several ways in order to different purposes. Although a good number of chemical of variable textile processing such as bleaching, scouring, dyeing, finishing and so forth, most obviously dyeing and finishing are the processes in which a variety of chemicals is widely used. Nonetheless, a large quantity of discharged wastewater turned to colored wastewater by the manufacturing of dyes in industry. But it now concerned of few causes like Toxicity, Accumulation and a few like them. If the toxicity increased and perfectly accumulated in nature, this process may be seemed like as a terrible view in environmental balance [1,2]. Textile dyes are ionized and organic, shows a strong affinity to the aqua solution and a bit on industry water, mostly used to be coloring during the manufacture of final product. Moreover, a lot of industries use finishing in the manufacturing process in leather, plastic, tanning and textile industry [3]. Furthermore, both of dye-manufacturing and textile finishing become a source of presumptive but only dye-manufacturing can discharge industrial waste-water without any disturbance.

Adsorption is a phantasm in which molecules of any state (gas or liquid) do focus at connecting surface with no reaction and the most effluent method for the removal of dyes from industrial waste water by treatment due to like ease of operation, little cost, efficiency and so on [2,4]. Activated carbon is a big potential for the treatment and used as an adsorbent. Other treatment methods can be operated during the process of wastewater of dyes removal not better than adsorption method with some factors such as cost, efficiency, operation system and so on. Besides, industrial wastewater for color removal can be treated biologically, chemically and physically [5]. As dyes are being produced in industry and later going for usage of the consumer product. Dyes have some common uses in business market–dyeing, soap, food, cosmetics,

and textiles and so forth [6]. Color containing dyes is essential to be removed due to the amount of dyes (below 2.0 ppm). It also creates a big impact on the environment. To the removal of textile dyes from industrial waste water, followed by Wastewater effluent treatment. The process of waste-water treatment can be elected by some requirements, for instance cost, characteristics and others [7]. The treatment methods are also divided into primary, secondary and tertiary such sedimentation, removal of constituents and so on [8]. Today's, over the study of environment, the treatment of textile dyes of wastewater in industry is more essential and the effluents of color from industries which is known as textile industries demands to be deliberated as a contaminant. A number of substances are uncertain, for example aromatic compounds, synthetic dyes, heavy metals in dyes effluents. Many common dyes are insoluble and mostly used for the treatment in textile industry for the high stability. Later, cotton and fabrics are obviously applied with more difficult operations during the treatment of industrial colors wastewater. A large number of textile industries are located in subcontinents for example India, Pakistan, Bangladesh, Sri-Lanka and some other countries. Now, this sector still has a big influence of scattering the latest design of any fabric throughout the world within a short period of time and great potential of unutilized materials which are undiscovered and inherently related to the textile productions like yarns, cottons, fabrics, garments and so forth.

---

*Corresponding author: Fazlur Rahman, Department of Petroleum and Chemical Engineering, Faculty of Engineering, Institute Technology Brunei, Jalan Tungku Link, Gadong, Bandar Seri Begawan, Brunei
E-mail: rahman. fazrahman.fazlur48@gmail.com

Denotation of textile dyes

A content or material can be called as natural or synthetic used to altering the color during the unit operation or connecting the color into the fabrics. Textile dyes are derived from sources (i.e., coal tar) and defined as colored compounds containing several chemicals with different concentrations. To view of industrial purpose, dyes are not directly being used and utilized.

Categorizing of textile dyes

By consisting data collections, it narrates every dye, consisting properties of physical and chemical. Gathering of information of some dyes mentioned as much as possible for environmental factor. Afterwards, the dyes are selected and have effect to the nature and health factor by developing activities. These dyes are propagated to the textile industry of dying process [9] (Figure 1).

- Sulphur dyes are applicable in alkaline solution. They widely used for cotton, viscous and staple fibers and so forth.

- Disperse dyes are insoluble in water. They have worldwide usage in textile industry, basically for synthetics fibers like polyester and cellulose acetate such as Di-acetate, Tri-acetate and others. This dyes also used for nylon. They are applied on the dye bath at high temperature range around 120-140°C.

- Direct dyes are the dyes, can be called as "substantive dyeing". Cotton used as natural fiber and viscous used as synthetic fiber. They also used for aqua solution. In solution, electrolysis and salts are available. Predominantly, direct dyes are used as the second dyes in worldwide. Direct dyes used for cotton, viscous.

- Azoic dyes are used for natural fibers, viscous, cellulosic fibers. They consist two soluble components to be formed colored molecules, are insoluble. The synthesis of azo dyes operated by two stages such as Diazotization [HNO_2] and Azo Coupling [Ar—H]

Nylon can be classified into some forms such as (nylon 6:6), (nylon 6:10) and nylon (6:11) that commercially used in the world market. Some of dyes are not associated to fiber on dyeing processing. Dyes are collaborated into three categories such as

- Cellulose fiber
- Synthetic fiber
- Protein fiber

Derivatives of textile dyes

Toxicity: Dyes are mainly induced by toxicity. Dyes are synthesized to various metals occupying such as chromium (Cr), Copper (Cu), Zinc (Zn), Cadmium (Cd) and more on are all violent. Dye molecules consist of two sort of molecule, might influence the toxicity of dye-colorant. The Toxicity could outcome by the elimination of metals to the nature or symbiotic action to raise the effect of toxics of metals. Study of toxicity is a vial of taking knowledge of toxic issues of materials, may be not well known and undecided. The easement of toxicity is performed by two tests, for example

- Sharp toxicity
- Enhancement toxicity

Firstly, Bacteria for example algae, protozoa are those who react to the sharp toxicity test. In the test, the organism is matured in the water-test where all the substances are present to necessary for growth.

Dyes	Examples	Structure of dyes
Acid Dyes	Acid violet 30	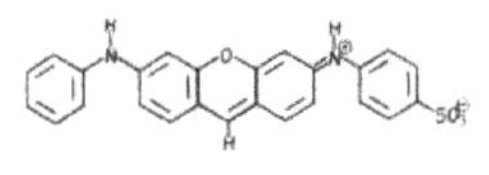
Azoic Dyes	Bismarck brown R	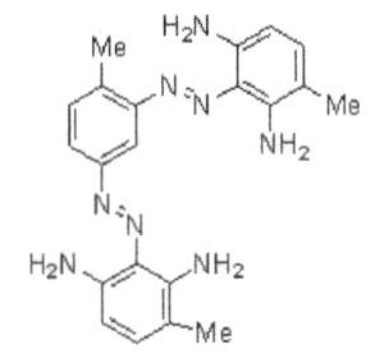
Basic Dyes	Basic Yellow 28	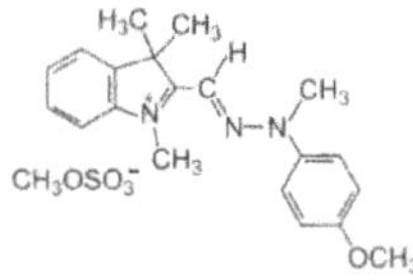
Disperse Dyes	Disperse Red 220	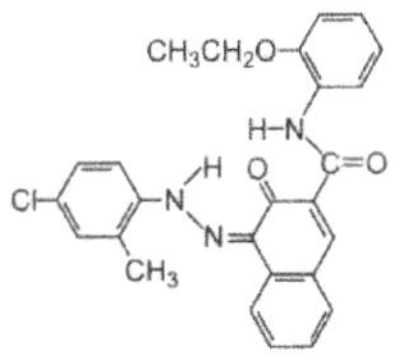
Reactive Dyes	Reactive Red 1	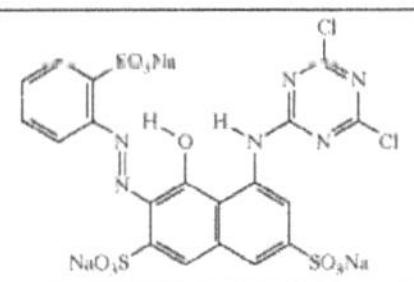
Vat Dyes	Vat blue 4	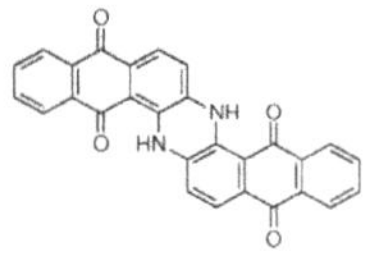
Pigment	Pigment Blue 15	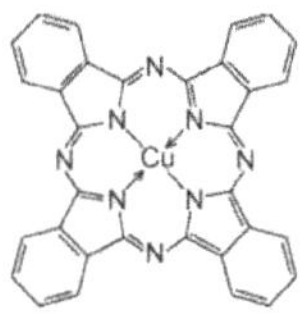

Figure 1: Some dyes and examples (structure) of dyes.

The number of organisms prompting in the sample, before and after growth, is estimated and the concentration of effluent required to affect with almost 50-60% of the organisms is found. Another test of the assessment of toxicity is enhancement toxicity. By using *Enterobacter aerogenes*, the enhancement toxicity test can be performed. Drinking water is contained the bacterium, developed with small growth as control. If any feasibility of increment growth is occurred to medium, the colonies are gained along with it. The reasons of toxicity may be determined through physical and chemical analysis by carrying out. The yield of dyes is also a big issue to the nature in terms of the event of toxicity [10]. By containing of disclosing of organisms are prototypical from the nature considering the time interval which are assertive to one or more components have many different concentrations, beneath the surroundings of environment are varied and grading the effects of poison. Thereby, tests which magnitude the concentration of a component, the appraisal of toxicity was executed. To be yielding a decisive effects, the time that is disclosure, is essential [11,12].

Mutagenicity: The effluents of toxic hold chemicals (i.e., mutagenic, carcinogenic) to several organisms. The genetic materials may be affected and also damaged over different types of chemicals. Many dyes are originated from known as carcinogens (i.e., benzidine). They are familiar to be acquired and seemed as a potential threat. In issue of the sources of mutagenic activity is inspected all over the parts of the globe since many years. Furthermore, expressing of azoic and nitro compounds like nitro-aminoazobenzenes dyes are contained along with Disperse violet 93, Disperse orange 37, and Disperse blue 373. The water consumption by human presented conferred mutagenic activity related to nitro-aromatics and aromatic amines compounds [13].

In purpose of dyes removal, few of processes can be applicable for the industrial waste water treatment. Names are mentioned below [14-19]:

- Expected treatment process
- Rehabilitation process
- Gushing removal process

In addition, expected treatment process are used by few technologies such as biodegradation, coagulation, flocculation for presenting onto treatment of wastewater whereas others process are also available for waste water treatment by process of rehabilitation or process of gushing removal. To be exploiting of rehabilitation process are also worn as wastewater treatment technologies like ion-exchange, oxidation, and separation by membrane and so on. Additionally, technologies of removal of gushing process are assigned on treating of industrial wastewater. For instance, selective oxidation process (advanced) is engaged to gushing removal process [20,21] (Table 1).

Biodegradation: Microbial-degradation, frugal decolorizing, bio-remission are included to the mode of biodegradation. They are fully devoted to the industrial effluents treatment due to several microorganisms, for instance that yeasts, bacteria, fungi, algae and so forth are capable to assemble and degrade different hazardous-wastes. In the operation of biodegradation, the activity of biomass of microbial is not only influential but microbial biomass is relevant to it. Lead (Pb), Chromium (Cr), Copper (Cu), Zinc (Zn), Arsenic (As), Cadmium (Cd) and so on are called heavy metals containing high concentration can have the effect is injurious on the action of microbial and also the microbial expansion. By adopting Fe (Iron) salts or Al (aluminum) drains the technology named coagulation. The treatment of textile derivations passed down by chemical coagulation completely. A major problem of coagulation is silt production, become a concern for carrying the process out. Photo-catalysis may be known as a new method of the advanced oxidation process, worn to clarify dye compounds. Many pros

are found in photo-catalytic degradation, like no more rudiments for further disposal in next. Photo-catalytic degradation is also applicable to the different dyes in different experiments [22].

Removal of textile dyes by techniques of treatment

Numerous methods are obviously relevant to the dyes removal of industrial waste water treatment. Among of them, it is feasible to identify the proper techniques are the best, can be described as below [3]:

- Chemical technique
- Physical technique
- Biological technique

A few of technologies have been applied for removal of textile dyes including above techniques, however, a device named microbial deterioration of color, is productive for clearing contaminant away from the nature [23]. By adsorption in lieu of degradation, some micro-organisms such as algae and fungi wash color out in effluents of industry. In consequence, colors abide in the environment. But under the surroundings that are assured, bacteria diminish colors [24]. Consistently, bacteria can lessen transitional by products like amines are aromatics adjacent to the enzyme of oxygenase and enzyme of hydroxylase [25].

- Physical technique is consisted of several methods such as sedimentation (clarification), screening, Nano filtration, reverse osmosis, electro dialysis and so forth. During the process of clarification, physical is a phenomena are allowed to be operated, relating to settling of solid by gravity. Sedimentation is one of the most physical treatment methods, used to attain treatment. In addition, another treatment method is aeration, by summing air to provide oxygen (O_2) to the wastewater. This phenomena also consists equalization and skimming. Due to high efficiency, A filtration method is renowned as Reverse Osmosis (RO) as well as can be suppressed that has drawn with good interest and thinking. The method can have a dormant of expelling of particles (over 90%) in solution and a prime might be applicable to the treatment of industrial wastewater. This method like reverse osmosis method moves on a membrane called semi-permeable membrane with very small pores (about 0.6 nanometer) and its function at a high rate of operation in this method. Since couple of years ago, some experiments had performed to do removal of dyes through treatment in industry in which can be required the rate of discharge up to the 93% after operation [26].

Kashefi and Bahrami [27] examined the removal of vat dyes (anthrasol brown IBR) using by the process of reverse osmosis. A survey is applied to the effects of the experiment operating with some key parameters such as temperature (T), feed concentration, and pH and so on. The experiment accepted use of 30 mg/L dye solutions where the final results demonstrated that the most ideal condition for the process was pH (= 4) and temperature (T = 25°C). The removal of colors was accomplished by following with the procedures through the membrane. Feed concentration was calculated beneath to some optimal conditions which disclosed that feed concentration is the largest and had a small positive supplement to the rate of color removal. In solution of feed, the effect of dye concentration is a specific parameter. Within the context of low concentration solutions a solution with a greater concentration brings greater elimination. Finally, many portions of solution through membrane and serial application of membranes are important aspects in promotion of rate of discharge and improvement of permeate quality in reverse osmosis process. As recognized, rate of discharge is attained with almost **94%** after completing through the membranes this experiment [27].

Technology	Pros	Cons
Biodegradation	Stable, profitable	Silt production high process is sluggish
Coagulation	Financially sufficient	Vending problem
Oxidation process	Quick process productive	Cost of energy is high
Ion–exchange	Effective small quantity of sludge produced	Generate concentrated Brine
Membrane separation	Is high	Pressure is high

Table 1: Pros and Cons of treatment by different technologies.

• Chemical technique is contained some methods like ozonation, neutralization, chlorination, and so on. Chemical treatment reposes of chemical reactions to develop the quality of water. Perhaps, chlorination is the most common chemical method. A strong oxidizing chemical named chlorine (Cl_2) used to kill bacteria and also the dissipation of wastewater. Neutralization is one of the most common methods, used industrial wastewater treatment. Neutralization consists of the addition of acid or base to modify pH levels back to neutrality. Ozonation is also one of the regular methods as chemical technique for using wastewater treatment. It is very competent for the deterioration of colors that are reactive, reduction of chemical oxygen demand, discharge of poisonous contaminants from the industrial dye effluents [28-32]. The main disadvantage of ozonation is duration of time. The period of time is halving of its total life. As the wastewater is remained in treatment, it can be lowered. Even if the condition is acidic for the ozonation, pH is required to the co-operation of treating industrial effluents for dyes removal [33]. Ozone (O_3) is more available to the environment and also used to be collected some data in experimental issues. It can be applied in gaseous state thoroughly.

Nabila et al. investigated for the sewage water treatment through the method of electrocoagulation by iron (Fe) anode. A good number of operational parameters of the treatment either expertise or not have consequences which were interrogated and progressed. By the current density, the initial pH, the volume of sodium chloride (NaCl) and initial concentration of color, the wastewater of excess using by iron (Fe) anode was prevailed. The optimum operating conditions of pH value was 7.6, 65 mA/cm² as current density, 30 min of electrolysis time, and amount of sodium chloride as supporting electrolyte was 1 g/L and gap distance of electrode was only 3 cm. The determination appears that the optimal current density would be constant at 65 mA/cm² that provides the highest removal of Suspended Solids (S.S), Chemical Oxygen Demand (COD), and Biochemical Oxygen Demand (BOD) where these values falls from 507 mg/L, 760 mgO_2/L, and 446 mgO_2/L to 5 mg/L, 98 mgO_2/L, and 77 mgO_2/l respectively. Furthermore, the amount of released iron climbed as the adapted current density rises, however the Suspended Solids (S.S), Chemical Oxygen Demand (COD), and Biochemical Oxygen Demand (BOD) were leveled off on more current density above 65 mA/cm², thereby 65 mA/cm² was preceded as the optimal adjusted current density for sewage electro coagulation. The electro-coagulation is the process which is a very favorable pre-treatment step for UF and RO process for the conversion of sewage water to irrigation water where it is high condition or drinking water which is decontaminated [34].

Stergiopoulos et al. discusses the electrochemical degradation treatment method of indigo carmine dye, that involves electrocoagulation, electro oxidation and advanced electrochemical oxidation using through the process of electro Fenton. The electrocoagulation process is implemented by electrodes of iron (Fe). The ambiguous electro oxidation process is by steady Titanium (Ti) or Platinum (Pt) and graphite electrodes in sodium chloride (NaCl) electrolyte solution, and the electro-Fenton process is done by electrodes of iron (Fe) and combined the amounts of hydrogen peroxide (H_2O_2). A variety of electrochemical processes are explained like electrocoagulation, electrochemical and electro oxidation where their efficiencies analyzed and assessed. The initial dye concentration of 100 mg l⁻¹ was quick and separated within a few minutes of electro-processing time effectively. The electro-Fenton treatment is the fastest, efferent and financial process operated at very low current densities nearly 0.33 and 0.66 mA cm⁻² and exhausting only 4.75×10-3 and 5.23 x 10-3 kWh m⁻³ of treated solution. The electrocoagulation

treatment with iron (Fe) electrodes and the electro oxidation process with Titanium (Ti) or Platinum (Pt) electrodes conducted at enforced current densities of 5 mA cm⁻² consumed 0.511 and 0.825 kWh m⁻³ of treated solution. The proposed procedure is a safe, economical method for discharge of indigo carmine dye from aqueous solutions and textile dye effluents. By employing the current densities of 0.33 and 0.66 mA cm⁻² **100%** the degradation of the dye was achieved in only 2 and 1 minutes of the method of electro-processing where consumption of energy was 4.75 x 10⁻³ and 5.23 × 10⁻³ kWh m⁻³ of treated dye solution [35] (Table 2).

• Fenton's reagent several methods are expressed by the chemical technique such as ozonation, Electrochemical, photochemical, electrochemical etc. However, Fenton's reagent as heating agent is comprised of decomposition. A mixture of hydrogen peroxide (H_2O_2) and ferrous ion (Fe^{2+}) is renowned as Fenton reagents. The technique of Fenton can be thermic with alternative [36]. It is so efficient of fading for many dyes. Afterwards, iron (Fe) (ll) or iron (Fe) (lll) oxalate ion, UV light and Hydrogen peroxide have received a focus of lightening for especially synthetic dyes [37-39]. But about of the silt generation is a major disadvantage of this technique through flocculation of molecules of dyes. Peroxide techniques are associated with UV and Fenton's reagent whereas the insertion of UV light is constrained. One more thing is the cost of Fenton's reagent, influenced to the improvement of this method. In this technique of photo-Fenton, through of some reactions like photo reduction of iron (Fe) (ll) to the ion of iron (Fe) (lll), activity of fenton, photolysis of hydrogen peroxide (H_2O_2), H_2O_2 are exploited quickly as far as possible. Hence, it can yield more OH^- radicals than method of photolysis or method of Fenton that is conventional [40,41]. By virtue of little depletion of hydrogen peroxide (H_2O_2), small price of removal of sediment, low dispersion of energy, inferior maintenance but flexibility is huge [42]. In contemplation of receiving the pros of burning power of Fenton's reagent exclude the desperation of Fe salts from solution, the process of exertion of hydrogen peroxide (H_2O_2) has been regarded. Additionally, in behalf of the reaction of Fenton, along with the chemisorption of Fe powder, it can produce as good for removal of textile dyes than Fe^{2+} (ferrous ions) or H_2O_2 (hydrogen peroxide) [43]. Over a wide range of pH (around 5-8) and sediment is not produced with comparing to the Fenton's reagent, the action of oxidation of H_2O_2 (hydrogen peroxide) gains merit of its ability of application.

• Biological technique is one of the most cost-effective treatment methods than others treatment methods. Biological treatment is awkward to the regular variation and toxoids whereas it has good affability in design and activity. It is overviewed to be proven that biological treatment is the way of eradication of color with tolerable. Some metals which are heavy are involved of the wastewater of textile industry. In addition, the effluents of textile industry enclose with complex form and also some metals that are ionic [44]. For the shake of capability to outgrow with little

Methods	Advantages	Disadvantages
Ozonation	Volume is unchanged, stayed in vapor-state	Life duration is short (20 min)
Photochemical	Production of silt is off	Metals, acids, halides Produced as by-product
Electrochemical	Disruption of non-haphazard compounds	Price of power supply is high
Electrocoagulation	Decolonization can be achieved	No failure of forming, metallic (OH) clouds in wastewater

Table 2: Advantages and disadvantages of some methods of chemical technique.

sludge, efficacy of price, benefaction of environment, some biological processes like bioaccumulation, bio colorization, and bio sorption are more supportive [45-47]. Because of the treatment of color effluents in industry, bioaccumulation and biosorption occupy good (potential) instead of methods are conventional. Moreover, they are known as the main technology in biological technique [48]. Biological systems are driven in situation of sites are contaminated, also called environmental benign were so costly and no contamination which is secondary. As the treatment of industrial color effluents, they can be said as the principle merits of biological technologies. Recently, for the treatment of color effluents related to industry, some necessary steps along with research have been carried out thoroughly on the methods of biological [49,50]. Efficiency with low degradation for many dyes, even no degradation efficiency is (suffered) to the process of degradation which might be as a demerit as well as it is not easy morally in process that is continuous [49,51,52]. By the biological materials which are passive, biosorption can be characterized as the static insolence of toxicants. As long as the treatment of industrial color effluents is largely harmful and continuous, the accommodating organism service is not desirable, can be said as merit of bioaccumulation and biosorption process. The use of biomass are dead is (flexile) to the situations of environment and concentrations of contaminant which run over of the current concern [13].

Shertate and Thorat inspected biodecolorization and degradation of textile diazo dye reactive blue-171 by- marinobacter sp.nb-6. Decolorization and Degradation of Reactive Blue 171 was achieved using by the adjusted marinobacter sp. NB-6 (Accession No. HF568873) detached from soil. The decolorization of dye Reactive blue 171 on 24 hours was up to 95.00% in nutrient broth consisting 8.0% NaCl and exhibited 93.11% decolorization in half strength nutrient broth having the concentration of sodium chloride (NaCl) which is equal. The percent decolorization of the dye was also executed by cell-free extract and was detected that the isolate can decolorize the dye 90.00% on 24 hours. The percent decolorization of the dye was disposed at 590 nanometer. The percent chemical oxygen demand (COD) debasement of the color by the segregate was 86.00%. The degradation products formed after degradation were analyzed by GC-MS analysis and it was found that degraded Reactive Blue 171 to the products having several molecular weights. Khalid et al. reported enhanced decolorization of azo dyes by Shewanella putrefaciens strain AS96 in presence of yeast extract as co-substrate under hypersaline condition. Rania et al. used Glucose, Sucrose, Starch and Sodium citrate as carbon sources among which Starch was best for decolorization of Crystal violet up to 96%. Gondaliya and Parikh reported the highest percentage decolourization 97.04% of Reactive Orange16 was obtained by Serratia marcescenswhen additional supplement of glucose (1 g/l) was added in nutrient broth [53].

Physicochemical approach: A number of different physical modes are available such as membrane filtration process. Further; the systems of physicochemical were expensive but feeble as well as not appropriate to wastewater of colors having a wide capacity [44]. The most typical process is membrane filtration process with some features. One of them is removal of all dyes by this. Another is its instant process but short area of space is needed. On the view of cost and volume, the cost is high and large volume cannot be handled by ability. By product of agricultural and industrial areas is done by non-ordinary adsorbent. It comprised to some advantages and dis-advantages. Firstly, adsorbent is active but not costly. Secondly, it provides smooth operation. Thirdly, it's also accessible in worldwide. The disadvantages are those such as toxic wastes are shifted from liquid to solid adsorbent as much as possible.

Adsorbent: The most effectual system is well-known as the adsorption process as established; the adsorption process is directly related to the wastewater treatment for industry. A good number of adsorbents are widely used such as activated carbon, seed of avocado pear, teak tree bark powder, natural clay, bentonite clay and so on. Activate carbon is not expensive, though its production creates trouble. Several types of by product can be feasible into activated carbon through conversion process from the area of agriculture and industry. Kind of thing is shell fiber of coconut, ash of fly (i.e., pearl millet husk, wheat straw, coconut straw etc.), corn rope, included to waste of wood and agriculture tended to produce activated carbon. The prerequisite of advantages for the removal of dyes process are some such as pore volume, high stability of heat, smooth production, and high stability of hydro thermal, no activity of catalytic and so forth. Many technologies already are prospered by researchers with some parameters like cost, efficiency, life cycle etc. by making the materials decolorizing from effusion of industry.

Removal of textile dyes by adopting adsorbents

3Carbon nanotubes as adsorbent for removal of dye: Carbon nanotubes (CNTs) are nanostructure shape having carbon atoms with some properties such as mechanical, chemical, thermal etc. Magnetic multi-wall carbon nanotubes (MMWCNTs) used to removal of cationic dyes from aqua solution [52-60]. Also single wall carbon nanotubes (SWCNTs), multi wall-carbon nanotubes (MWCNTs) and specially modified magnetic multi-wall carbon tubes (MMMWCNTS) used for different treatments on removal of dyes from industrial wastewater. Our effort is to look of the apply of the effective method for the treatment of industrial wastewater by removing of dyes by using SWCNTs or MWCNTs or MMMWCNTs.

Fernando et al. investigated Adsorption of a textile dye from aqueous solutions by carbon nanotubes. The adsorbents were mainly characterized by using N_2 adsorption/desorption isotherms, Raman spectroscopy and flashing and transmission electron microscopy. The effects of pH, agitation time and temperature on adsorption capacity were reviewed. In the acidic pH region, the adsorption of the dye was supportive using by both of adsorbents. The contact time to obtain equilibrium isotherms at temperatures like 298-323 K was fixed at 4 hours for both adsorbents. Because reactive blue 4 dyes, Liu isotherm model gave the best fit for the equilibrium data. The maximum sorption capacity for adsorption of the dye occurred at 323 K. Multi-walled carbon nanotubes (MWCNTs) and single-walled carbon nanotubes (SWCNTs) were better adsorbents for eliminating Reactive Blue 4 (RB-4) textile dye from aqueous solutions. The RB-4 dye interacted along with the MWCNT and SWCNT adsorbents at the solid or liquid interface while hanging in water. The equilibrium isotherm of the RB-4 dye was collected, and Liu isotherm model gave the best fit. Therefore, the maximum adsorption capacities were 502.5 and 567.7 mg g^{-1} for MWCNT and SWCNT. The enthalpy ($\Delta H°$) of adsorption implied which adsorption was the process that was an endothermic and the consequence of enthalpy was consistent with an electrostatic interaction of an adsorbent with the dye [37].

Activated carbon as adsorbent for removal of dye: Activated carbon is widely used as a large potential than other bio-adsorbents like expensive to the treatment for industry waste-water through method of adsorption. The adsorption of basic dyes (i.e., methylene blue) can be achievable by using activated carbon as an adsorbent. Experiments can be applied like batch experiment, equilibrium of tan is obtained almost 86 min with a variety of temperature variation in whatever place of an adsorbent such as activated carbon is worn. Isotherm models, for instance Langmuir and Freundlich are outfitted by adsorbent using as activated carbon [61].

Khraisheh et al. examined the Elucidation of Controlling Steps of Reactive Dye Adsorption on Activated Carbon. In this job, over the inspection of kinetics of adsorption, the rate-limiting traces of reactive dye adsorption onto FS-400 activated carbon were exemplified. Primarily, by these studies admitted that only 20% of the capacity of adsorption was available and obtained during the first 6 hours of mixing. The profiles of kinetic displayed that the adsorption process was maintained by diffusion that is external during the first 30 min of the reaction, after which internal diffusion restrained the process. What is more, the coefficients of external and internal diffusion and the rate of desorption declined after the period of interruption. Only 20% of the available ultimate capacity was achieved during the first 6 h of mixing with a dye solution. Interruption test results showed that adsorption of reactive dyes onto FS-400 was guarded by internal diffusion, with only a little effect from external diffusion [48].

Bentonite clay as adsorbent for removal of dye: The improvisation and adaption of physiochemical of Bentonite clay are apt of the removal of reactive dyes along with applications. At the beneath of upgrading condition with changing temperatue, Bentonite clay which comprised of properties for adsorption and approaching (reactive red 223) were observed at bottom of batch system during a period of time. Langmuir, Freundlich and Dubinin– Radushkevich isotherm models are fairly lighted by following of them. The upgrading conditions like amount of adsorbent, concentration of dye, temperature and so on were asses at the mark of no charge [62-65].

Tahir et al. discussed Physiochemical Modification and Characterization of Bentonite Clay and Its Application for the Removal of Reactive Dye. Thus the effective methods were ratified for the discharge of dyes and colorants from the textile effluents. In the current research, the discharge of textile dye reactive red 223 (RR 223) was operated by modified bentonite clay (MBC). The modification of bentonite clay was finalized beyond the manner of acid treatment. Moreover, the properties of adsorption of modified bentonite clay (MBC) towards RR 223 were prospected using by the batch method, on some temperatures (303-318 ± 2) K under the optimized conditions. The equilibrium data of adsorption were equipped in Langmuir, Freundlich and Dubinin-Radushkevich adsorption isotherm models and the values of the respective constants were evaluated by employing standard graphical method. Feasibility of adsorption process (RL) and sorption energy (Es) was also resolved. It completes the symbolic changes in the capacity of adsorption toward reactive dyes. Adsorption, desorption, zero point charge pH (pzc), thermodynamic and kinetics studies were continued to incline the validity of process. The experiments of adsorption were rush beneath the optimal conditions of amount of adsorbent, break time, basic concentration and various temperatures. The adsorption models like: Freundlich, D-R and Langmuir adsorption isotherm models were applied for the mathematical description of the adsorption equilibrium data. From the values of r2 it was suggested that Langmuir model is the best fitted model. Some parameters named as Gibbs free energy, enthalpy and entropy changes view that the adsorption of RR 223 onto MBC was instinctive and endothermic process. The kinetic data showed that the adsorption of RR 223 onto MBC followed the pseudo second order kinetics. The feasibility of adsorption and desorption process was surveyed in the study that is going to be running on. It was evaluated that inexpensive, available and active materials can be applied as adsorbents for the discharge of dyes. Certainly, low-cost adsorbents offer a lot of encouraging benefits for

commercial purposes with respect to exertion and derogation of waste. In addition, the alternation of adsorbent as well as the reformation of color was performed by desorption experiments. The sorption and desorption capacity of MBC was found to be 95.15% and 78% [63-68].

Isotherm model: The most common isotherm models are Langmuir and Freunlich which is used to the application of wastewater treatment of industry [69-74].

Langmuir adsorption isotherm: It expresses the production of adsorbate which is seemed as one layer on the adsorbent remaining the exterior surface. Additionally, more adsorptions does not access in further steps. So, the Langmuir model exemplifies the equilibrium division of metal ions occurred in between the phase of solid and liquid [37]. The Langmuir isotherm is accurate to the monolayer adsorption. The homogeneous energies of adsorption onto the surface are estimated with the model and the reincarnation of adsorbate in the surface plane is no more [69-75] (Table 3).

Freundlich adsorption isotherm: To depict the adsorption flavors for the different surface is worn by this isotherm model.

Dubinin–Radushkevich adsorption isotherm: Dubinin–Radushkevich isotherm is generally applied for the interpretation of the appliance of adsorption along with a energy named Gaussian dealing onto a surface, which is dissimilar [64-70]. More often, large solute activities equips this model effectively (Tables 4 and 5).

Conclusion

Several sorts of chemicals are interconnected to the textile industry such as removal, production of compounds and also the generation of some valuable products. Additionally, the outflow of textile industry is released as dyes while evaluated. The literature arrays a number of treatment methods are indispensable as well as regarded to the treatment of industrial wastewater for discharge of colors, though

Name of dyes	Name of adsorbents	Dye adsorption capacity (mg/g)	Isotherm model
Organic dyes	Surfactant modified coconut coir pith	76.3	Freundlich and Langmuir
Reactive blue 4	Multi-walled and single-walled carbon nanotubes	502.5 and 567	Liu
Reactive red 23	Activated carbon	59.88	Freundlich
Orange G	Activated carbon of Thespesia populnea pods	9.13	Freundlich and Langmuir
Reactive blue 171	Activated carbon	71.94	Freundlich
Congo red	Bagasse fly ash	11.89	Freundlich

Table 3: Dye adsorption capacities of different dyes.

Name of dyes	Name of adsorbents	Values of pH	Contact time (equilibrium)
Direct red	Calcined bone	2 – 12	60 min
Methylene blue	Montmorillonite clay	3 – 11	120 min
Organic dyes	Blue 21	4 – 10	180 min
Green anionic dye	Surfactant modified peanut husk	2 – 10	300 min
Reactive red	Activated carbon	4.5 – 6.0	340 min
Reactive blue	Multi-walled carbon nanotubes	2 – 10	360 min

Table 4: Different pH values and contact time of different dyes.

Isotherm model of Adsorption	Limitations of Isotherm model	Elemental Qualities
$q_e = \dfrac{q_{m,L}.K_L.C_e}{1+K_L.C_e}$	C_e = equilibrium concentration of adsorbate (mg/L) q_e = the amount of metal adsorbed per gram of the adsorbent at equilibrium (mg/g) q_{ML} = maximum monolayer coverage capacity (mg/g) K_l = Langmuir isotherm constant (L/mg)	Adsorption is consisted to many layers on surface which is diversified with handling of active ites of differential energy
$q_e = K_F.C_e^n$	K_F = Freundlich isotherm constant (mg/g) n = adsorption intensity C_e = equilibrium concentration of adsorbate (mg/L) Q_e = the amount of metal adsorbed per gram of the adsorbent at equilibrium (mg/g)	Adsorption is consisted of one layer having identical sites and no communication between adsorbed species
$E = \left[\dfrac{1}{\sqrt{2B_{DR}}} \right]$ $\varepsilon = RT\ell n \left[1 + \dfrac{1}{C_e} \right]$ $\ell nq_e = \ell n(q_s) - \left(K_{ad}\varepsilon^2 \right)$	q_e = the amount of metal adsorbed per gram of the adsorbent at equilibrium (mg/g) q_s = theoretical isotherm saturation capacity (mg/g) K_{AD} = Dubinin–Radushkevich isotherm Constant (mol²/kJ²) B_{DR} = isotherm constant R , T, Ce represents gas constant (8.314 J/mol.k) C_e = equilibrium concentration of adsorbate (mg/L)	A energy called 'Gaussian' allocated on surface, are composite, devoted to make difference between adsorption of physical and chemical by value of mean free energy

Table 5: Synopsis of elemental qualities and limitations of isotherm models.

of some technologies of processes (i.e., flocculation, ion-exchange, coagulation, biodegradation), techniques (i.e., physical, chemical, biological) and adsorbents (i.e., carbon nanotubes, activated carbon, bentonite clay) are explicitly connected. Almost all methods are not adequate to separate contaminants due to consisting of some constraints such as huge cost, absence of capability, formation of auxiliary contaminants, so forth. In physical technique of treatment, depends upon high energy and then forms many risky outgrowths. Also, in chemical technique of treatment, electrochemical is high cost of power supply. Furthermore, for the treatment process, biodegradation is slow and high sludge production. Although the treatment of biological technique is quiet cheaper than others, adsorption is the maximum cost-effective surrounded by all methods by admitting two major influential factors such as content of application and minor cost of employment. This method can be particularly carried through along with lot of adsorbents, but a few of them which are inferior in cost, obviously suitable economically and environmentally with their certain applications. Undoubtedly, it can be mentioned that none of any process or technique have more appliances like adsorption. Over view to all discussion, only adopting adsorbent for textile dyes removal might be the energetic mode for the treatment of textile dye effluents in industry in comparison to following techniques for the treatment. In near future, it can be devoted to the alternative color departments which are used in worldwide textile industries.

Acknowledgment

The correspondence author (MD Fazlur Rahman) is thankful to all those people who encouraged me to be approaching on making this literature on textile dyes removal through industrial wastewater treatment.

References

1. Mishra AK, Arockiadoss T, Ramaprabhu S (2010) Study of removal of azo dye by functionalized multi walled carbon nanotubes. Chemical Engineering Journal 162: 1026-1034.

2. Joshi M, Bansal R, Purwar R (2003) Colour removal from textile effluents. Indian Journal of Fibre & Textile Research 29: 239-259.

3. Karthik V, Saravanan K, Bharathi P, Dharanya V, Meiaraj C (2014) An overview of treatments for the removal of textile dyes. Journal of Chemical and Pharmaceutical Sciences 7: 301-307.

4. Damar Y (2012) Treatment of Textile Industry Wastewater by Sequencing Batch Reactor (SBR), Modelling and Simulation of Bio kinetic Parameters. International Journal of Applied Science & Technology 2: 302-318.

5. Elsagha A, Moradib O, Fakhrib A, Najafic F, Alizadehd R, et al. (2013) Evaluation of the potential cationic dye removal using adsorption by graphene and carbon nanotubes as adsorbents surfaces. Arabian Journal of Chemistry pp: 1-8.

6. Kamil AM, Abdalrazak FH, Halbus AF, Hussein FH (2014) Adsorption of Bismarck Brown R Dye onto Multiwall Carbon Nanotubes. Environmental Analytical Chemistry 1: 1-6.

7. Goosen MFA, Shayya WH (2000) Water Management, Purificaton, and Conservation in Arid Climates. CRC Press 2: 1-352.

8. Ramnath Lakshmanan (2013) Application of Magnetic nanoparticles and reactive filter materials for wastewater treatment. universitetsservice us-ab pp: 1-57.

9. Kyzas GZ, Γu J, Matis KA (2013) The Change from Past to Future for Adsorbent Materials in Treatment of Dyeing Wastewaters, Materials 6: 5131-5158.

10. Immich APS, De Souza AAU, De Souza SMAGU (2009) Removal of Remazol Blue RR dye from aqueous solutions with Neem leaves and evaluation of their acute toxicity with Daphnia magna. Journal of Hazardous Materials 164: 1580-1585.

11. Spraghe JB (1973) The ABC's of pollutant bioassay using fish. In: Biological Methods for the Assessment of Water Quality, American society for Testing and Materials. Philadelphia pp: 6-30.

12. Rand GM (1980) Introduction to Environmental Toxicology. Elsevier.

13. Batool S, Akib S, Ahmad M, Balkhair KS, Ashraf MA (2014) Study of Modern Nano Enhanced Techniques for Removal of Dyes and Metals. Journal of Nanomaterials pp: 20.

14. Kumar A, Chaudhary P, Verma P (2011) A comparative study on the treatment methods of textile dye effluents. Global journal of environmental research 5: 46-52.

15. Tsaia WT, Chang CY, Lina MC, Chiena SF, Suna HF, et al. (2001) Adsorption of acid dyeon to activated carbon prepared from agricultural waste bagasse by ZnCl2 activation. chemisphere 45: 51-58.

16. Manahan SE (2005) Fundamental of Enviromental Chemistry.

17. Adebayo, Otunola (2010) Assessment and Biological Treatment of Effluent from Textile Industry. African journal of biotechnology 9: 8365-8368.

18. Allen SJ, Oumanova BK (2005) Decolirization of Water/Wastewater Using Adsorption. Journal of the Univeristy of chemical technology and metallurgy 3: 175-192.

19. Abdelwahab O, El nemr A, El sikaily A, Khaled A (2005) Use of rice husk for adsorption of direct dyes from aqueous solution. Egyptian journal of aquatic research 31: 110-0354.

20. Nigam P, Armour G, Banat IM, Singh D, Marchant R (2000) Physical removal of textile dyes from e, uents and solid-state fermentation of dye-adsorbed agricultural residues. Bioresource Technology 72: 219-226.

21. Spellman FR (2000) Handbook of Water and Wastewater Treatment Plant Operations.

22. Beyene HD (2014) The potential of dyes removal from textile wastewater by using different treatment technology, a Review. International Journal of Environmental Monitoring and Analysis 2: 347-353.

23. Royer B, Cardoso NF, Lima EC, Ruiz VSO, Macedo TR, et al. (2009) Organ functionalized kenyaite for dye removal from aqueous solution. Journal of Colloid and Interface Science 336: 398-405.

24. Baath E (1992) Thymidine incorporation into macromolecules of bacteria extracted from soil by homogenization-centrifugation. Soil Biology and Biochemistry 24: 1157-1165.

25. Bader JL, Gonzalez G, Goodell PC, Ali AS, Pillai SD (1999) Chromiumresistant bacterial populations from a site heavily contaminated with hexavalent chromium. Water, Air, and Soil Pollution 109: 263-276.

26. Akbari A (2007) Polymeric nanomembranes and nanofiltration process.

27. Morteza MI, Bahrami F (2014) Removal of vat dyes from colored wastewater by reverse osmosis process, Bulletin of the Georgian National Academy of Sciences.

28. Liakou S, Pavlou S, Lyberatos G (1997) Ozonation of dyes. Water Science and Technology 141: 279-286.

29. Karahan O, Dulkadiroglu H, Kabdasli I, Sozen S, Babuna FG (2002) Effect of ozonation on the biological treatability of a textile mill effluent. Environmental Technology 23: 1325-1336.

30. Moraes SGD, Freire RS, Duran N (2000) Degradation and toxicity reduction of textile effluent by combined photocatalytic and ozonation processes. Chemosphere 40: 369-373.

31. Tehrani-Bagha AR, Mahmoodi NM, Menger FM (2010) Degradation of a persistent organic dye from colored textile waste water by ozonation. Desalination 260: 34-38.

32. Selcuk H (2005) Decolourization and detoxicification of textile waste water by ozonation and coagulation processes. Dyes and Pigments 65: 217-222.

33. Ahmet B, Ayfer Y, Doris L, Nese N, Antonius K (2003) Ozonation of high strength segregated effluents from a woolen textile dyeing and finishing plant. Dyes and Pigments 58: 93-98.

34. Hussien NH, Shaarawy HH, Shalaby MS (2015) Sewage Water Treatment via Electro coagulation Using Iron Anode. Asian Research Publishing Network (ARPN) 10: 8290-8299.

35. Dermentzis D, Giannakoudakis K, Sotiropoulos S (2014) Electrochemical Decolorization and Removal of Indigo Carmine Textile Dye From Wastewater. Global NEST Journal 16: 499-506.

36. Suty H, Traversay CD, Cost M (2004) Applications of advanced oxidation processes: present and future. Water science Tech 49 : 227.

37. Machado FM, Bergmannl CP, Lima EC, Adebayo MA, Fagan SB (2013) Adsorption of a textile dye from aqueous solutions by carbon nanotubes. Materials Research.

38. Swaminathan K, Sandhya S, Sophia AS, Pachhade K, Subrahmanyam YV (2003) Decolorization and degradation of H-acid and other dyes using ferrous–hydrogen peroxide system. Chemosphere 50: 619-25.

39. Bandara J, Morrison C, Kiwi J, Pulgarin C, Peringer P (1996) Degradation/ decoloration of concentrated solutions of Orange II. Kinetics and quantum yield for sunlight induced reactions via Fenton type reagents. Journal of Photochemistry and Photobiology A: Chemistry 99: 57-66.

40. Gogate PR, Pandit AB (2004) A review of imperative technologies for wastewater treatment II: hybrid methods. Advances in Environmental Research 8: 553-97.

41. Bertanza G, Collivignarelli C, Pedrazzani R (2001) The role of chemical oxidation in combined chemical-physical and biological processes: experiences of industrial wastewater treatment. Water sci Technol 44: 109-16.

42. Tang WZ, Chen RZ (1996) Decolorization kinetics and mechanisms of commercial dyes by H_2O_2/iron powder system. Chemosphere 32: 947-58.

43. Baath E (1998) Growth rates of bacterial communities in soils at varying pH: a comparison of the thymidine and leucine incorporation techniques. Microbial Ecology 36: 316-327.

44. Erdal S, Taskin M (2010) Uptake of textile dye reactive black-5 by Penicillium chrysogenum MT-6 isolated from cement-contaminated soil. African Journal of Microbiology Research 4 : 618-625.

45. Wang H, Zheng, Su JQ, Tian Y, Xiong XJ, et al. (2009) Biological Decolorization of the reactive dyes Reactive Black 5 by a novel isolated bacterial strain Enterobacter sp. EC3. Journal of Hazardous Materials 171: 654-659.

46. Aksu Z (2005) Application of biosorption for the removal of organic pollutants: a review. Process Biochemistry 40: 997-1026.

47. Malik PK (2004) Dye removal from wastewater using activated carbon developed from sawdust: adsorption equilibrium and kinetics. Journal of Hazardous Materials 113: 81-88.

48. Khraisheh MAM, Al-Degs YS, Allen SJ, Ahmad MN (2001) Examined the Elucidation of Controlling Steps of Reactive Dye Adsorption on Activated Carbon. Ind Eng Chem Res 4: 1651-1657.

49. Vijayaraghavan K, Yun YS (2008) Biosorption of C.I. Reactive Black 5 from aqueous solution using acid-treated biomass of brown seaweed Laminaria sp. Dyes and Pigments 76: 726-732.

50. Prasad MNV, De Oliveira Freitas HM (2003) Metal hyper accumulation in plants-biodiversity prospecting for phytoremediation technology. Electronic Journal of Biotechnology.

51. Stolz A (2001) Basic and applied aspects in the microbial degradation of azo dyes. Applied Microbiology and Biotechnology 56: 69-80.

52. Pearce CI, Lloyd JR, Guthrie JT (2003) The removal of color from textile wastewater using whole bacterial cells: a review. Dyes and Pigments 58: 179-196.

53. Shertate RS, Thorat PR (2013) Biodecolorization and Degradation of Textile Diazo Dye Reactive Blue-171 By- Marinobacter Sp.Nb-6 – A Bioremmedial Aspect. International Journal of Universal Pharmacy and Bio Sciences 3: 330-342.

54. Gupta VK, Kumar R, Nayak A, Saleh TA, Barakat MA (2013) Adsorptive removal of dyes from aqueous solution onto carbon nanotubes: A review. Advances in Colloid and Interface Science 193: 24-34.

55. Asad S, Amoozegar MA, Pourbabaee AA, Sarbolouki MN, et al. (2007) Decolorization of textile azo dyes by newly isolated halophilic and halotolerant bacteria. Bioresource Technology 98: 2082-2088.

56. Pandey A, Singh P, Iyengar L (2007) Bacterial decolorization and degradation of azo dyes. International Bio deterioration and Biodegradation 59: 73-84.

57. Moawad H, El-Rahim WMA, Khalafallah M (2003) Evaluation of biotoxicity of textile dyes using two bioassays. J Basic Microbiol 43: 218-229.

58. Gao HJ, Zhao SY, Cheng XY, Wang XD, Zheng LQ (2013) Removal of anionic azo dyes from aqueous solution using magnetic polymer multi-wall carbon nanotube nanocomposite as adsorbent. Chemical Engineering Journal 223: 84-90.

59. Geyikci F (2013) Adsorption of Acid Blue 161 (AB 161) Dye from Water by Multi-walled Carbon Nanotubes. Fullerenes Nanotubes and Carbon Nanostructures 21: 579-593.

60. Ghaedi M, Kokhdan SN (2012) Oxidized multiwalled carbon nanotubes for the removal of methyl red (MR): kinetics and equilibrium study. Desalination and Water Treatment 49: 317-325.

61. Qu S, Huang F, Yu S, Chen G, Kong J (2008) Magnetic removal of dyes from aqueous solution using multi-walled carbon nanotubes filled with Fe_2O_3 particles. Journal of Hazardous Materials 160: 643-647.

62. Machado FM, Bergmann CP, Fernandes THM, Lima EC, Royer B, et al. (2011) Adsorption of Reactive Red M-2BE dye from water solutions by multi-walled carbon nanotubes and activated carbon. Journal of Hazardous Materials 192: 1122-1131.

63. Tahir H, Sultan M, Qadir Z (2013) Physiochemical modification and characterization of bentonite clay and its application for the removal of reactive dyes. International Journal of Chemistry 5: 19-32.

64. Ver meulan TH, Vermeulan KR, Hall LC (1966) Fundamental. Ind Eng Chem 5: 212-223.

65. Arulkumar M, Sathishkumar P, Palvannan T (2011) Optimization of Orange G dye adsorption by activated carbon of Thespesia populnea pods using response surface methodology. J Hazard Materials 186: 827-834.

66. Almeida CAP, Debacher NA, Downs AJ, Cottet L, Mello CAD (2009) Removal of methylene blue from colored effluents by adsorption on montmorillonite clay. J Colloidal Interface Sci 332: 46-53.

67. Arami M, Limaee NY, Mahmoodi NM, Tabrizi NS (2005) Removal of dyes from colored textile wastewater by orange peel adsorbent: equilibrium and kinetic studies. J Colloid Interface Sci 288: 371-376.

68. Zhao B, Xiao W, Shang Y, Zhu H, Han R (2014) Adsorption of light green anionic dye using cationic surfactant-modified peanut husk in batch mode. Arabian Journal of Chemistry.

69. Dada AO, Olalekan AP, Olatunya AM, Dada O (2012) Langmuir, Freundlich, Temkin and Dubinin Radushkevich Isotherms Studies of Equilibrium Sorption of Zn^{2+} Unto Phosphoric Acid Modified Rice Husk. Journal of Applied Chemistry 3: 38-45.

70. Ghaedi M, Khajehsharifi H, Yadkuri AH, Roosta M, Asghari A (2012) Oxidized multiwalled carbon nanotubes as efficient adsorbent for bromothymol blue. Toxicological and Environmental Chemistry 94: 873-883.

71. Kuo CY, Wu CH, Wu JY (2008) Adsorption of direct dyes from aqueous solutions by carbon nanotubes: Determination of equilibrium kinetics and thermodynamics parameters. Journal of Colloid and Interface Science 327: 308-315.

72. Mall ID, Srivastava VC, Agarwal NK, Mishra IM (2005) Removal of congo red from aqueous solution by bagasse fly ash and activated carbon: Kinetic study and equilibrium isotherm analyses. Chemosphere 61: 492-501.

73. Pandey A, Singh P, Iyengar L (2007) Bacterial decolorization and degradation of azo dyes. International Biodeterioration and Biodegradation 59: 73-84.

74. Shah V, Verma P, Stopka P, Gabriel J, Baldrian et al. (2003) Decolorization of dyes with copper (II)/organic acid/hydrogen peroxide systems. Applied Catalysis B: Environmental 46: 287-292.

75. Kanawade SM, Gaikwad RW (2011) Removal of Methylene Blue from Effluent by Using Activated Carbon and Water Hyacinth as Adsorbent. International Journal of Chemical Engineering and Applications 2: 317-399.

Effects of Raising and Sueding on the Physical and Mechanical Properties of Dyed Knitted Fabric

Md. Touhiduzzaman[1], Rashid KMM[2] and Md. Syduzzaman[3]*

[1]*Department of Fabric Manufacturing Engineering, Bangladesh University of Textiles, Dhaka, Bangladesh*
[2]*Department of Yarn Manufacturing Engineering, Bangladesh University of Textiles, Dhaka, Bangladesh*
[3]*Department of Textile Engineering Management, Bangladesh University of Textiles, Dhaka, Bangladesh*

Abstract

For developing finishing methods and broadening their assortment, the research has explored raising and sueding methods in order to find out how these two finishing processes affect the aesthetic, dimensional and functional properties of dyed knitted fabric. In this work, three types of sample fabrics namely three thread fleece or Chief Value Cotton (CVC) fleece, two thread terry or organic fleece and 2x2 rib were used. CVC fleece was produced from yarns of count 30 Ne and 20 Ne respectively. Organic fleece and 2x2 rib were produced from 100% cotton. In this task, we studied six aspects of test results viz. the GSM test, shrinkage test, bursting strength test, spirality test, pilling test and Color fastness to wash. In addition, Wales per Inch (WPI), Course per Inch (CPI), stitch length, stitch density, course length was taken into consideration. The testing procedures were performed meticulously under ideal testing standards. The test results affirm significant effects regarding GSM, shrinkage, bursting and spirality test. No variation of pilling and color fastness of samples subjected to raising and sueding was noted.

Keywords: Knitted fabric; Raising; Sueding; Bursting strength; Spirality; Color fastness; GSM

Introduction

The final stage, amongst a series of processes in the manufacturing of textile clothing to impart the necessary decorative and functional characteristics in the end product involving both chemical and mechanical treatments is finishing. Finishing is best regarded as the final stage in the embellishment of the fabrics, most of which, as it comes from the loom or knitting machine, having an unattractive appearance, which persists, although to a less extent, even after dyeing or printing. A simple definition of finishing is the sequence of operations, other than scouring, bleaching and coloring, to which fabrics are subjected to after leaving the loom or knitting machine [1]. On the basis of processing technique, finishing can be segmented into two types viz. chemical and mechanical. Some mechanical finishing techniques are calendaring, beetling, embossing, raising sueding, and glazing [2]. This paper endeavors to discuss the two mechanical processes Raising and Sueding.

Raising is the technique whereby a surface effect is produced on the fabric that gives the fabric a brushed or napped appearance. It is achieved by teasing out the individual fibers from the yarns so that they stand proud of the surface [3]. The surface fibers are lifted up by means of sharp teeth imparting hairiness, softness and warmth. A velvety material surface is obtained by loosening a large number of individual fibers from the fabric. Natural teasels or metal cards are used to doing so. Raising gives a purely artificial character to the cloth. These effects are produced by ejecting only one end of the fibers residing on the outer surface of the yarn whereas the other end remains twisted within the thread construction so that the fibers are not completely detached from the thread. The thread receives considerable loss of strength, if the fibers totally separate from the yarn. So material tension needs to be carefully regulated.

Material tension is selected depending on the factors like staple length, fiber fineness, yarn twisting, fabric binding [4,5]. Raising can be performed in three states that include raising in loom state, raising after fabric preparation and raising after dyeing. There are two ways, wet and dry raising processes. Wet raising requires less time and gives laid pile while dry raising gives prominent pile and takes more time [6]. Wet technique requires less mechanical action whereas dry one needs repeated drum roller action. So moisture content of the fabric is also very important [7].

Sueding is technically known as peach finish mostly in woven manufacturing process. It's a mechanical finishing process in which a fabric is grazed on one or both sides to raise or create a fibrous surface. A sueding machine produces a very low pile on the fabric surface. The fabric surface is finished to get a feel of suede leather. Sueding machine can have a single cylinder sanding or multi-cylinder emerizing action. Multi cylinder emirizing is preferred over single cylinder process since it removes knots and loose fibers are not produced [8]. The abrasive material used for sanding or emerizing needs to be changed after certain number of circles of actions for uniform formation of pile. The abrasive grit size is determined by the fabric construction so that an optimum sueding effect can be obtained. Sueding effect can range from mild sueding to peach skin effect. By this action, a low pile is produced. It improves touch thus making the fabric surface softer and voluminous. It is very useful for synthetic material to beget a better textile quality.

Both raising and sueding improve the fabric appearance, give the fabric a softer, fuller hand. Masking of the fabric construction, subduing of coloration are also achieved. These improved aesthetics can increase the value of the fabric in the marketplace. Besides, both raising and sueding affect the dimensional as well as physical properties. At the same time, the processes being mechanical cause fiber damage. So finer count of yarn is used for fabric that requires raising and sueding. An assessment of the objective of raising- sueding can be drawn by studying the properties of raised and sueded fabric in comparison

*Corresponding author:** Md. Syduzzaman, Department of Textile Engineering Management, Bangladesh University of Textiles, 92 Shaheed Tajuddin Ahmed Ave, Dhaka 1208, Bangladesh, E-mail: sayeed33tex@gmail.com

with only dyed fabric. Fabrics of same structure at different processing stages show different values for GSM, bursting strength, pilling test, spirality and shrinkage upon implementing raising- sueding.

Materials and Methods

Three types of fabric samples such as three thread fleece or CVC fleece, two thread terry or organic fleece, 2 × 2 rib were taken to analyze. CVC fleece was made from yarn of count 30 Ne and 20 Ne. CVC Fleece or 3-thread fleece was produced from ring spun yarn in circular knitting machine (Machine gauge 20, Machine dia 30"). Organic Fleece or 2-thread fleece was produced from ring spun yarn in circular knitting machine (Machine gauge 24, Machine dia 30"). 2 × 2 Lycra rib was produced from ring spun yarn in circular knitting machine (Machine gauge 20, Machine dia 32").

CP Fleece or Single Action or European Style, "Normal" double drum machine has been used for CVC Fleece, organic Fleece and 2 × 2 Lycra rib. Here, European style- MC20/24 GF has been followed. As all type of fillets are not suitable for all type of fabrics, we had to be very careful to choose the right type of fillet for the right type of fabric. For CVC fleece fillet types were: Lower drum: 28/32 rubber-pile/counterpile and upper drum: wire 22 felt-3 counterpile/1 mushroom. For organic fleece or 100% cotton fleece fillet types were: Lower drum-wire 22 felt-pile/counterpile and upper drum-wire 22 felt-3 counterpile/1 mushroom. Lisa 4 Knit Plus machine has been used for getting sueding or peach effect on organic fleece and 2 × 2 Lycra rib (Table 1).

Test method

Wet raising process was done to avoid higher damage of yarn. Sueding was carried out with multi cylinder option. The standards that were followed in performing the tests are:

Fabric weight (GSM)	EN 12127
Pilling test	ISO 12945-1
Bursting strength test	ISO 13938-2
Spirality test	In House Method (1 wash)
Shrinkage test	In House Method (1 wash)
Color fastness to washing	ISO 105 C06

Results and Discussions

Visual appearance

The appearance of the samples when dyed and undergone raising and sueding appears prominently distinguishable to the naked eye.

From the Figures 1, 3, 5 and 7 we can see that the gray and dyed samples have distinctively visible intermeshing points with the loops clearly distinguishable. Then, in the Figures 2, 4 and 6 the distinction is replaced by a hairy surface with erect protruding fibres. It is also viewed that the raising process has abraded the surface more intensely than sueding.

Fabric type	Stitch length (mm)	Final color	Processing
CVC fleece or 3-thread fleece	4.46	Medium grey	Face side: None Back side: Raising
Organic fleece or 2-thread terry	2.27	Real black	Face side: Sueding Back side: Raising
2x2 lycra rib	3.20	Real black	Face side: Sueding Back side: Sueding

Table 1: Processing parameters of samples.

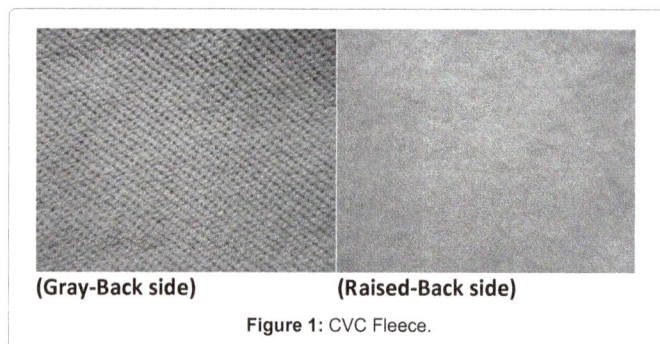

(Gray-Back side)   (Raised-Back side)

Figure 1: CVC Fleece.

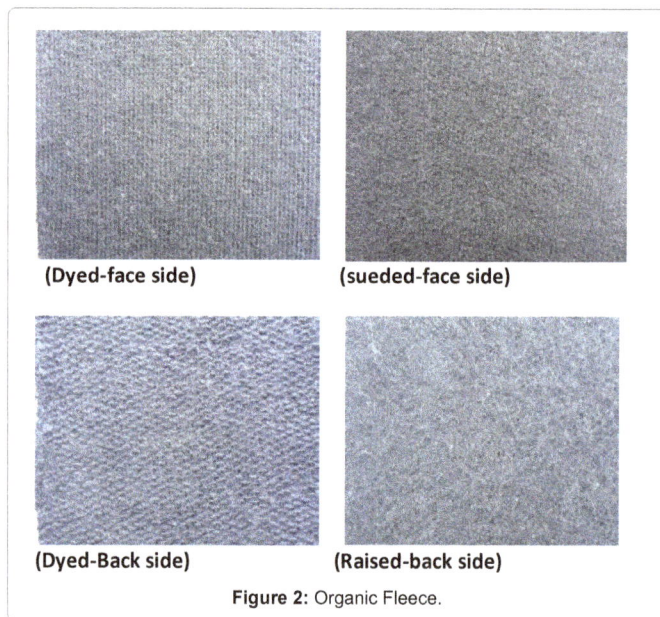

(Dyed-face side)   (sueded-face side)

(Dyed-Back side)   (Raised-back side)

Figure 2: Organic Fleece.

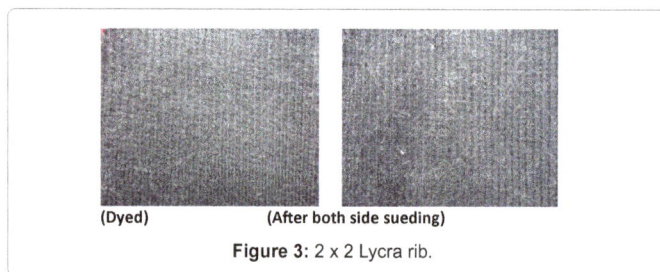

(Dyed)   (After both side sueding)

Figure 3: 2 x 2 Lycra rib.

Characteristic tests

The behavior of the samples under testing restraints allows us to characterize them with respect to bulk, softness, shrinkage etc. thus providing the scope to cast comments. Six tests have been used to bring forth the effects of raising and sueding. These are: Fabric weight (GSM), Pilling test, Bursting strength test, Shrinkage test, Spirality test, Color fastness to washing.

Fabric weight (GSM)

The weight of a fabric can be described in two ways, either as the 'weight per unit area' or the 'weight per unit length'. If either one of the two or the width of the cloth is known, the other can be calculated assuming the selvedge to have a negligible effect [9,10]. We opted for

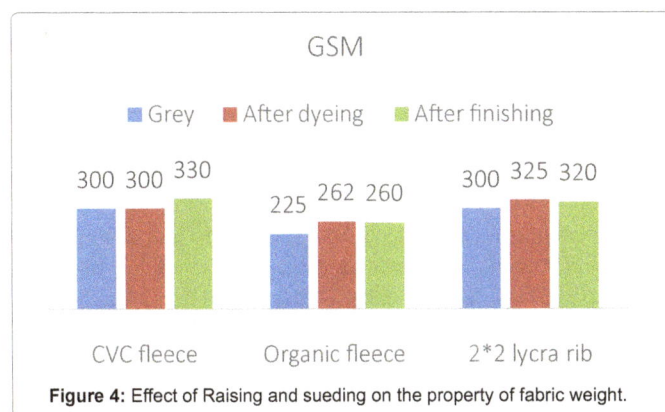

Figure 4: Effect of Raising and sueding on the property of fabric weight.

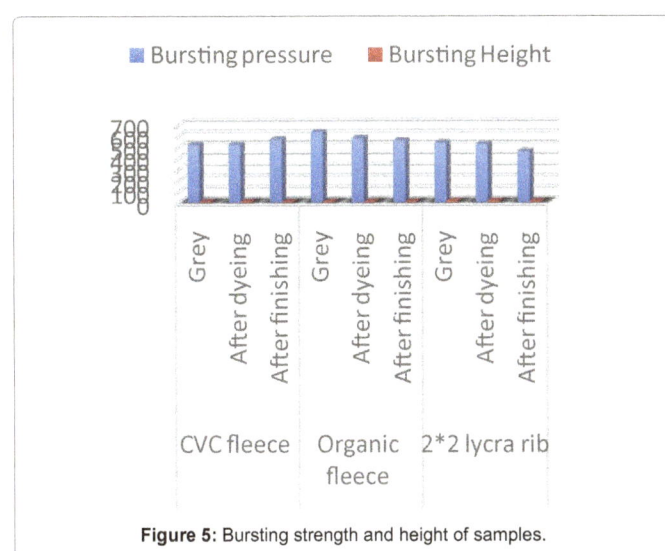

Figure 5: Bursting strength and height of samples.

GSM (Gram per Square Meter) for convenient understanding. EN 12127 method was used in this case. Five samples were taken by using GSM cutter. Then they were weighted on the electric balance. GSM of different stages of processing with and without raising- sueding has been calculated. From those data the average value was calculated.

CVC fleece was produced from dyed yarn. Since the fabric was not dyed further, the gray and dyed state refer to the same condition. So there is no variation in GSM in these states. When dyeing is performed to the yarn, flying fibers reduce, strength and mass of yarn is increased, the twisted fibers become more compact which leads to less raising action. As a result, increase in mass due to dye particles is more than decrease in mass due to fiber loss during raising. So GSM increased after finishing (raising) though there is fibre loss on the surface. Again, in case of organic fleece and Lycra rib, GSM increased notably after dyeing. Because of the absorption of the dye particles, the air and space in between the fibres, loops were replaced. The mass of dye directly contributed to the augmented GSM. Due to Finishing (raising or sueding or both), decrease in mass due to loss of fiber was more than increase of mass due to dye material.

Bursting strength test

The distending force applied at right angles to the plane of the fabric, under specified condition, which results in the rupture of a textile material is known as bursting strength [11]. Two parameters,

bursting pressure (in KPa) and bursting height (in mm) are attributed to the measure of strength. To do this, force is applied for a duration of 20 ± 3 seconds such that the material ruptures within this time. ISO 13938-2 standard was followed here.

Machine parameter for the test

Test Area	50 cm
Clamping	6.0 Bar
Pressure Rate	10 kPa/s
Pressure Drop	10 kPa
Diaphragm	1.0 mm Duraflex
Correction rate	2 kPa/s
Correction	35.2 kPa @ 46.1 mm

It is evident that CVC fleece experienced an increased pressure to rupture but its height has reduced. Yarn dyed fabric has more prominent loops. There is no accumulation in between the adjacent threads. After raising, ejected fibers took stand and the space between the adjacent threads was closed. So, higher pressure was needed to rupture the finished (raised) fabric. Organic fleece shows a gradual reduction in pressure but its bursting height increased after dyeing. When raised and sueded, the bursting pressure reduced because of the fabric becoming weaker. When the gray fabric was dyed, adjacent threads became more compact. After finishing, this compactness was opened. Due to the open end fibers, strength was reduced. In all these cases, duration of 20 seconds could be maintained. As we approach to Lycra rib, the reduction in bursting pressure after sueding is gigantic and the fabric ruptures long before 20 seconds is reached. Lycra is a highly extensible material. As it passes through various rollers and drums for sueding, its strength is reduced.

Shrinkage test

Shrinkage is the linear amount a fabric will contract warp wise (along wales) or filling wise (along course) when laundered. It is expressed as a percent of its original measurement. All fibers have some tendency to shrink, but this tendency is greatly increased if the fabric has been stretched in finishing [12]. Shrinkage is the process in which a fabric becomes smaller or extended than its original size, usually through the process of laundry. Cotton fabric suffers from two main disadvantages of shrinking and creasing during subsequent washing. When the fabric becomes smaller it is called negative shrinkage, in other case it is termed as positive shrinkage. Shrinkage is mainly due to yarn swelling and the resulting crimp increase. The largest amount of shrinkage is that represented by increase of crimp; yarn shrinkage takes a second place, being generally much less than increase in crimp, while fibre shrinkage is almost negligible [13]. Shrinkage is measured in two ways such as lengthwise and widthwise. For shrinkage test, 50 cm × 50 cm sample of each type of fabric was taken. In House Method (1 wash) resulted in the following measurements (Figure 6).

The CVC fleece shows major widthwise expansion whereas the organic fleece and lycra rib went through major lengthwise shrinkage. Since the processes of raising and sueding are mechanical and employ the fabric to be passed through rollers, dimensional integrity is compromised.

Spiraliy test

"Spirality" is a dimensional distortion in circular knitted fabric that arises from twist stress in the constituent yarns of plain fabric causing all loops to distort and throwing the fabric wales and courses into an

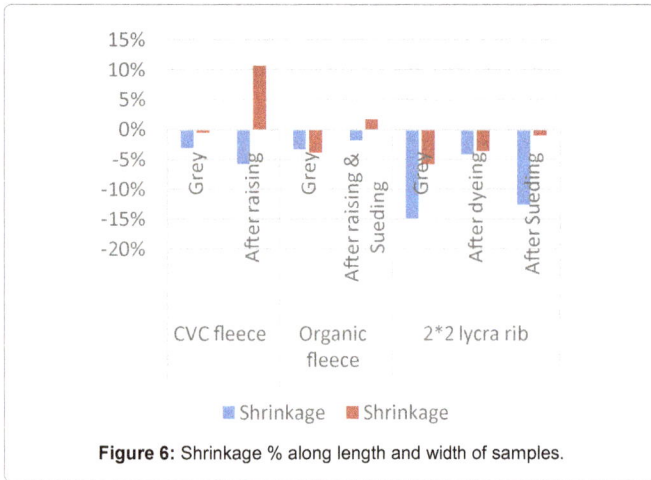

Figure 6: Shrinkage % along length and width of samples.

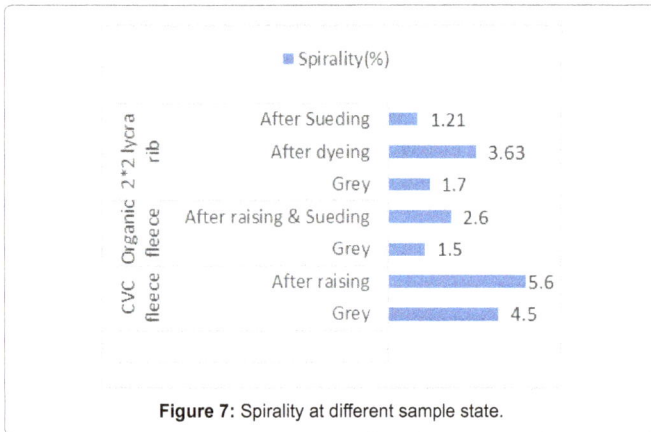

Figure 7: Spirality at different sample state.

angular relationship other than 90 degree. In House Method (1 wash) is used here. The length of left and right seam is measured and the average value for precision result is calculated. Now the distortion of the widths of the bag at the open end is measured. It is done at both end and calculated the average for the precision of the value. Now the spirality can be measured from the following equation: (Figure 7).

Spirality= (Distortion/length) × 100%

Both CVC and organic fleeces experienced increased spirality. Hence, dimensional stability and retaining its shape when worn are hampered. Lycra rib confronted more than double spirality after dyeing. After sueding, it was reduced again. This behavior is due to the presence of Lycra in the fabric.

Pilling test

Pilling is a fabric-surface fault characterized by little 'pills' of entangled fibre clinging to the cloth surface and giving the garment an unsightly appearance. The pills are formed during wear and washing by the entanglement of loose fibres which protrude from the fabric surface. After rubbing of a fabric it is possible to assess the amount of pilling quantitatively either by counting the number of pills or by removing and weighing them. Counting the pills and/or weighing them as a measure of pilling is very time consuming and there is also the difficulty of deciding which surface disturbances constitute pills. The more usual way of evaluation is to assess the pilling subjectively by comparing it with either standard samples or with photographs of

them or by the use of a written scale of severity. ISO 12945-1 was used as reference in this case.

Pilling scale

The pilling scale was explained at Table 2.

Pilling result

Severe pilling occurred in every case. As raising and sueding cause fibres to be raised and ejected, the free ends of the fibres can easily entangle to form pills. In case of 2 × 2 Lycra rib it showed exceptionality. Dyed rib fabric showed severe pilling but sueding affected that same fabric to show slight pilling (Table 3).

Color fastness test

Color fastness to wash is achieved by placing a specimen of textile, in contact with one or two specific adjacent fabrics and is mechanically agitated under described conditions of time and temperature in a soap solution, then rinsed and dried. The change in color of the specimen and the staining of the adjacent fabric are assessed with the grey scales. According to ISO 105 C06, DW multi-fibre fabric was used to recognize any kind of staining or color deterioration.

Grey scale for color change and staining

The grey scale for color change and staining was clearly explained at Table 4.

Color fastness result

Color fastness to wash depends on the selection of raw materials, dyes and chemicals. According to our experiments it can be said that the type of finishing process such as raising, sueding don't give any impact on coloration of the fabric (Table 5).

Rating	Surface evaluation
5	No pilling
4	Slight pilling
3	Moderate pilling
2	Severe pilling
1	Very severe pilling

Table 2: Pilling Scale.

Fabric Type		Grading Result	Remarks
CVC fleece	Grey	2	Severe pilling
	After finishing	2	Severe pilling
Organic fleece	Grey	-	
	After dyeing	2	Severe pilling
	After finishing	2	Severe pilling
2x2 lycra rib	Grey	-	
	After dyeing	2	Severe pilling
	After finishing	4	Slight pilling

Table 3: Pilling Result.

1	3
1/2	3/4
2	4
2/3	4/5
3	5

Here, 5= Best rating refers to no color change as well as no staining.
Table 4: Gray Scale.

Fabric Type		Color change	Staining
CVC fleece	Grey	4/5	4/5
	After raising	4/5	4/5
Organic fleece	Grey	4/5	4/5
	After dyeing	4/5	4/5
	After raising and Sueding	4/5	4/5
2x2 lycra rib	Grey	4/5	4/5
	After dyeing	4/5	4/5
	After Sueding	4/5	4/5

Table 5: Color fastness result.

Conclusion

It was observed that the appearance of the samples was directly affected. GSM was also found to be varied thus fill and volume of the fabric was changed. GSM of CVC fleece was increased but the other two samples saw the opposite consequence. Strength test elucidated that the samples lost strength to some extent. 2×2 Lycra rib was found to be heavily affected regarding strength. Shrinkage and spirality illustrated that the dimensional integrity was compromised whereas 2×2 Lycra rib suffered the most variation. CVC and organic fleeces resulted in a widthwise expansion. CVC and organic fleeces were found to be highly prone to pilling but 2×2 Lycra rib presented slight pilling. Color fastness to wash showed no impact. So fabrics could be raised and sueded without the concern of coloration and color fastness to wash being hampered.

References

1. Marsh JT (1966) An Introduction to Textile Finishing. 2nd eds. Open library, London.

2. Shenai VA (1990) Technology of Textile Finishing.

3. Palmar JW (1996) Textile Processing and Finishing Aids: Recent Advancements.

4. Rouette HK (2011) Encyclopedia of Textile Finishing Volume III.

5. Arora SM (1983) Modern Techniques of Textile Dyeing Bleaching and Finishing. Small Industry Research Institute.

6. Murphy WS (2000) Technology of Textile Finishing.

7. Haigh D (1971) Dyeing and Finishing Knitted Goods. Hosiery Trade Journal, England.

8. Manual of Raising, Shearing and Sueding of Mario Crosta.

9. Booth JE (1969) Principles of Textile Testing. Chemical Publishing, New York, USA.

10. Skinkle JH (1972) Textile Testing: Physical, Chemical and Microscopical. Chemical Pub Co, New York, USA.

11. ASTM International.

12. Johnson GH (1927) Textile fabrics: Their selection and care from the standpoint of use, wear, and launderability. Harper & Brothers, New York, USA.

13. Collins GE (2009) Fundamental principles that govern the shrinkage of cotton goods by washing. Journal of the Textile Institute Proceedings 30: 46-61.

Analysis of Sulfonated Anthraquinone Dyes by Electrospray Ionization Quadrupole Time-of-flight Tandem Mass Spectrometry

Min Li, Yufei Chen, David Hinks and Nelson R Vinueza*

Department of Textile Engineering, Chemistry and Science, North Carolina State University, Raleigh, NC 27695, USA

Abstract

A tandem mass spectrometric method using a commercial quadrupole–time-of-flight (QTOF) mass spectrometer is described for the identification of sulfonated anthraquinone type dyes, having a 1-amino anthraquinone-2-sulfonate backbone. A total of 9 anthraquinone dye model compounds were evaporated and ionized via negative-ion electrospray ionization (ESI). Ionization of the sulfonated anthraquinone compounds primarily results in the formation of deprotonated molecules, [M-H]-. Once ionized, the ions were subjected to collision-activated dissociation (CAD). The type of neutral molecules or ions cleaved during CAD facilitates identification of the original compound. In most cases, a loss of 64 amu was observed for all dyes and was confirmed to be SO_2 by high resolution mass spectrometry analysis. A unimolecular rearrangement of the sulfonate (SO_3) group was triggered by CAD that allowed loss of SO_2. Also, it was found that different group functionalities attached to the anthraquinone backbone (e.g., secondary aromatic amines and secondary alkyl amines) have specific fragmentation pathways that can be used to distinguish them under similar CAD conditions. For example, an anthraquinone having a secondary amine with an aromatic group attached to it (e.g., Acid Blue 25) can be differentiated from an anthraquinone having a secondary alkyl amine (e.g., Acid Blue 62) based on the product ions. The resultant fragmentation patterns could contribute to the identification of unknown dyes with similar chemical structures. The method was also successfully used in concert with targeted CAD for quantification purposes. The methodology presented here is the first stage in building a high resolution mass spectrometry dye database from the extensive uncatalogued Max Weaver Dye Library at North Carolina State University.

Keywords: Acid dyes; Anthraquinone; Mass spectrometry; Tandem mass spectrometry; Structural elucidation

Introduction

The ability to identify unknown dyes in complex mixtures is of great importance in the areas of food, environment, human health, and forensics [1-5]. For example, acid dye manufacturing constitutes one of the highest production worldwide, and is normally used on nylon, synthetic polyamides, wool, silk, paper, inks and leather [6]. The continuous development of new products requiring improved dyeing properties (i.e., leveling characteristics, washfastness, and lightfastness) requires the use of suitable analytical tools for structural identification of new dyes, their byproducts and degradation products, as well as synthetic impurities. However, the chemical structures of dyes are often protected by patents, making their characterization a challenge. In addition, the same dye type may have a different structure depending on the colorant manufacturer. For this reason, we set about to develop a suitable methodology for the analysis of these compounds, both qualitatively and quantitatively.

Mass spectrometry (MS) has evolved into an essential and powerful tool for mixture analysis [7,8]. The high sensitivity, specificity, and speed of MS provides rapid and useful molecular-level information regarding complex mixtures. The development of atmospheric pressure ionization techniques-such as electrospray ionization (ESI)-smoothed the coupling of high-performance liquid chromatography (HPLC)-an invaluable tool in mixture analysis-with MS [9,10]. This approach allows for the determination of the molecular weights and elemental composition of known and unknown analytes. In order to obtain detailed information on the molecular structures of these analytes, tandem mass spectrometry (MS/MS) is required [11-17]. MS/MS elucidates the structures of ionized compounds by their fragmentation reactions through collision-activated dissociation (CAD) [18-20].

The analysis of anionic dyes, such as sulfonated and sulfated dyes,

by MS has commonly used negative-ion ESI as an ionization source due to the polar character of the dyes [21-23]. The use of this ionization technique produces a series of deprotonated molecules, which can be very useful in the determination of the molecular weight (MW) of the dye [21]. The total number of acid groups can also be determined with negative-ion ESI based on the number of protons replaceable by sodium ions [24].

We report here the use of ESI in combination with a quadrupole-time-of-flight (QTOF) MS/MS as a way to determine specific fragmentation pathways of commercial sulfonated anthraquinone dyes for identification purposes. In this study, a series of sulfonated acid dyes containing the structure of 1-amino anthraquinone-2-sulfonate (Figure 1) were analyzed by HPLC coupled to ESI-MS and ESI-MS/MS with the purpose to investigate the featured fragment loss of sulfonated anthraquinone dyes, as well as the fragmentation mechanism by which these molecules disassociate. This will be part of the first phase of building a high resolution mass spectrometry dye database from the extensive uncatalogued Max Weaver Dye Library donated from Eastman Chemicals, with approximately 100,000 dyes, to North Carolina State University.

***Corresponding author:** Nelson R Vinueza, Department of Textile Engineering, Chemistry and Science, North Carolina State University 2401 Research Dr. Box 8301 Raliegh, NC 27695 , USA
E-mail: nrvinuez@ncsu.edu

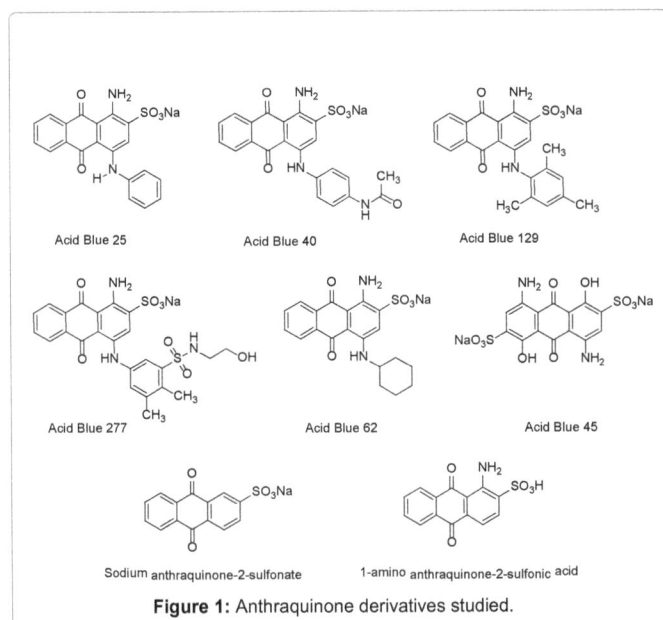

Figure 1: Anthraquinone derivatives studied.

No.	Name	Manufacturer	Molecular formula	MW[a]
1	Acid Blue 25	M. Dohmen	$C_{20}H_{14}N_2O_5S\ Na$	416.38
2	Acid Blue 45	Ciba Geigy	$C_{14}H_{10}N_2O_{10}S_2Na_2$	474.33
3	Acid Blue 40	M. Dohmen	$C_{22}H_{16}N_3O_6S\ Na$	473.43
4	Acid Blue 62	Classic Dyestuffs	$C_{20}H_{19}\ N_2O_5SNa$	422.43
5	Acid Blue 129	Sigma-Aldrich	$C_{23}H_{21}\ N_2O_5SNa$	460.48
6	Acid Blue 277	Ciba Geigy	$C_{24}H_{23}N_3O_8S_2Na$	567.57
7	1-amino anthraquinone-2-sulfonic acid	Sigma-Aldrich	$C_{14}H_9NO_5S$	303.29
9	Sodium anthraquinone-2-sulfonate	Sigma-Aldrich	$C_{14}H_7O_5S\ Na$	310.26

[a] Molecular Weight (MW)

Table 1: Sulfonated anthraquinone acid dyes and sulfonated anthraquinone model compounds.

Materials and Methods

Chemicals

Methanol and acetonitrile (LC-MS grade) were purchased from Honeywell & Burdick Jackson (Muskegon, MI, USA). The solvents were filtered through 0.22 µm Millipore filters (Whatman, GE Healthcare, UK). Ammonium formate (>99%, HPLC grade, Fluka, Switzerland), formic acid (~98%, MS grade), Sodium anthraquinone-2-sulfonate (**A1**), 1-Amino anthraquinone-2-sulfonic acid (**A2**) and Acid blue 129 were purchased from Sigma Aldrich (St Louis, MO, USA). The acid dyes (Table 1 and Figure 1) were supplied by Ciba Specialty Chemicals, M. Dohmen, and Classic Dyestuffs.

Sample preparation

Standard dye solutions were prepared at 1 mg/mL concentrations in HPLC grade 70:30 (v/v) methanol (MeOH)/ acetonitrile (CH_3CN). All solutions for analysis were prepared by adding 20 µL of standard dye solution to 980 µL of water (Milli-Q).

Instrumentation

Liquid chromatography/mass spectrometry (LC/MS): The experiments were carried out using an Agilent (Santa Clara, CA)

Accurate Mass 6520 Q-TOF mass spectrometer equipped with an ESI source operating in negative-ion mode ((-)ESI) and coupled with an Agilent 1260 SL HPLC system. The chromatography runs were performed using an Agilent Zorbax Eclipse Plus C_{18} column (2.1 × 50 mm, 3.5 µm) with a Zorbax Eclipse Plus C_{18} narrow bore guard column (2.1×12.5 µm, 5 µm). The mobile phase used for separation consisted of a mixture of 20 mM ammonium formate and 0.01% formic acid in H_2O (A) and 70:30 MeOH/CH_3CN (B). The flow rate was 0.5 mL/min with an injection volume of 3 µL. Negative ESI source parameters were as follows: nebulizer pressure, 35 psi; capillary voltage, 4000 V; drying gas flow, 12 L/min at 350 °C; and fragmentor voltage, 110 V. The instrument was operated in 4 GHz high resolution mode. Instrument control, data aquisistion, and analysis were performed using Agilent MassHunter Workstation Acquisition and Agilent MassHunter Qualitative Analysis B.06.00.

Collision activated dissociation (CAD): MS/MS experiments were performed on the collision cell (hexapole) of the QTOF by selecting the ion of interest with an isolation width of 1.3 Da (narrow) and colliding the ion with nitrogen gas (99.9995%) with collision energy of 40 V. All fragment ions were guided to the TOF mass analyser for detection.

Targeted MS/MS QTOF quantitation for acid blue 25: Quantitation of Acid Blue 25 (AB25) was achieved by using the fragment ion (m/z 329.0926) abundance generated from the corresponding deprotonated dye molecule. Five calibration solutions of AB25 were prepared (20.0, 40.0, 60.0, 80, and 100.0 µg/mL in 70:30 MeOH/CH_3CN) for calibration. A solution containing 50.0 µg/mL was analyzed by this MS/MS method and validated with a traditional liquid chromatographic method. All experiments were run at least three times for reproducibility purposes.

Results

A total of nine anthraquinone compounds were examined via (-) ESI/MS/MS. Most analytes primarily formed stable deprotonated molecules upon negative-ion ESI when using the 70:30 MeOH/CH_3CN as the solvent. The deprotonated molecules were subjected to CAD to obtain structural information. High-resolution measurements were carried out to verify the identities of the neutral molecules lost upon fragmentation. A detailed discussion on each of the anthraquinone derivatives studied is provided below.

Anthraquinones model compounds

A1, was the simplest anthraquinone derivative studied; it contained a sulfonate group at the 2 position with no other substituents on the aromatic system. Under negative-ion ESI conditions **A1** forms an abundant deprotonated molecule [M-H], which fragments upon CAD by loss of SO_2-confirmed by exact mass measurements (Table 2) to generate a unique fragment ion with a mass-to-charge ratio (m/z) of 223.0429. The loss of this neutral SO_2 molecule suggested a rearrangement of the sulfonate group (SO_3), which is in agreement with previous theoretical [25] and experimental [26] studies on aromatic systems. The fragment ion after the SO_3 rearrangement is considered a phenoxide anion and is useful for identification of similar structures.

A2, having an amino group at the ortho-positon to the sulfonate group, was found to undergo loss of SO_2 upon CAD of the deprotonated ion, giving a single fragment ion with m/z 238.0512 (Table 2). This fragment ion also generates a phenoxide anion, which was confirmed by high resolution MS.

In order to understand the fragmentation efficiency of these two

Analyte (MW)	MS (m/z)[a]	MS/MS CAD fragment ions (m/z)
A1 (310.26)	[M-H]$^{-b}$ (287.0022)	**287.0022 – SO$_2$ (223.0409)**
A2 (303.28)	[M-H]$^-$ (302.0129)	**302.0129 – SO$_2$ (238.0512)**
Acid blue 25 (416.38)	[M-H]$^-$ (393.0529)	**393.0529 – SO$_2$ (329.0926)** 393.0529 – NHC$_6$H$_5$ (301.0944) 393.0529 – SO$_2$ - NHC$_6$H$_5$ (329.0926)
Acid blue 40 (474.33)	[M-H]$^-$ (450.0763)	**450.0763 – SO$_2$ (386.1151)** 450.0763 – SO$_2$ – C$_2$H$_3$O(386.1151) 450.0763 – SO$_2$ – C$_8$H$_9$N$_2$O(237.0429) 450.0763 – SO$_2$ – C$_8$H$_8$NO(252.0540)
Acid blue 45 (473.43)	[M-H]$^-$ (428.9781)	**428.9781– SO$_2$ (365.0052)** 428.9781– C$_6$H$_{12}$N (289.0043) 428.9781– SO$_2$ – SO$_2$ (285.0527)
Acid blue 62 (422.43)	[M-H]$^-$ (399.1009)	**399.1009 –C$_6$H$_{11}$ (316.0155)** 399.1009 – SO$_2$ (289.0060) 399.1009 – SO$_2$ – C$_6$H$_{11}$ (252.0533)
Acid Blue 129 (460.48)	[M-H]$^-$ (435.1020)	**435.1020– SO$_2$ (371.1412)** 435.1020– SO$_2$ – C$_6$H$_{11}$ (252.0548) 435.1020– SO$_2$ – C$_7$H$_{12}$N (225.0464)
Acid blue 277 (567.57)	[M-H]$^-$ (544.0846)	544.0846 – SO$_2$ (480.1213) **544.0846 – C$_2$H$_6$NO$_3$S (420.0780)** 544.0846 – SO$_2$ – C$_2$H$_6$NO$_3$S (356.1165) 544.0846 –C$_{10}$H$_{15}$N$_2$O$_3$S (301.0045) 544.0846 – SO$_2$ –C$_{10}$H$_{15}$N$_2$O$_3$S (237.0427)

[a]The m/z value is the monoisotopic mass. [b]Deprotonated molecule, the sodium (Na+) cation under negative-ion ESI conditions will not be added to the m/z value. The m/z value of deprotonated molecule will differ by one or two sodium ions and/or one hydrogen atom from the total molecular weight of the dye.

Table 2: Ions formed upon ESI as well as the product ions formed upon CAD experiments for sulfonated dyes.

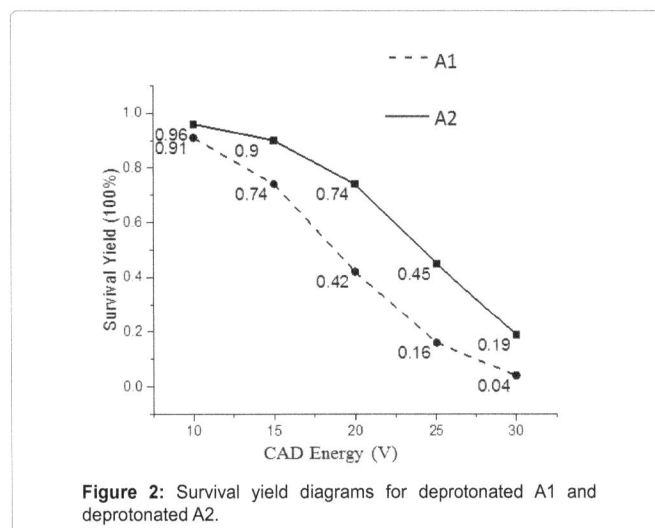

Figure 2: Survival yield diagrams for deprotonated A1 and deprotonated A2.

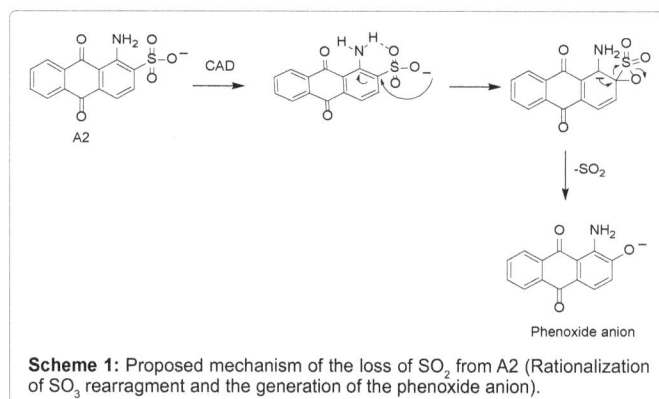

Scheme 1: Proposed mechanism of the loss of SO$_2$ from A2 (Rationalization of SO$_3$ rearragment and the generation of the phenoxide anion).

molecules, the two compounds were fragmented at the same CAD energies (ranging from 10 to 35V in 5V intervals). The Survival Yield (SY) methodology [27-29] was used as defined in Eqn. 1:

$$SY = \frac{I_p}{I_p + \sum I_f} \qquad (1)$$

where I_p is the intensity of the precursor ion (deprotonated molecule) and $\sum I_f$ is the sum of fragment intensities. A higher SY suggests the compound requires higher energy to be fragmented. Comparison of the SY between the compounds **A1** and **A2** is shown in Figure 2. The results suggest that the **A2** requires higher energy to fragment.

A possible explanation of the extra energy needed for fragmentation of **A2** can be related to hydrogen bonding between the hydrogen atom of the amino group and the oxygen atom of the sulfonate group. This can form a six-membered ring that stabilizes the deprotonated

molecule. A proposed mechanism of the sulfonate rearrangement is depicted in Scheme 1. The formation of an epoxide that leads to the cleavage of the SO$_2$ molecule has been proven as a favorable pathway by theoretical studies [25].

Acid dyes containing a secondary aromatic amine: Acid blue dyes 25, 40, 129 and 277 form abundant deprotonated molecules [M-H]$^-$ upon ionization via negative-ion ESI. All deprotonated acid dyes fragment upon CAD with a main fragment ion loss of SO$_2$ and a minor fragment ion loss of the secondary aromatic amine (NHC$_6$H$_5$, Table 2). These results suggest that the lowest energy pathway of fragmentation is the rearrangement of the sulfonate group to lose SO$_2$ as compared to the loss of the secondary aromatic amine. This characteristic fragmentation pathway is useful in determining possible substituents present on the anthraquinone backbone (Figure 3).

Acid Blue 62 (AB62) generates an abundant deprotonated molecule upon ionization via negative-ion ESI. CAD experiments showed that deprotonated AB62 dissociated via loss of alkyl group C$_6$H$_{11}$ as the main fragment ion (m/z 316.0155) followed by loss of SO$_2$ (yielding an anion with a m/z of 289.0060, Table 2). These results suggested that the lowest fragmentation energy pathway is the loss of the cyclohexyl group from the deprotonated acid blue as compared to the loss of SO_2 (Figure 4). Furthermore, this dissociation pathway is different from the ones observed for acid dyes containing secondary aromatic amines as substituents on the anthraquinone backbone.

Acid Blue 277 forms an abundant deprotonated molecule upon ionization via negative-ion ESI. This ion fragments via loss of $C_2H_6NO_3S$ to generate the main fragment ion with a m/z 420.0780 followed by the loss of SO_2 (m/z 480.1213, Table 2). This result suggests that the lowest energy path of fragmentation is loss of alkyl groups attached to aromatic ring of the secondary aromatic amine.

The CAD studies of the nine anthraquinone derivatives showed an interesting loss of SO_2, which can generate a very stable phenoxide anion useful for quantification by MS/MS. AB25 was selected to test the efficiency of quantitation by this method. The fragment ion with m/z 329.0926, generated upon CAD of the deprotonated ion (m/z 393.0529), was selected for calibration and sample quantification. The calibration curve is showed in Figure 5.

A solution of AB25 with a concentration of 50.0 μg/mL was

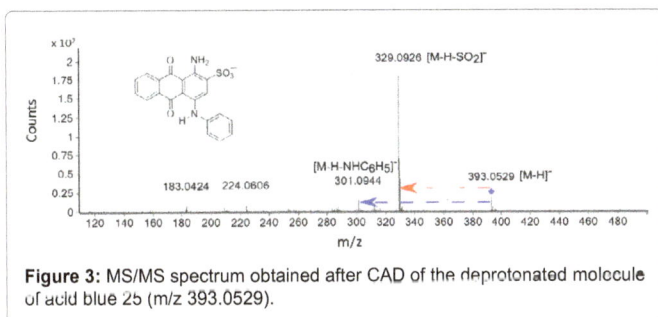

Figure 3: MS/MS spectrum obtained after CAD of the deprotonated molecule of acid blue 25 (m/z 393.0529).

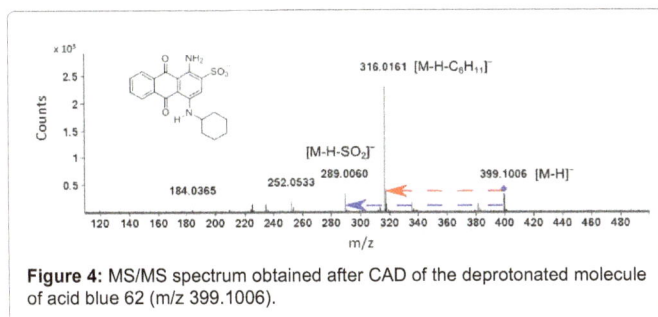

Figure 4: MS/MS spectrum obtained after CAD of the deprotonated molecule of acid blue 62 (m/z 399.1006).

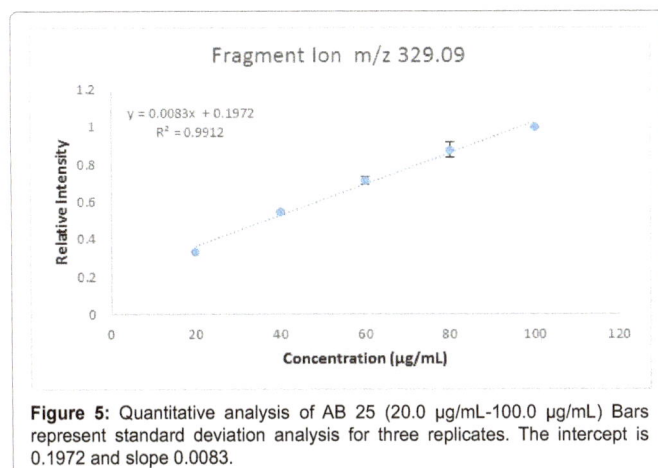

Figure 5: Quantitative analysis of AB 25 (20.0 μg/mL-100.0 μg/mL) Bars represent standard deviation analysis for three replicates. The intercept is 0.1972 and slope 0.0083.

quantified by HPLC and by targeted MS/MS. The results of the AB25 solution were for HPLC 50.77 μg/mL and for Targeted MS/MS 54.64 μg/mL. The percent error for the MS/MS method was 7.62% compared to the HPLC quantification value. The overall results strongly suggest that the phenoxide anion generated by loss of SO_2 by CAD is an exceptional ion for performing MS/MS quantification.

Conclusions

The deprotonated molecules formed upon negative-ion ESI were subjected to CAD events revealing fragmentation patterns that facilitated the identification of the anthraquinone derivatives. A loss of SO_2 was observed for all the sulfonated anthraquinone dyes, which indicated a characteristic rearrangement of -SO_3 functional group to form a stable phenoxide fragment ion. It became clear that this was the dominant fragment ion when the dye contained a secondary aromatic amine substituent (e.g., Acid Blue 25). In the presence of alkyl substituent groups on the anthraquinone backbone, the SO_3 rearrangement was not the main fragmentation pathway. Quantification by QTOF MS/MS of a sulfonated anthraquinone was feasible when the stable phenoxide anion was selected as observed in the AB25 experiments.

Overall, the fragmentation of acid dyes with similar group functionalities was analyzed. This provided tools for fingerprinting anthraquinone dyes and generating a database of dyes containing sulfonated anthraquinone structures. The results of this study will be applied on the Max Weaver Dye Library for structure characterization.

Acknowledgement

This work was supported by North Carolina State University startup fund.

References

1. Edlund PO, Lee ED, Henion JD, Budde WL (1989) The determination of sulfonated azo dyes in municipal wastewater by ion spray liquid chromatography tandem mass spectrometry. Biomed Environ Mass Spectrom 18: 233-240.

2. Fuh M, Chia K (2002) Determination of sulphonated azo dyes in food by ion-pair liquid chromatography with photodiode array and electrospray mass spectrometry detection. Talanta 56: 663-671.

3. Holcapek M, Jandera P, Prikryl J (1999) Analysis of sulphonated dyes and intermediates by electrospray mass spectrometry. Dyes and Pigments 43: 127-137.

4. Oliveira DP, Carneiro PA, Sakagami MK, Zanoni MVB, Umbuzeiro GA (2007) Chemical characterization of a dye processing plant effluent—identification of the mutagenic components. Mutation Research/Genetic Toxicology and Environmental Mutagenesis 626: 135-142.

5. Pan H, Feng J, He G, Cerniglia CE, Chen H (2012) Evaluation of impact of exposure of sudan azo dyes and their metabolites on human intestinal bacteria. Anaerobe 18: 445-453.

6. Hunger K (2003) Industrial Dyes: Chemistry, Properties, Applications. Weinheim, Germany: Wiley-VCH.

7. Lecchi P, Zhao J, Wiggins WS, Chen T, Yip PF, et al. (2009) A method for monitoring and controlling reproducibility of intensity data in complex electrospray mass spectra: A thermometer ion-based strategy. J Am Soc Mass Spectrom 20: 398-410.

8. Habicht SC, Vinueza NR, Archibold EF, Duan P, Kenttämaa HI (2008) Identification of the carboxylic acid functionality by using electrospray ionization and ion-molecule reactions in a modified linear quadrupole ion trap mass spectrometer. Anal Chem 80: 3416-3421.

9. Prakash C, Shaffer CL, Nedderman A (2007) Analytical strategies for identifying drug metabolites. Mass Spectrom Rev 26: 340-369.

10. Ma SK, Chowdhury SB, Alton K (2006) Application of mass spectrometry for metabolite identification. Curr Drug Metab 7: 503-523.

11. Auld J, Hastie DR (2009) Tandem mass spectrometry and multiple reaction monitoring using an atmospheric pressure chemical ionization triple quadruple

mass spectrometer for product identification in atmospherically important reactions. International Journal of Mass Spectrometry 282: 91-98.

12. Bandu ML, Watkins KR, Bretthauer ML, Moore CA, Desaire H (2004) Prediction of MS/MS data. 1. A focus on pharmaceuticals containing carboxylic acids. Anal Chem 76: 1746-1753.

13. Liu Y, He J, Zhang R, Shi J, Abliz Z (2009) Study of the characteristic fragmentation behavior of hydroquinone glycosides by electrospray ionization tandem mass spectrometry with optimization of collision energy. Journal of Mass Spectrometry 44: 1182-1187.

14. Lopez LL, Tiller PR, Senko MW, Schwartz JC (1999) Automated strategies for obtaining standardized collisionally induced dissociation spectra on a benchtop ion trap mass spectrometer. Rapid Communications in Mass Spectrometry 13: 663-668.

15. Habicht SC, Vinueza NR, Amundson LM, Kenttamaa HI (2011) Comparison of functional group selective Ion–Molecule reactions of trimethyl borate in different ion trap mass spectrometers. J Am Soc Mass Spectrom 22: 520-530.

16. Amundson LM, Owen BC, Gallardo VA, Habicht SC, Fu M, et al. (2011) Differentiation of regioisomeric aromatic ketocarboxylic acids by positive mode atmospheric pressure chemical ionization collision-activated dissociation tandem mass spectrometry in a linear quadrupole ion trap mass spectrometer. J Am Soc Mass Spectrom 22: 670-682.

17. Amundson LM, Gallardo VA, Vinueza NR, Owen BC, Reece JN, et al. (2012) Identification and counting of oxygen functionalities and alkyl groups of aromatic analytes in mixtures by positive-mode atmospheric pressure chemical ionization tandem mass spectrometry coupled with high-performance liquid chromatography. Energy Fuels 26: 2975-2989.

18. McLuckey SA (1992) Principles of collisional activation in analytical mass spectrometry. J Am Soc Mass Spectrom 3: 599-614.

19. Sleno L, Volmer DA (2004) Ion activation methods for tandem mass spectrometry. Journal of Mass Spectrometry 39: 1091-1112.

20. Mayer PM, Poon C (2009) The mechanisms of collisional activation of ions in mass spectrometry. Mass Spectrom Rev 28: 608-639.

21. Ballantine JA, Games DE, Slater PS (1997) The use of diethylamine to determine the number of sulphonate groups present within polysulphonated alkali metal salts by electrospray mass spectrometry. Rapid Communications in Mass Spectrometry 11: 630-637.

22. Volna K, Holcapek M, Suwanruji P, Freeman HS (2006) Mass spectrometric analysis of sulphonated dyes based on diaminobiphenyls. Coloration Technology 122: 22-26.

23. Rafols C, Barcelo D (1997) Determination of mono-and disulphonated azo dyes by liquid chromatography–atmospheric pressure ionization mass spectrometry. Journal of Chromatography A 777: 177-192.

24. Holcapek M, Jandera P, Zderadicka P (2001) High performance liquid chromatography–mass spectrometric analysis of sulphonated dyes and intermediates. Journal of Chromatography A 926: 175-186.

25. Ben-Ari J, Etinger A, Weisz A, Mandelbaum A (2005) Hydrogen-shift isomerism: Mass spectrometry of isomeric benzenesulfonate and 2-, 3-and 4-dehydrobenzenesulfonic acid anions in the gas phase. Journal of Mass Spectrometry 40: 1064-1071.

26. Binkley RW, Flechtner TW, Tevesz MJ, Winnik W, Zhong B (1993) Rearrangement of aromatic sulfonate anions in the gas phase. Org Mass Spectrom 28: 769-772.

27. Kuki A, Nagy L, Zsuga M, Keki S (2011) Fast identification of phthalic acid esters in poly (vinyl chloride) samples by direct analysis in real time (DART) tandem mass spectrometry. International Journal of Mass Spectrometry 303: 225-228.

28. Kertesz TM, Hall LH, Hill DW, Grant DF (2009) CE 50: Quantifying collision induced dissociation energy for small molecule characterization and identification. J Am Soc Mass Spectrom 20: 1759-1767.

29. Kuki A, Shemirani G, Nagy L, Antal B, Zsuga M, et al. (2013) Estimation of activation energy from the survival yields: Fragmentation study of leucine enkephalin and polyethers by tandem mass spectrometry. J Am Soc Mass Spectrom 24: 1064-1071.

5

Optimizing Effects of Cots Shore Hardness on Cotton Yarn Properties at Ring Frame

Bagwan AS[1*], Policepatil R[2] and Pawar S[1]

[1]Mukesh Patel School Of Technology, Management and Engineering, Shirpur, Dhule, India
[2]Spentex Pvt. Ltd. Baramati, Pune, Maharashtra. India

Abstract

The effect of eight different spinning front and back line cots (Synthetic rubber cot) varying only in shore A hardness (65°,83°) on 100% cotton 30's ring spun yarn has been investigated. The change in cotton yarn properties like mass uniformity, unevenness percent, Imperfection levels (in all class) with progressive change in shore A hardness has also been reported. The count and process parameter's from opening and cleaning machines that covers blow room and carding then breaker and finisher drawing, speed frame and up to ring spinning kept identical. As one progress from lesser shore hardness (65°) to higher shore hardness (83°) the yarn unevenness percent and imperfection levels gradually increases.

Keywords: Cotton; Spinning; Yarn

Introduction

Yarn quality is essential to the economic success of spinning plants. International competition and market requirements dictate the necessity to produce quality yarns at an acceptable price [1,2].

In general yarn quality is influenced by:

- Quality of raw material

- Opening and cleaning operations at Blow room and Carding

- Speeds and Settings kept at various stages of yarn production and its functions.

- Process control techniques and parameters kept at spinning

- Humidification, (temperature and humidity)

- labour force training and their skills.

- Maintenance of production equipment and vital components.

Drafting components have a significant influence on yarn quality and production costs in ring spinning. Especially spinning top roller covers i.e., cots and drafting aprons [3]. These are the main components of the drafting mechanism and certainly it has more influence on the quality of the yarn produced Cots are used in draw frame, comber, speed frame and ring frame, whereas aprons are used only in speed frame and ring frame. The purpose of cots is to provide uniform pressure on the fibre strand to facilitate efficient drafting and use of aprons help to have better grip and control on fibres particularly floating fibres [4]. A front line cot in ring spinning should also offer sufficient pulling force to overcome drafting resistance. Mathematically, Force of pulling required at front line cot Frictional resistance between fibers and Force exerted by the aprons on fibers [5-7].

Essential characteristics of a spinning cot

The raw material Compounds on the basis of special rubber in the hardness range of approx. 65 to 83 Shore A hardness are used as coating raw materials [8].

The composition of the raw material determines the characteristics of the cot such as

- Shore A hardness of the rubber cot

- Resilience properties, low Compression set values and elasticity

of the cot.

- Surface Characteristics like grip offered on fibre strands.

- Abrasion resistance.

- tensile strength

- Swelling resistance

- Color - These characteristics should fulfill the following demands made on a top roller cover

- Good fiber guiding

- No lap formation

- Long working life

- Good ageing stability

- Minimal film formation

Normally synthetic top roller cots are available in cylindrical form. The technical specifications of a top roller cot are i) Bare roller diameter BRD ii) Finished outer diameter FOD iii) Width or Length iv) construction like Alufit or PVC core and v) Shore A hardness. Shore hardness is one of the main properties of top roller cot and varies for different types of fibre, application etc. (Figure 1).

Literature Review

Shore hardness

Generally Shore hardness of a rubber cot is measured by using an

---

*Corresponding author: Abdul salam Bagwan, Centre For Textile Functions, Mukesh Patel School Of Technology, Management and Engineering, Shirpur, District – Dhule, India, E-mail: abdulsalaambagwan@gmail.com

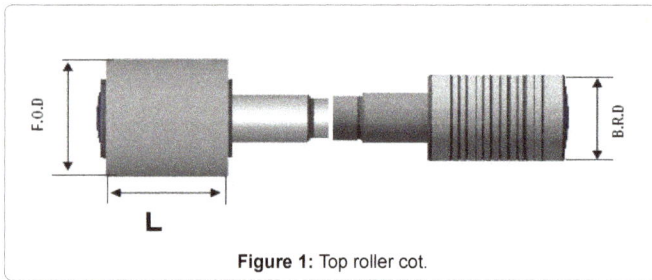

Figure 1: Top roller cot.

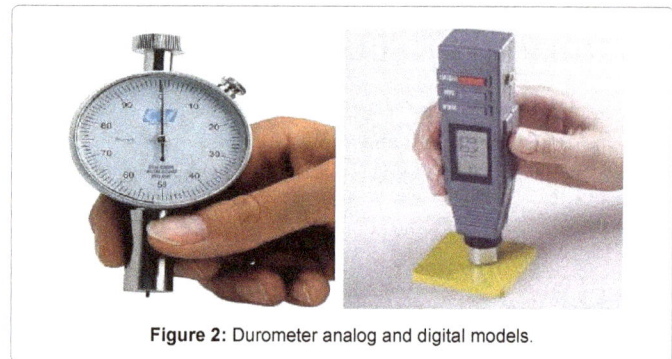

Figure 2: Durometer analog and digital models.

instrument called 'Durometer' and the value is expressed in A scale. Cots are available in wide shore hardness ranging from 63° to 90° shore.

Definition of shore hardness

Hardness may be defined as the resistance to indention under conditions that do not puncture the rubber. It is called elastic modulus of rubber compound. These tests are based on the measurement of the penetration of the rigid ball into the rubber test piece under specific conditions. The measured penetration is converted into hardness degrees. Shore ADurometer is used for measuring soft solid rubber compounds. Other scales are also used like Shore D which is used to measure the hardness of very hard rubber compounds including ebonite. The main drawback is in reproducibility of results by different operators. So, a practical tolerance of 5° is acceptable. As per the ASTM (D 2240 – Defines apparatus to be used and its sections such as diameter, length of the indentor, force of spring and D 1415 –Defines specimen size), DIN, BRITISH and ISO Standards following test conditions have been laid for measuring.

Shore a hardness of rubber products (Figures 2 and 3)

1. The specimen should be at least 6 mm in thickness.

2. The surface on which the measurement made should be flat.

3. The lateral dimension of the specimen should be sufficient to permit measurements at least 12 mm from the edges.

Mathematically, Arc of contact or the nipping length made by top roller cot with fluted roller (I) is inversely proportional to the shore hardness of the rubber cot. In general, Lower the shore hardness higher will be the contact area with steel bottom roller better so that there will be positive control on fiber's strand producing the yarn with better mass uniformity, lesser imperfection levels. Under Identical condition a cot measuring 65° Shore Hardness will make larger arc of contact with steel bottom than a cot measuring 83° Shore hardness.

Experimental Work

The Experimental work flow is shown in (Figure 4)

Material Methods

In this investigation 100% MCU -5 cotton was chosen as raw material with the following fiber parameters and 30 s Ne combed count was produced at ring spinning. In the production process, cotton processed through blow room, carding, breaker draw frame, unilap, finisher draw frame followed by speed frame and ring frame. Table 1 indicates the properties of cotton used in production of 30's cotton yarn. At ring frame altering the positions of front shore hardness and back shore hardness 30's yarn produced. Obtained yarn tested for unevenness, Rkm, Elongation, Imperfections on Uster Uneveness tester (UT-5) and Uster tensorapid tester (UT-3) in order to assess the yarn properties of 30's count.

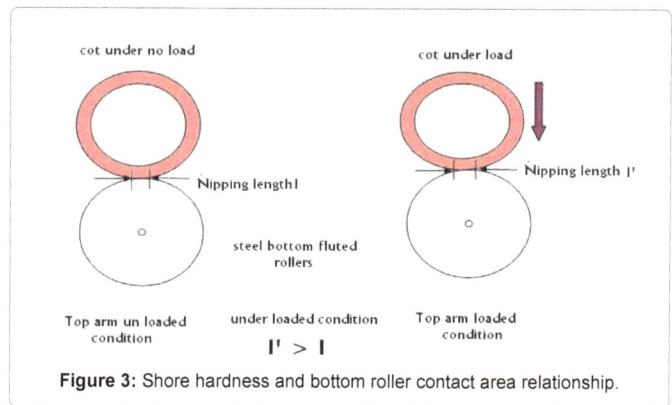

Figure 3: Shore hardness and bottom roller contact area relationship.

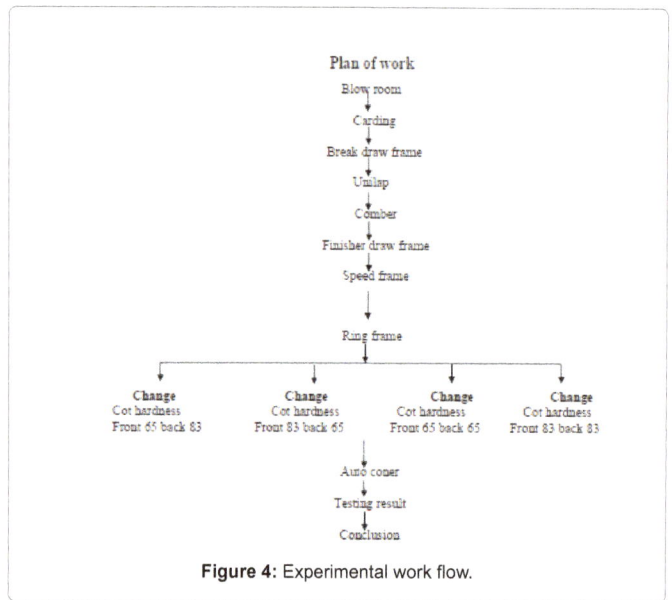

Figure 4: Experimental work flow.

HVI test data

Fibre parameters are shown in the Table 1.

Result and Discussions

From Table 2 and Figure 5 The present investigation summarized that, The effect different spinning front and back line cots (Synthetic rubber cot) varying only in shore A hardness (65°, 83°) on 100% cotton 30's ring spun yarn has been investigated. The change in cotton yarn properties like mass uniformity, unevenness percent, Imperfection levels (in all class) with progressive change in shore hardness has also

2.5 % Span Length in mm	30.70	Bundle Strength at 3 mm Gauge	23.5 gms / Tex
50 % Span Length in mm	13.70	FibreMicronaire	3.8 µgs / Inch
Raw Material Trash %	3.3 %	Short Fibre Content by (n)	27.8 %
Short Fibre Content by (w)	10.3 %	Maturity Ratio	0.88
Immature Fibre Content	6.2 %	Neps / Gram	106

Table 1: Shows fiber parameters selected in manufacturing of cotton yarn.

Count	30 Ne	30 Ne	30 Ne	30 Ne
Cot Shore Hardness	F 65/ B 83	F83/ B 65	F65/B 65	F 83/ B 83
Count	29.73	29.59	29.54	29.14
U%	8.97	9.55	8.79	9.54
CVm%	11.07	12.09	11.12	12.08

Table 2: Effect of shore hardness on 30's count.

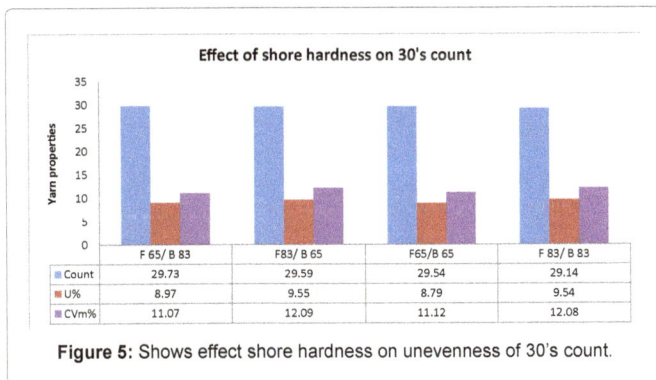

Figure 5: Shows effect shore hardness on unevenness of 30's count.

Count	30 Ne	30 Ne	30 Ne	30 Ne
Cot Shore Hardness	F 65/ B 83	F83/ B 65	F65/B 65	F 83/ B 83
Thin -30%	503.3	875.5	442.5	796.3
Thik +35%	156.3	318.5	139.8	300
Neps +140%	310	348.8	292	78.8
Total Sensitivity	969.6	1484.8	874.3	1174.6

Table 3: Effects of different shore hardness on Imperfections of 30's count.

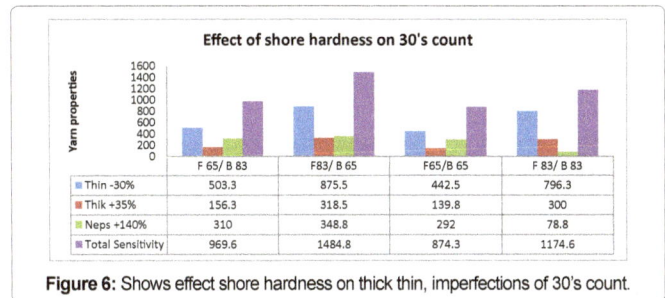

Figure 6: Shows effect shore hardness on thick thin, imperfections of 30's count.

Count	30 Ne	30 Ne	30 Ne	30 Ne
Cot Shore Hardness	F 65/ B 83	F83/ B 65	F65/B 65	F 83/ B 83
SH	0.92	0.95	0.89	0.92
UTR 3 Results	–	–	–	–
RKM	17.23	17.55	17.76	18.03
CV%	6.54	7.61	6.27	6.65
Elongation	4.33	4.3	4.37	5.2
CV%	7.93	9.96	9.77	7.03

Table 4: Effects of different shore hardness on Rkm and elongation of 30's count.

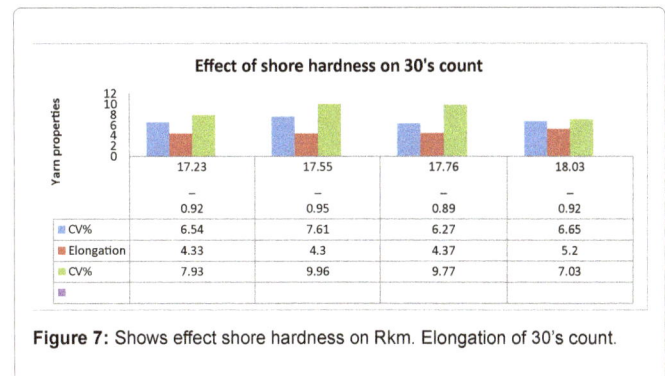

Figure 7: Shows effect shore hardness on Rkm. Elongation of 30's count.

been reported.

Unevenness (U percent)

From Table 3 and Figure 6 summarized that, among the three shore hardness, yarn obtained F65, B65, shows improvement in Upercent, RKm, IPI, of 30's yarn, this is due to the fact that lower shore hardness cot helps for increase in area of contact with the fluted bottom roller, which significantly shortens the uncontrolled area between apron to cot nipping point.

Among the all shore hardness third F 65/B65 shows optimum quality because it produce less thick, thin, neps this is obtained due to Lower the degree of shore hardness, higher the softness of rubber compound and vice –versa. Even though softer cots under normal spinning conditions produces better yarns with better mass uniformity and IPI levels

Yarn strength Rkm and IPI

From Table 4 and Figure 7 summarized that, among the three shore hardness, yarn obtained with F65, B65, shows improvement in RKm due to Lower the degree of shore hardness, higher the softness of rubber under normal spinning conditions produces better yarns gripping and twist realization in yarn which results in increase in strength and improvement in IPI alue.

Among the shore hardness {65, 83°} the third combination found more suitable for cotton processing of 30's count because it would give optimum strength and elongation. From the data it was inferred that F65/B65 gives optimum result in the production of 30's count.

The effect of nine different spinning front line cots (Synthetic rubber cot) varying only in shore A hardness (65°, 83°) on 100% cotton ring spun yarn has been investigated. The change in cotton yarn properties like mass uniformity, unevenness percent, Imperfection levels (in all class) with progressive change in shore hardness has also been reported.

Conclusions

Present investigation summarized that

- In cotton yarns, with increase in shore hardness form 65° to 83°, the co-efficient of variation of yarn mass CV (m) percent and Yarn Unevenness, U (m) percent increased

- Imperfection level and yarn unevenness percent usually increases with increase in shore hardness. This is due to the fact that lower shore hardness cot helps for increase in area of contact with the fluted bottom roller, which significantly shortens the uncontrolled area between apron to cot nipping point.

- Lower the degree of shore hardness, higher the softness of rubber compound and vice –versa. Even though softer cots under normal spinning conditions produces better yarns with better mass uniformity and IPI levels.

Scope of the study

This investigation is done with respect to one particular yarn count (30 s combed yarn), further studies can be conducted with different count range, raw material and other yarn quality parameters can be tested and the relationship can be further extended with respect to cots shore hardness.

Acknowledgements

The authors acknowledged valuable support received from The Director MPSTME, Shirpur and Associate Dean (Textile Technology) MPSTME, shirpur The Principal, Centre for textile functions, MPSTME Shirpur and Chief executive, [2] Spentex Pvt. Ltd. Baramati, District –Pune, Maharashtra, INDIA.

Reference

1. Singh AK, Agarwal V (1997) Effect of Shore Hardness of Cots on yarn Quality. Indian Textile Journal.

2. Btra J, Sitra, Atira, Nitra (2012) Technical report for the project Measures for Meeting Quality Requirements of cotton Yarns for Export Sponsored by Ministry of Textiles.

3. Kane CD, Ghalsaso SG (1992) Studies on Ring Frame Drafting Part. Indian Textile Journal.

4. Parn PK (1999) Latest Trends in Cots & Aprons. Journal of Textile Association.

5. Nawaz M, Butt MA (1996) effect of different drafting parameters on yarn regularity. Pakistan Textile Journal 45: 29 -31.

6. Russell SJ, Dobb MG (1995) the effect of rubbing on the structural and tensile properties of woolen slubbings. J Fibre Textile Inst 86: 415- 424.

7. Sambandhan PKS, Sudhangaran C (1983) Aprons and Cots. Ind J Fibre Textile Res 94: 91-97.

8. Subramanian V, Mohammed A (1991) Effects of Apron Spacing and Break Draft on Double Rove Yarn Quality in Short-Staple Spinning. Textile Res J 61: 280- 285.

6

Predicting the Tensile and Air Permeability Properties of Woven Fabrics Using Artificial Neural Network and Linear Regression Models

Ghada Ali Abou-Nassif*

Fashion Design Department, Design and Art Faculty, King Abdul Aziz University, Jeddah, Saudi Arabia

Abstract

The objective of this paper is to investigate the predictability of some of woven fabric properties using artificial neural network (ANN) and regression models. For achieving this purpose, a neural network with three layers was adopted. The regression model was of type a multiple – linear regression one. The independent variables were weft yarn count, twist multiplier and weft density; and the dependent ones were tensile strength, breaking extension and air permeability of the woven fabrics. The ANN and regression models were assessed using the Root means square error (RMSE) and the coefficient of determination (R2-value). The findings of this study revealed that ANN is superior to regression model in predicting the woven fabric properties.

Keywords: Neural networks; Back-propagation; Woven fabrics properties; Tensile strength; Air permeability

Introduction

Artificial neural network

Artificial Neural Networks (ANNs) are algorithmic structures derived from a simplified concept of the human brain structure. They belong to the Soft Computing family of methods, along with fuzzy logic/fuzzy control algorithms and genetic algorithms [1]. They all share an iterative, non-linear search for optimal or suboptimal solutions to a given problem, without the presupposition of a model of any type for the underlying system or process [2]. Various different ANN types have already been successfully employed in a wide variety of application fields [3]. Major ANN functionalities are: function approximation which can be exploited in system input-output modeling and prediction, and pattern recognition and classification problems [4].

The structure or 'architecture' of an ANN contains a number of nodes, called neurons, organized in a number of layers and interconnected to form a network. Each neuron receives connections from other neurons and/or itself, each with an associated weight. The interconnectivity defines the topology of the ANN. The weights represent information being used by the neural network model to solve a problem. One of the central issues in neural network design is to utilize systematic procedures (a training algorithm) to modify the weights directly from the training data without any assumptions about the data's statistical distribution [5].

There are different kinds of topologies and training algorithms but the multi-layered feed-forward neural network with back-propagation learning algorithms is more popular and commonly used [6]. In this structure, the neurons are located in layers and from one layer to another one connected with each other with links to carry the signals between them. There is a weight for each connection link which acts as a multiplication factor to the transmitted signal. An activation function such as linear or sigmoid is applied to each neuron's input to determine the output signal. Usually a feed forward neural network consists of several layers of nodes, one input layer, one output layer and some hidden layers in between.

The training of a neural network by back-propagation involves three stages: The feed-forward of the input training pattern, the calculation and back-propagation of the associated error, and the adjustment of the weights. The calculation of error vector to adjust the weights is done according to the calculated mean square error (MSE) form the difference between actual and predicted outputs according to the following relationship.

$$\text{MSE} = \frac{1}{N} \sum_{i=1}^{N} (y_i - x_i)^2 \tag{1}$$

Where N=the number of observations, y_i=the neural network predicted values, and x_i=the actual target values.

In the backward pass, this error signal is propagated backwards to the neural network and the synaptic weights are adjusted in such a manner that the error signal decreases with each iteration process. Thus, the neural network model approaches closer and closer to producing the desired output. The corrections necessary in the synaptic weights are carried out by a delta rule, which is expressed by the following equation.

$$\Delta W_{ji(n)} = -\eta \left[\frac{\partial(MSE)}{\partial W_{ji(n)}} \right] \tag{2}$$

Where $W_{ji(n)}$ is the weight connecting the neurons j and i at the nth iteration; $\Delta W_{ji(n)}$, the correction applied to $W_{ji(n)}$ at the nth iteration; and η, a constant known as learning rate [7,8].

Prediction of yarn and woven fabric properties

Generally, modeling and prediction of yarn and fabric properties based on fiber properties, fabric characteristics and process parameters have been considered by many researchers. Over the years, one of the first approaches has been the use of mechanistic models. In this category, some studies such as the work of Alsaid A Almetwally [9] and Pinar [10] to predict cotton yarn strength and spinning quality, Hearle, El-Behery and Thakur [11] in relation to structural mechanics

---

*Corresponding author: Ghada Ali Abou-Nassif, Fashion Design Department, Design and Art Faculty, King Abdul Aziz University, Jeddah, Saudi Arabia
E-mail: dr_ghada2013@yahoo.com

of fibers, yarns, and fabrics, the prediction of yarn strength by Frydrych [12], and works performed by Vitro et al., [13] Kim and El-sheikh [14] and Yong Ku Kim [15] and El-sheikh [16] can be mentioned. However, statistical regression models for this purpose have been used by some researchers, namely El-Mogahzy [17], Hunter [18] and Alsaid Ahmed Almetwally [19,20] and M M Mourad [21].

The limitation of mechanistic and statistical regression models was described in previous works [22,23]. ANNs, genetic algorithm, and fuzzy set theory are presented as attractive alternatives for predictive modeling. ANN algorithms have been used by many researchers for modeling the different textile processes, especially for predicting different kinds of yarn and fabric properties. Alsaid Ahmed Almetwally [5] compared between artificial neural networks and regression models in relation to the tensile properties of cotton/spandex core spun yarns. The results of his study revealed that ANN has better performance in predicting comparing with multiple linear regressions. ANNs were also used to predict the stress strain curve of the woven fabrics [24].

Fabric hand is a property that combines the mechanical properties of a fabric with the sensory perception of the fabric by the humans when they touch it. Fabric hand was predicted by using artificial neural networks by Youssefi and Faez [25], Hui et al. [26], and Matsudaira [27]. Drapability is far the most complex mechanical property of the fabrics and it is essential for many applications of the textile fabrics. The prediction of the drape has been made using ANNs [28].

Materials and Methods

In this study eighteen fabric samples were woven. All fabric samples were produced from Egyptian cotton of type Giza 80. The properties of cotton used to spin the weft and warp yarns from which the fabric samples are woven, were listed in Table 1. All warp yarns were made with the same English count, but the weft yarns were spun from three different counts. Each weft yarn also has two twist multipliers. Each fabric sample was woven with three different densities for the weft yarns. Generally, fabric samples used in this study were woven on Rapier weaving machine of model Picanol Gamamax with the following particulars:

Warp yarn count: 30/1 Ne

Weft yarn counts: 24/1 Ne, 30/1 Ne, and 36/1 Ne.

Warp yarn density: 108 ends/inch.

Weft yarn densities: 50, 60, and 70 ppi.

Twist multiplier of warp yarns: 4.

Twist multipliers of weft yarns: 3.8 and 4.2.

Weave structure: plain 1/1.

Warp width: 176.8 cm.

Number of harness: 6

Parameter	Value
Micronaire value	4.67
Mean length (mm)	28.7
Uniformity Index (%)	82.4
Strength (g/tex)	34.1
Elongation (%)	7.8
Maturity %	86
Fineness (mtex)	185

Table 1: Characteristics of cotton used to make the fabric samples.

Reed count: 20.5 dent/cm

Fabric width: 170 cm (Table 1).

Since the variation in the fabric samples were conducted in the weft direction, then testing the fabric properties were carried out in the same direction. Before testing, all fabric samples were conditioned and then tested under standard condition of 20 ± 2°C and 65 ± 5% relative humidity. Tensile strength and breaking extension measurements of the fabric samples were executed by standard strip method on an Instron 4411 Tester (Instro Inc., USA) in accordance with ASTM D1682; ten individual readings for each fabric sample were taken and averaged. The air permeability was measured by shierly Air permeability Tester according to ASTM D737-04; ten readings for each fabric sample were recorded and averaged.

Neural network design

Architecture: First, the size of the network must be determined by the number of hidden layers and the number of neurons in these layers. A network with three layers is sufficient for most practical applications. The number of input neurons normally corresponds to the number of input variables of the process to be modeled. In selecting the output neurons, note that it is generally inadvisable to train a network for several tasks simultaneously. In this study we have selected a network with three layers: an input layer with three neurons, hidden layer with 5 neurons, and an output layer with one neuron corresponding to one dependent variable subject to the analysis at a time.

Learning strategy and transfer functions: Before learning, the whole experimental data were segregated into training, validation, and testing patterns: 60% of the patterns were randomly selected for training, 20% were for testing the neural network, and the remaining 20% of patterns were for validating the model's performance [5]. Training is an important feature of neural networks. The objective of the training process is to minimize the squared error between the network output and the desired output. This is done by adjusting the connection weights across the network. The validation set is a part of the data used to tune network topology or network parameters other than weights. For example, it is used to define the number of units to detect the moment when the neural network performance started hidden to deteriorate. To choose the best network (i.e. by changing the number of units in the hidden layer), the validation set is used. Whereas, the test set is a part of the input data-set used to test how well the neural network will perform on new data. The test set is used after the network is ready (trained), to test what errors will occur during future network application. This set is not used during training and thus can be considered as consisting of new data entered by the user for the neural network application. The learning method is back-propagation with Levenberg–Marquardt algorithm. Three of the most commonly used transfer functions are linear, sigmoid, and tanh:

Linear: $f(x)=x$ (3)

Sigmoid: $f(x)=\dfrac{1}{(1+e^{-x})}$ (4)

Tanh: $f(x)=\dfrac{(e^{x}-e^{-x})}{(e^{x}+e^{-x})}$ (5)

The topology architecture of feed-forward three-layered back-propagation neural network is illustrated in Figure 1.

Performance of the neural network

The performance of the ANN was assessed using the root mean

square error (RMSE), and the coefficient of determination according the following formulas:

Root mean square error: RMSE= $\sqrt{\dfrac{1}{N}\sum_{i=1}^{N}(y_i - x_i)^2}$ (6)

Coefficient of determination= $\dfrac{SS_{reg}}{SS_{total}}$ (7)

Where N=the number of observations, y_i=the neural network predicted values of fabric properties, x_i=the actual values of fabric properties.

Statistical regression

In this study, multiple-linear regression was used for developing three predictive models of fabric tensile strength, fabric breaking extension, and fabric air permeability. In this concern, the same sets of data used for evaluating ANN model were used in a multiple linear regression algorithm. Independent variables were weft yarn count (Ne), weft density (ppi) and the twist multiplier of weft yarns. The following linear regression models correlate the fabric properties of woven fabrics with the independent variables, i.e. weft yarn count, weft density, and twist multiplier.

Tensile strength, kg=97.3 – 2.3 × weft yarn count + 5.3 × twist multiplier + 0.6 × weft density

Breaking extension, %=31.2- 0.2 × weft yarn count + 1.1 × twist multiplier – 0.14 × weft density

Air permeability, (cm³/cm².sec)=45.4 + 0.4 × weft yarn count + 5.2 × twist multiplier – 0.96 × weft density

As seen from the regression models above, twist multiplier of weft yarns affected all fabric properties positively. However, weft yarn density affected breaking extension and air permeability negatively and affected the fabric tensile strength positively. Whereas weft yarns count

influenced the tensile strength and breaking extension negatively; while air permeability is affected positively. The statistical analysis also revealed that yarn count accounts for 76% of the variability in fabric tensile strength and that twist multiplier accounted for 7%, while weft density accounted for 52% of the variability in fabric tensile strength. With respect to breaking extension, it was determined that weft yarn count, twist multiplier and weft density accounted for 50%, 11% and 55% respectively of the variability in fabric breaking extension. Finally, it was found that weft yarn count, twist multiplier and weft density accounted for 23%, 12% and 92%, respectively of the variability in fabric air permeability.

Results and Discussion

In this study, we have adopted a three-layer neural network consisting of a three-input layer, a 5-neuron hidden layer, and a one-neuron output layer at a time, focusing on fabric tensile strength (kg), breaking extension (%), and air permeability (cm³/cm².sec). The neural network learning model is in accordance with the experiment data of three inputs and three targets, which are listed in Table 2. The general view of the neural network used in this study is shown in Figure 1. The RMSE and coefficient of determination (R^2) were used to judge the performance of ANN and regression models (Table 2).

Tensile strength

Tensile strength has been accepted as one of the most important attributes of woven fabrics. It is the main characteristic that distinguishes it from non-woven and knitted fabric. The strength of a woven fabric depends not only on the strength of constituent yarns, but also on the yarn and fabric structure and many other factors [29].

The actual and predicted tensile strength values of woven fabric samples, according to the variation of the independent variables, using ANN and regression models and its performance are presented in Tables 3 and 4 and in Figure 2, respectively. The results revealed that ANN model gives the best performance with the least RMSE and highest R^2 values. The values of the root mean square error were 3.1 and zero for regression and ANN models, respectively. The coefficient of determination values of regression and ANN models were found to equal 0.87 and 1, respectively. This signifies that an artificial neural network model fits the data very well and it is better than the regression model in predicting the tensile strength of the woven fabrics.

From Table 3, it can be seen that the minimum and maximum absolute errors for regression model are 1.6 and 15.5%, while the corresponding values for ANN model are zero and 6.1%. This confirms that ANN model has a better predictive performance than regression analysis to predict the tensile strength of woven fabrics under study.

Figure 2 compares the predicted outputs of the ANN model and

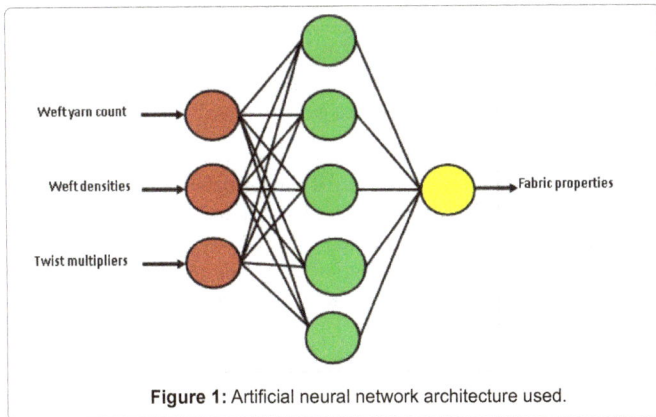

Figure 1: Artificial neural network architecture used.

Sample No.	Inputs			Targets		
	Yarn count (Ne)	Twist multiplier	Weft density (ppi)	Tensile strength (kg)	Breaking extension (%)	Air permeability (cm³/cm².sec)
1	24	4.2	50	39.33333	24.66667	30.33333
2	30	3.8	50	33	22.00000	28.66667
3	30	3.8	70	55.33333	19.33333	11.00000
4	30	4.2	60	41.66667	21.66667	22.00000
5	36	3.8	60	25	21.00000	19.66667
6	36	3.8	70	31.66667	17.66667	13.33333
7	36	4.2	70	36	18.33333	13.66667

Table 2: Learning data of the back-propagation neural network for fabric tensile properties and air permeability.

Actual values	Multi-linear regression		Artificial neural network	
	Predicted value	Absolute error (%)	Predicted value	Absolute error (%)
54.66667	56.83333	3.963402	54.55063	0.212268
39.33333	45.44444	15.53672	39.33333	0
51.33333	54.72222	6.601734	48.21225	6.080026
48.33333	47.55556	1.609179	48.33333	0

Table 3: Tensile strength predicted by multi-linear regression and neural network.

Statistical parameters	Regression	ANN
Coefficient of determination, R^2	0.87	1
Root mean square error, RMSE	3.072428	0.000

Table 4: Comparison of prediction performance of ANN and regression models for tensile strength.

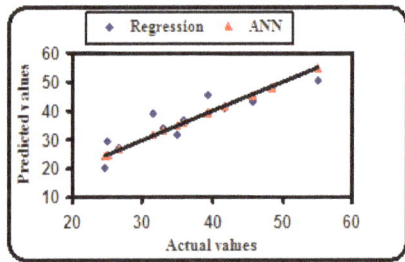

Figure 2: Actual against predicted values using ANN and regression models for tensile strength.

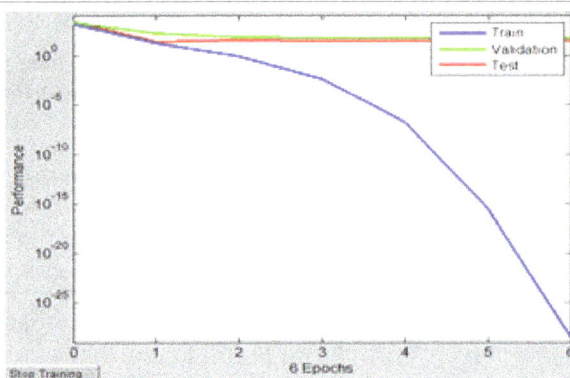

Figure 3: The performance plot of the ANN used to predict the fabric tensile strength.

multi-linear regression model with the experimentally actual tensile strength of woven fabrics. It is observed that ANN model reproduced the trend in the measured values better than regression model; and the ANN model has more capability of making quantitatively accurate tensile strength predictions. The correlation coefficient between actual and predicted yarn tensile strength using ANN and regression models were 1 and 0.94, respectively (Tables 3 and 4) (Figures 2 and 3).

Figure 3 describes the performance training algorithm for the ANN used to predict the tensile strength of the woven fabrics. This is a useful diagnostic tool to plot the training, validation, and test errors to check the progress of training. This figure describes that the test set error and the validation set error has similar characteristics, and it does not appear that any significant over fitting has occurred. The training stopped after 6 iterations because the validation error increased.

Breaking extension

Equally important to the fabric strength is its ability to extend under load. When the fabric is subjected to tension in one direction, the extension takes place in two main phases. The first phase is decrimping or crimp removal in the direction of the load. The removal of the crimp is accompanied by a slow rate of increase of the load. The second phase is the extension of the yarn during which the fabric becomes stiffer; the stiffness depends mainly on the character of the yarn. The more is the crimp in the yarn, the more extensible is the fabric [29,30].

Actual and predicted values of breaking extension of the woven fabrics under study and their predicted errors are listed in Table 5. A Comparison between the performances of the ANN and multiple linear-regression models on the data-sets is provided in Table 6.

It can be noticed that ANN model achieved an average RMSE of 0.0006. Comparing this value to the regression model with an average RMSE of 0.1.2, suggested that the ANN model highly superior to the regression one.

It is also observed that the artificial neural network model has a capability of achieving a very good fit to the measured breaking extension values as noticed by high R^2 value of 0.9999. While the R^2 value associated with the multi linear – regression model was lower to large extent (R^2=0.67), which means that regression model failed to interpret 37% of the total variation in the breaking extension values of the woven fabric under study. The poor performance of the regression model in predicting the breaking extension values may be ascribed to this model may be nonlinear.

As seen from Table 5, the predicted errors associated with ANN models were all close to zero, while that associated with regression one range between 6 and 12%, which confirms the high-ability of ANN model to predict the breaking extension values of woven fabrics. Figure 4 illustrated the predicted breaking extension values of ANN and regression models. From this figure it can be seen that a good agreement between actual and predicted values for ANN model assuring that the predicted errors nonsexist. Whereas the predicted values in the regression model are randomly distributed around the regression line. This confirms that ANN model can perform good prediction with very minor errors.

The performance of the artificial neural network used to predict the breaking extension at the validation, test and training stages were depicted in Figure 5. It can be seen that validation and testing errors have nearly the same characteristics. This type of ANN attains their performance goal after five iterations (Table 6) (Figures 4 and 5).

Actual values	Multi-linear regression		Artificial neural network	
	Predicted value	Absolute error (%)	Predicted value	Absolute error (%)
18.33333	20.37963	11.16161	18.33342	0.000489
23	22.24074	3.301130	22.99990	0.000399
24.66667	23.15741	6.118608	24.66657	0.000382
20	21.29630	6.4815	20.00003	0.000156

Table 5: Breaking extension predicted by multi-linear regression and neural network.

Statistical parameters	Regression	ANN
Coefficient of determination, R^2	0.67	0.9999
Root mean square error, RMSE	1.2479	0.0006

Table 6: Comparison of prediction performance of ANN and regression models for breaking extension.

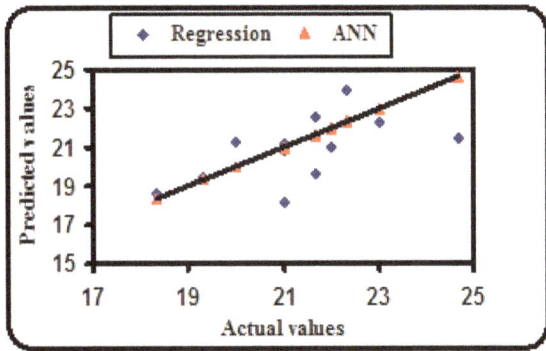

Figure 4: Actual against predicted values using ANN and regression models for breaking extension.

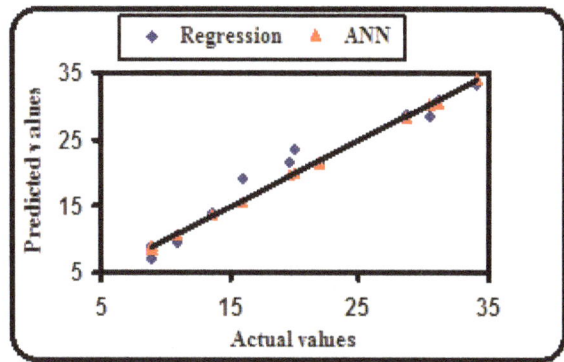

Figure 6: Actual against predicted values using ANN and regression models for air permeability.

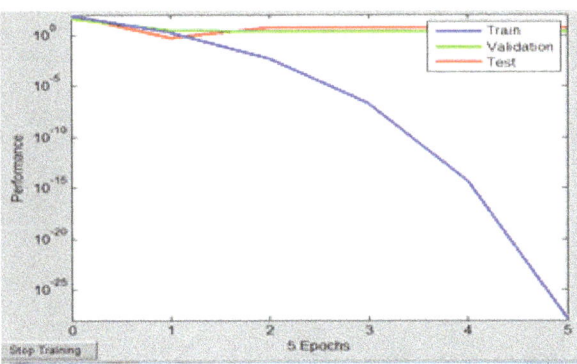

Figure 5: The performance plot of the ANN used to predict the fabric breaking extension.

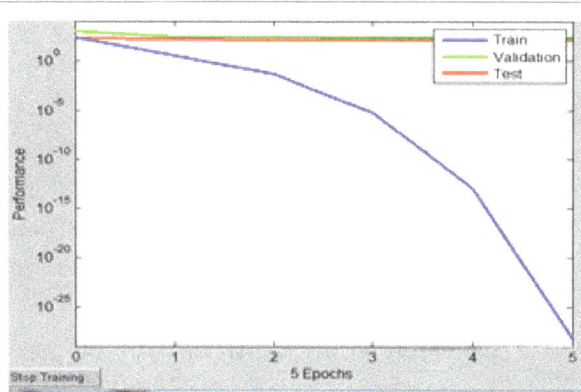

Figure 7: The performance plot of the ANN used to predict the fabric air permeability.

Actual values	Multi-linear regression		Artificial neural network	
	Predicted value	Absolute error (%)	Predicted value	Absolute error (%)
11	9.518518	13.46802	10.7	2.835158
22	21.25926	3.367	21.8	1.04382
30.3333	28.50926	6.013325	30.2	0.286504
34	33.34259	1.933559	34.1	0.29391

Table 7: Air permeability predicted by multi-linear regression and neural network.

Air permeability

The air permeability is a very important factor in the performance of some textile materials. Especially, it is taken into consideration for clothing, parachutes, sails, vacuum cleaner, fabrics for air bags and industrial filter fabrics. The air permeability is mainly dependent upon the fabric's weight and construction (thickness and porosity).

The actual and predicted values of air permeability of woven fabrics and their absolute errors for ANN and regression models were listed in Table 7. From this table, it can be observed that the absolute errors of the air permeability predicted by ANN model are highly lower compared to those predicted by regression model. The predicted errors of the air permeability associated by ANN model ranges between 0.3 and 3%, whereas the corresponding errors associated with regression model falls between 2 and 13%. The lower predicted errors accompanied by ANN model give a high degree of confidence in this type of prediction models.

Statistical parameters	Regression	ANN
Coefficient of determination, R^2	0.95	1
Root mean square error, RMSE	4.4776	1.08807

Table 8: Comparison of prediction performance of ANN and regression models for air permeability.

The R^2 and RMSE values of air permeability of the woven fabrics under study predicted by ANN and regression models are presented in Table 8. The average R^2 and RMSE values predicted via ANN model are 1 and 1.09, respectively. While such values predicted using regression model are 0.95 and 4.5 for R^2 and RMSE, respectively. The coefficient of determination values show that ANN model explained about 100% of the variations in the air permeability, whereas the variations explained by the regression model approaches 95%.

The air permeability values of the woven fabrics predicted by ANN and the regression models were depicted in Figure 6. From this figure it can be observed that the agreement between the actual values of air permeability and that predicted by ANN model is very high; while the agreement between the corresponding values predicted by the regression model is lower to some extent. The correlation coefficient between actual values of air permeability and that predicted by ANN model is approximately 1; while the correlation coefficient between the corresponding values for regression model equals to 0.95. Therefore, ANN model can perform good prediction with least error compared to regression one (Tables 7 and 8) (Figures 6 and 7).

Figure 7 depicts the plot of the performance of the ANN used to predict the fabric air permeability at training, validation and testing stages. It is shown that this neural network attains their optimum performance after 5 epochs, whereas the training and validation errors have the same characteristics.

Conclusion

In this study, tensile properties and air permeability of woven fabrics were predicted using ANN and regression models. The independent variables in the predicted models were weft yarn count (Ne), twist multiplier and weft density (ppi). Dependent variables which predicted by ANN and regression models were tensile strength, breaking extension and air permeability. The findings of this study revealed that R^2 values associated with ANN models for all fabric properties were very high compared to the R^2 values accompanied the regression models. Also, the RMSE values with ANN models are lower than those associated with regression ones. All absolute errors of the fabric properties predicted by ANN model were highly lower than that errors predicted by the regression model. Good agreement between predicted and actual values of tensile strength, breaking extension and air permeability of the woven fabrics proves that the neural network is an effective technique. With the help of the neural network, we can predict woven fabric properties easily, effectively and accurately.

References

1. Zadeh L (1994) Soft computing and fuzzy logic. IEEE Software 11: 48-56.

2. Keeler J (1992) Vision of neural networks and fuzzy logic for prediction and optimization of manufacturing processes, In: Applications of artificial neural networks III. 1709: 447-456.

3. Haykin S (1998) Neural networks: A comprehensive foundation. Prentice Hall, New York.

4. Lippman RP (1987) An introduction to computing with neural nets. IEEE ASSP Magazine 4-22.

5. Ahmed AA, Idrees HF, Hebeish Ali Ali (2014) Predicting the tensile properties of cotton/ spandex core-spun yarns using artificial neural networks and regression models. The Journal of the Textile Institute 105: 1221-1229.

6. Ogulata SN, Sahin C, Ogulata RT (2006) The prediction of elongation and recovery of woven bi-stretch fabric using artificial neural network and linear regression models. Journal of Fibers and Text in East Euro 14: 46-49.

7. Beltran R, Wang L, Wang X (2004) Predicting spinning performance with an artificial neural network model. Text Res J 74: 757-763.

8. Majumdar A, Majumdar PK, Sarkar B (2005) Application of linear regression, artificial neural network and neuro-fuzzy algorithms to predict the breaking elongation of rotor-spun yarns. Indian J Fibers Text Res 30: 19-25.

9. Ahmed, AA Mourad MM, Ali HA, Mohamed AR (2015) Comparison between physical properties of ring-spun yarn and compact yarns spun from different pneumatic compacting systems. Indian Journal of Fibers and Textile Research 40: 43-50.

10. Pinar D, Osman B, Rizvan E (2006) Prediction of strength and elongation properties of cotton/polyester blended yarn. Fibers and Textiles in Eastern Europe 14: 18-21.

11. Hearle JWS, El-Behery HM, Thakur VM (1961) The mechanics of twisted yarns: Theoretical developments. Journal of the Textile Institute Transactions 52: T197-T220.

12. Frydrych I (1992) A new approach for predicting strength properties of yarn. Textile Research Journal 62: 340-348.

13. Vitro C, Paul C (2009) Yarn irregularity parameterization using optical sensors. Fibers and Textiles in Eastern Europe 17.

14. Lucas LJ (1983) Mathematical fitting of modulus-strain curves of poly (ethylene terephthalate) industrial yarns. Textile Research Journal 53: 771-777.

15. Kim YK, El-sheikh A (1984) Tensile behavior of twisted hybrid fibrous structures. Part II: Experimental studies. Textile Research Journal 54: 534-543.

16. Kim AYK, El-Shiekh A (1984) Tensile behavior of twisted hybrid fibrous structures. Part I: Theoretical investigation. Textile Research Journal 54: 526-534.

17. El Mogahzy YE (1988) Selecting cotton fiber properties for fitting reliable equation to HVI data. Textile Research Journal 58: 392-397.

18. Hunter L (1988) Prediction of cotton processing performance and yarn properties from HVI test results. Melliand Textilber 4: 229-232.

19. Ahmed AA, Mourad MM (2014) Effects of spandex drawing ratio and weave structure on the physical properties of cotton/spandex woven fabrics. The Journal of the Textile Institute 105: 235-245.

20. Mourad MM, Ahmed ES, Ahmed AA (2011) Core spun yarn and the secret behind its popular appeal. Textile Asia 41-43.

21. Fan J, Hunter L (1998) A worsted fabric expert system. Part II: An artificial neural network model for predicting the properties of worsted fabrics. Textile Research Journal 68: 763-771.

22. Ramesh MC, Rajamanickam R, Jayaraman S (1995) The prediction of yarn tensile properties by using artificial neural networks. The Journal of the Textile Institute 86: 459-469.

23. Hadizadeh M, Jeddi AAA, Tehran MA (2009) The prediction of initial load-extension behavior of woven fabrics using artificial neural network. Text Res J 79: 1599-1609.

24. Youssefi M, Faez K (1999) Fabric Handle Prediction Using Neural Networks, Proceedings of the IEEE-EURASIP Workshop on Nonlinear Signal and Image Processing (NSIP'99), Bogaziçi University Printhouse, Antalya, Turkey.

25. Hui CL, Lau TW, Ng SF, Chan KCC (2004) Neural network prediction of human psychological perceptions of fabric hand. Text Res J 74: 375-383.

26. Matsudaira M (2006) Fabric handle and its basic mechanical properties. J of Textile Engineering 52: 1-8.

27. Fan J, Newton E, Au R, Chan SCF (2001) Predicting garment drape with a fuzzy-neural network text. Res J 71: 605-608.

28. Eid MY, Alsalmawy A, Almetwally AA (2010) Performance of woven fabrics containing spandex. Indian Textile Journal 120: 1-22.

29. Mansour SA, Morsy AE, Almetwally S (1998) Twist loss on Air – Jets: Effects on Various Fabric Properties. The Indian Textile Journal 41: 1-12.

30. Tugrul O (2006) Air permeability of woven fabrics. Journal of Textile and Apparel, Technology and Management 5: 1-10.

7

Development of Directionally Oriented Compressive Weft Knitted Fabrics

Cruz J*, Sampaio S and Fangueiro R

Centre for Textile Science and Technology, University of Minho, Portugal

Abstract

Polyamide/elastane knitted fabrics with directionally oriented compressive ability was studied. Knitting designs with coursewise, diagonal, walewise, and both coursewise and walewise compressive orientations were created and analysed at four different angles (0º, +45º, -45º and 90º) and up to 50% elongation. At the same stretching and same direction (angle), the tensile force was higher for samples from Serie B (only the elastane was producing miss stitches) than for samples from Serie A (polyamide and elastane yarns were simultaneously producing miss stitches), in most of the cases. There are higher differences between tensile force values of samples from Serie B with the different number of miss stitches, however, very little difference in tensile force values is seen between samples from Serie A with the different number of miss stitches, in most of the cases. Also the tensile force is the highest when samples from Serie B with walewise compressive orientation are stretched at 90º direction, the tensile force value was two times and half higher than that of the same compressive orientation but from Serie A. At 0º direction, the force values of the samples with coursewise compressive orientation from Serie A were higher than that of samples from Serie B due to miss loops force orientation. At the +45º and -45º directions, the force values of the two Series are similar, showing that the designs and the yarns used are less important due to force orientation. Fabrics with diagonal compressive orientation have force values similar among each sample, in all four directions, the diagonal structure enables the control of fabric anisotropy. The prediction of force values in relation to the number of miss stitches sequences and in relation to the different angles are made using regression model. This analysis helps engineer compressive devices with adjustable and directionally oriented stretch structures.

Keywords: Elastic behaviour; Directionally oriented compression; Knitted fabrics

Introduction

The main application of compression garment is for treating venous insufficiency, for example, treatment of varicose veins. However, it presents many other applications such as for treating burns, fatigue, dislocations, muscle fatigue, sprains and low blood pressure. It is also used and recommended for long flights to avoid the occurrence of deep vein thrombosis and postpartum recovery [1,2].

Moreover, since no body part is a uniform cylinder, pressure exerted by a garment with a given tension is not uniform and is distributed differently over the various areas of the body for any given person. Clearly concave areas of the body do not make contact with the pressure garment and therefore no pressure is exerted on them [3].

The level of compression is governed by the garment size as well as the amount of fabric stretching. Fabrics used for compression garments are usually engineered with stretchable structure and containing elastomeric material to achieve highly stretchable appropriate compression [4].

In today's athletics environment, many athletes wear compression garments. Elastic garments for sports and outer wear play an important role in optimizing an athletic performance by providing freedom movement, minimizing the risk of injury or muscle fatigue, and reducing friction between body and garment. In the absence of body motion, many garments provide apparent comfort. However the moment the physical movement is made, the comfort performance level changes, and that change could be significant. During movement, different parts of the body stretch very differently, and the amount of stretch will vary differently in each direction [5].

However, most compression garments do not provide targeted support to specific body areas, but rather overall compression. As a result, compression garments have inherent unwanted push-pull effects that reduce the garment ability to maintain optimal compression

support for complex moving muscles. Some compression garments provide specific targeted higher compression zones by sewing particular compression fabric panels in dedicated areas and adjoining the panel shapes at the seams. However, the amount of targeted support zones is always limited due to the practical construction complications that will always arise from sewing too many seam lines in the garment. In addition, an abundance of seams naturally leads to a higher propensity towards the possibility of skin chaffing for the wearer, regardless even if the seams are flat-locked [6].

Single or double covered elastomers with different stretch moduli and linear densities are usually used in knitted fabrics to achieve specific stretch and pressure scales. The degree of stretch and recovery of knitted fabrics depends on the amount of elastic material incorporated and on the stitch construction of the fabrics [7,8]. The understanding on how to optimise the stretch potential in pattern design is, in relative terms, still in its infancy. Comprehensive study detailing all aspects of an objective approach to stretch pattern development has not been done so far [9]. Furthermore, the possibility to design knitted fabrics with directionally oriented compressive ability is also not yet been done.

The objective is to find ways of improving the compressive properties of knitted fabrics at different fabric orientations by the engineering design of structures (pattern designs). Various authors [10-12] stated that the tensile behaviour of knitted fabrics are directly derived from

---

*Corresponding author: Cruz J, Centre for Textile Science and Technology, University of Minho, Portugal, E-mail: julianacruz@det.uminho.pt

the loop configuration and the yarn properties. However, these analysis were mostly limited to knitted fabrics subjected to biaxial (coursewise and walewise) stresses. The coursewise compressive orientation is expected to be manily influenced by the yarn properties and the diagonal, walewise, and both coursewise and walewise compressive orientations, is expected to be mainly influenced by the knitted pattern. Based on these assumptions, in this paper, different knitted fabrics were developed with coursewise, diagonal, walewise and both coursewise and walewise compressive orientations based on their pattern designs. The influence of different structural designs on the elastic behaviour of knitted fabrics composed by elastane and polyamide yarns have been thoroughly investigated and discussed. Different knitting designs have been created and analysed at four different angles (0º, +45º, -45º and 90º). Furthermore, the impact of the knitting yarn and the number of miss loops on knitted fabrics elastic behaviour was also studied for all the samples at the four different angles selected.

Materials and Methods

Polyamide and elastane knitted fabrics have been produced for investigating the elastic behavior affected by varying the number and position of miss loops. Polyamide 6.6 yarn produced with 44 dtex linear density, with 34 filaments (PA 44/34), and elastane yarn produced in a conventional double covering machine with linear density of elastane core of 156 dtex, with polyamide 6.6 double cover 78 dtex with 23 filaments (EA 156 D PA78/23) were used.

24 fabric samples were produced on a circular knitting machine with 13 inches cylinder and E24 gauge. Half of the samples were prepared from knitted constructions where the polyamide and the elastane yarns were simultaneously producing miss stitches, designated by Serie A, and the remaining half samples solely comprised constructions where only the elastane yarn was producing miss stitches, designated by Serie B.

By varying the number and the position of the miss loops different structural designs were engineered with predictable orientations in order to provide compression in that directions. Coursewise, diagonal, walewise and both coursewise and walewise orientations were created. The directionally oriented compressive knitted fabrics (DOCKFs) produced were classified into four groups:

- DOCKF_C fabrics with coursewise orientation - produced from miss stitches sequences on adjacent needles (1 to 3 successive miss loops) on same course (samples: 1A, 1B, 2A, 2B, 3A and 3B);

- DOCKF_D fabrics with diagonal orientation (-45º) - produced from miss stitch sequences on adjacent needles with one needle space (1 to 3 miss stitches) (samples: 4A, 4B, 5A, 5B, 6A and 6B);

- (3) DOCKF_W fabrics with walewise orientation - produced from the effect of the miss stitches sequences on the same needle (1 to 3 successive miss loops) (samples: 7A, 7B, 8A, 8B, 9A and 9B);

- (4) DOCKF_C &W fabrics with both coursewise and walewise orientations, produced from miss stitches sequences on the same course and miss stiches sequences on the same wale (samples: 10A, 10B, 11A, 11B, 12A and 12B).

Their corresponding loop configurations are illustrated in Figure 1 and as can be seen in Figure 1 structure 1A is the same as 7A, 1B and 7B are the same, 4A and 10A are the same and 4B and 10B are also

the same, and this is due to the groups' choice that was done for this research. Table 1 shows the dimensional properties of the 24 DOCKFs produced.

To investigate the impact of the different structural designs on the elastic behaviour of DOCKFs, tests were carried out according to NP EN ISO 13934 standard. The fabrics elastic behaviour were evaluated in a Hounsfield H10KS-UTM universal testing machine, at 100 mm/min crosshead speed using 10 cm gauge length, in four directions: 0º, +45º, -45º and 90º directions of the DOCKFs specimens (Figure 2). Strip specimens were stretched up to 50% elongation and force and elongation were recorded and averaged for the 24 samples. These fabrics were dry relaxed before mechanical testing in lab conditioning at 22ºC and 65% RH for more than 24 hours. Such extension percentages are chosen as simple body movements stretch the knitted garment by up to 50% [13].

Results and Discussion

Load capacity of DOCKFs up to 50% elongation

Fabrics used for compression garments are usually engineered to achieve highly stretchable appropriate compression (pressure exerted by a garment with a given tension) [4]. Inlay yarn and laid-in stitches are the key knitting elements in the fabrication of compression textiles, and play a critically important role in managing and controlling high pressure magnitudes [9]. A need still exists for an effective prevention and treatment device that provides mild stretchable compression on different directions (simultaneously at different parts of the body), while, at the same time, provides comfort and wearability all day long. Therefore, in this study, the use of polyamide/elastane structures which contain miss and normal loops were chosen to produce knitted fabrics with directionally oriented mild stretchable compression. The tensile force of the elastic DOCKFs up to 50% extension has been studied and the results are presented in Table 2, Figures 3-5.

Table 2 displays the tensile strength of the different group samples for each angle at 50% extension. Figure 3 shows the force-extension curves of the four different group samples up to 50% extension. Figure 4 shows the correlations among samples DOCKF_C, DOCKF_D, DOCKF_W and DOCKF_C&W for force in relation to the number of miss stitches at 50% elongation; Figure 5 shows the correlations among samples DOCKF_C, DOCKF_D, DOCKF_W and DOCKF_C &W for force in relation to the different angles (directions) used for the tensile tests at 50% extension.

Figures 4 and 5 just include results with values of the Pearson correlation coefficient (R^2) higher than 0, 80 (strongly correlated) [14].

Influence of knitting yarns and number of miss stitches sequences on the stretch behaviour

The knitting designs and the yarns used can be adjusted to provide the correct level of surface pressure in specific parts of the body. Through selection of the knitting yarns and the stitch used, 24 samples with directionally orientated compressive ability were developed. DOCKFs where the polyamide and the elastane yarns were simultaneously producing miss stitches, designated by Serie A, and the DOCKFs where only the elastane yarn was producing miss stitches, designated by Serie B, were compared.

Tensile test results are given in Table 2 and compressive force-extension curves are shown in Figure 3. Figure 3 presents typical load-extension characteristic curves for weft knitted fabrics [15]. The tensile test results of Figure 3 reveal a two stage deformation process, at the

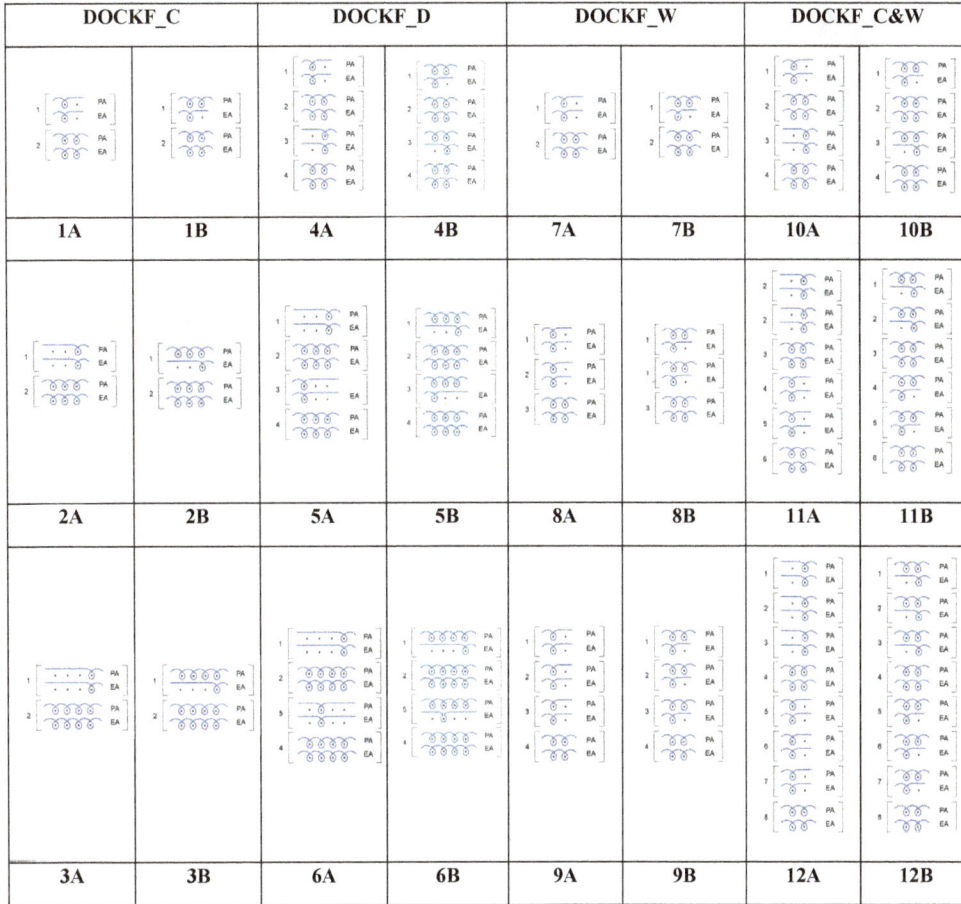

Figure 1: Schematic diagram of different designed knitted fabrics.

Samples		Loop length (mm)		Wales/cm	Courses/cm	Thickness (mm)	Aerial mass (g/m²)
		PA	EA				
DOCKF_C	1A	2.9	2.5	15	30	1.7 ± 0.1	489 ± 9
	1B	3.1	2.5	13	25	1.4 ± 0.0	376 ± 2
	2A	3.3	2.9	13	25	1.3 ± 0.1	365 ± 4
	2B	3.0	2.5	15	30	2.2 ± 0.1	484 ± 2
	3A	3.0	3.1	13	24	1.1 ± 0.1	370 ± 8
	3B	2.7	2.6	18	20	2.3 ± 0.2	606 ± 5
DOCKF_D	4A	2.9	3.1	13	26	1.3 ± 0.0	385 ± 1
	4B	2.6	3.0	13	24	1.6 ± 0.0	483 ± 6
	5A	2.7	2.8	13	26	1.6 ± 0.0	380 ± 4
	5B	2.8	2.6	14	20	2.0 ± 0.1	556 ± 9
	6A	2.9	3.2	13	26	1.4 ± 0.1	460 ± 9
	6B	3.2	2.8	13	25	1.9 ± 0.1	455 ± 4
DOCKF_W	7A	3.1	2.5	13	25	1.4 ± 0.0	376 ± 2
	7B	2.9	2.5	15	30	1.7 ± 0.1	489 ± 9
	8A	2.7	3.2	13	24	1.3 ± 0.0	361 ± 5
	8B	2.7	2.3	15	28	2.0 ± 0.1	492 ± 1
	9A	2.8	2.8	13	26	1.3 ± 0.1	355 ± 1
	9B	2.3	2.2	15	27	2.0 ± 0.1	537 ± 3
DOCKF_C&W	10A	2.9	3.1	13	26	1.3 ± 0.0	385 ± 1
	10B	2.6	3.0	13	24	1.6 ± 0.0	483 ± 6
	11A	2.7	3.2	13	26	1.4 ± 0.1	395 ± 2
	11B	2.7	2.6	13	25	1.9 ± 0.1	495 ± 9
	12A	2.6	3.0	13	25	1.3 ± 0.1	393 ±1
	12B	2.8	3.1	15	26	2.1 ± 0.2	529 ± 9

Table 1: Dimensional properties of the DOCKFs used in the study.

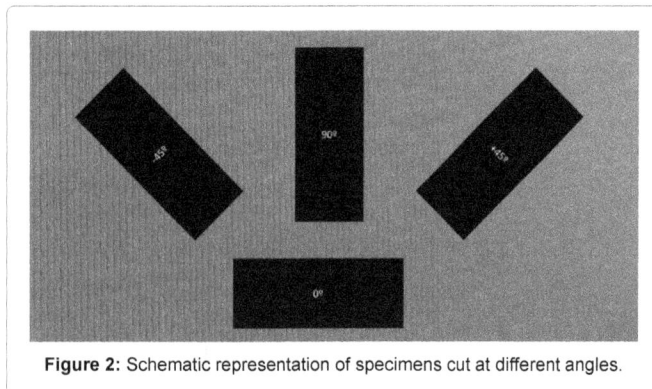

Figure 2: Schematic representation of specimens cut at different angles.

first stage (up to approximately 10% extension) during the application of a tensile load, the loops change their shape in order to accommodate the applied load. The fabric structure allows large deformation where the yarns are straightening without yarn elongation itself. At the second stage, it can be seen that the relationship between load and extension is almost linear when the fabric is stretched (from 10% up to 50% extension), where the elongation of the straightened yarn takes place. In general, from fabric samples in Figure 3, it can be seen that at the same stretching and direction (angle), the stretching force is higher for samples from Serie B than for samples from Serie A, in most of the cases. This might be due to the higher loop density of these fabrics of Serie B, the number of yarns that bear the tensile force in the samples from Serie B is more than that of the samples from Serie A (Table 1). Results also show that there are higher differences between tensile force values of samples from Serie B with different number of miss stitches. However, very little difference in tensile force values is seen between samples from Serie A with different number of miss stitches, in most of the cases. Also the tensile force is the highest when samples from Serie B are stretched at 90° direction, the highest tensile force value is verified for the sample from Serie B with walewise compressive orientation (DOCKF_W) and with three miss stitches sequences and it is two times and half higher than that of the same compressive orientation fabrics from Serie A. The lowest tensile forces are when samples from Serie B are stretched at 0° direction for fabrics with coursewise compressive orientation (DOCKF_C). When the force is applied to a fabric at 0° direction the stretching force is higher for fabrics with

more miss stitches (from Serie A) because they will resist stronger due to direct stretch of some polyamide yarns (without loops to deform). At +45° and -45° directions, the tensile force of samples from Serie B have higher values than Serie A, but not too much, although a small structure effect of the fabric still exists, this may be ignored as it is less important due to force orientation.

Analysing each group of samples in more detail, it is possible to observe from Figure 3a that Sample 3A from DOCKF_C group, as expected, presents the best performance at 0° direction, due to the positive influence of the number of miss loops. When a knitted fabric is extended along its coursewise direction, the yarns will bear most of the force, and the higher number of miss stitches sequences (especially the ones from polyamide yarn) in this fabric might have blocked earlier the stretching, requiring higher force to stretch. In this way, Sample 3A with coursewise compressive orientation may provide more sustained compression to some target positions at 0° direction, for the production of compression textiles.

From Figure 3b, it is observed that all samples from DOCKF_D group present similar tensile force (samples from Serie A as well as samples from Serie B), in all four directions. In this way, it can be concluded that the diagonal structure enables the control of fabric anisotropy. However, the maximum value of tensile force (13,00 N) was obtained by stretching Sample 6B, at 90° direction, providing more sustained compression to some target positions at 90° direction, for the production of compression textiles.

From Figure 3c, it can be observed that the maximum value of tensile force (24,92 N) for DOCKF_W Group was obtained by stretching the Sample 9b at 90° direction. This sample presents high loop density with higher number of miss stitches, providing better compression behaviour. In this way, Sample 9B with walewise compressive orientation may provide more sustained compression to some target positions at 90° direction, for the production of compression textiles.

From Figure 3d, the maximum value of tensile force (21,79 N) for DOCKF_ C&W Group was obtained by stretching Sample 12B at 90° direction. This behaviour can be explained by the higher loop density and number of miss stitches to resist the fabric stretching. This sample also shows very good performance at 0° direction (15,75 N). Sample 12B with both coursewise and walewise compressive orientations may

Samplegroups	Samples	Serie A				Serie B			
		Force (N) at 0°	Force (N) at 45°	Force (N) at -45°	Force (N) at 90°	Force (N) at 0°	Force (N) at 45°	Force (N) at -45°	Force (N) at 90°
DOCKF_C	1	10.83 (±17%)	11.15 (±6%)	11.94 (±12%)	9.09 (±29%)	9.77 (±13%)	10.30 (±9%)	11.77 (±5%)	13.56 (±6%)
	2	10.38 (±5%)	10.03 (±14%)	10.75 (±7%)	10.17 (±8%)	9.00 (±26%)	10.11 (±7%)	10.36 (±5%)	14.89 (±18%)
	3	13.76 (±9%)	9.22 (±5%)	10.65 (±10%)	9.540 (±9%)	7.70 (±7%)	9.22 (±5%)	10.22 (±7%)	13.75 (±14%)
DOCKF_D	4	10.00 (±0%)	9.33 (±6%)	10.27 (±7%)	10.33 (±3%)	11.87 (±2%)	11.50 (±4%)	9.90 (±2%)	11.07 (±6%)
	5	10.68 (±2%)	10.17 (±3%)	11.76 (±17%)	8.87 (±1%)	10.60 (±2%)	10.53 (±2%)	10.00 (±14%)	11.81 (±21%)
	6	10.53 (±4%)	9.32 (±6%)	9.85 (±1%)	9.50 (±8%)	12.90 (±4%)	11.47 (±2%)	10.88 (±3%)	13.00 (±8%)
DOCKF_W	7	10.83 (±17%)	11.15 (±6%)	11.94 (±12%)	9.09 (±29%)	9.79 (±13%)	10.30 (±9%)	11.77 (±5%)	13.56 (±6%)
	8	11.44 (±13%)	10.44 (±17%)	11.22 (±13%)	10.44 (±21%)	13.80 (±32%)	11.04 (±6%)	11.27 (±13%)	18.66 (±24%)
	9	11.75 (±9%)	13.40 (±8%)	12.24 (±3%)	9.94 (±6%)	12.73 (±10%)	13.33 (±8%)	12.31 (±7%)	24.92 (±7%)
DOCKF_C&W	10	10.00 (±0%)	9.33 (±6%)	10.27 (±7%)	10.33 (±3%)	11.87 (±2%)	11.50 (±4%)	9.90 (±2%)	11.07 (±6%)
	11	10.53 (±1%)	9.78 (±3%)	10.43 (±6%)	10.00 (±0%)	14.36 (±5%)	13.27 (±5%)	12.64 (±4%)	15.47 (±4%)
	12	11.39 (±4%)	9.80 (±1%)	10.33 (±3%)	9.54 (±6%)	15.75 (±1%)	15.019 (±0%)	14.67 (±2%)	21.79 (±1%)

Table 2: Tensile test results at 50% elongataion.

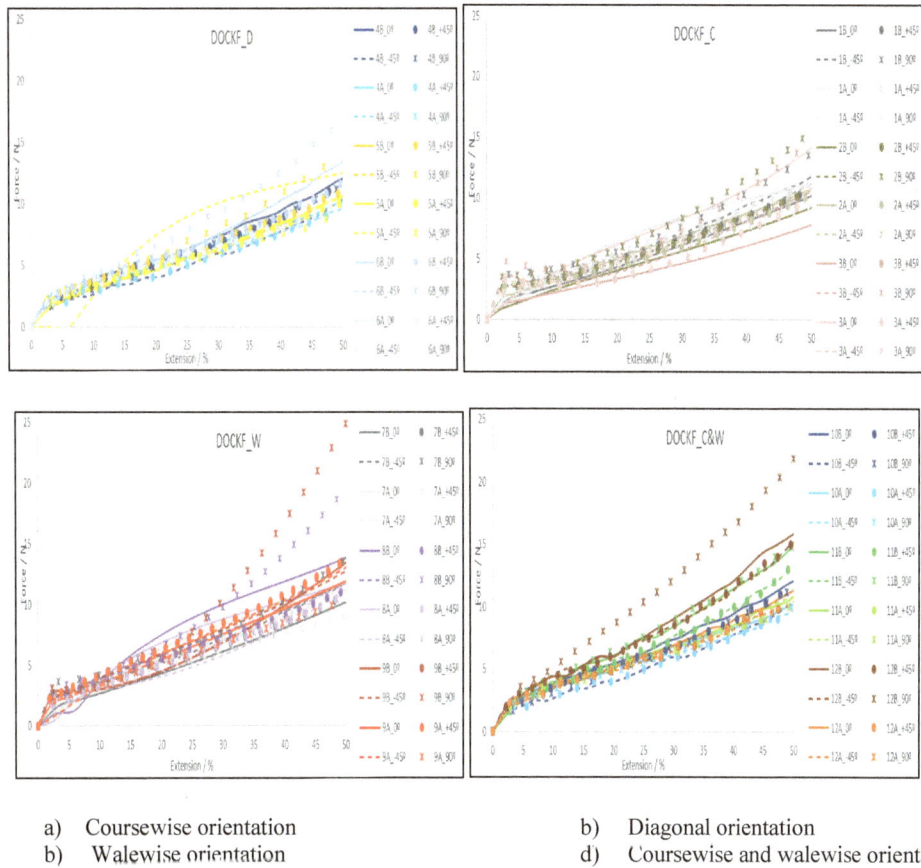

a) Coursewise orientation   b) Diagonal orientation
b) Walewise orientation   d) Coursewise and walewise orientation

Figure 3: Force-extension curves of DOCKFs.

provide more sustained compression to some target positions at 90° and at 0° directions, for the production of compression textiles.

The results obtained showed that the developed structural designs influenced the elasticity of the knitted fabrics, and oriented compressive structures were created, furthermore a combination of knitted structures with different compressive orientations can be created, on one unique support device, to give comfortable support to different parts of the body at the same time.

Influence of number of successive miss stitches (1, 2, and 3 miss loops) on the stretch behaviour

Figure 4a (samples from DOCKF-C Group - coursewise effect) shows the dependence of the number of successive miss loops sequences on the fabrics compressive behaviour when tested at + 45°, -45° and at 0°, i.e. for samples represented in this figure the increase on the number of miss stitches (nms) is inversely proportional to the tensile force. This might be occurring because at + 45° and -45° it is mainly the structure of the fabric that resists the stretching. The increase in the number of miss stitches leads to a decrease in the loop density of these fabrics decreasing their stretchability. The same behaviour is observed at 0° for Serie B, where only the elastane is producing miss stitches. In this case, when this fabric is extended along its coursewise direction, the yarns will support most of the force, however straight polyamide yarn (miss stitches) are not presented to block the stretching. It is possible to predict the tensile force values of designed knitted fabrics with compressive coursewise orientation when varying the number of successive miss stitches, when these fabrics are stretched along +45°,

-45° and courswise direction (0°). The influence of this parameter on tensile force (F) of DOCKF_C from Serie B, tested at + 45°, -45° ant at 0° is described by the linear equations 4.1, 4.2 and 4.3 given in Figure 4a. The influence of this parameter on tensile force (F) of DOCKF_C from Serie A tested at + 45° and -45° is described by the linear equations 4.4 and 4.5 given in Figure 4a.

For the DOCKF_D Group (diagonal effect), Figure 4b shows that the dependence of the number of successive miss loops sequences on the tensile force is only verified for fabrics from Serie B when tested at -45°, i.e. for fabrics represented in this figure the increase on the number of miss stitches (nms) is inversely proportional to the tensile force. This might be occurring because at - 45° it is mainly the structure of the fabric that resists the stretching; the decrease in the loop density due to the increase in the number of miss stitches, leads to less stretch resistance. It is possible to predict the tensile force values of designed knitted fabrics with compressive diagonal orientation when varying the number of successive miss stitches on the design pattern, when these fabrics are stretched along - 45° direction. The influence of this parameter on tensile force (F) of DOCKF_D from Serie B, tested at -45°, is described by the linear equation 4.6 given in Figure 4b.

Knitted fabrics of DOCKF_W group (walewise effect), represented in Figure 4c, indicates that the increase number of successive miss loops sequences on same needles did have influence on the tensile force (F) performance of fabrics from Serie B tested at + 45° and 90°, and of fabrics from Serie A at 0°. The tensile force of Serie B samples tested at +45° and at 90° is correlated with the number of miss stitches (nms) sequences. The increase on the number of miss stitches is

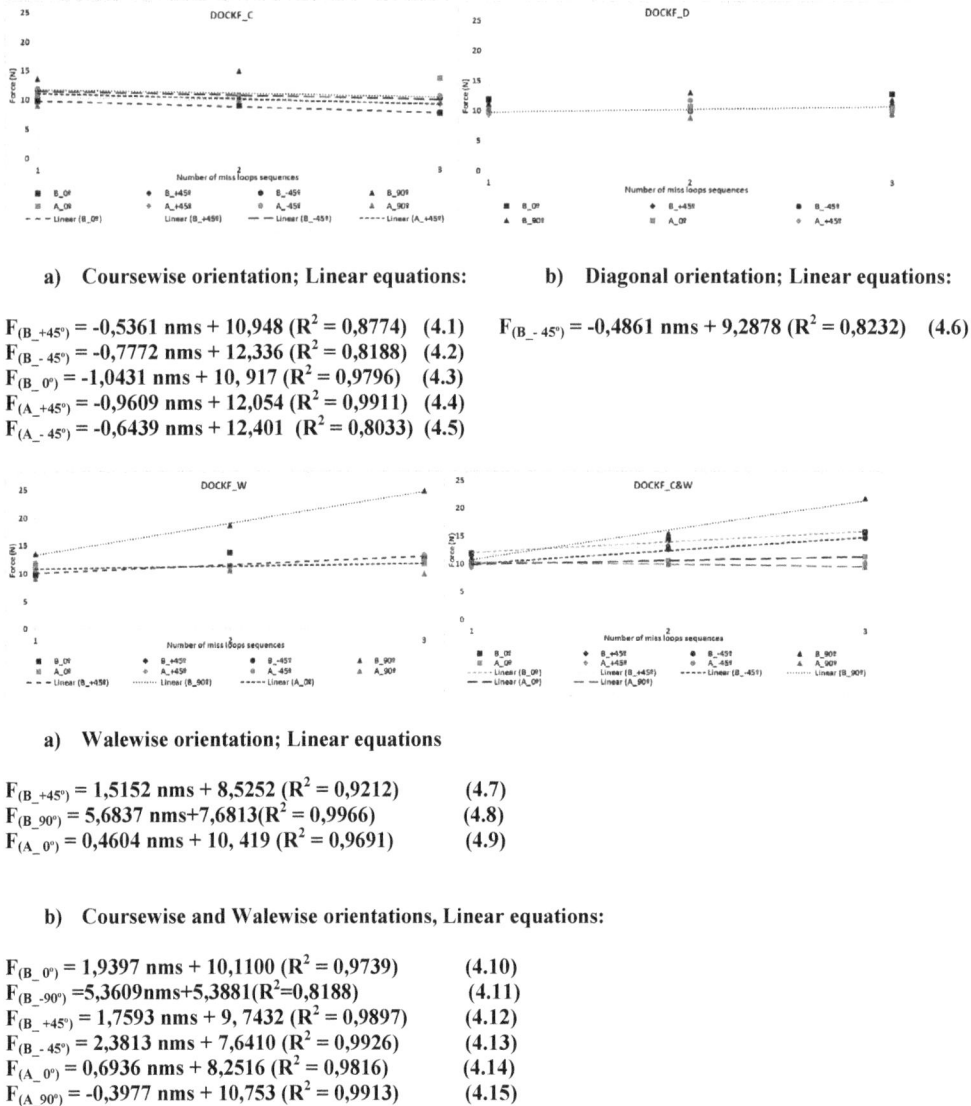

a) **Coursewise orientation; Linear equations:**

$F_{(B\_+45°)} = -0,5361 \text{ nms} + 10,948 \ (R^2 = 0,8774) \quad (4.1)$

$F_{(B\_-45°)} = -0,7772 \text{ nms} + 12,336 \ (R^2 = 0,8188) \quad (4.2)$

$F_{(B\_0°)} = -1,0431 \text{ nms} + 10,917 \ (R^2 = 0,9796) \quad (4.3)$

$F_{(A\_+45°)} = -0,9609 \text{ nms} + 12,054 \ (R^2 = 0,9911) \quad (4.4)$

$F_{(A\_-45°)} = -0,6439 \text{ nms} + 12,401 \ (R^2 = 0,8033) \quad (4.5)$

b) **Diagonal orientation; Linear equations:**

$F_{(B\_-45°)} = -0,4861 \text{ nms} + 9,2878 \ (R^2 = 0,8232) \quad (4.6)$

a) **Walewise orientation; Linear equations**

$F_{(B\_+45°)} = 1,5152 \text{ nms} + 8,5252 \ (R^2 = 0,9212) \quad (4.7)$

$F_{(B\_90°)} = 5,6837 \text{ nms} + 7,6813 (R^2 = 0,9966) \quad (4.8)$

$F_{(A\_0°)} = 0,4604 \text{ nms} + 10,419 \ (R^2 = 0,9691) \quad (4.9)$

b) **Coursewise and Walewise orientations, Linear equations:**

$F_{(B\_0°)} = 1,9397 \text{ nms} + 10,1100 \ (R^2 = 0,9739) \quad (4.10)$

$F_{(B\_-90°)} = 5,3609 \text{nms} + 5,3881 (R^2 = 0,8188) \quad (4.11)$

$F_{(B\_+45°)} = 1,7593 \text{ nms} + 9,7432 \ (R^2 = 0,9897) \quad (4.12)$

$F_{(B\_-45°)} = 2,3813 \text{ nms} + 7,6410 \ (R^2 = 0,9926) \quad (4.13)$

$F_{(A\_0°)} = 0,6936 \text{ nms} + 8,2516 \ (R^2 = 0,9816) \quad (4.14)$

$F_{(A\_90°)} = -0,3977 \text{ nms} + 10,753 \ (R^2 = 0,9913) \quad (4.15)$

Figure 4: Force versus number of miss stitches at 50% elongation for DOCKFs.

directly proportional to the tensile force of the fabrics from Serie B, leading to an increase of the loops density which influences positively the stretch resistance. For Serie A samples tested at 0° direction a good correlation has been also found. The number of miss stitches is directly proportional to the tensile force, when fabrics with compressive walewise orientation are extended along the coursewise direction, the yarns will support most of the force. The predicted tensile force of Serie B at +45° and at 90° are given in the equations 4.7 and 4.8 respectively, and the predicted tensile force of Serie A at +0° is given in the equations 4.9 (Figure 4c).

For DOCKF_C&W Group (combination of coursewise and walewise effects), Figure 4d shows that the increase number of successive miss loops sequences did affect the tensile force (F) of the fabrics from Serie A when tested at 0° and 90°. This behaviour demonstrates the importance of the polyamide yarn in resisting the stretching at 0° and the importance of the design structure in resisting the stretch at 90°. The increase number of successive miss loops was directly proportional to the tensile force at 0° and inversely proportional at 90°. The increase number of successive miss loops sequences did also affect the tensile

force of the fabrics from Serie B. For these samples, when an increase in the number of successive miss loops sequences is verified, an increase in the tensile force in all directions is observed. The higher increment is noticed at 90° direction, the force value increased 47% from 1 miss loop to 3 successive miss loops sequences, showing again that the design structure plays a key role on the elongation along the walewise direction. The predicted tensile force of DOCKF_C&W at 0° and at 90° are given in the equations 4.10 and 4.11 respectively. The predicted tensile force of DOCKF_C&W at all directions are given in the equations 4.12, 4.13, 4.14 and 4.14, in Figure 4d.

Influence of the force direction on the stretch behaviour

Fabric characteristics can be engineered to enhance the tensile properties only in the required orientations. The impact of different knitting designs orientations on tensile force of knitted fabrics at different angles (da) were studied and can be seen in Figure 5.

In DOCKF_C Group (coursewise orientation) the results show (Figure 5a) that only two samples, Sample 1A and Sample 2A, present good correlations. The best result represented was obtained at -45°

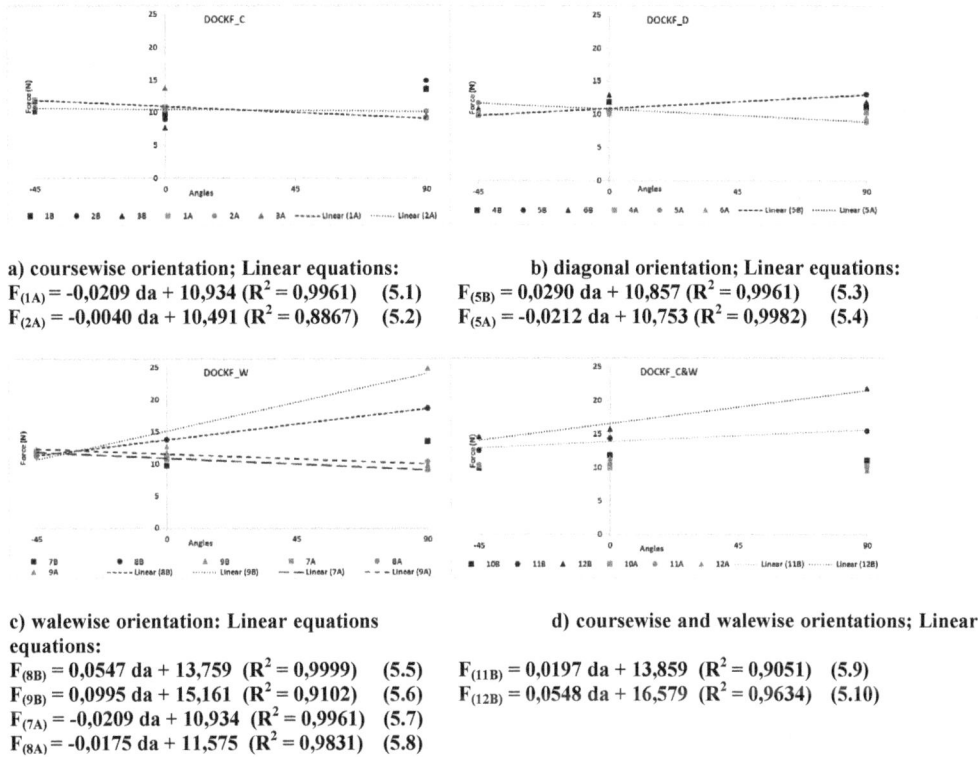

a) coursewise orientation; Linear equations:
$F_{(1A)} = -0,0209 \, da + 10,934 \, (R^2 = 0,9961)$ (5.1)
$F_{(2A)} = -0,0040 \, da + 10,491 \, (R^2 = 0,8867)$ (5.2)

b) diagonal orientation; Linear equations:
$F_{(5B)} = 0,0290 \, da + 10,857 \, (R^2 = 0,9961)$ (5.3)
$F_{(5A)} = -0,0212 \, da + 10,753 \, (R^2 = 0,9982)$ (5.4)

c) walewise orientation: Linear equations:
$F_{(8B)} = 0,0547 \, da + 13,759 \, (R^2 = 0,9999)$ (5.5)
$F_{(9B)} = 0,0995 \, da + 15,161 \, (R^2 = 0,9102)$ (5.6)
$F_{(7A)} = -0,0209 \, da + 10,934 \, (R^2 = 0,9961)$ (5.7)
$F_{(8A)} = -0,0175 \, da + 11,575 \, (R^2 = 0,9831)$ (5.8)

d) coursewise and walewise orientations; Linear equations:
$F_{(11B)} = 0,0197 \, da + 13,859 \, (R^2 = 0,9051)$ (5.9)
$F_{(12B)} = 0,0548 \, da + 16,579 \, (R^2 = 0,9634)$ (5.10)

Figure 5: Force for the different angles (0°, +45°, -45°, and 90°) for DOCKFs.

for both samples; the tensile force of these samples slightly increased. The force value of DOCKF_C fabrics, from Serie A with 1 and with 2 miss loops sequences, present good correlation within the different directions tested. The predicted tensile force values of DOCKF_C from Serie A are given in the equations 5.1 and 5.2 (Figure 5a). It is possible to predict the tensile force values of designed knitted fabrics with compressive coursewise orientation from Serie A, with 1 and 2 miss stitches sequences, when these fabrics are stretched along any direction.

In the group of DOCKF_D (coursewise orientation) it can be seen from Figure 5b that only two samples were representative. Represented sample 5A and sample 5B had slightly better performance at 45° direction and at 90° direction respectively. The tensile force of DOCKF_D samples with two miss stitches sequences and produced by Serie A and Serie B has good correlation with the different angles (da) and the predicted linear equations 5.3 and 5.4 are given in Figure 5b. It is possible to predict the tensile force values of designed knitted fabrics with compressive diagonal orientation from Serie A with 2 miss stitches sequences and from Serie B with 2 miss stitches sequences when these fabrics are stretched along any direction.

For DOCKF_W group (walewise effect) illustrated in Figure 5c the following conclusions might be drawn: for Serie B samples (8B and 9B) the increase tensile force is directly proportional to the increase in the angles direction performances (-45°, 0°, + 45° and 90°) indicating that the best performance was obtained at 90° in accordance with the orientation of the structure design. As for serie A samples (7A and 8A) the increase in tensile force is inversely proportional to the increase in the angles direction performances (-45°, 0°, + 45° and 90°) indicating that the best performance was at - 45° and it is not in accordance with the orientation of the structure design. The tensile force of DOCKF_D

samples with two and three miss stitches sequences from Serie B and samples with one and two miss stitches from Serie A has good correlation with the different angles and the predicted linear equations 5.5, 5.6, 5.7 and 5.8 are given in Figure 5c. It is possible to predict the tensile force values of designed knitted fabrics with compressive walewise orientation from Serie B with 1 and 2 miss stitches sequences and from Serie A with 1 and 2 miss stitches sequences when these fabrics are stretched along any direction.

From representative samples of DOCKF_C &W group (Figure 5d), it can be seen that only two samples from serie B are represented, samples 11B and 12B have the best performance at 90° direction tests, 15,47 N and 21,79N respectively. The force value of DOCKF_C&W fabrics from Serie B with 2 and with 3 miss loops sequences has good correlation with the different directions tested. The tensile force values of DOCKF_C&W from Serie B are given in the equations 5.9 and 5.10 (Figure 5d). It is possible to predict the tensile force values of designed knitted fabrics with both compressive coursewise and walewise orientation from Serie B with 2 and with 3 miss stitches sequences when these fabrics are stretched along any direction.

The relationships presented in Figures 4 and 5 allows to predict the level of compression required, and design gradual and oriented compression garments by selecting the fabric structure with the right number of miss stitches and right fabric compressive orientation.

Conclusions

In this study, the influence of different structural designs on the elastic behaviour of seamless knitted fabrics composed of elastane and polyamide yarns were thoroughly investigated and discussed. Furthermore, the possibility to design knitted fabrics with directionally oriented compressive knitted fabrics (DOCKFs) was also studied.

Different knitting designs with coursewise (DOCKF_C), diagonal (DOCKF_D), walewise (DOCKF_W), and with both coursewise and walewise (DOCKF_C&W) compressive orientations were created and analysed at four different angles (0°, +45°, -45° and 90°) and up to 50% elongation. The impact of the knitting yarn and stitch type on the stretch behaviour of DOCKFs was noticed for all the samples from DOCKF_C group, from DOCKF-D group, from DOCKF_W group and from DOCKF_C&W group. It was found that in general, fabric constructions where only the elastane was producing miss stitches (Serie B) have higher tensile force values than fabrics from Serie A, when stretched at -45°, at + 45° and at walewise direction (90°). The loop density of the DOCKFs samples, was affected mainly by the type of stitch produced, and it played an important role in the tensile force values of the samples. There were higher differences between tensile force values of samples from Serie B with the different number of miss stitches, however, very little difference in tensile force values was seen between samples from Serie A with the different number of miss stitches, in most of the cases. The prediction of force values in relation to the number of miss stitches sequences and in relation to the different angles studied are made using regression model. A combination of knitted structures with different compressive orientations can be created, on one unique support device, to give comfortable support to different parts of the body at the same time.

Acknowledgments

This work is financed by FEDER funds through the Competitivity Factors Operational Programme - COMPETE and by national funds through FCT – Foundation for Science and Technology within the scope of the project POCI-01-0145-FEDER-007136.

References

1. Cruz J, Fangueiro R, Soutinho F, Ferreira C, Andrade P, et al. (2010) Study on the Compressive Behaviour of Functional Knitted Fabrics Using Elastomeric Materials. Autex 2010 World Textile Conference, Vilnius, Lithuania.

2. Cruz J, Carvalho R, Sohel R, Fangueiro R, Coutinho G, et al. (2012) Influence of Structural and Process Parameters on Mechanical Behaviour of Elastane Yarns. The 41th Textile Research Symposium, Universidade do Minho, Guimarães, Portugal.

3. Atiyeh BS, El Khatib AM, Dibo SA (2013) Pressure garment therapy (PGT) of burn scars: evidence-based efficacy. Ann Burns Fire Disasters 26: 205-212.

4. Wang L, Felder M, Cai J (2011) Study of properties of medical compression garment fabrics. Journal of Fibre Bioengineering and Informatics 4: 15-22.

5. Senthilkumar M, Anbumani N (2014) Dynamic Elastic Behavior of Cotton and Cotton / Spandex Knitted Fabrics. Journal of Engineered Fibers and Fabrics 9: 93-100.

6. Galluzzo G, Regan P, Strong R (2013) Seamless circular or warp knitted compression garment with targeted anatomical musculature support. Patent Application 0254971-A1, USA.

7. Anon (1990) Lycra-the fitness fiber. Textiles 19: 58-61.

8. Watkinsa P (2011) Designing with stretch fabrics. Indian Journal of Fibre and Textile Research 36: 366-379.

9. Liu R, Lao TT, Wang S (2013) Impact of Weft Laid-in Structural Knitting Design on Fabric Tension Behavior and Interfacial Pressure Performance of Circular Knits. Journal of Engineered Fibers and Fabrics 8: 96-107.

10. MacRory BM, McCraith JR and McNamara AB (1975) The Biaxial Load-Extension Properties of Plain, Weft-Knitted Fabrics - A Theoretical Analysis. Textile Research Journal 45: 746-760.

11. MacRory BM, McNamara AB (1967) Knitted Fabrics Subjected to Biaxial Stress - An Experimental Study. Textile Research Journal 37: 908-911.

12. Hong H, de Araújo M, Fangueiro R and Ciobanu O (2002) Theoretical Analysis of Load-Extension Properties of Plain Weft Knits Made from High Performance Yarns for Composite Reinforcement. Textile Research Journal 72 991-996.

13. Ališauskienė D, Mikučionienė D (2012) Influence of the Rigid Element Area on the Compression Properties of Knitted Orthopaedic Supports. Fibres and Textiles in Eastern Europe 20: 103-107.

14. Zou KH, Tuncali K, Silverman SG (2003) Correlation and Simple Linear Regression. Radiology 227: 617-628.

15. de Araújo M, Fangueiro R, Hong H (2003) Modelling and Simulation of the Mechanical Behaviour of Weft-Knitted Fabrics for Technical Application. Autex Research Journal 3: 166-172.

8

Study of Performance Characteristics of Fabrics Coated with PVC based Formulations

Asagekar SD*

Textile and Engineering Institute, Ichalkaranji, Maharashtra, India

Abstract

The cotton base fabric with plain weave was coated with different coating formulations for exploring the best possible formulation for achieving reasonably good waterproofing and breathability. The coated fabrics were manufactured with Polyvinyl Chloride (PVC) resin, mixture of PVC and Polyurethane (PU) and PVC mixed with Polyethylene Glycol (PEG)/Ethylene Glycol (EG) additives. The properties of coated fabric were studied for two solutions of coagulation bath. The characteristics of coating film were studied by subjecting the coated fabric samples to water treatment. The film porosity was found to be high for all cases, corresponding to low rate of coagulation. Also it was observed that on washing the porosity of film was increased. The fabric coated with mixture of PVC and PU formulation found to give best breathability (around 60 to 75% of control fabric) with water penetration resistance exceeding 150 cm head of water.

Keywords: Cotton fabric; Coating resins; Knife over roll coating; Wet coagulation; Water penetration resistance; Water vapor permeability

Introduction

The waterproofing and breathability are the essential requirements of fabrics for the applications such as rain wears, fabrics for foul weathering, defence wears, marine wears etc. Earlier the attempts were made to develop such fabrics by application of water repellent finishes but could sustain low water pressure and the focus was shifted to physically coating the fabric surface by polymeric materials. Various resins such as Polytetrafluoroethylene, Polyvinyl chloride, Polyacrylates, Polyurethancs etc. Saunders [1] has been used as coatings. The coating layer does prevent the entry of water but it traps the body heat and moisture in the microclimate which makes the wearer uncomfortable [2]. In dealing with thermo-physiological comfort, which relates the maintenance of constant body temperature, can be achieved by transfer of body heat through conduction, convection, radiation as well as evaporative cooling [3,4]. Especially the moisture transmission can be achieved by developing the microspores in coating layer through which the diffusion of moisture takes place [5].

In last few decades the constant efforts were taken to develop the waterproof breathable fabric which can fulfil the necessary requirement at the economic cost. At present many such fabrics are commercially available in the market for different applications with various brands. The rubber, polyurethane etc. [6] have been used on larger scale due to their good adhesive and mechanical properties. In this study an attempt was made to develop waterproof breathable fabric with cotton as base fabric, with PVCresin based coatingformulations. PVC resin was used because it becomes soft, flexible on adding plasticizer and remains thermoplastic. In one formulation PVC was mixed with PU.

Experimental

Materials

Substrate: 100% cotton fabric with the following specifications was taken as base fabric (Table 1).

The following commercial or laboratory grade reagents, chemicals were used

- Polyvinyl Chloride (PVC) resin-commercial grade
- Polyurethane adhesive (PL-2)-commercial grade
- Polyethylene Glycol (PEG) –laboratory reagent
- Ethylene Glycol (EG)-Laboratory reagent
- DioctylPthalate (DOP)-laboratory reagent
- Dimethyl Formamide (DMF)-solvent

Coating formulations

- PVC solution in DMF-33 gm of DOP mixed with every100 gm of solid PVC. The mixture was dissolved in DMF to get 16% PVC solution.
- The effect of water soluble additives was studied by adding 3% PEG/EG in PVC solution.
- 16% PU solution in ethyl acetate (commercial sample) PL-2 adhesive (viscosity 12 poise) from Pidilite industries limited. PVC and PU was mixed in 60:40 proportion.

Methodology

Coating method: Coating was carried out on Benz coating machine. It is laboratory machine with number of controls. Coating head consists of a blade, rubber blanket supported by rolls. The clearance between knife and fabric was varied to give different weight add-ons. The fabric after coating was passed through coagulation bath wherein the microporous structure was expected to develop and in the subsequent step the fabric was dried at 120°C for 3 minutes. The coagulation was carried out in 100% water and mixture of DMF and water in the ratio 50:50. The process of coating, coagulation and drying was continuous. The coating was carried out at three levels of knife clearance.

*Corresponding author: Asagekar SD, Textile and Engineering Institute, Ichalkaranji, Maharashtra, India, E-mail: sdasgekar@gmail.com

Characterization

Coating viscosity: The coating viscosity was measured by Brooke Field Viscometer. The viscosity was measured as the torque required rotating the spindle through polymer solution.

Add on (%): The percentage add-on of coated fabric was measured by weighing the fabric of given area before and after coating. The difference was expressed in percentage with respect to the weight of base fabric.

Bending length: The bending length of fabric was measured on Shirley stiffness tester. The test was carried out according to ASTM D1388-14E01 standards.

Air permeability: The air permeability of coated fabric was measured by textest instrument, with suction principle, at the pressure difference of 2000 Pa. The test was carried out according to ASTM D737-04 (2012) standards.

Water penetration resistance (WPR): The water penetration resistance of fabric was measured by Shirley hydrostatic head tester. The waterpenetration resistance of fabric was measured in terms of head of water the fabric can sustain. The test was carried out according to ASTM D3393-91(2014) standards.

Water vapor permeability (WVP): The water vapor permeability was measured by Ludlow method. In principle the cell measures the humidity generated under controlled conditions as a function of time. This is based on the application of the gas permeability equation and the ideal gas law. The test was carried out according to ASTM F1868-98 standards.

Launderability: The integrity of coated film was determined by subjecting the fabric to laundering. The test was carried out according to ASTM D4265-14 standards. The performance of the fabric was evaluated after laundering in terms of add-on (%), water vapor permeability, water penetration resistance, bending length. Qualitative assessment was also made to see the distortion or puckering of the coated film.

Results and Discussion

The characteristics of the coated fabrics were measured and the results were analyzed which have been presented as follows.

Effect of coating with PVC

Coagulation in water: The cotton fabric was coated with 16% PVC and then passed through coagulation bath containing water and finally dried. The characteristics of fabric are summarized in Table 2.

All the samples, excepting corresponding to 40% add on, show excellent WPR, which exceeds 150 cm head of water. On the other hand, air permeability is drastically reduced from 2630 l/m²/sec of control fabric to 3.34 at highest level of add-on. This is expected because control fabric has open structure as against the coated fabric. The air permeability seems to decrease with increase in add-on but only marginally. The WVP also decreased with coating level. For all cases it varies from around 25% to 40% with respect to the control fabric. Bending length found to increase with add-on, reflecting the increase in stiffness with level of coating.

On laundering, marginal weight loss was observed (around 3-5% of the actual add-on). The increased WVP indicate that some residual solvent might have come out during washing, producing additional micro-pores. After washing, the fabric became relatively softer as indicated by decrease in bending length.

Coagulation in water/DMF mixture: The effect of rate of coagulation was studied by passing the coated fabric through coagulation bath containing 50:50 mixtures of water and DMF. The results are summarized in Table 3. It may be noted from Table 2, that all the fabric samples crossed water head of 150 cm, meaning add-on levels in excess of 50% produce strong uniform film. The WVP values show marginal increase in comparison with that of previous case (Table 1) which shows the increased micro-porosity on slow coagulation. Also there is significant increase in air permeability with respect to the previous case. Interestingly even bending length is reduced with slow rate of coagulation.

On washing treatment, slight loss in weight was recorded due to removal of solvent traces. The removal of solvent traces improved WVP and softness. All the fabric samples passed WPR test crossing 150 cm head of water indicate that the integrity of film was maintained even after washing.

Effect of PEG as additive in PVC

Coagulation in water: The effect of water soluble polymer additive was studied by adding 3% PEG in PVC solution. The fabric was then coated withmixture and subsequently passed through the coagulation bath containing water. The results are summarized in Table 4.

It may be noted that the water-vapor permeability and the air permeability have been found to decrease with add-ons. All the coated fabrics pass the water penetration test. Bending length too increased with add-on. But in comparison with PVC coated samples the bending length is quite low. This may be due to the presence of PEG which may act as plasticizer. However, the expected improvement in WVP was not observed in this set of experiments. The reason could be that coagulation time may not be sufficient for effective removal of PEG. On washing a considerable increase in WVP was observed. The bending length was further reduced after washing.

Coagulation in water/DMF mixture: In this set of experiments, the fabric was first coated with mixture (PVC+3% PEG) and then passed through 50:50 water/DMF bath and dried. The results are summarized in Table 5.

It may be noticed from Table 5, that slow rate of coagulation does improve the properties, in particular WVP values which confirms the improvement in porosity. The bending length and air permeability are also improved as a result of slow coagulation. On washing a significant weight loss of around 8-10% of the actual add-on was observed. It may be related mainly to removal of large PEG molecules entrapped in the film during coagulation. All the fabric samples passed the WPR test even after washing, indicating that the film was intact even after removal of PEG.

Effect of EG additive

Coagulation in water: The fabric was coated with a mixture containing PVC and water soluble additive Ethylene Glycol (EG). The coated fabric was then passed through the coagulation bath containing water. The results of the experiment are summarized in Table 6.

The fabric coated with this mixture show improved WVP and bending length in comparison with PVC coated fabric. But its performance is almost similar to that of fabric coated with PVC and PEG mixture. This indicates that the addition of EG does not affect much on performance of fabric compared with the PVC+ PEG coated

Fabric type	GSM	Ends/inch	Picks/inch
Cotton with plain weave	120	40	29

Table 1: Fabric type.

Add-on (%)	WPR (cm head of water)	WVP (gm/m²/day)	Bending length (cm)	Air permeability (l/m²/sec)
40.0(38.6)	120(109)	291(382)	4.40(4.20)	5.46
62.0(59.7)	>150	214(236)	4.60(4.35)	4.30
74.0(71.6)	>150	189(229)	5.20(5.00)	3.34
control	--	730	1.65	2630

*Quantity in the bracket indicates the corresponding figures after washing.

Table 2: Performance of PVC coated fabric with coagulation in water.

Add-on(%)	WPR (cm head of water)	WVP (gm/m²/day)	Bending length (cm)	Air permeability (l/m²/sec)
54.8(53.4)	>150	289(330)	4.10 (4.00)	8.80
61.0(59.3)	>150	220(283)	4.30(4.10)	8.00
76.6(74.2)	>150	193(253)	4.50 (4.30)	6.70
control	--	730	1.65	2630

* Quantity in the bracket indicates the corresponding figures after washing.

Table 3: Performance of PVC coated fabric with coagulation in water/DMF.

Add-on(%)	WPR (cm head of water)	WVP (gm/m²/day)	Bending length (cm)	Airpermeability (l/m²/sec)
48.0(46.4)	>150	269 (330)	3.10 (2.95)	7.10
65.0(62.6)	>150	244 (286)	3.40 (3.20)	5.20
76.0(73.2)	>150	220 (271)	3.65 (3.40)	4.32
control	--	730	1.65	2630

*Quantity in the bracket indicates the corresponding figures after washing.

Table 4: Effect of PEG additive on the performance of PVC coated fabric with coagulation in water.

Add-on (%)	WPR (cm head of water)	WVP (gm/m²/day)	Bending length (cm)	Airpermeability (l/m²/sec)
52.0(48.9)	>150	313 (398)	2.75 (2.65)	7.46
61.0(57.5)	>150	275 (330)	2.90 (2.80)	6.60
73.0(68.9)	>150	239 (288)	3.20 (3.00)	6.20
control	--	730	1.65	2630

*Quantity in the bracket indicates the corresponding figures after washing.

Table 5: Effect of PEG additive on the performance of PVC coated fabric with coagulation in water/DMF.

Add-on(%)	WPR (cm head of water)	WVP (gm/m²/day)	Bending length (cm)	Airpermeability (l/m²/sec)
54.0(51.6)	> 150	272(341)	3.20(3.10)	6.70
65.0(62.7)	> 150	240(289)	3.40(3.25)	5.86
72.0(69.8)	> 150	208(248)	3.55(3.45)	4.20
control	--	730	1.65	2630

*Quantity in the bracket indicates the corresponding figures after washing.

Table 6: Effect of EG additive on the performance of PVC coated fabric with coagulation in water.

fabric. As usual the WPR exceeds the 150 cm mark of water head. On washing, around 3 to 5% weight loss was observed. This has improved the WVP by around 20%. The WPR resistance value continues to show better performance after washing, as the water head remained above 150 cm.

Coagulation in water/DMF mixture: The effect of reduced rate of coagulation was studied by passing the coated fabric through mixture of DMF and water. The performance characteristics of fabric coated

with PVC/EG mixture at reduced rate of coagulation is summarized in Table 7. As expected the reduced rate of coagulation has considerably improved WVP. The bending length which varied from 3.10 to 2.70 cm, represents a good handle. All the fabric samples pass the WPR test. The WVP values have been found to increase further on washing. The bending length also shows improvement on washing. The film remained intact and continuous which can be confirmed by high values of WPR even after washing.

Coating with PVC/PU mixture

In this set of experiment PVC was mixed with PU so that the benefit of excellent adhesiveness of PU along with possibility of improvement of porosity can be explored. In preliminary experiments the composition was optimized by taking different combinations of PVC and PU. It was observed that for equivalent add-ons the combination of 60:40 PVC/PU coat show excellent results as far as the WVP and WPR are concerned. Therefore, it was decided to use 60:40 mixtures for the study.

Coagulation in water: Cotton fabric was coated with 60:40 mixtures of PVC/PU resins and passed through the coagulation bath containing water. The results are summarized in Table 8. It can be observed from the Table 7 that the properties are greatly improved. The WPR resistance continues to show excellent figures as it crossed 150 cm mark. The WVP found to increase drastically in comparison with previous cases. Also bending length shows excellent figures varying between 2.60 to 2.90 cm in comparison with earlier experiments. This shows the excellent hand of the fabric coated with this mixture. The properties are improved considerably on washing treatment.

Coagulation in water/DMF mixture: The mixture of PVC/PU coated fabric was passed through the coagulation bath containing water/DMF mixture and then dried. The results are presented in

Add-on (%)	WPR (cm head of water)	WVP (gm/m²/day)	Bending length (cm)	Airpermeability (l/m²/sec)
53.0(50.8)	>150	344(426)	2.70 (2.60)	6.90
64.6(62.9)	>150	290(347)	2,95(2.80)	6.50
71.7(69.4)	>150	254(287)	3.10(3.00)	5.90
control	--	730	1.65	2630

*Quantity in the bracket indicates the corresponding figures after washing.

Table 7: Effect of EG additive on the performance ofPVC coated fabric with coagulation in water/DMF.

Add-on(%)	WPR (cm head of water)	WVP (gm/m²/day)	Bending length (cm)	Airpermeability (l/m²/sec)
59.0(58.0)	>150	429(536)	2.60 (2.50)	6.70
69.0(66.9)	>150	327(400)	2.50 (2.60)	6.74
77.0(74.2)	>150	300(351)	2.90 (2.70)	4.80
control	--	730	1.65	2630

*Quantity in the bracket indicates the corresponding figures after washing.

Table 8: Performance of PVC/PU coated fabric with coagulation in water.

Add-on(%)	WPR (cm head of water)	WVP (gm/m²/day)	Bending length (cm)	Air permeability (l/m²/sec)
58.0(56.3)	>150	464(548)	2.35 (2.30)	7.50
68.0(65.7)	>150	368(467)	2.40(2.35)	6.80
76.2(73.9)	>150	340(423)	2.50(2.40)	6.10
control	--	730	1.65	2630

*Quantity in the bracket indicates the corresponding figures after washing.

Table 9: Performance of PVC/PU coated fabric with coagulation in water/DMF.

Table 9. It is observed that the WVP is improved further because of slow rate of coagulation. For the equivalent add-ons the WVP for this case is improved by around 10% compared with that for high rate of coagulation.

On washing the WVP is further increased. For add-on level of around 58% the WVP is increased to 548 gm/m^2/day which is nearly 75% of the WVP of control fabric, and certainly fulfils the breathability requirements. The air permeability and the bending length follow the trend and show the improved performance. The bending length dropped further on washing indicating the improved fabric hand. As far as the WPR is concerned all samples crossed 150 cm water head.

Conclusion

All the coated fabric samples give excellent WPR even at low levels of add-ons.

All the coated fabric samples at slow rate of coagulation show better performance characteristics. Washing treatment had no effect on film integrity as well as fabric film bonding. Thecoated samples containing PEG and EG additives in PVC, show loss of fabric weight indicating the effective removal of additives with improved film porosity. The water vapor permeability of PVC/PU coated fabric at slow coagulation condition is exceptionally good which is the major success of the study.

References

1. Saunders KJ (1988) Organic polymer chemistry.springer publications, Netherlands.

2. Anon (1987) Fibres and Fabrics: Waterproof and comfortable Textile Month 28: 37.

3. Gohlke DJ (1981) Improved analysis of comfort performance in coated fabrics. Journal of Industrial Textiles 10: 209-224.

4. Keighley JH (1985) Breathable fabrics and comfort in clothing. Journal of Industrial Textiles 15: 89-104.

5. Lomax GR (1985) The Design of Waterproof, Water Vapor-Permeable Fabrics. Journal of Industrial Textiles 15: 40-66.

6. Damewood JR (1980) The Structure-property relationships of Polyurethanes designed for coated fabrics. Journal of Industrial Textiles 10: 136-150.

9

Synthesis, Characterization and Application of Nano Cellulose for Enhanced Performance of Textiles

Chattopadhyay DP* and Patel BH

Department of Textile Chemistry, Faculty of Technology and Engineering, The Maharaja Sayajirao University of Baroda, Vadodara, India

Abstract

In the present investigation cellulose nano whisker is separated from industrial waste viscose rayon fibre, characterized by SEM images and FTIR spectroscopy. The size and size distribution of these nano crystals have also been examined using particle size analyzer; the average size of the particles is found to be 348 nm. The findings support the size and shape of the synthesized nano cellulose particles. These nanoparticles have been applied to polyester fabric by padding technique and manifested the improved physical and thermal properties. The dyeing behaviour of the treated fabrics with direct dye has also been studied and the build-up of dyes, measured as colour strength in terms of K/S values, reported. The higher K/S values are obtained when the cellulose nano is anchored in the fibre matrix, i.e. when the fibre is pre-treated and dyed with direct dyes. Improved colour strength with good resistance towards soaping is obtained after treatment of fabrics with nano cellulose.

Keywords: Absorbency; Cellulose nano; Direct dye; Physical property; Thermal property

Introduction

Textile materials made from natural fibers have played an important role in the life of human beings from time immemorial and still are widely used in the modern textiles industry for their unique properties as high quality textile materials. Due to the variation in staple length, the natural fibers with short staple length can't be used to spin yarns. Consequently, natural fibers such as wool, silk, cotton or hemp are wasted during processing and final usages. A new way of reusing these fibers has large marketing potential because of their excellent intrinsic properties. Meanwhile, not only the textile industry, but many other industries like the bio-medical industries need such bio-compatible materials [1-3].

Fine/super fine powder prepared from protein or cellulose fiber is generally known as nano-whiskers, which can impart various functional properties not only to the textiles but also contribute significantly in the field of electronics and medicines. Some potential applications of nano cellulose in the field of paper and paperboard applications as dry strength agent, surface strength agent or nanocoatings/nanobarriers, bio-nanocomposites, food applications, cosmetics/skin creams, medical/pharmaceutical applications, hygiene/absorbent products, emulsion/dispersion applications and oil recovery applications. Many researchers have reported new method of synthesis nano-whiskers and their application in bio-technological and bio-medical fields [4-7].

In this paper, nano scale cellulose polymers were prepared from viscose rayon yarns by a novel technique. The prepared cellulose nano whiskers were characterized for their size, shape and chemical composition. Nano scale cellulose was applied to polyester textiles by padding technique. Changed in physico-chemical characteristics and thermal behaviour of the new polyester fiber incorporated with nano cellulose were analyzed by SEM, FTIR, Image analyzer and computer colour matching system, the thermal behaviour of polyester cellulose nano composite were analyzed using DSC.

Materials and Experimental Methods

Material

Pure polyester woven fabric with specification as mentioned in Table 1 was used. The fabrics was cleaned with 2% sodium carbonate and 5% nonionic detergent at 70°C temperature for 15 minute then again washed and neutralized before used.

Experimental methods

Nano-cellulose was prepared by treating waste viscose rayon fibers with freshly prepared solution of sodium zincate.

Preparation of sodium zincate solution: Sodium zincate was prepared by adding 180 gms of NaOH to 200 ml of water then 80 gms of ZnO was gradually added with constant stirring. The solution was kept for 24 hours in a container. Finally, the solution was filtered using Whatman No.1 filter paper to get sodium zincate solution.

Preparation of nano cellulose: In this study, suspensions of nanocrystals were prepared from waste viscose rayon fibers the scheme for the preparation of nano cellulose is illustrated in the following Figure 1.

The waste viscose rayon fibers were ground to smaller than 20 mesh powder. Ground viscose rayon fiber powder was mixed with sodium zincate in a ratio of 1:9 (g/ml). A reaction temperature of 50°C was maintained for the diffusion of sodium zincate into the amorphous region of the fibers resulting in a subsequent cleavage of the glycosidic bonds. After 1 hour the particles were neutralized by glacial acetic acid solution. The suspension was washed and further filtered by Whatman No.1 filter paper. The colloidal suspension was evaporated aconverted in powder form. The powder was washed with distilled water and dried.

Characterization of nano cellulose particles: The particle size and size distribution of the cellulose nano were analyzed on the particle size analyzer (Malvern Instrument, MAL501131, DTS version 5.03,

*Corresponding author: Chattopadhyay DP, Department of Textile Chemistry, Faculty of Technology and Engineering, The Maharaja Sayajirao University of Baroda, Vadodara, India, E-mail: dpchat6@gmail.com

ample	Material Specification						
	Count/Denier		Ends/inch	Pick/inch	Type weave	Wt.gm/ sq.m.	Thickness (mm)
	Warp	Weft					
100% polyester	128d	146d	90	72	Plain	109.7	0.21

Table 1: Specifications of polyester fabric.

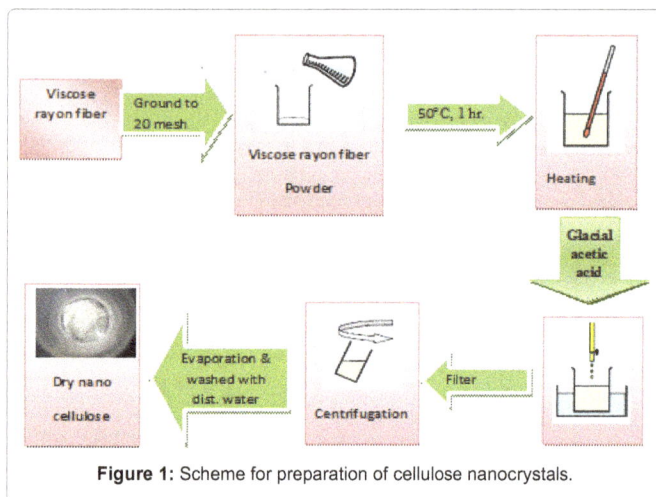

Figure 1: Scheme for preparation of cellulose nanocrystals.

U.K.). The morphology of cellulose nano nanoparticles was examined on scanning electron microscope (SEM) (Model JSM5610LV, version 1.0, Jeol, Japan). Chemical composition of prepared nano powder was analyzed by FTIR Spectroscopy Nicolet is10 FT-IR Spectrometer (Thermo Scientific).

Application of nano cellulose by *Pad–dry–cure method*: Polyester fabric samples were padded with varying concentrations of nano cellulose suspension viz., 1 gpl, 5 gpl, and 10 gpl. For 1 gpl solution, 0.1 gm nanoparticle was added in 100 ml liquor with 5 gm lissapol L surfactant. The mixture was then stirred using magnetic stirrer at 250 rpm for 30 minutes at 50°C temperature. Likewise all concentration solution was prepared. Polyester fabric samples (size: 40cm × 30cm) were immersed in padding liquor at room temperature for 10 minutes and then passed through a two bowl laboratory padding mangle, which was running at a speed of 15 rpm with a pressure of 1.75 Kg/cm^2 using 2-dip-2-nip padding sequence at 70% expression for polyester fabric. The padded substrates were dried at 80°C. The dried samples were cured in a preheated curing oven at 180°C temperature for 60 seconds.

Testing and analysis

Fabric characterization: The morphology of cellulose nanoparticles deposited on polyester fabric was observed by SEM. The samples were also observed on microscope at 100 X magnification. The images at selected places of the specimen were captured by digital camera attached to the microscope. These images were transferred to image analyzer in computer. Image analyses of these samples were carried out using Image-Pro ^{Plus}, Version 4.1 Software of Media Cybernetics, USA. The presence of cellulose in the polymer structure was detected by FTIR Spectroscopy. The thermal characterizations of polyester fabric and nano cellulose were analyzed using Differential Scanning Calorimetric measurements (DSC) (Model 6000 from PerkinElmer, Singapore) in the temperature range 50°C to 300°C. Nano cellulose and polyester-nano cellulose composites were scanned on an X-ray diffractometer (X'Pert-Pro, PAN Analytical, Singapore). Cu Kα radiation at 45 Kv and 40 mA was utilized and scanned for 2θ between 5°C and 40°C.

Physical testing: Before physical testing the samples were dried and conditioned at 65± 2% RH and 27 ± 2°C temperature.

Determination of tensile properties: 2 cm x 8 cm fabric samples were tested at 100 mm/min traversing speed for the determination of breaking load, breaking elongation, stress and strain. The test was performed as per B.S. 2576:1959

Determination of crease recovery angle: The test specimen was folded and compressed under controlled condition of defined force to create a folded angle, the specimen was suspended in an instrument for a controlled recovery and the recovery angle was measured. The test was performed as per AATCC test method 66-2003.

Determination of bending length: The stiffness in terms of bending length of nano treated and untreated samples were measured as per AATCC Test Method 115-2005 using Prolific stiffness tester (India).

Determination of absorbency by wicking test: Wicking behavior of the treated and untreated samples were evaluated as per T-PACC standard method.

Evaluation of water permeability: These test methods provide procedure for determining the hydraulic conductivity (water permeability) of textiles materials in terms of permittivity under standard testing conditions in uncompressed state. The test was conducted using ASTM D 4491 (Constant Head Method) water permeability test method.

Evaluation of air permeability: The air permeability of treated and untreated polyester fabric samples were measured on Metefem air permeability tester as per ASTM D 737 test method. The result of the test measured reported in m^3/h/m^2 to three significant digits.

Dyeing of nano cellulose treated and untreated fabric samples

Mild scoured polyester fabric samples (size: 40cm × 30cm) were immersed in the padding liquor containing 3 gram and 5 gram direct dye, 2 gram sodium carbonate and 5 gram glauber's salt in 100 ml liquor. Polyester sample were entered in above liquor at room temperature for 10 minutes and then passed through a two bowl laboratory padding mangle, which was running at a speed of 15 rpm with a pressure of 1.75 Kg/cm^2 using 2-dip-2-nip padding sequence at 70% expression for polyester fabric. The padded substrates were dried at 80°C. The dried samples were cured in a preheated curing oven at 180°C temperature for 60 seconds.

Evaluation of dyed samples

Measurement of colour strength value (*K/S Value*): The dyed samples were assessed for *K/S* values using computer colour matching system (illuminant D65/100 observer, Spectra scan 5100 RT, Spectrophotometer, Premier Colourscan Instrument, India).

Fastness tests: The light fastness of the dyed samples was tested on Fad-o-meter (FDA-R, Atlas, U.S.A.) after partially exposing the samples to the xenon arc lamp for 16 h and graded for the colour change with the ratings. The wash fastness of the samples was performed as per ISO-2 tests using launder-o-meter (Digi.wash, Paramount Scientific Instruments., India). Samples were also evaluated for the rating in terms of colour change

Results and Discussion

This section of the paper discussed the results of preparation of nano cellulose particles and their application to polyester fabrics by pad-dry-cure techniques. The prepared nanoparticles were characterized using particle size analyzer, the morphology of the particles was observed using SEM. The nano cellulose polymers were using FTIR. The functional properties of nano treated fabrics were tested as per the standard methods of testing.

The rod-like particles that were produced as a result of treatment were dried and again washed with distilled water and dried. Figure 2 represent organization and separation of nano cellulose from the fiber.

Characterization of prepared nano cellulose

The dried powder of nano cellulose prepared is in the insect photographs shown in Figure 3. The analysis of the sample of cellulose powder dispersed in water by particle size analyzer showed a narrow and sharp peak at around 348 nm diameter.

Figure 4 shows the scanning electron micrographs of prepared nano cellulose particles deposited on carbon coated aluminum sheet. It can be seen from the figure that the shape of prepared nano cellulose particles was rod-like. The breaking of the cellulose chain, which contains high order crystalline regions, connected with low order amorphous regions, which appear like individual rods. These rod-like particles are commonly called as whiskers. FTIR spectrum of the nano cellulose powder is shown in Figure 5 and X-ray diffraction pattern of nano cellulose are shown in Figure 6.

The absorption in the region of 3600-3100 cm^{-1} was due to the stretching of -OH group and at 3000 to 2800cm^{-1} to the CH stretching. The band observed at 1642 cm^{-1} across from the H-O-H bending of the absorbed water. The symmetric C-H bending occurred at 1400 cm^{-1}; the FTIR absorption band at 1430 cm^{-1}, assigned to a symmetric CH$_2$ bending vibration, decreases. This band is also known as the

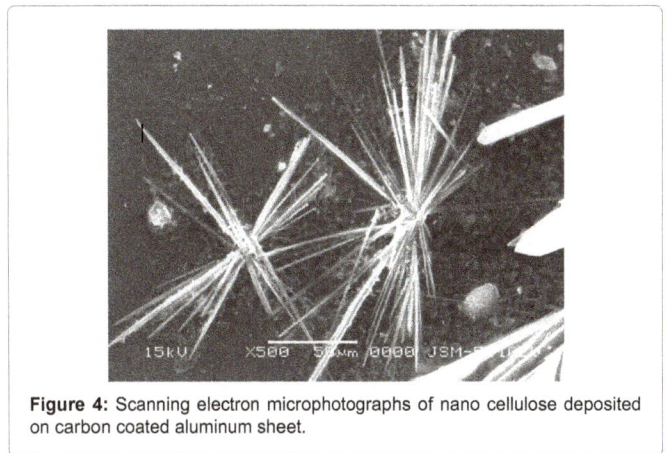

Figure 4: Scanning electron microphotographs of nano cellulose deposited on carbon coated aluminum sheet.

"crystallinity band", indicating that a decrease in its intensity reflects reduction in the degree of crystallinity of the samples. The main characteristic peaks were detected at absorption band 898 cm^{-1}, assigned to C-O-C stretching at β-(1→4)-glycosidic linkages, is designed as an "amorphous" absorption band. The IR spectra confirmed the presence of amorphous microcrystalline celluloses structure.

Characterization of polyester/nano cellulose composite

The dispersion of nanoparticles on the surface of the fiber and their penetration in the polymer matrix were examined by image analyzer (100 X). Cross sectional and longitudinal view of the sample were prepared in laboratory as per AATCC Test method 20-2005. (AATCC Technical manual, vol. 81, 2006, pp-40.). The prepared cross sectional and longitudinal sections were further stained with direct dye and examined under image analyzer. The images captured by image analyzer are shown in Figure 7.

Cellulose particles applied on 100% polyester fabric, the treated fabric was then stained with a direct dye (Congo Red BDC) to highlight the cellulose particles. Deposition of cellulose particles is seen on the surface of polyester fiber, from the longitudinal view of the polyester fiber presented in Figure 7a. Dispersion of these particles in the polymer matrix is also observed from the Figure 7b which represents the cross sectional view of polyester fiber.

Figure 8 shows the SEM images of polyester fiber surfaces after the nano-cellulose treatment. The Micro photographs captured at different magnifications i.e 500 X, 1000 X, 1500 X and 2000 X show that the fiber surface is covered with nano-cellulose particles after treatment.

The FTIR spectra of the polyester fabric before and after nano cellulose treatment are illustrated in IR spectra Figure 9a and 9b respectively. The peaks in the IR spectra of the polyester loaded with nano cellulose and untreated fabric appeared in the range of 600-4000 cm^{-1}. The waves were assigned as follows: 1715 cm^{-1} (C=O), 1409 cm^{-1} (aromatic ring), 1331 cm^{-1} and 1021 cm^{-1} (carboxylic ester or anhydride), and 1021 cm^{-1} (O=C-O-C or secondary alcohol), 967 cm^{-1} (C=C), 869 cm^{-1} (five substituted H in benzene). The peak at 1409 cm^{-1} corresponded to the aromatic ring. It was the characteristic absorption peak of PET. The peak at 1715 cm^{-1} was assigned to the ester group.

No significant changes in the spectra were observed after nano cellulose treatment for polyester portion appears over the course. But the spectra also confirmed the presence of cellulose from the absorption peaks in the region of 3600-3100 cm^{-1} due to the stretching of –OH group; at 3000 to 2800 cm^{-1} to the CH stretching; 1642 cm^{-1} across from

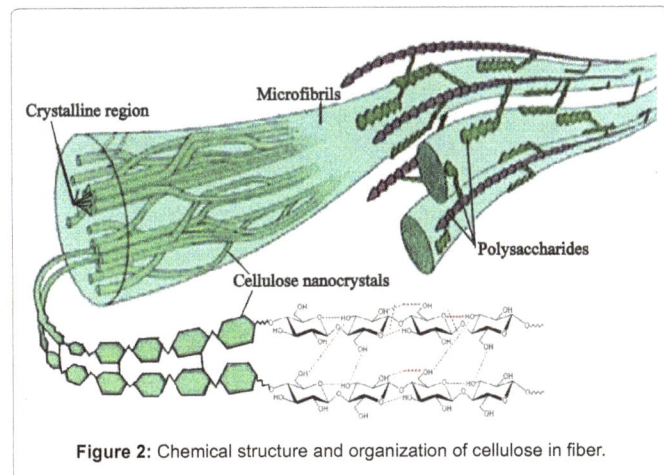

Figure 2: Chemical structure and organization of cellulose in fiber.

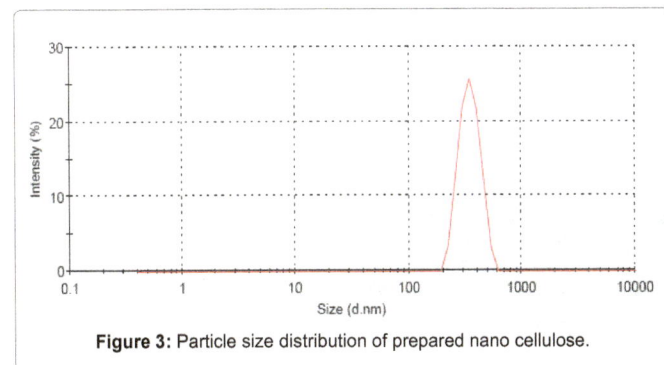

Figure 3: Particle size distribution of prepared nano cellulose.

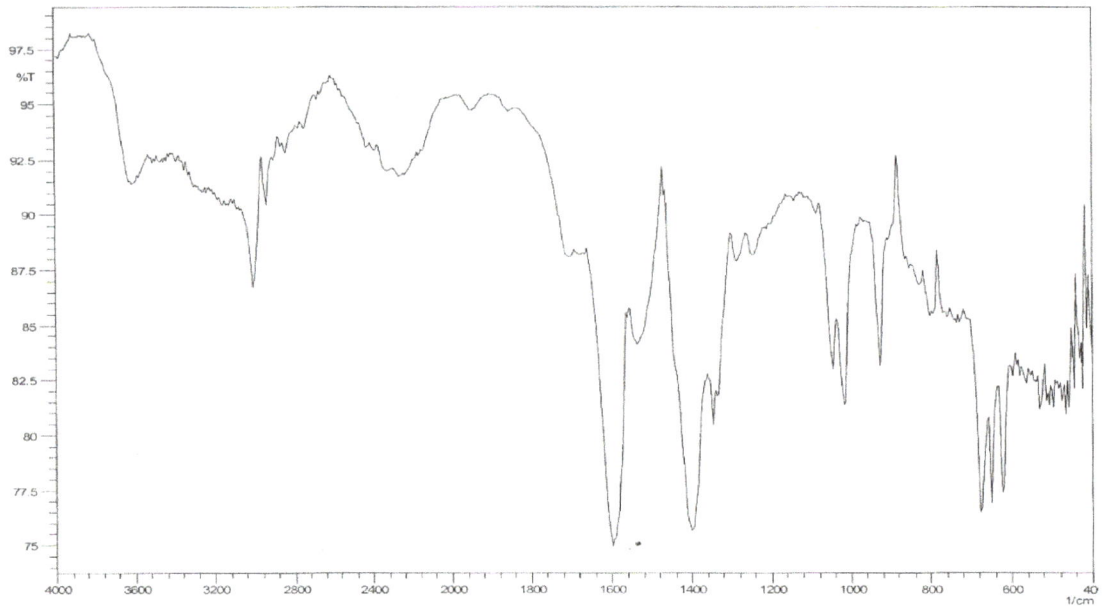

Figure 5: IR Characterization absorption peak of prepared nano cellulose powder.

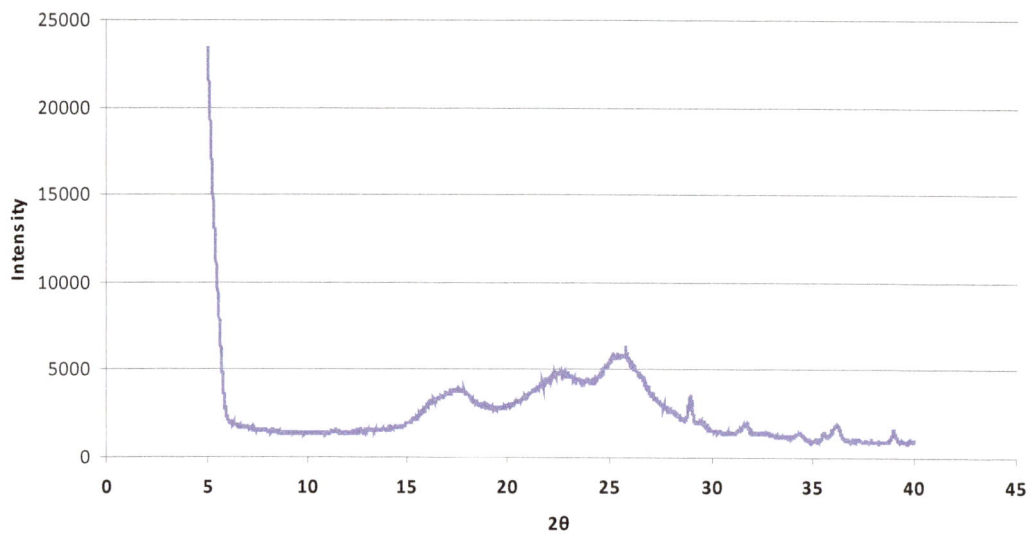

Figure 6: X-ray diffraction pattern of nano cellulose.

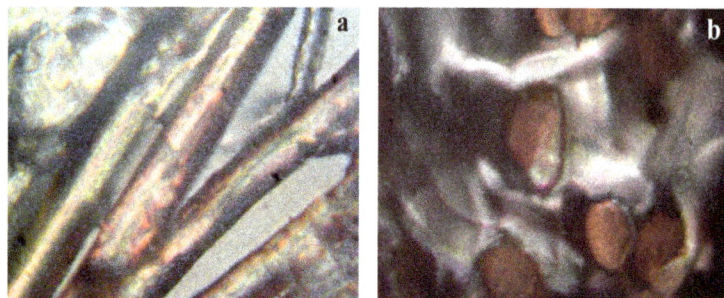

Figure 7: Images (100X) of polyester substrate treated with nano-cellulose and stained with congored direct dye (a) longitudinal and (b) cross sectional view.

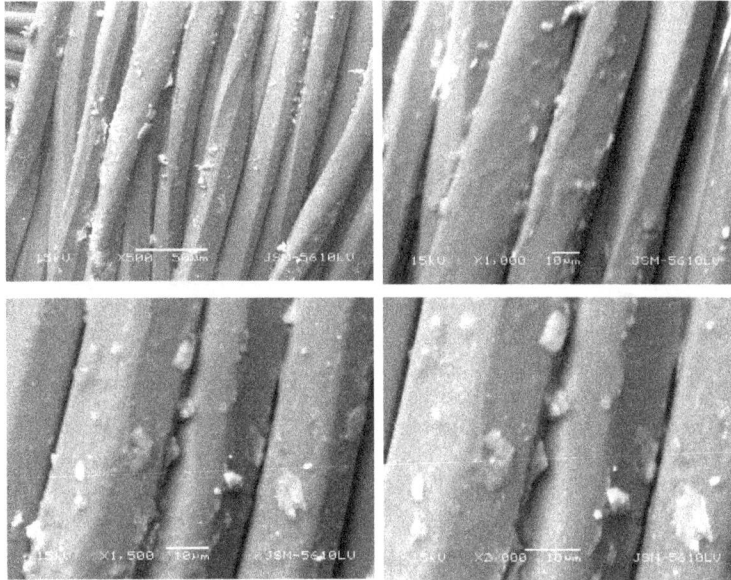

Figure 8: Polyester fiber surface morphology after treatment with the nano-cellulose treatment.

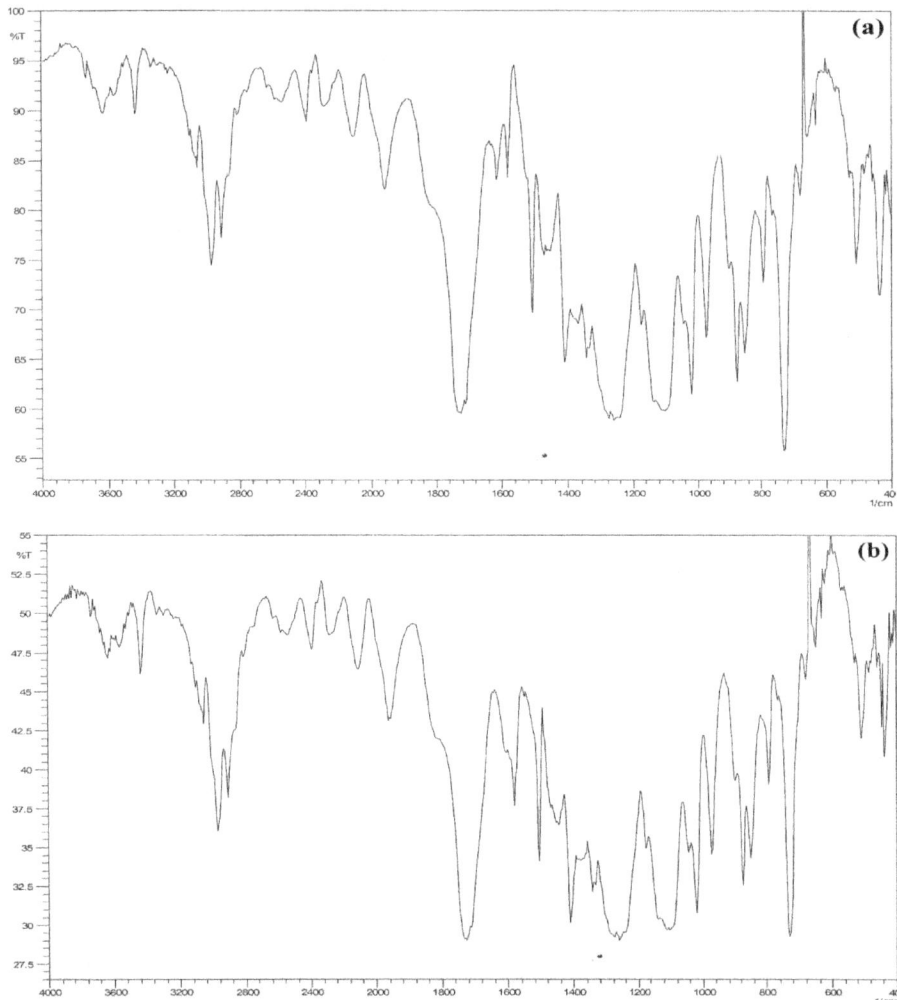

Figure 9: FTIR spectra of (a) pure polyester fabric and (b) polyester fabric treated with nano-cellulose.

the H-O-H bending of the absorbed water; symmetric C-H bending occurred at 1430 cm⁻¹; at absorption band 898 cm⁻¹, assigned to C–O–C stretching.

Thermal analysis

The thermal reactivity of synthesized nano powder (C), polyester fabric treated (B) and untreated (A) were then studied. The onset temperatures for samples were determined using Differential Scanning Calorimetric measurements (DSC). The sample A and B showed the onset temperature at 250.32°C and 250.73°C respectively. The slight increase in onset temperature can be attributed due to the incorporation of cellulose nano into the polyester fiber matrix; however, the effect is not very significant. For the nano cellulose powder the onset temperature at 54.29°C. The reactivity of the untreated PET was then compared with the PET treated with cellulose nano powder, which revealed little difference in reactivity as recorded by DSC. The various onset temperatures for samples are shown in Table 2 and DSC curves registered in the 50-300°C temperature range are shown in the Figure 10.

To study the effect of nano cellulose treatment on thermal degradation of polyester fabric was analyzed. The endothermic peaks characteristic of nano cellulose, occurred at 57.59°C on the DSC curves is presented in Figure 11. A shift of the maximum temperature of the nano treated sample to higher values may be observed with the decrease of the crystallinity degree. Thus, in the case of untreated polyester fabric A, the maximum temperature of the peak appears at 251.72°C, followed by the PET sample treated with nano cellulose B at 253.18°C noticed for the 50-300°C region, which corresponds to the incorporation of nano cellulose in polyester fiber matrix. This behavior is explained by the fact that the thermal degradation reaction starts in the amorphous domain of the cellulosic materials by statistical degradation of cellulose. In the present case, the amorphous content increases from pure polyester to polyester/nano cellulose material.

Effect of nano cellulose on physical properties of polyester fabric

Treated and untreated polyester fabrics were evaluated for the change in physical properties in terms of breaking load and crease recovery angle.

Effect on tensile strength: The treated and untreated polyester

Compound	Onset temperature in °C	Endothermic peak °C	ΔH, J/g	Area of the endothermic peak, mJ
A- Pure PET	250.32	251.72	67.77	67.77
B- PET/Nano-cellulose	250.73	253.18	67.28	45.75
C- Nano - cellulose powder	54.29	57.59	121.27	179.48

Table 2: DSC results.

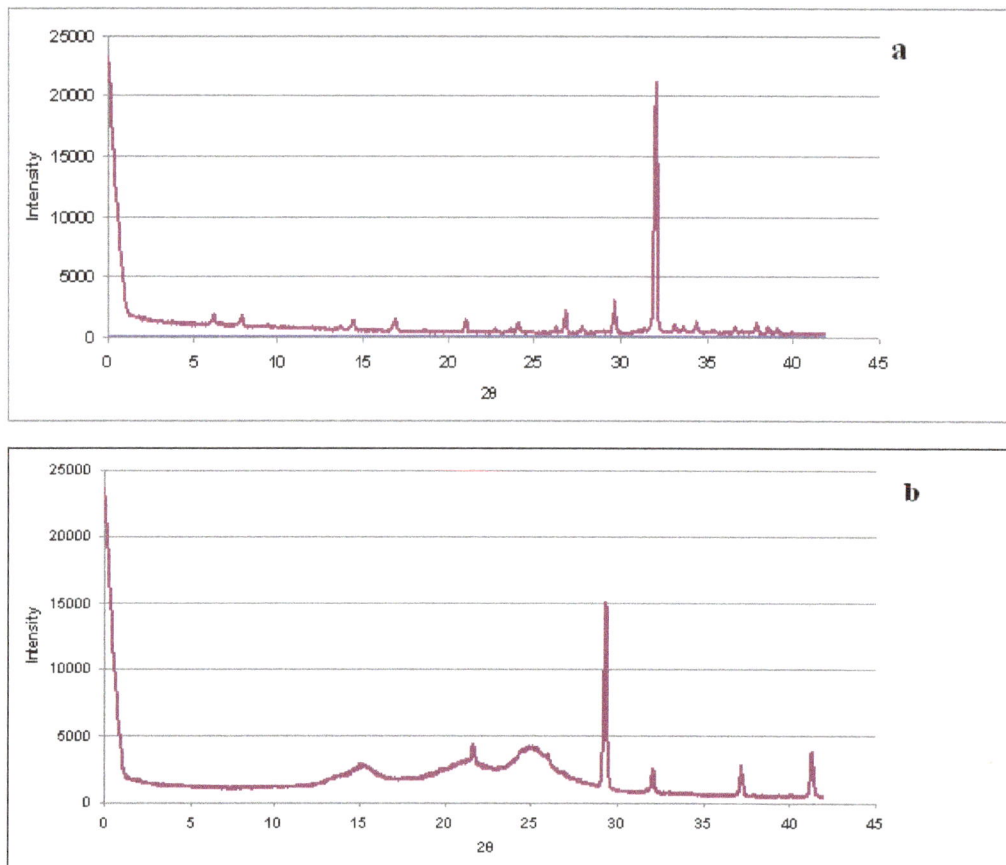

Figure 10: X-ray diffraction pattern of a) pure polyester and b) polyester treated with nano cellulose.

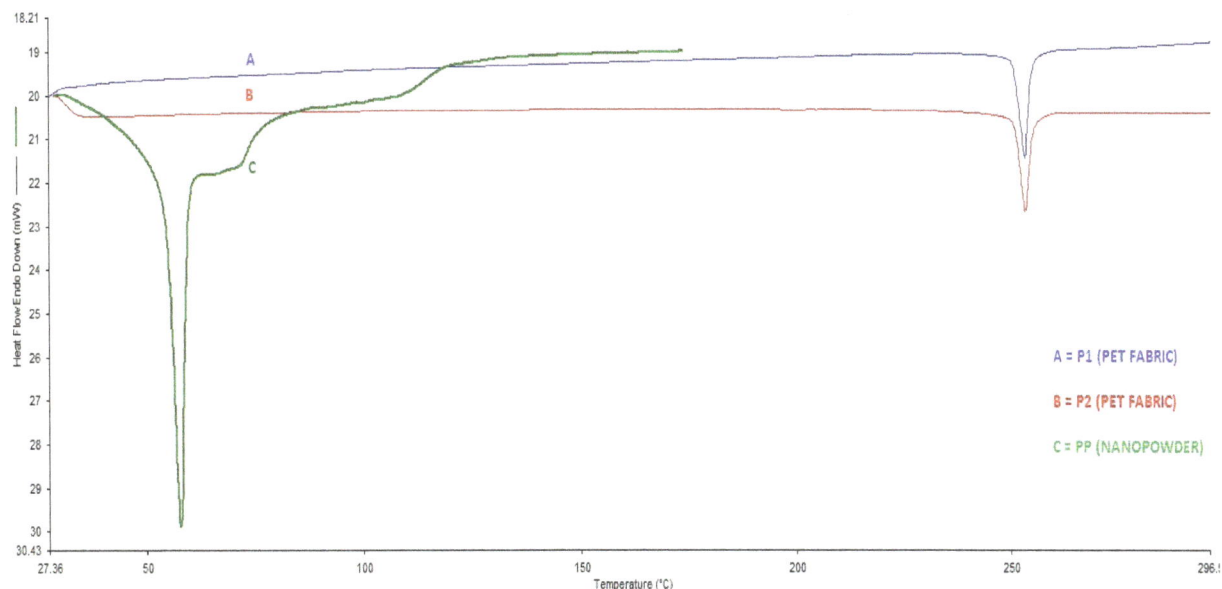

Figure 11: DSC curves registered in the 50-300 °C temperature range for untreated polyester (A), polyester treated with nano-cellulose (B) and nano-cellulose powder (C).

samples were tested to evaluate change in tensile strength; the results are shown in Table 3. The results show that the application of nano cellulose particles to polyester fiber causes an improvement in the load bearing capacity of the fiber. It may be due to the more amount of nano cellulose diffused in the polymer matrix.

Effect on crease recovery: The result presented in Table 4 shows minor improvement in crease recovery angle of the treated samples. The nano cellulose particles because of their small size can enter in between the polymer molecules and perhaps act as filler or cross linking agent. The crease recover angle of fabric was improved with increase in the concentration of nanoparticle. The improvement in physical properties is due to the mechanical interlocking caused by the mechanical anchoring of the nano cellulose in the intermolecular pores.

Effect of nano cellulose on water absorbency of polyester fabric

The absorbency of polyester fabric treated with nano cellulose was measured by drop test and wicking behavior test. Tables 5 and 6 show results of absorbency of the fabric using drop test and wicking height.

It is found from the results shown in Table 4 that the water absorbency of polyester fabric treated with 1 gpl concentration of nano cellulose take less time for the absorption of water indicates improvement in absorbency. But as the concentration of nano cellulose was increased to 5 gpl and 10 gpl, the water droplet took more time to absorb water drop indicating reduction in absorbency of nano treated material.

The improvement in hydrophilicity of polyester treatment due to low dosage of nano cellulose treatment is because of the inherent hydrophilicity of cellulose. But interestingly as the concentration of nano cellulose was increased it started hindering the penetration of the water molecules as they start acting as nano whiskers and do not allow the water drop to be accommodated within the interpolymeric spaces.

Effect of nano cellulose on water permeability of polyester fabric

Table 7 shows reduction in water permeability through the polyester fabric sample treated with nano cellulose compared to untreated sample, it may be attributed due to the resistance offered by the nanoparticle present in the polymer matrix towards the flow of water through the fabric. It can also be seen that as the concentration of nanoparticle in fabric was increased, the permeability of water was reduced.

Effect of nano cellulose on air permeability of polyester fabric

The treated polyester fabric were tested for air permeability and compared with air permeability of untreated polyester sample. The results presented in Table 8 show that the air permeability of polyester fabric treated with nano cellulose was reduced compared with the untreated sample. It may be due to the resistance of the nanoparticles present in the polymer matrix towards the flow of air through the fabric sample.

Effect of nano cellulose on dyeing of treated fabric with direct dye

The dyeing of polyester fabric treated with 5 gpl nano cellulose using pad-dry-cure method and subsequently dyed with direct dye (Congo red BDC) using pad-dry-cure method. The results in terms of colour strength and colour coordinate values are reported in Table 9. From the table it can be seen that the polyester fabric dyed without nano cellulose treatment is stained only. It can be observe from the table that as the concentration of dye in the dye bath increases the K/S value of the sample also increases in case of treated polyester fabric subsequently dyed with direct dye. The treated and dyed samples were further soaped with 5% (v/v) non ionic detergent at 70°C temperature for 15 minutes. The results shows that the loss in K/S values of these samples still remain higher than the samples dyed without the nano treatment. The dyeing of nano cellulose treated polyester fabric with direct dye may be attributed due to the presence of cellulose particles in the polyester structure.

Conclusions

- FTIR spectral analysis results confirm the formation of nano cellulose from the waste viscose rayon fiber by treatment with sodium zincate.

- The z-average size of the separated cellulose nano whiskers is found to be 348 nm. The SEM images indicate that cellulose crystals produced are of rod-like shape.

- The treatment with cellulose nano whiskers improves the

Sample	Tensile Strength (kgf)	
	Bre.load(kgf)	Extention(mm)
Control Polyester	105.6	63.07
Polyester treated with 1 g/L nano cellulose	106.7	63.03
Polyester treated with 5 g/L nano cellulose	108.3	64.77
Polyester treated with 10 g/L nano cellulose	110.5	66.05

Table 3: Effect of nano cellulose treatment on tensile strength of sample.

Sample	Crease recovery angle (°)
Control Polyester	148
Polyester treated with 1 g/L nano cellulose	157
Polyester treated with 5 g/L nano cellulose	158
Polyester treated with 10 g/L nano cellulose	162

Table 4: Crease recovery angle of nano cellulose treated polyester fabric.

Polyester fabric treated with nano cellulose (grams/liter)	Time (sec)
Untreated sample	20.10
1	12.0
5	76.0
10	110

Table 5: Effect on absorbency of polyester fabric due to the nano cellulose treatment.

Polyester fabric treated with nano cellulose (grams/liter)	Wicking height(mm)		
	1min	5 min	10 min
Untreated sample	20	40	55
1	22	44	57
5	15	29	41
10	11	22	31

Table 6: Effect of nano cellulose treatment on wicking height (mm) of polyester fabric.

Polyester fabric treated with nano cellulose (grams/liter)	Ψ-Water Permeability (S⁻¹)
Untreated sample	0.3253
1	0.3223
5	0.3087
10	0.2965

Table 7: Effect of nano cellulose on water permeability of polyester fabric.

Polyester fabric treated with nano cellulose (grams/liter)	Air permeability (m³/m²/h)
Untreated sample	223.14
1	215.33
5	211.47
10	195.21

Table 8: Effect of nano cellulose treatment on air permeability of polyester fabric.

Concentration of dye (gpl)	sample	K/S	Sample	L*	a*	b*
50	Treated with 5 gpl nano cellulose	23.84		66.81	29.66	7.51
	Soaped	14.44		66.51	27.97	6.84
	Without nano cellulose treatment	6.05		66.06	20.15	6.80
30	Treated with 5 gpl nano cellulose	21.71		66.83	29.48	7.65
	Soaped	5.15		65.94	23.78	5.74
	Without nano cellulose treatment	2.29		66.34	14.05	7.77

Table 9: Colour strength and co-ordinate values of polyester fabric pretreated and dyed with direct dye (Congo red BDC).

breaking load and crease recovery angle with almost no effect on rigidity of the material.

- Cellulose nano treatment to polyester also alters the thermal property, It is found that incorporation of cellulose nano slightly increase the onset temperature and The endothermic peak of nano cellulose treated polyester, occurred at 253.18°C.

- Cellulose nano treatment improves absorbency and reduces water and air permeability of polyester fabric.

- Nano cellulose treatment enhances the colour strength of polyester fabric dyed with direct dyes and also improves the fastness towards soaping.

References

1. Havancsak K (2003) Nanotechnology at present and its promises for the future. Mat Sci Forum 414: 85-94.

2. Hengstentaerg J, Mark H (1929) Crystalline Materials. Journal of Crystallography. 69: 271.

3. Gardner, Douglas (2008) Adhesion and Surface Issues in Cellulose and Nanocellulose. Journal of Adhesion Science and Technology 22: 545-56.

4. Atalla RH (1987) The Structures of Cellulose. ACS Symposium Series.

5. Plunguian M (1943) Cellulose Chemistry. Chemical Publishing Co Inc New York.

6. Ranby BG (1951) The colloidal properties of cellulose micelles. Discussions Faraday Society 11: 158-164.

7. Ciolacu D, Ciolacu F, Popa VI (2011) Amorphous cellulose–structure and characterization. Cellulose Chem Technol 45: 13-21.

10

Mechanical Properties of Geotextiles after Chemical Aging in the Agriculture Wastewater

Alsalameh KA*, Karnoub A, Najjar F, Alsaleh F and Boshi A

Department of Textile and Spinning, Faculty of Mechanical Engineering, University of Aleppo, Syria

Abstract

Aging of geotextile, which is widely used as reforming medium in structures, attracted a great deal of attention in recent years, as it is very important to the stability of the whole work. Especially, the prediction of geotextile's aging-time has become one of the focuses nowadays. In this study, four types of nonwoven geotextiles was used in tests, which are heat bounding, needle punched, chemical adhesive, and supporting by thread. The modified EPA 9090 test method was applied to compare the chemical resistance in pH 8 for agricultural wasting water in Syria. The immersion conditions are 30~90 days under 25°C and 50°C respectively. On other hand, chemical resistance of these nonwoven geotextiles was estimated by the average retentions of mechanical properties before/after exposure in the above chemical solution. However, the relied mechanical properties are grab tensile response, trapezoidal tear strength, and CBR puncture strength testes. In addition, we have compared between specimens in related to pore size volume, thickness, weight per m2, and row material. Transmissivity of geotextile for drainage were slightly decreased in pH8 solution. Finally, needle punched nonwoven geotextile has the best resistance to the tensile, tear, and puncture before and after aging.

Keywords: Geotextiles; Chemical resistance; Average retentions of tensile strength; Transmissivity

Introduction

Recently, the use of non-woven fabrics for the purposes of geological is increased. The reason for this enjoyment of characteristics suited to these purposes, exceeds the characteristics of woven and knitted fabrics, most of the studies are on plastics, and few are on fibers. Koerner [1], studied systematic investigations on the aging-time of polypropylene fibers at different temperatures have been made Moreover, a particular emphasis has been laid on how to build up the equation of geotextile's aging-time, which was based on Arrhenius equation (K=A e-E/RT). The experiments were respectively carried out at 120°C, 125°C, 130°C and 135°C by means of oven accelerated aging test. Then the lifetime of the fibers at normal temperature could be sscalculated according to the equation. where τ f is the final durability period; K is competitive multiplication coefficient; τ r is reference durability period; τr is reference temperature, 150°C; τi is using temperature (K); Kj is multiplication coefficient at it; Fi is time fraction of °Ci, Fi=1. Moreover, the effects of changing critical value on the equation were elucidated. Furthermore, the effect of soil's acidity (pH = 5) and basicity (pH = 9), pure water and copper ion in the water on the aging-time was discussed. The results showed that acid and alkali made the fiber's lifetime decrease about 13% and water make the fiber's lifetime decrease about 20%, while copper ion shorten the aging-time of the fiber more than 54%. Acid, alkali, metal ion would shorten the lifetime of PP fiber, and the effect of metal ion is the highest, the effect of water is the second, acid and alkali is the lowest. Under the pressure the aging rate of PP geotextile would be accelerated. This study also indicated that fiber grade anti-aging PP chip could be spun at conventional temperature; plastic and flat fiber grade would be spun at high temperature. However, high spun temperature would make the antiager consume and decompose, which will shorten the geotextile's lifetime. Therefore, the antiager and the spin ability of resin were very important. As there are different effect factors in different environment, experiment should be done based on particular natural conditions [2,3].

In another research the effectiveness of layered-geotextile protection layers comprised of combinations of nonwoven needle-punched, woven slit-film, and nonwoven heat-bonded geotextiles to minimize strains in landfill geomembranes has been examined. Results from physical experiments were reported where a sustained 700-N force was applied to a 28-mm-diameter machined steel probe on top of the protection layer, which was above a 60-mm-diameter, 1.5-mm-thick high-density polyethylene geomembrane and a 50-mm-thick compressible clay layer. The experiments are intended to simulate the physical conditions in a medium-sized landfill with an average vertical stress of 250 kPa and to capture the mean response with nominal 50-mm coarse gravel above the geomembrane. Screening tests were first conducted for up to 100 h at temperatures up to 55°C to evaluate three different combinations of layered geotextiles. Of those examined, the combination with a low-slack, heat-bonded geotextile above and below a thick, nonwoven, needle-punched geotextile as its central core was found to provide the lowest strains. A time-temperature superposition method was then developed and validated as a means to predict the long-term effectiveness of the most promising layered-geotextile composite. Last, long-term predictions of tensile strain were made and compared with proposed allowable limits. Despite the encouraging results from the short-term screening tests, even the most promising layered-geotextile composite is not recommended as a protection layer to limit long-term geomembrane strains for the particular force, particle size, and materials examined because the predicted strain after 100 years at 22–55°C of ~10% exceeds the range of currently proposed limits of 3-8% [4].

During the revision of Technical Specification for Application of Geotextile in Marine Works (JTJ239-98) published by the Ministry

---

*Corresponding author: Khawla Almohamad Alsalameh, Department of Textile and Spinning, Faculty of Mechanical Engineering, University of Aleppo, Syria
E-mail: khawlasalama90@gmail.com

of Communications of China, artificially accelerated ageing tests in laboratory, natural insolating tests, measurement of underwater ultraviolet radiation energy, ageing tests of buried geotextile in sandy soil and tests of specimens from practical engineering works were carried out for the monographic research on ageing resistance of geotextile. The paper is the summary of the test results, which can be of the reference for designers and contractors [5-7].

The studies in 1970 [8], were showed that nonwoven geotextiles were used for the first time in an earth dam. The geotextile acted as a filter for the toe drain and on the upstream slope below the rip-rap. In 1992, specimens were taken from both locations and performance tests were conducted in the laboratory and the main results of the hydraulic behavior of the geotextile filter in association with the soil of the damhae been presented [9]. Also microscopic analyses are presented and, as the filter is considered to be performing well, selected filter criteria are checked and the effectiveness of layered-geotextile protection layers comprised of combinations of nonwoven needle-punched, woven slit-film, and nonwoven heat-bonded geotextiles to minimize strains in landfill geomembranes is examined [10]. Results from physical experiments are reported where a sustained 700-N force was applied to a 28-mm-diameter machined steel probe on top of the protection layer, which was above a 60-mm-diameter, 1.5-mm-thick high-density polyethylene geomembrane and a 50-mm-thick compressible clay layer. The experiments are intended to simulate the physical conditions in a medium-sized landfill with an average vertical stress of 250 kPa and to capture the mean response with nominal 50-mm coarse gravel above the geomembrane. Screening tests were first conducted for up to 100 h at temperatures up to 55°C to evaluate three different combinations of layered geotextiles. Of those examined, the combination with a low-slack, heat-bonded geotextile above and below a thick, nonwoven, needle-punched geotextile as its central core was found to provide the lowest strains [11]. A time-temperature superposition method was then developed and validated as a means to predict the long-term effectiveness of the most promising layered-geotextile composite. Last, long-term predictions of tensile strain were madse and compared with proposed allowable limits. Despite the encouraging results from the short-term screening tests, even the most promising layered-geotextile composite is not recommended as a protection layer to limit long-term geomembrane strains for the particular force, particle size, and materials examined because the predicted strain after 100 years at 22–55°C of ~10% exceeds the range of currently proposed limits of 3-8%.

Finally, Answers to the problem of durability of geotextiles according to the French experience have been given particularly in papers by Sotton et al. presented at the Las Vegas conference in 1988 [12]. In addition, in contributions by Leclercq at the RILEM seminar on long-term behavior of geotextiles, held near Paris in 1986. More recently additional information has been obtained [13].

This paper will summarize the results that have been presented earlier and give new results obtained from recent measurements.

Materials and Methods of Search

Geotextile's aging-time has become one of the focuses nowadays. Therefore, that it is important to know the chemical resistance of nonwoven geotextiles.

The main mechanical properties that have been considered in this research were tensile, tear, puncture and air pockets for four types of non-woven fabrics made in different manufacturing ways (thermal bonding, needle punching, chemical paste, and sewing by supportive thread). Table 1 shows the different types of geotextile used in this work.

Geotextile types

Table 1 shows the brief idea about geotextile types.

Types of fibers of row materials used in geotextile specimens

The float solution used in chemical aging is agricultural wasting water taken from Syrian irrigation projects which consist of the elements shown in Table 2.

Chemical composition of used agricultural wastewater

Table 3 shows Agricultural wastewater has prepared of following ingredients.

Weighting of specimens

The weight of the different types of specimens was defined using electronic and accurate balance DNAUS, manufacturing by Adventurer corporate, it depends on measuring circulatory piece with exact scaling, through contingents of area we cane defining the weight per square meter. By following modified ASTM D5261 test method.

Thickness of specimens

Specimens are different in thickness, we measured the thickness of them, and the results were located in Table 4.

Specimen name	Specimen type	Specimen photo
A	Heat bounding nonwoven	
B	Needle punched nonwoven	
C	Chemical adhesive nonwoven	
D	Nonwoven with supporting thread	

Table 1: Geotextile types.

Mg++	24.64	So₄⁻	13.14
Ca++	27.70	Cl-	50.54
Na+	31.76	Hco₃⁻	26.06
K+	0.17	Co₃⁻	0

Table 2: Types of fibers of row materials used in geotextile specimens.

Specimen	Type of raw material of specimens
A	Polypropylene 100%
B	Polyamide 50% + Polypropylene 50%
C	Cotton 67% + Polyester 33%
D	Hemp 80% + Polyamide 20%

Table 3: Chemical composition of used agricultural wastewater.

Pore size volume defining device of specimens

The pore size volume of specimens has been defined using shaker device with sieves, manufacturing by CISA corporate, by following the modified ASTM D4751 test method. The principle of experiment depends on applying group of sieves that have sequential volume holes on the shaker device, we will put the type on the top of sieves, and we put multifarious volume sand above the type. Next, we set shaker device to work for a period in order to filter the sand grains through the type pore size, and then check up the highest volume for that grains, which is the same of pore size volume (Figure 1).

Method of chemical aging

All of specimens were immersed within path water device (Figure 2), in agricultural wastewater, depending on test method EPA 9090.

Mg^{++}	24.64	So_4^-	13.14
Ca^{++}	27.70	Cl-	50.54
Na^+	31.76	Hco_3^-	26.06
K^+	0.17	Co_3^-	0

Table 4: Thickness of specimens.

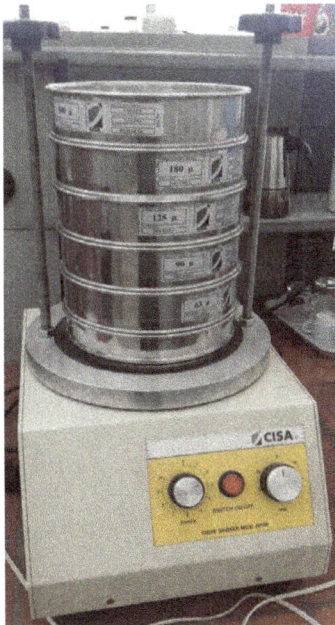

Figure 1: Pore size volume defining device of specimens.

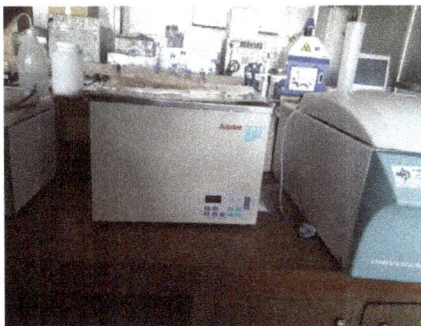

Figure 2: Path water for chemical aging test.

Grab tensile response, trapezoidal tear strength, and CBR puncture strength testes

The utmost mechanical properties for geotextile are grab tensile response, Trapezoidal tear strength and CBR puncture strength. The difference between previous tests is the changing of the jaws grade and distance between them as the norm for each test.

The modified ASTM D4632 test method for grab tensile response, the modified ASTM D4533 test method for trapezoidal tear strength, the modified ASTM D6241 test method for CBR puncture strength.

Since the specimens are different in thickness, the break force [N] must divide on thickness and width to get the stress. In that way the comparison between specimen is true (Figure 3).

Results and Discussion

Mass per unit

Weight test of specimens was repeated 10 times for each type. The average values of mass per unit area are illustrated in the Table 5.

Apparent opening size test

Table 6, shows the sizes of apparent opening for all specimens, after repetitions 10 times.

We noticed that holes volume in type B is the highest because manufacturing way, it is a needle punched nonwoven, that allows to layers fabric to stay as it before. While the holes in type B are smaller than type A because this type manufactured by Heat bounding nonwoven. Type C is the lowest holes volume because this type manufactured by Chemical adhesive nonwoven way. While the holes in type D are bigger than type C because this type manufactured by Nonwoven with supporting thread.

Grab tensile resistance

Grab tensile test was done on specimens by the following parameters in Table 7, according to related test method.

Grab tensile test was done before chemical aging also after 30, 60, and 90 immersion days at 25°C. Resulted stresses [N/mm²] were shown in Table 8.

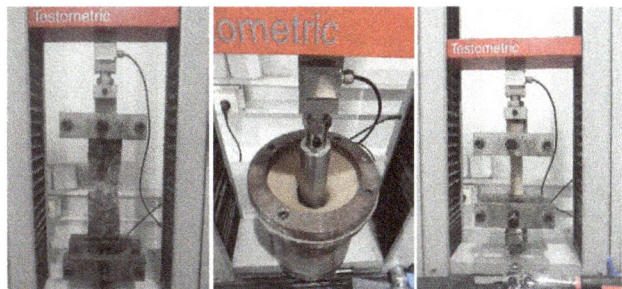

Figure 3: Grab tensile response, Trapezoidal tear strength, CBR puncture strength.

Specimen	Thickness [mm]
A	0.76
B	1.69
C	2.8
D	7.4

Table 5: Mass per unit area of specimens.

Previous result could be in illustrative form in Figure 4, to show the behavior of each specimen under chemical aging conditions at 25°C.

Similarly, specimens were exposure to grab tensile test in the same conditions but at 50°C. Besides, force break was divided on thickness and width. Stresses [N/mm²] were shown in Table 8 and Figure 5.

Before the chemical aging, specimen B has the highest tensile resistance with 20 to 10% more than the rest, followed by D then A and C, which (A and C) have the same tensile resistance before aging.

After the chemical aging, all of specimens lose several amount of their tensile resistance, due to the conditions of aging. Nevertheless, specimen B still has the highest tensile resistance, because it lose just 19% of its resistance, so it is the best specimen against the rest. However, specimen D lose more than 44% of its tensile resistance to be in the behinds. Specimen A lose 24% while C lose 33% of its resistance, as shown in Figure 6.

Trapezoidal tear strength

Trapezoidal tear test parameters were shown in Table 9, according to related test method (Table 10).

Specimen	Weight of 1 m²
A	412.06 gr
B	314.02 gr
C	817.30 gr
D	971.27 gr

Table 6: opening size test.

Specimen	Test speed [mm/min]	Specimen thickness [mm]	Specimen width [mm]	Distance between jaws [mm]	Test repeats
A	10	0.76	25	100	10
B	10	1.69	25	100	10
C	10	2.8	25	100	10
D	10	7.4	25	100	10

Table 7: Analogical and nominal specification of specimens for Grab Tensile tests.

Specimen	Before aging [N/mm²]	After 30 days of aging [N/mm²]	After 60 days of aging [N/mm²]	After 90 days of aging [N/mm²]
A	0.781763	0.721713	0.691716	0.634321
B	1.028188	1.012184	0.925188	0.823288
C	0.766923	0.636233	0.600223	0.587923
D	0.898662	0.728662	0.67998732	0.6134567

Table 8: Tensile test results of specimens in 25°C.

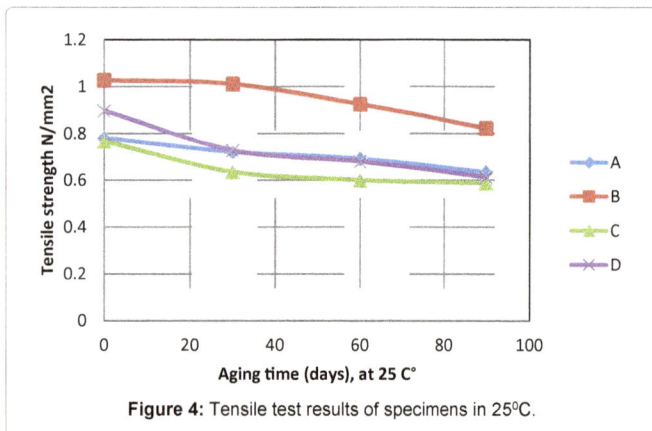

Figure 4: Tensile test results of specimens in 25°C.

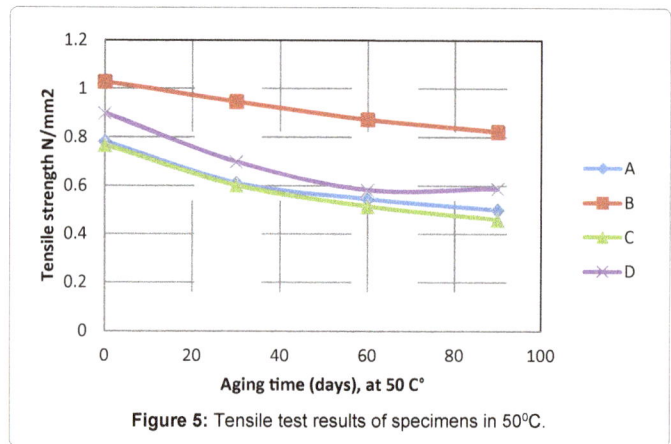

Figure 5: Tensile test results of specimens in 50°C.

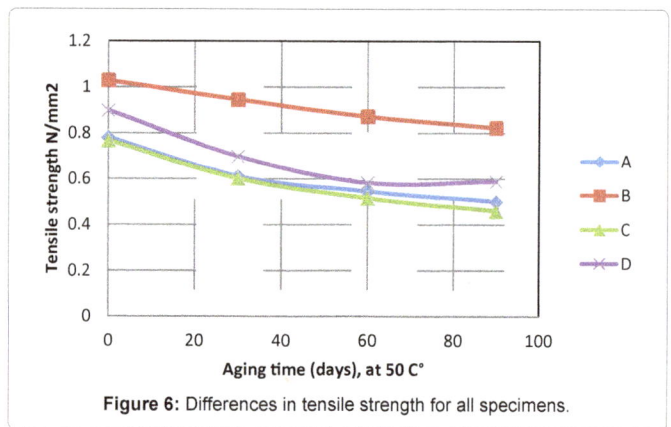

Figure 6: Differences in tensile strength for all specimens.

Specimen	Before aging [N/mm²]	After 30 days of aging [N/mm²]	After 60 days of aging [N/mm²]	After 90 days of aging [N/mm²]
A	0.781763	0.6118662	0.5448732	0.501067
B	1.028188	0.9456384	0.872488	0.823008
C	0.766923	0.601233	0.515023	0.459873
D	0.898662	0.698453	0.582648	0.588979

Table 9: Tensile test results of specimens in 50°C.

Test repeats	Specimen	Test speed [mm/min]	Specimen thickness [mm]	Specimen width [mm]	Distance between jaws [mm]	Test repeats
A	300	15	0.76	75	200	10
B	300	15	1.69	75	200	10
C	300	15	2.8	75	200	10
D	300	15	7.4	75	200	10

Table 10: Analogical and nominal specification of specimens for trapezoidal tear tests.

Specimens were exposure to trapezoidal tear strength before chemical aging in addition to after aging with 30, 60, and 90 immersion days at 25°C. Resulted stresses in [N/mm²] were shown in Table 11 and Figure 7.

Specimens were aged in the same conditions but at 50°C, after that it were tested with trapezoidal tear test. Finally, stresses [N/mm²] were shown in Table 12.

Previous result could be in illustrative form in Figure 8, to show the behavior of each specimen under chemical aging conditions (Figure 8).

Withal, specimen B is the best specimen against the rest in

trapezoidal tear test, followed by C, D, and A respectively, Because B has the highest trapezoidal tear resistance before the aging and even after that.

Figure 8, shows that all specimens lose a convergent amount of their resistance to trapezoidal tear. However, they lose 24%, 23%, 22%, and 18% for A, B, D, and C respectively. Nevertheless, specimen B still the best specimen (Figure 9).

CBR puncture strength

CBR Puncture test has special parameters according to interdependent test method, parameters were shown in Table 13.

CBR Puncture test was applied on all specimens before chemical aging and after aging with 30, 60, and 90 immersion days at 25°C. Resulted stresses in [N/mm²] were shown in Table 14.

To see the behavior of each specimen under chemical aging conditions, previous result could be in illustrative form in Figure 10.

Similarly, specimens were exposure to puncture test in the same conditions but at 50°C. Stresses [N/mm²] were shown in Table 15.

Figure 10, clearly shows the behavior of each specimen under chemical aging conditions at 50°C (Figure 11).

Before the chemical aging, specimen A has a puncture resistance higher than specimen B with 8%. While specimen C has the lowest puncture resistance. Finally, the puncture resistance of specimen D higher than C with 30%.

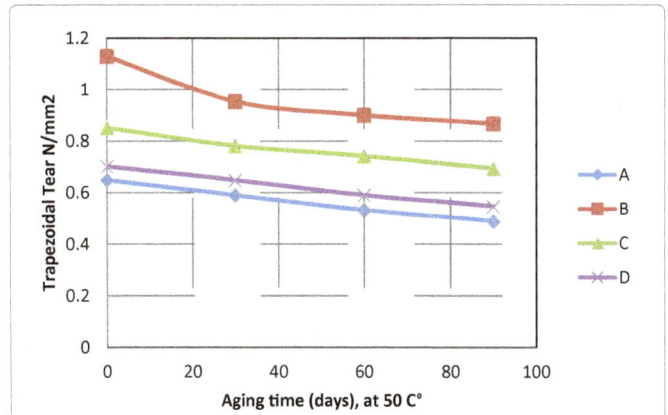

Specimen	Before aging [N/mm2]	After 30 days of aging [N/mm2]	After 60 days of aging [N/mm2]	After 90 days of aging [N/mm2]
A	0.64994	0.60211	0.55345	0.50123
B	1.129412	0.976001	0.923412	0.897121
C	0.852199	0.802199	0.771909	0.723459
D	0.702717	0.678901	0.623067	0.587012

Table 11: Trapezoidal Tear test results of specimens in 25°C.

Figure 7: Trapezoidal tear test results of specimens in 25°C.

Specimen	Before aging [N/mm²]	After 30 days of aging [N/mm²]	After 60 days of aging [N/mm²]	After 90 days of aging [N/mm²]
A	0.64994	0.59101	0.53345	0.48923
B	1.129412	0.954001	0.900411	0.867121
C	0.852199	0.782099	0.741804	0.693459
D	0.702717	0.648901	0.591067	0.547012

Table 12: Trapezoidal Tear test results of specimens in 50°C.

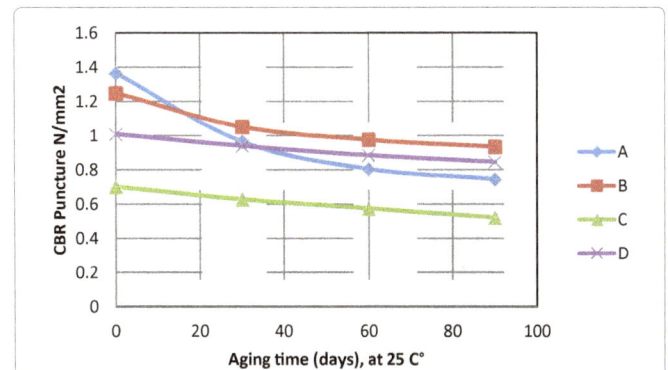

Figure 8: Trapezoidal Tear test results of specimens in 50°C.

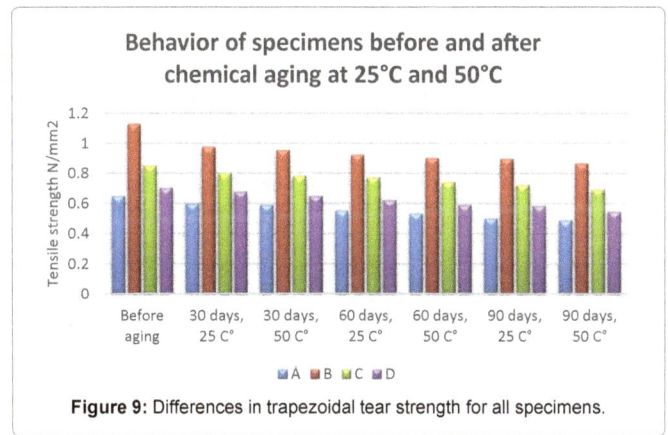

Figure 9: Differences in trapezoidal tear strength for all specimens.

Specimen	Test speed	Specimen diameter	Test cape diameter	Test repeats
A, B, C, D	100 mm/min	150 mm	50 mm	10 tests

Table 13: Analogical and nominal specification of specimens for CBR Puncture tests.

Specimen	Before aging [N/mm2]	After 30 days of aging [N/mm2]	After 60 days of aging [N/mm2]	After 90 days of aging [N/mm2]
A	1.36555	0.96525	0.80432	0.74436
B	1.248179	1.051179	0.97525	0.93325
C	0.701026	0.628106	0.574026	0.521103
D	1.008193	0.940725	0.885215	0.845105

Table 14: CBR Puncture Strength test results of specimens in 25°C.

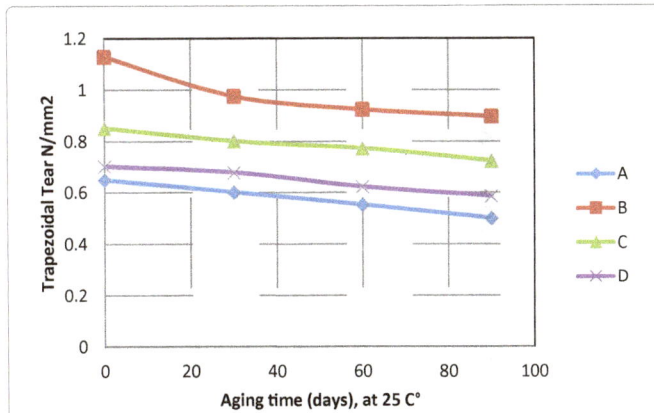

Figure 10: CBR Puncture Strength test results of specimens in 25°C.

After the chemical aging with 30 days only, all of specimens lose the highest amount of their puncture resistance, after that they lose a little amount of their puncture resistance.

After 90 days of chemical aging, specimens A, B, and D have a similarly puncture resistance but less than before aging with 37%, 25%, and 10% respectively. While, specimen C still in behinds by losing 16% of its puncture resistance (Figure 12).

Discussion

Specimen B

Specimen B is the best one of specimens in the three tests grab tensile, tear, and puncture, before and after aging. Regarding the situation before aging, it is made by needle-punched method; this method helps the layers of nonwoven fabric to enlacement with each other that causes to increase the resistance of this specimen against the three tests. On other hand, correlative layers prevent the tear slot to stretching, especially for trapezoidal tear test.

While for the situation after aging, this specimen has the largest pore size, which leads to the best immersion in agriculture wastewater. In addition, the raw material of this specimen (polypropylene and

Specimen	Before aging [N/mm2]	After 30 days of aging [N/mm2]	After 60 days of aging [N/mm2]	After 90 days of aging [N/mm2]
A	1.36555	0.90325	0.74505	0.60325
B	1.248179	1.010172	0.98425	0.92225
C	0.701026	0.579711	0.534026	0.480026
D	1.008193	0.900725	0.811215	0.735005

Table 15: CBR Puncture Strength test results of specimens in 50°C.

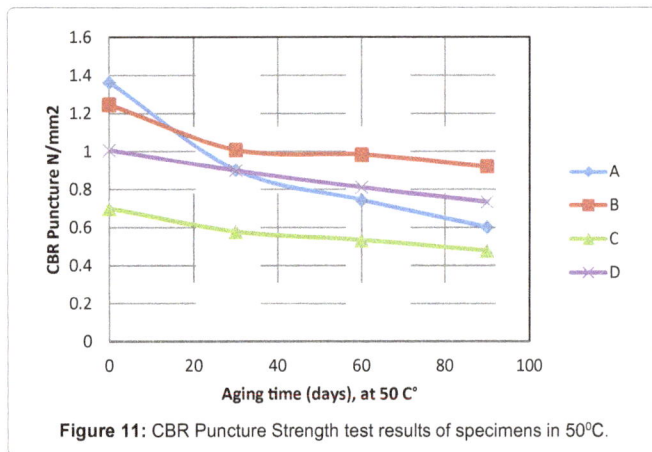

Figure 11: CBR Puncture Strength test results of specimens in 50°C.

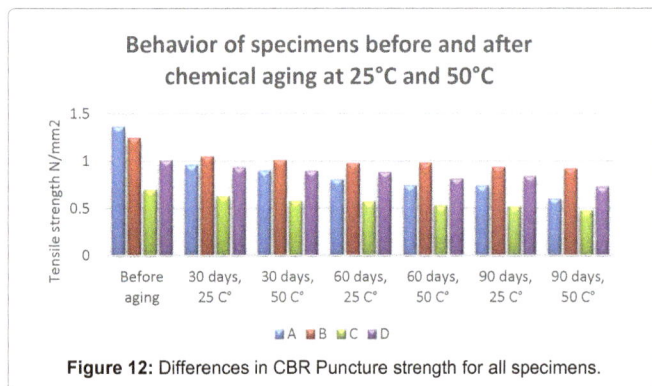

Figure 12: Differences in CBR Puncture strength for all specimens.

polyamide) did not interact with the agriculture wastewater. That explains the best results of this specimen after aging, in addition to the microscope photo to it, in Figure 13, which clarifies that is not any change on it, before and after the chemical aging.

Specimen D

Specimen D has has a resistance to tensile and puncture less than B with 13%, 23% respectively before aging, and 25%, 9% respectively after aging at 25°C, while 41%, 20% respectively after aging at 50°C. This specimen located in the second level next than specimen B in related to tensile and puncture tests, that is because it is made of hemp, which has a heavy qualitative weight, it is clearly shown in Table 5. On the other hand, layers of this specimen also enlacement with each other in a good way by supporting thread.

But in tear test it has a low resistance less than B with 39% approximately before and after aging at 25°C, and 50°C. The low resistance against tear test is because the direction of supporting thread, it is horizontal while the test is vertical, which leads to break the specimen quickly, as it is shown in Figure 14.

Specimen A

Specimen A is the best specimen just in puncture test before aging, that is because in puncture test the resistance of specimen depends on friction between nonwoven fabric, this specimen is a heat bounding specimen, therefore this specimen has the lowest thickness that means it has high friction between layers, so it has high resistance against

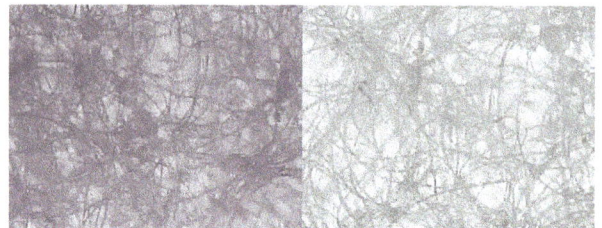

Figure 13: Microscope photo of needle punched specimen before and after aging.

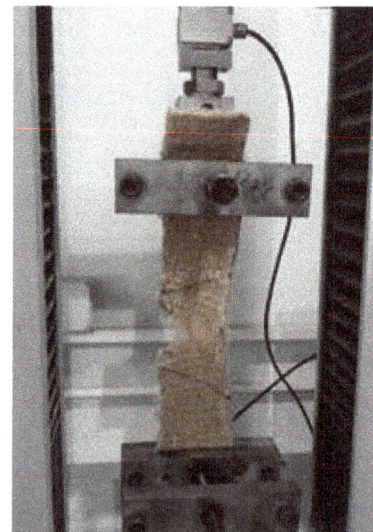

Figure 14: Supporting thread specimen during tear test.

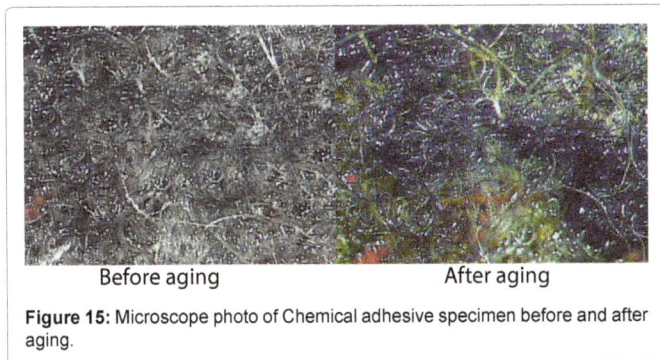

Figure 15: Microscope photo of Chemical adhesive specimen before and after aging.

puncture. While after aging, it lose a lot of its resistance especially at 50°C, to be less than B with 20%, 35% respectively at 25°C, 50°C for puncture test, also less than B with 23%, 40% respectively at 25°C, 50°C for tensile test. To explain previous result, specimen's pore size was measured by microscope, before and after aging particularly at 50°C. Figure 14, clearly displays the increment of pore size, this is due to the high temperature of aging, which leads layers to detachment of each other, so break the specimen easily under puncture and tensile tests.

For tear test, it has a resistance less than B with 44% approximately before and after aging. This low tear resistance because manufacture method (heat bounding), this method make the nonwoven fabric as one layer, it leads to be in the lowest thickness, that causes to stretching the tear slot quickly.

Specimen C

Specimen C is the worth specimen particularly after aging in related to tensile and puncture tests with 45% approximately less than B, that is because it made of natural fibers (cotton), which interacted with agriculture wastewater and lost most of its resistance. In addition to the interaction between immersion liquid and adhesive of specimen, which cause to dissolve a lot of adhesive, then layers of nonwoven fabric of specimen will disport of each other, which leads the specimen to be weak, as it clearly is shown in Figure 14. However, tear resistance of this specimen quite a bit, it is less than B with 24% before aging, and 20% after aging, this is because the random installing of specimens' filaments, which retards stretching the slot of tear test (Figure 15).

Conclusion

Geotextile nonwoven characteristics are different to each other because generally:

- Manufactured way
- Kind of raw material
- Pore size volume

Following conclusion were made after assessing the experimental results and after effecting tests (tensile, tear, and penetration) on types we notice and compare results:

- The type B, which is a needle punch specimen, was better in all situations.
- Chemical aging was affected in bad manner with higher temperature.

References

1. Koerner RM (2005) Designing with Geosynthetics. 5TH edtn, Prentice-Hall, Eaglewood Cliffs NJ.

2. Holtz RD (1997) Geosynthetic Engineering. BioTech Publish Ltd, Richmond, US.

3. Baker TL (1997) Proceedings of '97 Geosynthetics Conference.

4. Salman A (1997) Proceedings of '97 Geosynthetics Conference.

5. Koerner GR, Hsuan GY, Koerner RM (1998) Journal of Geotechnical and Geoenvironmental Engineering.

6. Artires O, Gaunet S, Bloquet C (1997) Geosynthetics International.

7. Inglod TS (1994) The Geotextiles and Geomembranes Manual. Elsevier Oxford 1: 229-242.

8. Koerner RM (1989) Durability and Aging of Geosynthetics. Elsevier 3: 65-109.

9. Koerner RM, Lord AE Jr, Halse YH (1988) Geotextiles and Geomembranes.

10. Jeon HY, Cho SH, Mun MS, Park YM, Jang JW (2005) Assessment of chemical resistance of textile geogrids manufactured with PET high-performance yam. Polymer Testing 24: 339-345.

11. Brchman R, Sabir A (2013) Long-Term Assessment of a Layered-Geotextile Protection Layer for Geomembranes. J Geotech Geoenviron Eng 139: 752-764.

12. Faure YH, Farkouh B, Delmas P, Nancey A (1999) Geotextiles and Geomembranes 17: 353-370.

13. Koerner GR, Koerner RM (1992) Geosynthetics in Filtration, Drainage and Erosion Control.

11

Designing and Development of Denim Fabrics: Part 1 - Study the Effect of Fabric Parameters on the Fabric Characteristics for Women's Wear

Kumar S[1], Chatterjee K[1], Padhye R[2] and Nayak R[2]*

[1]*Technological Institute of Textiles and Sciences Bhiwani, India 127021*
[2]*School of Fashion and Textiles, RMIT University, Brunswick, Australia 3056*

Abstract

The performance of a garment during its usage is very important for consumers. Performance as such is a very wide term and may range from satisfying the requirements during its use to the durability. Comfort is also considered by many consumers today as one of the performance requirements. For stretchable denim fabric, the fit related comfort lies in the ability of the material to be stretched when a load is applied according to body movements and retain to its original length. While designing the stretch denim, fabric weight and weave plays important role for comfort, performance and fashion. In this study, the effect of fabric parameters such as areal density, Lycra content and weave on characteristics of stretchable denim fabrics were investigated. Various properties such as thickness, tensile strength, flexural rigidity, stretch and recovery properties and air permeability of the fabrics were evaluated. The test results revealed that increasing the fabric weight increased fabric warp tensile strength, compressibility, stretch and recovery, whereas the flexural rigidity, weft tensile strength and air permeability were decreased. As Lycra contents in fabric increased, fabric thickness was increased which resulted in higher flexural rigidity of fabric. Fabric construction and weave also influenced the fabric flexural rigidity and air permeability related to performance and comfort of stretchable denim fabric.

Keywords: Stretch denim; Lycra; Fabric compression; Air permeability; Stretchability; Elastic recovery

Introduction

Denim, the favourite fabric of the youngsters has indeed come a long way. The consumer's choice, although unstable and unpredictable, it has remained almost the same while selecting denim for their fashion item [1-4]. The scope for denim wear is increasing tremendously every year and its worldwide market share has increased unpredictably in the last few decades. Recently the fashion trend is moving from denim to stretch denim (denim with Lycra) [5]. Stretch denim usually incorporates an elastic component (such as elastane) into the fabric to allow a degree of stretchability in garments [6,7]. Denim is a heavy woven fabric made from 100% cotton coarse indigo dyed warp and grey weft yarn [8-10]. The traditional denim is rather hard and high density fabrics with high mass per unit area. Twill weaves such as three-up-one-down (3/1) and two-up-one-down (2/1) are predominantly used for denim construction [11-13]. Denim is available in attractive indigo blue shades and is made for a variety of applications and in a wide range of qualities. Denim is comfortable, fashionable, affordable and durable for which it is popular in all the age groups.

Denim is available in different weights ranging from 200-300 g/m² which are categorized as light denim to 300-600 g/m², known as heavy denim. Denim's durability lies in the combination of the yarn and the weave. The consumer's today need durability and comfort in their fashion items including denim [14-17]. Twill weaves have good abrasion resistance, meaning the fabric will absorb a lot of friction before it breaks apart. The reason for such great durability is the way the yarns are woven together: one set of yarns floats over another in 2-5 (generally 2-3) sets of yarns at regular intervals to create a diagonal textured fabric surface. It is these yarn "floats" that absorb the abrasion. Denim has always been used for very durable outdoor work clothing because of its weight, rigidity and thickness. Denim is a good choice for casual jackets, skirts and jeans. In recent years, the advancements in garment-finishing techniques have led to easy processing and subsequently its use has broadened into different lifestyles. The clothing made of denim can be often sold in high prices depending on its fit, ornamentation, finishing and brand [18].

The stretchable denim fabrics give the elasticity to fabric so that it closely fit to body without restricting the body movement [19-21], hence providing wear comfort [22-24]. Lycra yarn is added to denim to increase its stretch and recovery properties [25]. Generally, adding 1-5% of Lycra with cotton will stretch the fabric over the body providing a more comfortable fit. For example, Rahman [26] studied the effect of spandex ratio on fabric physical and mechanical properties such as: breaking strength, breaking extension, shrinkage and fabric growth [27-30]. The findings of this study revealed that the ratio of Lycra had a significant influence on the physical properties of woven denim fabrics. Özdil [7] had studied the stretch and bagging properties of denim fabrics containing different amounts of elastane. The test results revealed that increasing the amount of elastane in denim fabric enhanced comfort properties related to stretch.

Core spun yarn can also be used as filling in which core part is Lycra filament and sheath fibres cotton, to improve the stretchability of denim [7,25,31,32]. The performance and comfort factors of these garments during use are very important. Generally, the comfortable stretching of fabrics according to body movements as well as recovery after stretching, are the desirable properties. In recent years, due to the demand for more comfortable clothing, elastane-containing denim fabrics are becoming increasingly popular. Some of the earlier studies focused on the effects of Lycra on physical and stretch properties of the denim fabrics. Some recent studies focused on the anti-bacterial

***Corresponding author:** Rajkishore Nayak, School of Fashion and Textiles, RMIT University, Brunswick, Australia 3056, E-mail:rajkishore.nayak@rmit.edu.au

properties of denim fabric [33]. However, limited research has been done on the effect of Lycra content and weaves on the performance and comfort properties of the denim fabric.

Hence, this research aims to investigate the effect of fabric areal density, Lycra content and weave on stretch and recovery properties in addition to the physical and comfort properties. For this purpose, different properties of denim fabrics introduced with varying amounts of elastane incorporated into core spun yarns in the weft direction were measured. The comfort properties related to stretch, performance factors and comfort related to air permeability were measured. Hence, the effect of amount of Lycra on fabric thickness, flexural rigidity, tensile strength, and stretch and recovery properties were evaluated. The effect of weave on the physical and stretch properties of the fabric was also investigated.

Experimental

Materials

Four denim fabrics were developed with varying amount of stretch (depending on the Lycra content) in the present study by keeping the GSM and weave same. The samples were abbreviated as L0, L1, L2 and L3 with 0, 1, 1.5 and 2% of Lycra, respectively. The specifications of denim fabrics are shown in Table 1.

Numerical notations for different denim designs, such as 3/1, denote what each warp yarn is doing relative to the filling yarns that it is interlacing with. In this case, each warp yarn is going "over" three picks and then "under" one pick. This would be verbally stated as "3 by 1" twill or "3 by 1" denim. At the next end or warp thread to the right, the same sequence is repeated but advanced up one pick. This advancing upward sequence continues, giving the characteristic feature of twill lines. In this case, the twill line is rising to the right and the fabric is classified as right hand twill (RHT) weave (Figure 1a). If the twill line is made to rise to the left, then the design is left hand twill (LHT) (Figure 1b). Broken twills are designed by breaking up the twill line at different intervals thus keeping it from being in a straight line.

For a more pronounced twill line in a denim fabric, the direction of twist in the warp yarn should be opposite to the twill direction in the fabric. For example, if "Z-twisted" yarn is woven into right hand twill, the twill line is less pronounced. If "S-twisted" yarn is woven into the same fabric, then the twill line is more pronounced. It must be remembered that only Z-twisted yarns are formed in open-end yarns, while ring-spun yarns have either "Z" or "S" twist. For this reason, open-end yarn can be used in left hand twills when a more pronounced twill line is desirable. Having the twist direction opposite from the direction of the twill line also tend to make the fabric handle a little softer.

Methods

Followings methods were used for testing the fabric samples. The details of the procedure and calculation of each testing is described below.

Conditioning: All the denim fabrics were evaluated in the greige form for the physical and other properties. Unless otherwise mentioned, the test specimens were conditioned at standard atmosphere of $20 \pm 2°$ C temperature and $65 \pm 5\%$ relative humidity (RH) before performing any of the tests.

Fabric weight and thickness: The thickness of stretchable denim fabric was measured using a thickness tester (Karl Schroder KG) under 0.5 kPa pressure following BS EN ISO 9073-2: 1997 standard. Thread density was measured by using a pick glass as per ISO 7211/2-1984. The weight per unit area (g/m²) was measured by using a circular specimen of 100 cm² area following ISO 3801-1977 standard test method.

Compressibility and compression recovery: For the assessment of fabric compressibility and fabric compression recovery, fabric thickness was measured at higher pressures (2.0 and 4.0 kPa). Compressibility and compression recovery percentage was obtained using the following formula:

Compressibility (%) $= 100 \times (t_1 - t_2) / t_1$

Where, t_1=measured thickness at 0.5 kPa, t_2=measured thickness at 2.0 kPa

Compression recovery (%) $= 100 \times (t_3 - t_2) / (t_1 - t_2)$

Where, t_1=measured thickness at 0.5 kPa, t_2=measured thickness t at 2.0 kPa and t_3=measured thickness at 4.0 kPa.

Flexural rigidity: The flexural rigidity is the measure of fabric stiffness. Denim fabrics samples were tested for flexural rigidity as per ASTM D1388-08. The principle of stiffness was based on the principle of cantilever bending of the fabric under its own mass. The test was done on a Shirley stiffness tester. Five samples of 20 cm × 2.5 cm were cut from the fabric each in warp and weft direction. Ten readings were taken for each fabric sample in each of the warp and weft direction. The specifications of the stiffness testers is described in Table 2.

Fabric flexural rigidity was calculated by the following formula:

Flexural rigidity (G)=W × C³ mg-cm

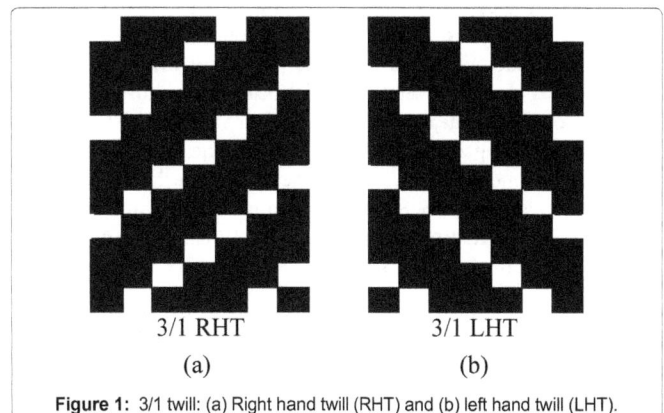

Figure 1: 3/1 twill: (a) Right hand twill (RHT) and (b) left hand twill (LHT).

Fabric Code	Lycra content (%)	Fabric weight (GSM)	Fabric thickness (mm)	Weave	Thread density (threads/cm)		Yarn linear density (Ne)	
					Warp	Weft	Warp	Weft
L0	0.0	320	0.631	3/1 RHT	28	26	14	12
L1	1.0	320	0.654	3/1 RHT	28	26	14	12
L2	1.5	320	0.728	3/1 RHT	28	26	14	12
L3	2.0	320	0.792	3/1 RHT	28	26	14	12

Table 1: Specifications of denim fabrics with different Lycra content.

Where, W is fabric weight in g/cm^2, C is bending length in cm.

Breaking strength: Breaking strength and elongation were tested on a universal tensile testing machine from Instron® at a traverse of 300 mm/min. Fabric samples with 25 mm × 150 mm dimension were used and ASTM D 5035-11 (Standard test method for breaking force and elongation of textile fabrics (strip method)) was followed for the test. Five readings were taken for each fabric sample and the average values were reported.

Stretch properties: Fabric stretch properties are related with fabric extension and recovery when in use. Fabric stretch properties were tested as per ASTM D 3107-07. Test specimens of 65 mm × 560 mm were cut from the fabric. Specimens from weft direction were hung on the apparatus after marking a 250 mm index in the central part of each specimen. A 1.8 kg load, which was hung according to the fabric weight in the bottom hanger, was applied to the sample three times and after the fourth application; the marked distance was measured. The samples were hung for 30 minutes, and the distance was measured once again. After that fabric samples were removed from the testing apparatus and were relaxed for 1 hour. Fabric stretch properties after cyclic loading were measured [34]. The total number of ten cycles was required at fixed load of fabric. Sample strip length was same as used in stretch testing of fabric. Fabric stretch, growth and elastic recovery values were calculated from these measured outcomes, as follows:

$$\text{Fabric stretching (\%)} = \frac{B - A}{A} \times 100$$

$$\text{Fabric growth (\%)} = \frac{C - A}{A} \times 100$$

$$\text{Elastic recovery (\%)} = \frac{B - C}{B - A} \times 100$$

Where, A is the distance marked between the upper and lower part of the fabric (250 mm), B is the distance between the marked points after hanging the sample for 30 minutes with the load (in mm) and C is the distance between the marked points after 1 hour of relaxation.

Air permeability: Air permeability is a measure of how well the fabric allows air flow through it under a differential pressure between the two surfaces. Air permeability is defined as the volume of air in millilitres, which is passed in one second through 100 mm^2 of a fabric at a pressure difference of 10 mm head of water. During the test, the specimen was clamped over an air inlet of the apparatus and air was sucked through it by means of a pump. The air valve is then adjusted to give a pressure drop across the fabric of 10 mm head of water and the air flow is then measured using a flow meter. Five specimens were used each with a test area of 508 mm^2 (25.4 mm diameter) and the mean air flow in cubic centimetre per square centimetre per second was calculated from the five results.

Statistical analysis: The difference between the average results for each test was estimated using the one-way analysis of variance (ANOVA) using Excel 2013 at the $p \leq 0.001$ level. The difference between the fabrics was significant when the F value was higher than $F_{critical}$. $F_{critical}$ is the value which the test results must exceed to reject the null hypothesis.

Test parameters	Values
Sample size	20 cm × 2.5 cm
Angle of inclination of intersecting plane to horizontal	41.5°
Load on test specimen	10 ± 2 g/cm

Table 2: Specifications of Stiffness Tester.

Results and Discussion

Effect of lycra content on fabric properties

Compressibility and compression recovery: Fabric compressibility is one of the important parameters affecting the fabric mechanical properties. The compressional properties affect fabric's softness, fullness, smoothness and are related to fabric handle [35]. Fabric compressional properties depend on surface properties of fibre and yarns, lateral compressional properties of fibre and yarns, and the fabric structure. The effect of Lycra content on compressibility and compression recovery properties of denim fabric is graphically shown in Figure 2. It can be observed that the increase in Lycra content significantly affected the compressibility and compression recovery properties of the denim fabric samples. As the Lycra content increased, fabric compressibility increased. This can be attributed to the spring like behaviour of the Lycra fibre and its tendency to recover to its original dimensions after the load is being removed. The compression recovery of the denim fabrics also showed similar trend as compressibility. In addition with the increased Lycra content, the fabric thickness was increased as shown in Table 1, which resulted in the increase of the compressibility and compression recovery properties. From ANOVA analysis it was observed that there is a significant difference in fabric compressibility and compression recovery of the denim fabrics having different Lycra content (F=1290.92, $p \leq 0.001$).

Flexural rigidity: Fabric flexural rigidity is a measure of the fabric stiffness. A fabric with higher flexural rigidity will be stiffer. The stiffness affects the fabric drape and tactile comfort. The fabric with very high stiffness may not be comfortable to wear and cannot bend as per the body contours. The effect of Lycra content on the flexural rigidity of the stretchable denim fabric is shown in Figure 3. It can be observed that the flexural rigidity of denim fabrics increased with the increase in the Lycra content both in the warp and weft direction. The filament structure of the Lycra made the yarn stiffer, which finally led to increased stiffness. Furthermore, it can be observed that as the as the thickness of the fabric was higher with higher Lycra content, this resulted in higher flexural rigidity. It is evident that the fabric handle becomes stiffer as the Lycra content in the structure of the fabric was increased. The higher flexural rigidity in warp direction can be attributed to the weft yarns containing Lycra having coil like structure and its tendency to recover to its initial dimension. The warp yarn has to come closer, which makes contraction in the fabric. Therefore, the larger flexural rigidity was observed in warp direction than weft direction of fabrics. From statistical analysis (one-way ANOVA) it can be observed that there is a significant difference

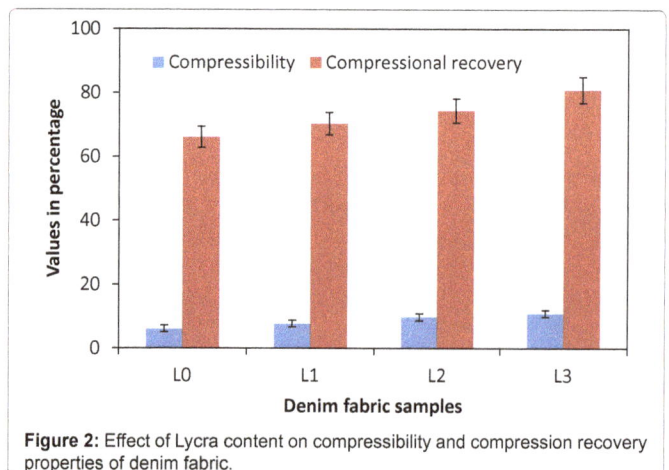

Figure 2: Effect of Lycra content on compressibility and compression recovery properties of denim fabric.

between the fabric flexural rigidity of the denim fabric samples having different Lycra content (F=101220.81, p ≤ 0.001).

Breaking strength: The breaking strength of a fabric is a measure of its performance during use [36]. The fabric used for the construction of particular clothing should be able to withstand the fatigue applied to it. A garment with insufficient breaking strength may fail during use, which can lead to the rejection of the garment. The effect of Lycra content on breaking strength in the direction of both warp and weft are shown in Figure 4. It can be observed that the braking strength of the fabric gradually increased in the warp direction with the Lycra content, whereas it decreased significantly in the weft direction with Lycra content. The increase in the breaking strength in the warp direction can be attributed to higher fabric assistance provided by the weft yarns with increased Lycra content. As the Lycra percentage was gradually increased, the additional amount of Lycra provided higher assistance to the warp yarn, resulting in increased warp-wise tensile strength. However, the decrease of the breaking strength in the weft direction can be attributed to the low strength of Lycra compared to cotton. Although Lycra is highly stretchable, it does not contribute to the strength of the fabric. It was observed that more Lycra content in fabrics reduced the fabric breaking strength than 100% cotton based fabrics (L0). This is one of the negative attribute of Lycra fibres as the stretchability is achieved with the compensation of the fabric strength. From statistical analysis (one-way ANOVA) it was observed that there is a significant difference in the breaking strength of the three experimental fabrics having different Lycra content (F=4013.55, p ≤ 0.001).

Stretch properties: Fabric stretchability indicates the property of the fabric that facilitates the body part movements. A fabric with higher stretch may follow the body movement easily. However once the force is being removed, the fabric should return to its original dimensions. The fabric containing Lycra are well known for their good stretchability and stretch recovery characteristics. Figure 5 shows the effect of Lycra content on stretchability and recovery properties of denim fabrics. It can be observed that the stretchability was increased with the Lycra content. As Lycra yarn possesses higher extensibility and elastic recovery, the increase in the stretchability was observed. The stretch recovery was also increased with increase of the Lycra content in the fabric due to the presence of Lycra. From ANOVA analysis it was observed that there is a significant difference in fabric stretch properties of the three denim fabrics having different Lycra content (F=408.02, p ≤ 0.001).

Air permeability: The air permeability of a fabric is the ability of the fabric to allow the atmospheric air to flow thorough the fabric and reach the skin. Depending on the usage of the fabric, the air permeability values are determined. For example, high air permeability may be a desired for clothing in hot climate. However, it may be negative in cold climates. The effect of Lycra content on air permeability of denim fabric is shown in Figure 6. It can be observed that the Lycra content has a profound effect on fabric air permeability. The higher value of air permeability is observed in the fabric with lower value of Lycra content and it decreased as the Lycra content was increased. As Lycra content increased in the fabric, contraction of the woven fabric was more, which made the fabric more compact and thicker, resulting in higher resistance to air flow. Hence, the higher amount of Lycra content can help in achieving higher stretchability, with reduced air permeability. The reduced air permeability may result in the lower water vapour resistance and hence cause discomfort to the wearer. However, as the fabrics are used in single layers the water vapour resistance, which is generally in the lower range for denim will be changing marginally [37]. From statistical analysis it was observed that there is a significant

difference in fabric air permeability of the three denim fabrics having different Lycra content (F=3366.718, p ≤ 0.001).

Conclusions

The results obtained in this study indicated that the amount of Lycra has a significant influence on physical and elastic properties of denim fabrics. Fabric tensile strength was decreased in weft direction with Lycra content, while fabric tensile strength was increased in warp direction because of the higher fabric assistance by the weft yarns. As the fabric thickness increased, fabric compressibility, compression recovery and flexural rigidity increased. The compressibility, compression recovery increased as the Lycra content was increased in the fabric. The flexural rigidity was also increased with the increase in the amount of Lycra in the denim fabric. Furthermore, As the Lycra content was

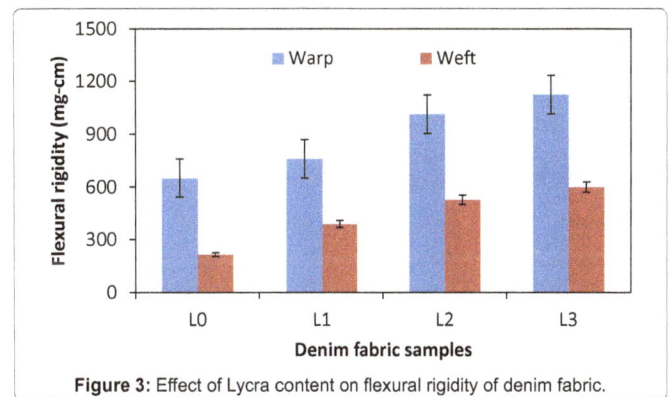

Figure 3: Effect of Lycra content on flexural rigidity of denim fabric.

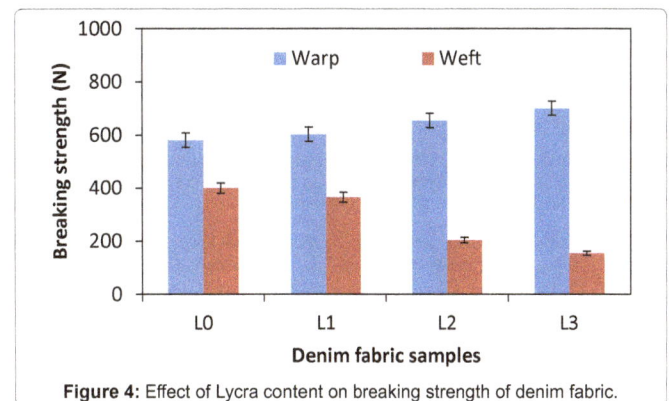

Figure 4: Effect of Lycra content on breaking strength of denim fabric.

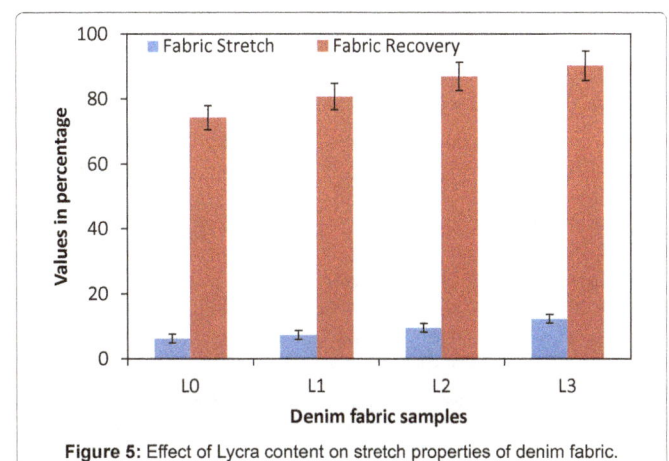

Figure 5: Effect of Lycra content on stretch properties of denim fabric.

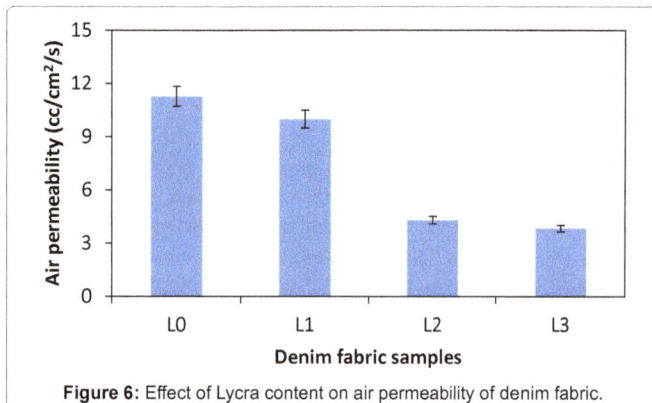

Figure 6: Effect of Lycra content on air permeability of denim fabric.

increased the air permeability and decreased significantly due to the fact that the fabrics become thicker and more compact with the increase in Lycra content in the woven fabrics. Statistical analysis proved that the difference between the results for the three fabric were significant for all the properties. As per the survey results, denim was preferred to be the first choice in the women's wear for style and fashion, which will be published in Part 2 of the paper.

References

1. Zervent Ünal B (2012) The prediction of seam strength of denim fabrics with mathematical equations. Journal of the Textile Institute 103: 744-751.

2. Nayak R, Padhye R (2015) Garment Manufacturing Technology. Elsevier.

3. Nayak R, Padhye R (2014) Introduction: the apparel industry, in Garment Manufacturing Technology. Elsevier.

4. Lee IY, Jeong EG, Kim SR, Bengelsdorff C, Kim SD (2015) Effects of biowashing and liquid ammonia treatment on the physical characteristics and hand of denim fabric. Coloration Technology 131: 192-199.

5. Kumar V, Nayak R (2014) Sewing performance of PV and PES air-jet textured sewing threads in denim fabrics. Journal of Textile and Apparel, Technology and Management 8: 1-12.

6. El-Ghezal S, Babay A, Dhouib S, Cheikhrouhou M (2009) Study of the impact of elastane's ratio and finishing process on the mechanical properties of stretch denim. The Journal of Textile Institute 100: 245-253.

7. Özdil N (2008) Stretch and bagging properties of denim fabrics containing different rates of elastane. Fibres and Textiles in Eastern Europe 16: 66.

8. McLoughlin J, Hayes S, Paul R (2015) Cotton fibre for denim manufacture. Denim: Manufacture, Finishing and Applications.

9. Rahman O, Jiang Y, Liu WS (2010) Evaluative criteria of denim jeans: A cross-national study of functional and aesthetic aspects. The Design Journal 13: 291-311.

10. Behera B, Chand S, Singh TG, Rathee P (1997) Sewability of denim. International Journal of Clothing Science and Technology 9: 128-140.

11. Raina M, Gloy Y, Gries T (2015) Weaving technologies for manufacturing denim. Denim: Manufacture, Finishing and Applications.

12. Adanur S, Qi J (2008) Property analysis of denim fabrics made on air-jet weaving machine part I-Experimental system and tension measurements. Textile Research Journal 78: 3-9.

13. Glassner A (2002) Digital weaving-1, Computer Graphics and Applications. IEEE 22: 108-118.

14. Nayak R, Padhye R (2014) The care of apparel products, in Textiles and fashion: Materials, design and technology. Elsevier 799-822.

15. Nayak R (2009) Comfort properties of suiting fabrics. Indian Journal of Fibre and Textile Research 34: 122-128.

16. Morris M, Prato H (1981) Consumer perception of comfort, fit and tactile characteristics of denim jeans. Textile Chemist and Colorist 13: 24-30.

17. Wu J, Delong M (2006) Chinese perceptions of western-branded denim jeans:

a Shanghai case study. Journal of Fashion Marketing and Management: An International Journal 10: 238-250.

18. Nayak R, Padhye R, Wang L, Chatterjee K, Gupta S (2015) the role of mass customisation in the apparel industry. International Journal of Fashion Design, Technology and Education 8: 162-172.

19. Nayak R, Padhye R, Dhamija S, Kumar V (2013) Sewability of air-jet textured sewing threads in denim. Journal of Textile and Apparel Technology and Management 8: 1-11.

20. 20. Kan C, Yuen C (2009) Evaluation of the performance of stretch denim fabric under the effect of repeated home laundering processes. International Journal of Fashion Design Technology and Education 2: 71-79.

21. Mourad M, Elshakankery M, Almetwally AA (2012) Physical and stretch properties of woven cotton fabrics containing different rates of spandex. Journal of American Science 8:(4)

22. Nayak R, Padhye R, Wang L (2015) How to Dress at Work, in Management and Leadership-A Guide for Clinical Professionals. Springer 241-255.

23. Chatterjee K, Das D, Nayak R, Kavita (2011) Study of handle and comfort properties of Poly-khadi, handloom and powerloom fabrics. Man-Made Textiles in India 39: 351-358.

24. Crowther E (1985) Comfort and fit in 100% cotton-denim jeans. Journal of the Textile Institute 76: 323-338.

25. Nayak R, Padhye R, Gon DP (2010) Sewing performance of stretch denim. Journal of Textile and Apparel Technology and Management 6: 1-9.

26. Rahman O (2011) Understanding consumers' perceptions and buying behaviours: Implications for denim jeans design. Journal of Textile and Apparel Technology and Management 7: (1).

27. Lee JT, Kim MW, Song YS, Kang TJ, Youn JR (2010) Mechanical properties of denim fabric reinforced poly (lactic acid). Fibers and Polymers 11: 60-66.

28. Haghighat E, Etrati SM, Najar SS (2013) Modeling of needle penetration force in denim fabric. International Journal of Clothing Science and Technology 25: 361-379.

29. Hua T, Tao XM, Cheng KPS, XU BG, Huang XX (2013) An experimental study of improving fabric appearance of denim by using low torque singles ring spun yarns. Textile Research Journal.

30. Sariisik, Merih (2004) Use of Cellulases and Their Effects on Denim Fabric Properties. AATCC review 4:(1).

31. Lou C, Chang CW, Lin JH, Lei CH, Hsing WH (2005) Production of a polyester core-spun yarn with spandex using a multi-section drawing frame and a ring spinning frame. Textile Research Journal 75: 395-401.

32. Jaouachi B,Hassen MB, Sahnoun M, Sakli F (2010) Evaluation of wet pneumatically spliced elastic denim yarns with fuzzy theory. The Journal of Textile Institute 101: 111-119.

33. Sumithra M, Raaja NV (2012) Antibacterial efficacy analysis of Ricinus communis, Senna auriculata and Euphorbia hirta extract treated on the four variant of denim fabric against Escherichia coli and Staphylococcus aureus. Journal of Textile Science and Engineering.

34. Mukhopadhyay A, Nayak R, Kothari V (2004) Extension and recovery characteristics of air-jet textured yarn woven fabrics. Indian Journal of Fibre and Textile Research 29: 62-68.

35. Nayak R, Chatterjee KN, Tanwar A, Gon DP (2007) Handle and comfort properties of Polyester/Viscose suiting fabrics. Man-Made Textiles in India 50: 288-292.

36. Khedher F, Dhouib S, Msahli S, Sakli F (2009) The influence of industrial finishing treatments and their succession on the mechanical properties of denim garment. AUTEX Research Journal 9: 311

37. Nayak R, Kanesalingam S, Houshyar S, Vijayan A, Wang L, et al. (2016) Effect of repeated laundering and Dry-cleaning on the thermo-physiological comfort properties of aramid fabrics. Fibers and Polymers 17: 954-962.

12

Development of Antibacterial Silk Sutures Using Natural Fungal Extract for Healthcare Applications

Parthiban M[1]*, Thilagavathi G[2] and Viju S[2]

[1]Department of Fashion Technology, PSG College of Technology, Coimbatore, India
[2]Department of Textile Technology, PSG College of Technology, Coimbatore, India

Abstract

The purpose of the study was to treat silk sutures with natural fungal pigment namely *Thermomyces* of different concentrations 1.5%, 2.0% and 2.5%, and to analyse its effect on the properties of silk sutures such as tenacity, knot strength, friction and antimicrobial activity. The result showed that the pigment concentration in the selected range has no significant effect on friction, tenacity and knot strength of silk sutures. Antimicrobial test results showed that as the pigment concentration increases the antimicrobial activity also increases against both *E. coli* and *S. aureus* bacteria. At 2.5% concentration, a zone of inhibition of 10 mm and 14 mm are observed against *E. coli* and *S. aureus* respectively. Silk suture treated with optimum concentration of the natural fungal pigment is appropriate to retard the exponential growth of *S. aureus*, a gram-positive bacterium and *E. coli* a gram-negative bacterium and hence silk sutures can be developed with the required characteristics for healthcare applications.

Keywords: Pigment; Fungal; Antimicrobial; Bacterium

Introduction

Wound closure using suture materials is an integral part of the surgical process. Sutures are natural or synthetic textile biomaterials widely used in wound closure, to ligate blood vessels and to draw tissues together [1]. Sutures consist of a fibre or fibrous structure with a metallic needle attached at one of the fibre ends and they can be classified into two broad categories namely absorbable and non-absorbable sutures. The most crucial requirements of suture materials are physical and mechanical properties, handling properties, biocompatibility, and antimicrobial nature [2]. Till date, there is no single suture material which can fulfill all the crucial requirements of sutures [3]. The present surgeon has several choices of suture material available and he may choose them based on availability and his familiarity.

Silk, a natural non-absorbable suture material has been used as biomedical suture for centuries due to its advantageous characteristics. However, one of the major problems associated with the silk is its poor microbe resistance characteristics. Several researchers have used different antimicrobial agents onto silk sutures to impart microbe resistance characteristics. Researchers have also used silver doped bioactive glass powder to coat silk surgical suture [4]. Recently, studies on the effect of chitosan coating on the characteristics of silk sutures [5]. Another study on tetracycline coating on silk sutures was carried out and they investigated the effect of tetracycline treatment on silk suture properties [6].

Recently, antimicrobial finishing of textiles using microbial dyes have received greater attention as they require less labour, land, and cost effective solvents for extraction as opposed to higher plant materials. In this study, silk sutures are treated with *Thermomyces*, a natural fungal extract and its effect on the properties of silk sutures such as antimicrobial activity, friction, tenacity and knot strength are studied.

Materials and Methods

Materials

Bombyx mori silk filaments were used for the study. All other chemicals were analytical grade and used as received.

Methods

The process of development of silk suture is shown in the Figure 1. The following section discusses each step in detail.

Isolation of silk fibroin

The raw silk filaments were immersed in an aqueous solution of 0.1% (w/v) sodium carbonate at 98°C - 100°C for 30 minutes, to remove sericin [7]. The sericin free silk filaments were subsequently washed with copious water to remove sodium carbonate. The silk filaments were subsequently dried and conditioned at an atmosphere of 27°C and 65% relative humidity for 48 hours.

Extraction of fungal pigment from *Thermomyces*

Czapek yeast broth was prepared and *Thermomyces* culture was inoculated and incubated as stationary cultures for 5 days. After incubation the grown up culture was filtered to separate the fungal biomass from the broth. The supernatant broth was filtered using a sterilized muslin cloth to remove the fungal mat. To the filtrate, one volume of 95% (v/v) methanol was added and the mixture was kept on a rotary shaker for 30 min at 150 rpm and at 35°C. The mixture was then centrifuged at 5000 rpm for 15 min. The same process was repeated for removal of fungal biomass and the filtrate was filtered through a preweighed 47 mm filter paper. Next, the absorption spectrum was observed at 300-600 nm using spectrometer (U-2000, Hitachi Ltd., Tokyo, Japan.) The purified pigments were kept in a Buchi rotary evaporator and lyophilized (Lyobeta 35) to obtain the yellow pigment in a powder form [8]. The sample of the extracted pigment is shown in Figure 2.

*Corresponding author:** Parthiban M, Department of Fashion Technology, PSG College of Technology, Coimbatore, India,
E-mail: parthi_mtech@yahoo.com

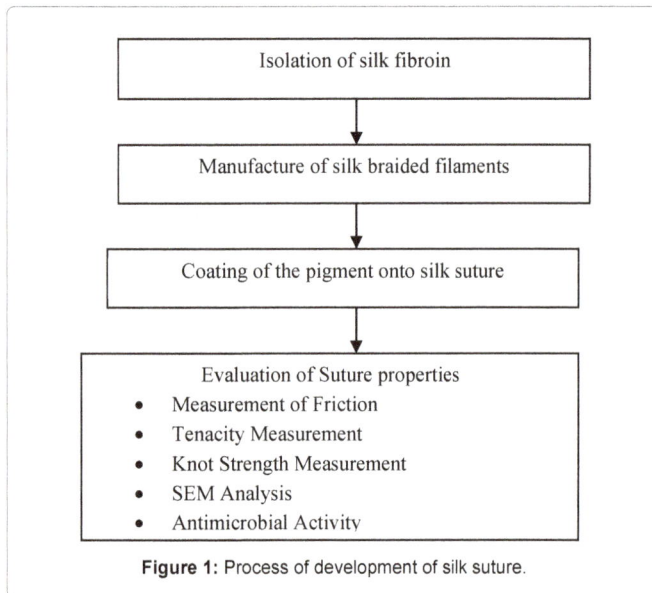

Figure 1: Process of development of silk suture.

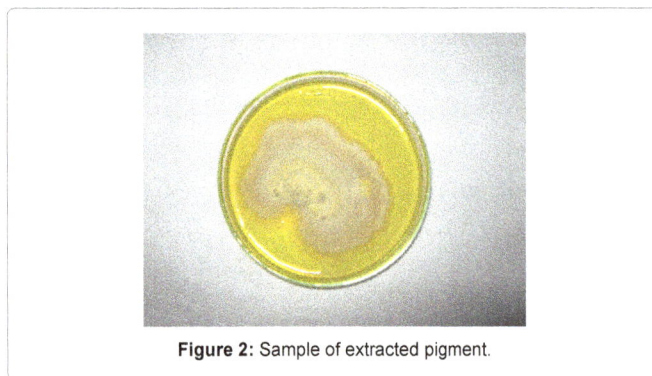

Figure 2: Sample of extracted pigment.

Manufacture of braided silk filaments

Braided silk filaments were manufactured using 16 sericin free silk yarns through a circular braiding machine with 16 carrier arrangement. The count of the braided silk suture was 124 tex.

Extraction of *Thermomyces* and coating onto silk braided filaments

Pigments from *Thermomyces* of different concentrations 1.5%, 2.0% 2.5% (w/v) were prepared as per the standard procedure. The silk filaments were then immersed in pigments of different concentrations for 24 hours at room temperature. The silk sutures were then dried at room temperature for 24 hours.

Friction measurement

The frictional properties of braided silk sutures were measured using Lawson Hemphill Dynamic Friction Tester. The treated and untreated braided silk sutures were tested for friction at a constant sliding speed of 150 m/min and at an input tension of 60 cN. For all the friction testing, 180 degree wrap angle was maintained and for each set of experiment 20 tests were conducted.

Tenacity and knot strength measurement

Tensile Tester (Instron Make) was used to measure the tenacity and knot strength of silk sutures. The braided silk sutures were tested

for tenacity and knot strength at a gauge length of 150 mm and extension rate of 90 mm/min. In the knot strength measurement, knot was formed with square knot method. For each test method at least 20 readings were taken.

SEM analysis

The surface characteristics of silk filaments were studied using Scanning Electron Microscope (JSM 6390) after coating them with pigment.

Antimicrobial activity evaluation

The *Thermomyces* treated and untreated silk filaments were evaluated for their antimicrobial activity using Agar Diffusion Test (SN 195920- 1992) and Shake Flask Method (AATCC 100). All the tests were carried out in triplicate and the results were averaged.

Statistics

Datas were expressed as mean ± standard deviation. Statistical analysis was carried out using F- test. A value of $P<0.05$ was considered to be statistically significant.

Results and Discussions

Effect of natural fungal treatment on tenacity and knot strength

Tenacity of suture material is significant for the practitioner making a knot. If the material is too weak and the knotting force is stronger than tensile strength of suture material, suture can easily break while tightening the knot. Therefore it is essential to know the tensile properties of sutures [9]. The Figures 3 and 4 show the effect of pigment concentration on tenacity and knot strength of silk sutures respectively. From the Figures 3 and 4 it is observed that the pigment treated and untreated silk sutures exhibits similar tenacity and knot strength and there is no significant difference between the properties of

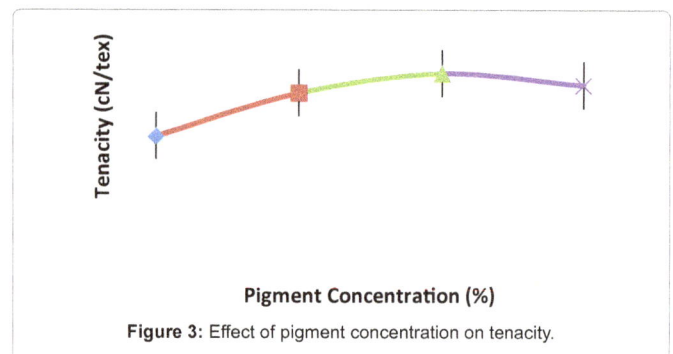

Figure 3: Effect of pigment concentration on tenacity.

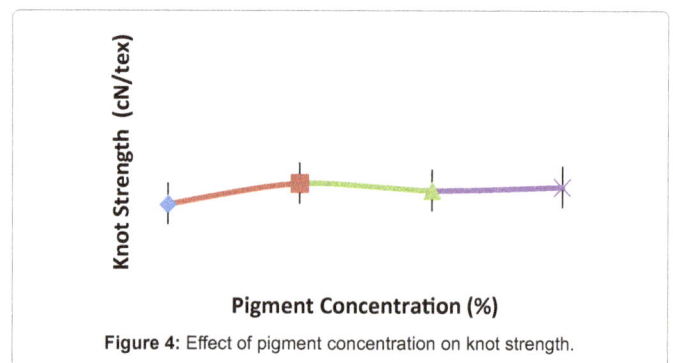

Figure 4: Effect of pigment concentration on knot strength.

treated and untreated samples. This shows that the incorporation of the pigment into the silk sutures has no significant effect on both tenacity and knot strength.

Furthermore, it is observed that for the untreated and all the pigment treated sutures, the knot strength is lower than the tenacity as shown in Figures 3 and 4. The failure of strands occurred at the knot rather than along the suture strand, indicating the knot itself causes an area of high stress concentration [10]. Several factors may contribute to failure occurring at the knot rather than along the suture [11]. First, breakage at the knot may be caused by forces being oriented at the knot at an acute angle to the suture axis. Second, the suture yarn in the knot region may be weakened during knot construction and during loading [12]. Third, tightening of the knot and the friction between yarns in the knot may contribute to the failure [13].

Effect of natural fungal treatment on friction

Braided silk structures are difficult to pass through tissue because of tissue drag and thus they cause a greater extent of tissue damage. Also, braided sutures provide higher frictional values than the monofilament sutures [14]. Hence braided sutures are given special surface coatings to reduce friction. The common form of characterizing the frictional properties of yarns and filaments is the coefficient of friction [15]. The frictional properties of pigment treated silk sutures are shown in the Figure 5. From Figure 5, it is observed that the pigment concentration in the selected range has no significant effects on frictional force. The results suggest that pigment treated silk suture does not generate more friction than the untreated silk sutures.

Effect of natural fungal treatment on antibacterial activity

Bacterial species are capable of colonizing different surfaces and proliferating on them, forming adherent biofilms. This could represent a major problem for implantable suture material. Experimental and clinical data indicate most of the wound infections begin around material left within the wound, and that the incidences of post-surgical complications are directly related to the degree of contamination at the time of material placement. Measuring for reducing the risk of surgical site infections include surgical technique, appropriate antimicrobial agent, adjunctive strategies for reducing wound contamination and promoting wound healing [16].

In this study two different types of bacteria are selected namely *S. aureus*, a gram-positive bacterium and *E. coli* a gram-negative bacterium. *S. aureus* causes skin and tissue infections, septicemia, endocarditis and meningitis whereas *E. coli* causes urinary tract and wound infections [17]. The antimicrobial activity of pigment treated silk sutures tested using agar diffusion method is given in the Figures 6 and 7.

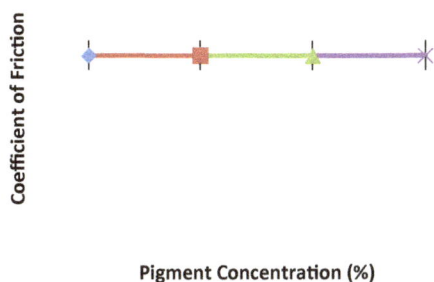

Figure 5: Effect of pigment concentration on friction coefficient.

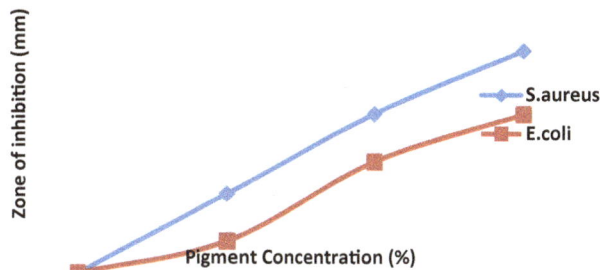

Figure 6: Zone of inhibition as a function of pigment concentration.

(a)   *S.aureus*

(b)   *E.coli*

Figure 7: Antibacterial activity of the fungal treated silk suture against *S.aureus* and *E.coli*.

From the Figures 6 and 7, it can be seen that silk sutures treated with pigment exhibit excellent antimicrobial activity against both *E. coli and S. aureus*. This may be due to the intrinsic antimicrobial characteristics of the pigment. The most accepted mechanism for microbial inhibition by fungal pigment is the presence of anthraquinone compounds, carboxylic acids and pre-anthraquinones [18]. The results also showed that pigment treated silk suture exhibits lower antimicrobial activity against gram negative bacteria *E. coli* than gram positive bacteria *S. aureus*. These results are correlated with the data published previously by the researchers. Moreover, the size of zone is higher for the suture with higher pigment concentration%, which is essential to prevent infectious diseases.

Morphology of pigment free and pigment loaded silk sutures

The SEM micrographs of the pigment free and 2% pigment loaded silk sutures are shown in Figures 8 and 9 respectively. From Figure 9, it is clear that pigment particles are well deposited on the surface of braided silk suture.

(a) X100   (b) X1400

Figure 8: SEM micrograph of the pigment free braided silk sutures.

(a) X100   (b) X1400

Figure 9: SEM micrograph of the 2 % pigment loaded braided silk sutures.

Conclusions

Silk suture produced was treated with natural fungal extract at optimum concentration and the effects of natural fungal treatment on the suture properties were studied. The result showed that the tenacity and knot strength of silk braided sutures increased compared to the untreated silk suture. The frictional properties of both the fungal treated silk suture and the untreated silk suture were determined by the dynamic coefficient of friction and there is a slight reduction in frictional value found in the treated silk suture compared to the untreated silk suture. The uniform deposition of natural fungal pigment on to the surface of the silk braided suture was confirmed by Scanning Electron Microscopy. The antibacterial activity of fungal treated silk braided suture at optimum concentration against *S. aureus* and *E. coli* is found to be good compared to the untreated silk suture. The result suggests that the silk suture treated with optimum concentration of the natural fungal pigment is appropriate to retard the exponential growth

of *S. aureus*, a gram-positive bacterium and *E. coli* a gram-negative bacterium and hence silk sutures can be developed with the required characteristics for healthcare applications.

References

1. Chu CC, Fraunhofer JAV, Greisle HP (1997) Wound closure biomaterials and devices. 1: 35- 49 CRC Press New York, USA.

2. Saxena S, Ray AR, Kapil A, Pavon-Djavid G, Letourneur D, et al. (2011) Development of a New Polypropylene-Based Suture: Plasma Grafting, Surface Treatment, Characterization, and Biocompatibility Studies. Mol Biosci 11: 373-382.

3. Pillai CKS, Sharma CP (2010) Absorbable Polymeric Surgical Sutures: Chemistry, Production, Properties, Biodegradability, and Performance. J Biomater Appl 25: 291-366.

4. Blacker JJ, Nazat SN, Bocc AR (2004) Development and characterisation of silver-doped bioactive glass-coated sutures for tissue engineering and wound healing applications. Biomaterials 25: 1319-1329.

5. Viju S, Thilagavathi G (2013) Effect of chitosan coating on the characteristics of silk-braided sutures. J Ind Tex 42: 256-268.

6. H Liu H, Ge Z, Wang Y, Toh SL, Sutthikhum V, et al. (2007) Modification of sericin-free silk fibers for ligament tissue engineering application. J Biomed Mat Res App Bio 82B: 129-138.

7. Parthiban M, Thilagavathi G (2012) Optimuzation of process parametres for coloration and antibacterial finishing of wool fabric using natural fungal extract. Ind J Fib Tex Res 37: 257-264.

8. Kim JC, Lee YK, Lim BS, Rhee SH, Yang HC (2007) Comparison of tensile and knot security properties of surgical sutures. J Mater Sci Mater Med 80: 332-338.

9. Tera H, Aberg C (1976) Tissue strength of structures involved in musculo aponeurotic layer sutures in laparotomy incisions. Acta Chirurgica Scandinavica 142: 349-355.

10. Heward AG, Laing RM, Carr DJ, Niven BE (2004) Tensile Performance of Nonsterile Suture Monofilaments Affected by Test Conditions. Tex Res J 74: 83-90.

11. Kim J, Yong K (2007) Conformal growth and characterization of hafnium silicate thin film by MOCVD using HTB (hafnium tertra-tert-butoxide) and TDEAS (tetrakis-diethylamino silane). Journal of Materials Science: Materials in Electronics 18: 391-395.

12. Abdessalem SB, Debbabi F, Jedda H, Elmarzougui S, Mokhtar S (2009) Tensile and Knot Performance of Polyester Braided Sutures. Tex Res J 79: 247-252.

13. Gupta BS, Wolf KW, Postlethwait RW (1985) Effect of suture material and construction on frictional properties of sutures. Surgical Gynecology Obstet 161: 12-16.

14. Zurek W, Frydrych I (1963) Comparative Analysis of Frictional Resistance of Wool Yarn. Tex Res J 63: 322-335.

15. Harnet JC, Guen EL, Ball V, Tenenbaum H, Ogier J, et al. (2009) Antibacterial protection of suture material by chlorhexidine-functionalized polyelectrolyte multilayer films. J Mat Sci Mater Med 20: 185-193.

16. Giridev VR, Venugopal J, Sudha S, Deepika G, Ramakrishna S, et al. (2009) Dyeing and antimicrobial characteristics of chitosan treated wool fabrics with henna dye. Carbohyd Poly 75: 646-650.

17. Velmurugana P, Kamala-Kannana S, Balachandarb V, Lakshmanaperumalsamyc P, Chaea J, et al. (2010) Natural pigment extraction from five filamentous fungi for industrial applications and dyeing of leather. Carbohyd Poly 79: 262-268.

18. Sharma D, Gupta C, Aggarwal S, Nagpal N (2012) Pigment extraction from fungus for textile dyeing. Ind J Fibre and Tex Res 37: 68-73.

13

Characterization of Electrostatic Discharge Properties of Woven Fabrics

Perumalraj R*

Bannari Amman Institute of Technology, Sathyamangalam, Erode, India

Abstract

In this research work, the woven fabric samples of cotton, polyester PC blend, glass and silk materials have been selected to analyze the electrostatic discharge behavior of various woven fabrics using electrostatic discharge tester. The electrostatic discharge is mainly depends upon various controlled factors of electrostatic discharge tester hence, the controlled factors of number of rubbing cycle (20, 30 and 40 strokes), pressure (100,150 and 200 gms), speed (10, 20 and 30 mpm) were considered and analyzed the electrostatic discharge behaviors of various woven fabrics under various temperature and relative humidity using Box Behnken design and regression analysis. It was found that the glass and silk woven fabric have greatest tendency to give up electrons and gain a positive electrical charge but in case of polyester woven fabric have the greatest tendency to attract electrons and gain a negative electrical charge. It was observed that glass woven fabrics have more electrostatic discharge value than the polyester, PC blend, silk and cotton woven fabrics and it was also found that higher number of rubbing cycle, pressure and speed have significant effects on the electrostatic discharge properties of various types of woven fabrics.

Keywords: Electrostatic discharge; Box-Behnken method; Tribo electric effect; Static electricity; Regression analysis

Introduction

Electrostatic discharge (ESD) is the release of static electricity when two objects come into contact. Materials are made of atoms that are normally electrically neutral because they contain equal numbers of positive charges (protons in their nuclei) and negative charges (electrons in "shells" surrounding the nucleus). The phenomenon of static electricity requires a separation of positive and negative charges. When two materials are in contact, electrons may move from one material to the other, which leaves an excess of positive charge on one material, and an equal negative charge on the other. When the materials are separated they retain this charge imbalance [1-4]. During textile manufacturing process, there is a potential of static charge generation when fibers are extruded, and yarns are woven or knitted, and finished. Fibers, yarns, or fabrics are rubbed with guides, rollers or tension devices on the machinery and this operation of contact and separation continuously occur throughout the process [5-8]. This gives many spinners and weavers much trouble in terms of productivity, and can lead to malfunction of electronic equipment [9-12]. Static problems in textile industry have become more serious problems in synthetic fibre production [13-19]. There are many factors that affect charge generation such as environment (temperature, humidity), structural (polymer type, structure of fabric) and working factors (fabric speed, tension, and contact area between fabric and machine parts, material type that is in contact with fabric [20,21]. From the literature survey, there is no extensive study about the electrostatic discharge properties of various types of woven fabric. Hence In this research work, the woven fabric samples of cotton, polyester PC blend, glass and silk materials have been selected to analyze the electrostatic discharge behavior of various woven fabric using electrostatic discharge tester. The electrostatic discharge is mainly depends upon various controlled factors of electrostatic discharge tester hence, the controlled factors of number of rubbing cycle (20, 30 and 40 strokes), pressure (100,150 and 200 gms), speed (10, 20 and 30 mpm) were considered and analyzed the electrostatic discharge behaviors of various woven fabrics under various temperature and relative humidity using Box Behnken design and regression analysis.

blend, silk and glass with a sample size of 10 inch × 5 inch were selected to analyze the electrostatic charge behavior of woven fabric using electrostatic discharge tester. All the selected woven fabrics were made with plain weave and same ends per inch (30), picks per inch (30). The electrostatic charge properties are mainly depends upon various controlled factors and uncontrolled factors namely number of rubbing cycle, pressure, speed, temperature and relative humidity (RH). In order to analyze and optimize the process parameters of electrostatic discharge tester, the following three factors were considered to analyze the electrostatic discharge properties of textile materials i.e., number of cycles (X1), pressure(X2) and speed (X3) by using Box-Behnken design as shown in the Table 1. The varying values of rubbing cycle(X1) are 20, 30 and 40 strokes; the varying values of pressure(X2) are 100,150 and 200 gms. The varying values of speed (X3) are 10, 20 and 30 mpm under temperature (27°C) and RH (50%) as shown in the Table 1. In electrostatic discharge tester, the sample is fixed between clamp1 and 2. The required parameters are fed into the electrometer before hand and the arm consisting the weight at the top and Teflon roller at the bottom is moved over the fabric in a linear reciprocating to and fro manner over the fabric. Finally the linear reciprocating movement is stopped after the predetermined values are reached, where the rubbing strokes is sensor by the proximity sensor and the static charge produced on the fabric can be measured by the electrostatic sensor. As per the coulombs law the force, pressure, speed were directly proportional to the amount of charge produced. The range of static charge that can be measured in the range of 1-15 kilovolts. When Teflon roller is moved over the fabric, the electrostatic charge is generated on the fabric, and then static charge can be sensed by electro static sensors. The electrostatic discharge values of various woven fabrics were analyzed using Box

Materials and Methods

The woven fabric samples of cotton, polyester, polyester/cotton

***Corresponding author:** Rathinam Perumalraj, Bannari Amman Institute of Technology, Sathyamangalam, Erode, India
E-mail: raj134722002@gmail.com

Factor level combination			
Expt Run	**X1**	**X2**	**X3**
	20	100	20
	40	100	20
	20	200	20
	40	200	20
	20	150	10
	40	150	10
	20	150	30
	40	150	30
	30	100	10
	30	200	10
	30	100	30
	30	200	30
	30	150	20
	30	150	20
	30	150	20

Table 1: Box-Behnken method.

Behnken design and regression analysis.

Results and Discussion

The static electricity of woven fabrics is mainly depends upon the various factors like speed, pressure and number of rubbing cycles, temperature and RH. In general, when the temperature is high, the static charge generation will be more but when humidity is high, the static charge generation will be less. The various type of woven fabrics have a tendency of either giving up electrons and become positive (+) in charge (or) attracting electron and become negative (-) in charge when brought in contact with other materials. The glass and silk woven fabric materials have tend to give up electrons and gain a positive electrical charge when brought in contact with other materials but the polyester woven fabrics have tend to attract electrons and gain a negative electrical charges when brought in contact with other materials. In case of cotton woven fabric do not tend to get to attract or giving up electrons when brought in contact or rube with other materials. The tribo electric effect (tribo electric charging) is a type of contact electrification in which sudden materials become electrically charged after the come in to contact with other materials through friction. Any two materials are come into contact and then separate, the electron to be exchange. After coming in contact, the chemical bond is formed between parts of the two surfaces, called adhesion, and electrical charges move from one material to other material in order to equalize the electrochemical potential.

Electrostatic discharge properties of polyester woven fabrics

It was observed from the Figures 1-3, that the polyester woven fabric have more static charge generation range from 1.11 to 2.60 kilovolts under various controlled factors of rubbing cycles, pressure and speed as per the Box Behnken experimental design. The polyester woven fabrics have more static discharge value than cotton and silk woven fabrics. It is mainly due to the moisture regain value of polyester woven fabrics (0.4%MR). The polyester woven fabric have less moisture absorption properties and retain more static charge than the natural fibres and also the polyester material are hydrophobic in nature which hardly allows moisture through polyester woven fabric material. Hence polyester woven fabric does not able to dissipate or conduct the electrostatic charges. It was also observed that there is no significant difference of electrostatic charge of polyester woven fabrics

under high and low humidity conditions. It was understood that there was a significant effect between number of rubbing cycle and pressure

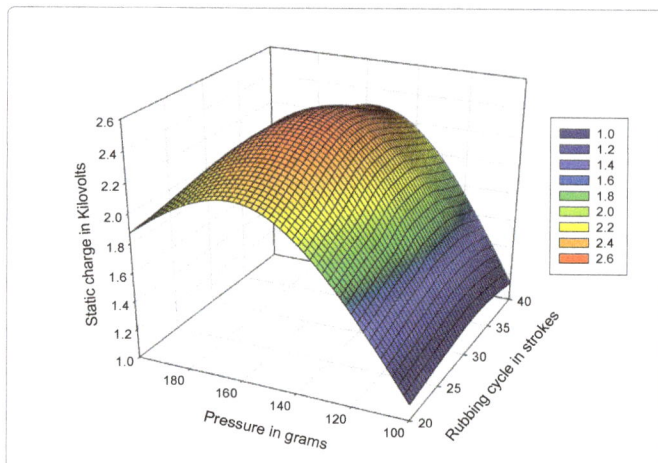

Figure 1: Effect of pressure and rubbing cycle on electrostatic discharge properties of polyester fabrics.

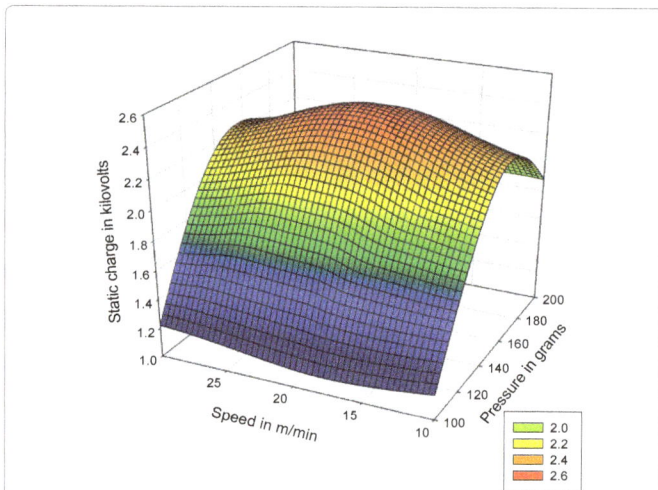

Figure 2: Effect of pressure and speed on electrostatic discharge properties of polyester fabrics.

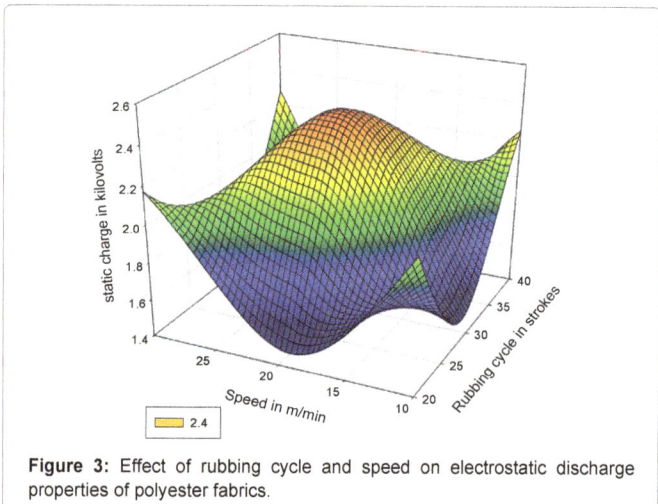

Figure 3: Effect of rubbing cycle and speed on electrostatic discharge properties of polyester fabrics.

and speed on electrostatic discharge properties of polyester woven fabrics. It was also found that higher number of rubbing cycle, pressure and speed had significant influences on the electrostatic discharge properties.

Electrostatic discharge properties of cotton woven fabrics

It was observed from Figures 4-6, that the cotton woven fabric have less electrostatic discharge value range from 1.11-1.53 kilovolts under the various controlled factors of number of rubbing cycle, speed and pressure as per the Box Behnken experimental design. The cotton woven fabrics have less electrostatic discharge than Polyester, PC blends, silk and glass woven fabrics. It is mainly due to moisture regain value of cotton materials. The cotton woven fabric moisture regain value is 8% but the polyester woven fabric moisture regain value is only 0.4%. The cotton fibers have 70% of crystalline region and rest of region is amorphous and cotton fibre is very absorbent owing to the counter less polar O-H group in its polymer, these attract water molecules. Hence this water conducts and dissipates the static electrical charge. There is a significant difference between cotton, PC blends and glass woven fabrics of electrostatic discharge properties. This cotton fabric moisture will able to contain and helps to distribute/dissipate the electrostatic charge very easily. However the observed moisture particle is actually conducting the electrical static charge and not the cotton fibre itself. It

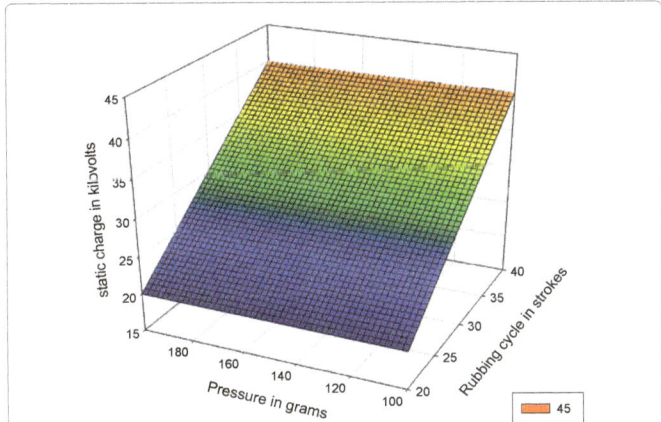

Figure 6: Effect of rubbing cycle and speed on electrostatic discharge properties of cotton fabrics.

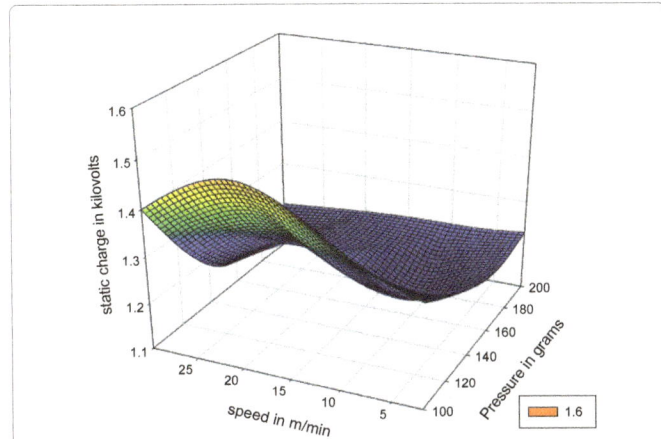

Figure 7: Effect of pressure and rubbing cycle on electrostatic discharge properties of PC blend fabrics.

was also observed that the cotton woven fabrics have less static in high humidity conditions because they observed water from the atmosphere and this water conduct and helps to dissipate and distribute the electrical static charge and also cotton fabrics are ineffective at dissipating static electric charge at very low level RH. It was understood that there was a significant effect between number of rubbing cycle and pressure and speed on electrostatic discharge properties of cotton woven fabrics. It was also found that higher number of rubbing cycle, pressure and speed had significant influences on the electrostatic discharge properties

Electrostatic discharge properties of PC blend fabrics

It was observed from the Figures 7-9, that the PC blend woven fabrics have static discharge value range from 1.12-1.85 kilovolts under various control factors of rubbing cycles, pressure and speed as per the Box Behnken experimental design. It was found that the PC blend fabrics have less electrostatic discharge generation value than the polyester woven fabric, but higher than the cotton woven fabrics. It was understood that there was a significant effect between number of rubbing cycle and pressure and speed on electrostatic discharge properties of PC blend woven fabrics. It was also found that higher number of rubbing cycle, pressure and speed had significant influences on the electrostatic discharge properties.

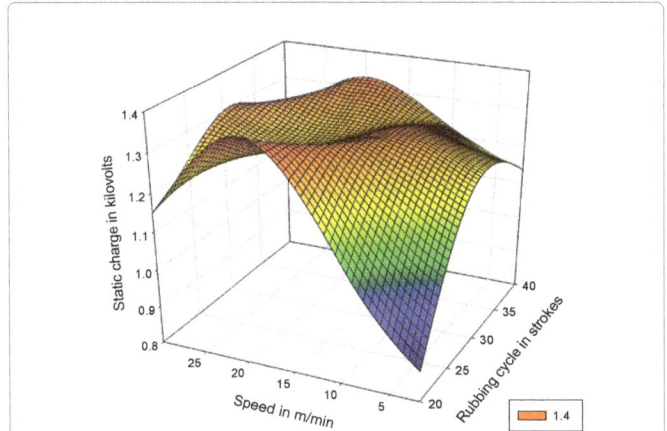

Figure 4: Effect of pressure and rubbing cycle on electrostatic discharge properties of cotton fabrics.

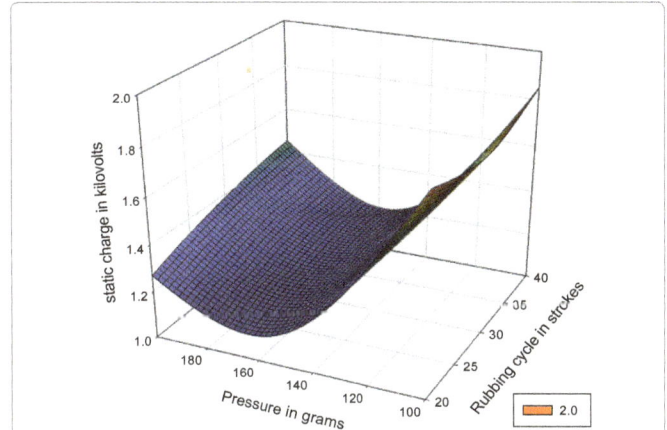

Figure 5: Effect of pressure and speed on electrostatic discharge properties of cotton fabrics.

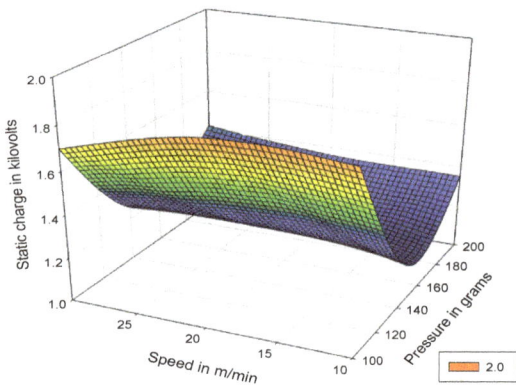

Figure 8: Effect of pressure and speed on electrostatic discharge properties of PC blend fabrics.

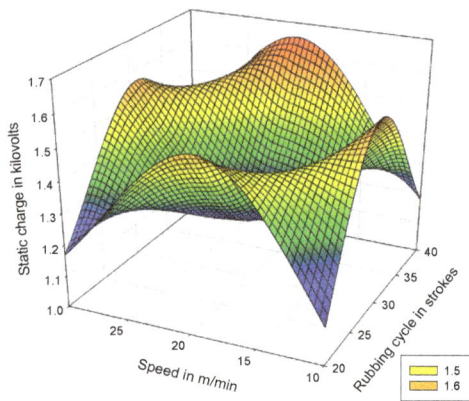

Figure 9: Effect of rubbing cycle and speed on electrostatic discharge properties of PC blend fabrics.

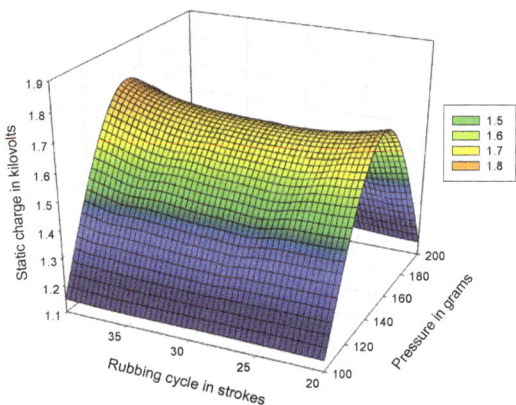

Figure 10: Effect of pressure and rubbing cycle on electrostatic discharge properties of Silk fabrics.

The electrostatic discharge properties of silk woven fabrics

It was observed from the Figures 10-12, that the silk woven fabric have electrostatic discharge generation value range from 1.1-1.82

kilovolts under the various controlled factors of number of rubbing cycles, speed and pressure as per the Box Behnken experimental design. The silk woven fabric less electrostatic discharge than polyester woven fabric, it is mainly due to the moisture regain value of silk woven fabric (11%). Hence the silk woven fabrics have conduct and dissipate the electrostatic charge easily than the polyester woven fabric. It was understood that there was a significant effect between number of rubbing cycle and pressure and speed on electrostatic discharge properties of PC blend woven fabrics. It was also found that higher number of rubbing cycle, pressure and speed had significantly influence the electrostatic discharge properties.

The electrostatic discharge properties of glass woven fabrics

It was observed from Figures 13-15, that the glass woven fabric have very high electrostatic discharge value range from 1.5-2.88 kilovolts under various controlled factors of rubbing cycles, speed and pressure as per the Box Behnken experimental design. It was found that the glass woven fabric have more electrostatic discharge value than polyester, PC blend, silk and cotton woven fabrics. Since, the glass fibre woven fabric does not absorb the moisture and water; hence there is more accumulation of static electric charge over the surface of the glass woven fabrics than other woven fabric materials and also there is no any conducting or dissipating or distribution of static electric charge.

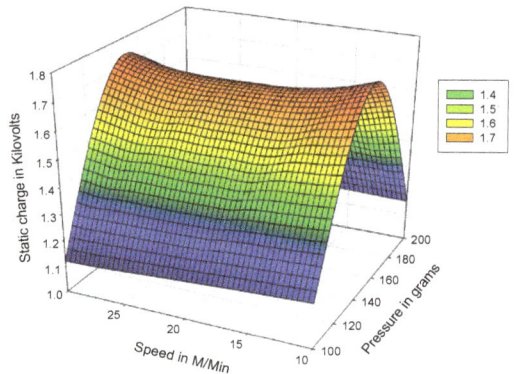

Figure 11: Effect of pressure and speed on electrostatic discharge properties of Silk fabrics.

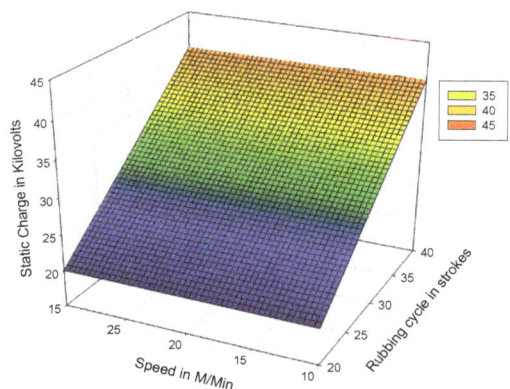

Figure 12: Effect of rubbing cycle and speed on electrostatic discharge properties of Silk fabrics.

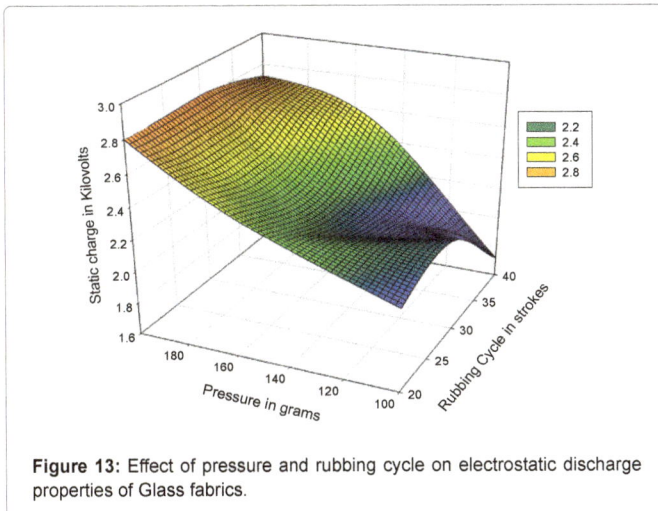

Figure 13: Effect of pressure and rubbing cycle on electrostatic discharge properties of Glass fabrics.

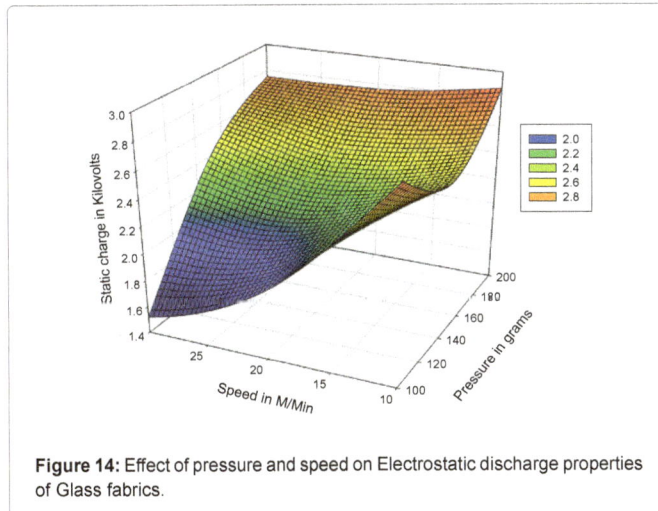

Figure 14: Effect of pressure and speed on Electrostatic discharge properties of Glass fabrics.

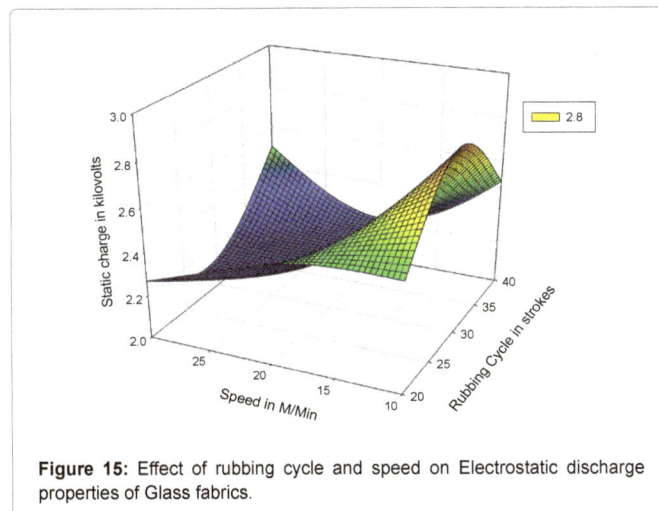

Figure 15: Effect of rubbing cycle and speed on Electrostatic discharge properties of Glass fabrics.

In general, the static electricity is the accumulation of electrical charges on the surface of the material, usually an insulator or non-conductor of electricity. It called "static" because there is no current flowing, as there is in alternative current (AC) or direct current (DC) electricity.

Sl. No.	Process parameter	Optimum level
	Rubbing cycle	+
	Pressure	0
	Speed	+

Table 2: Optimum level process parameter of electrostatic discharge tester.

The glass and silk fibre woven fabrics have the greatest tendency to give up electrons and gain a positive electric charge, but in case of polyester woven fabric have the greatest tendency to attract electrons and gain a negative electrical charge.

The regression equations were formed using co-efficient obtained in the Box Behnken experimental design to assess the static electricity charge of the textile woven fabrics. Regression equation analysis for various woven fabrics is equated as y1=-198.310232a-2.51277x1+1.719121x2+0.918667x3 where y1 indicates the average of static charge generated, x1, x2 and x3 are the three factors and levels i.e. rubbing cycle, pressure and speed. The above equation is formed in this exercise and appeared to fit in a better way with the actual value obtained at all level, expressed by the higher correlation values.

The three optimum process parameters obtained from data and graphs were analyzed for best occurrence of each process parameter and once set of process parameter is obtained to find out the (i) optimum level of each variable,(ii) best trial. The process parameters for optimum results of electrostatic discharge are shown in the Table 2.

Conclusion

The electrostatic properties of various woven fabrics have mainly depends upon the control factors of number of rubbing cycle, pressure and speed of electrostatic discharge tester. The following conclusion can be drawn from this study.

•	The glass woven fabric have more static charge generation than polyester, PC blend, silk and cotton woven materials.

•	The polyester woven fabrics have more static electrostatic charge accumulation over the surface of the fabrics than the cotton, PC blend, silk and cotton.

•	The glass and silk fibre woven fabrics have greatest tendency to give up electrons and gain a positive electrical charge.

•	The polyester fibre woven fabric have greatest tendency to attract electrons and gain a negative electrical charge.

•	The optimum process parameters were analyzed and it was found that rubbing cycle (+), pressure (0) and speed (+) were obtained for effective measurement of electrostatic discharge properties of various types of woven fabrics.

References

1.	Slade PE (1998) Antistats. Handbook of Fiber Finish Technology. New York, USA.

2.	Bailey, Adrian G (2001) The Charging of Insulators. Journal of Electrostatics 51-52: 82-90.

3.	Castle GSP (1997) Contact Charging between Insulators. Journal of Electrostatic 40-41: 3-12.

4.	Arita Y, Shiratori SS, Ikezaki K (2003) Methods for the Detection and Visualization of Charge Trapping Sites in Amorphous Parts in Crystalline Polymers. Journal of Electrostatics 57: 263-271.

5.	Taylor DM, Secker PE (1994) Industrial Electrostatics: Fundamentals and Measurements. John Wiley and Sons New York, USA.

6. Gompf R (1988) Standard test method for evaluating triboelectric charge generation and decay. NASA Report.

7. Chubb JN (1990) Instrumentation and standards for testing static control materials. IEEE Trans Ind App 26: 1182-1187.

8. Buhler C, Calle C, Clements S, Ritz M, Starnes J (2006) Test methodology to evaluate the safety of materials using spark incendivity. Journal of Electrostatics 64: 744-751.

9. Chubb JN (1999) The assessment of materials by tribo and corona charging and charge decay measurement. Inst Physics 'Electrostatics 163: 329-333.

10. Chubb JN (1996) Corona charging of practical materials for charge decay measurements. Journal of Electrostatic 37: 53-65.

11. (2000) IEC 61340-2-1 Measurement methods in electrostatics – Test method to measure the ability of materials and surfaces to dissipate static electric charge. International Electrotechnical Commission.

12. Chubb JN, Malinverni P (1992) Comparative studies on methods of charge decay measurement. Journal of Electrostatics 30: 273-283.

13. Chubb JN (1995) Dependence of charge decay characteristics on charging Parameters. York, USA.

14. Jean C (1987) Electrostatics: Principles, problems and Applications. Adam Hilger.

15. Berezin AA (1995) Electrification of Solid Materials. Handbook of Electrostatic Processes, New York, USA.

16. Morton WE, Hearle JWS (1993) Physical Properties of Textile Fibers. Manchester UK.

17. Vosteen, William E (1984) A Review of Current Electrostatic Measuring Techniques and Their Limitations. Electrical Overstress Exposition.

18. Noll CG (1995) Electrostatic Charge Elimination Techniques. Handbook of Electrostatic Processes. New York, USA.

19. Kacprzyk R, Stec C (1997) Measurement of the Surface Charge Density on Moving Webs. Journal of Electronics 40-41: 455-461.

20. Seaver AE (1995) Analysis of Electrostatic Measurements on Non-Conducting Webs. Journal of Electrostatics 35: 231-243.

21. Durkin WJ (1995) Dangers in Interpreting Electrostatic Measurements on Plastic Webs. Journal of Electrostatics 35: 215-229.

14

Spectrophotometric Investigation of the Interactions between Cationic (C.I. Basic Blue 9) and Anionic (C.I. Acid Blue 25) Dyes in Adsorption onto Extracted Cellulose from *Posidonia oceanica*

Douissa NB*, Dridi-Dhaouadi S and Mhenni MF

Research Unity of Applied Chemistry and Environment, Department of Chemistry, Faculty of Sciences, University of Monastir, 5019 Tunisia

Abstract

Extracted cellulose from *Posidonia oceanica* was used as an adsorbent for removal of a cationic (Basic blue 9, BB) and anionic textile dye (Acid blue 25, AB) from aqueous solution in single dye system. Characterization of the extracted cellulose and extracted cellulose-dye systems were performed using several techniques such as Fourier transform infrared spectroscopy (FTIR), X-ray photoelectron spectroscopy (XPS), zeta potential and Boehm acid–base titration method. Adsorption tests showed that the extracted cellulose presented higher adsorption of BB than AB in single dye system, revealing that electrostatic interactions are responsible, in the first instance, for the dye–adsorbent interaction. In single dye systems, the extracted cellulose presented the maximum adsorption capacities of BB and AB at 0.955 and 0.370 mmol.g^{-1}, respectively.

Adsorption experiments of AB dye on extracted cellulose saturated by BB dye exhibited the release of the latter dye from the sorbent which lead to dye-dye interaction in aqueous solution due to electrostatic attraction between both species. Interaction of BB and AB dyes were investigated using spectrophotometric analysis and results demonstrated the formation of a molecular complex detected at wavelengths 510 and 705 nm when anionic (AB) and cationic (BB) dye were taken in equimolar proportions. The adsorption isotherm of AB, taking into account the dye-dye interaction was investigated and showed that BB dye was released proportionately by AB equilibrium concentration. It was also observed that AB adsorption is widely enhanced when the formation of the molecular complex is disadvantaged.

Keywords: Extracted cellulose; Single dye system; Extracted cellulose saturated by BB dye; Dye-dye interactions; Molecular complex

Introduction

Dyes are complex aromatic substances essential in many industries such as textile, paper, paint, leather and pharmaceutical. Compared with natural dyes [1], synthetic dyes are superior in terms of color availability, easiness to use, quick-setting and ensured by accurate formulas [2]. However, most of the dyes are stable to light, heat and many oxidizing agents, and more difficult to be biodegraded. With the introduction of strict environmental legislation, effluents containing dyes require proper treatment prior to discharge, not only for their high chemical oxygen demand (COD), suspended solids and toxic breakdown products, but also for color, which is not only highly visible and undesirable, but also harm the environment and cause health problems to humans and aquatic animals.

In this context, several methods of decolourization were used such as biological treatment [3], coagulation-flocculation [4] chemical and electrochemical oxidation [5-7]. However, these treatments are generally expensive and their effectiveness is sometimes asked. Sorption onto solid surface, namely adsorption, has proven to be an effective and cheaper process for removing pollutants from textile effluents due to its simplicity and high efficiency. However, the conventional solid retention, usually activated carbons [8] remains for the professionals a relatively expensive material [9]. Recent studies have been conducted to propose alternative sorbents.

Posidonia oceanica (P.oceanica), a local biomass abundant on the coasts of Tunisia has shown its effectiveness in removing organic pollutants (textile dyes [10-13]) and inorganic (heavy metals [10,11]). The cellulosic material namely cellulose and hemicellulose, presented the major constituent of the *Posidonia oceanica* (62%) with a relatively high lignin (27%) [14]. During this work, firstly, the extracted cellulose

from the marine plant is used to remove cationic (Basic blue 9, BB) and anionic dye (Acid blue 25, AB) from aqueous solution in single dye system. Secondly, adsorption of an anionic dye (AB) onto extracted cellulose previously saturated by cationic dye (BB) was investigated to understand the adsorption mechanism and the dye-dye interactions on one hand and between the dyes and the adsorbent on the other hand.

Experimental

Adsorbent preparation

Posidonia oceanica's balls were collected from Tunisian coasts. They were washed with tap water to remove sand and other solids, dried in sunlight, crushed into small pieces (45-1000 µm) using a grinder (AM80 Nx2), then washed repeatedly with distilled water and finally dried in an oven at 60°C for 24 h. The cellulose extraction was performed using the follow protocol [15]. The powdered material (20 g) was first treated with 100 mL of a 3 mol.L^{-1} sodium hydroxide solution at 100°C for 4 h. This alkaline treatment removes both lignin and hemicellulose. The residue was collected by filtration, washed with

---

*Corresponding author: Najoua Ben Douissa, Research Unity of Applied Chemistry and Environment, Department of Chemistry, Faculty of Sciences, University of Monastir, 5019 Tunisia
E-mail: najoua_bendouissa_2000@yahoo.fr

distilled water, and then dried in an oven for 16 h at 50°C. In order to eliminate the pectin as well as some residuals lignin that gives brown colour to the biomass, conventional bleaching treatment was used. In fact, 10 g of the cellulose were agitated with NaClO solution (100 mL; pH 10) at 50°C for 90 min. After filtration, the residue was washed thoroughly with distilled water until the filtrate was neutral, dried in sunlight for two days, and then stored in plastic bottles for further use.

The second adsorbent of this study is extracted cellulose from *Posidonia oceanica* saturated by the basic dye BB. Indeed, the cellulosic material (5 g.L^{-1}) is brought into contact with a solution of dye [BB] = 1 g.L^{-1} in the aqueous dye solution pH (without addition of acid or base) (pH = 6.2 ± 0.4). The whole is subjected to continuous agitation of 150 rpm (Heidolph Vibramaxe 100) for 24 h. The extracted cellulose is filtered and washed thoroughly with distilled water until the filtrate became clear water. Residual concentrations of BB in all filtrates solutions were determined using the colorimetric method at the maximum wavelength of BB dye (655 nm). Thereafter, the adsorbed amount of BB on extracted cellulose was calculated as 0.586 mmol.g^{-1}.

Characterization of materials

Acidic sites on extracted cellulose were determined by the acid–base titration method proposed by Boehm [16]. Aqueous solutions of NaHCO$_3$ (0.05 mol.L^{-1}), Na$_2$CO$_3$ (0.05 mol.L^{-1}) and NaOH (0.05 mol L^{-1}) were prepared. A volume 25 mL of these solutions was added to 0.5 g of extracted cellulose, shaking for 48 h and then centrifuged. The excess of base was then determined by back titration using NaOH (0.05 mol L^{-1}) and HCl (0.025 mol.L^{-1}). All titrations were carried out at room temperature. The number of acidic sites was determined under the assumptions that NaOH neutralizes carboxylic, lactonic, and phenolic groups; that Na$_2$CO$_3$ neutralizes carboxylic and lactonic groups; and that NaHCO$_3$ neutralizes only carboxylic groups. The test was repeated at least three times.

The Zeta potential of extracted cellulose was measured with Zeta Meter System 3.0. 100 mg of extracted cellulose were added to 100 mL of distilled water. After pH adjustment, mixing for 2 h and centrifugation of the sorbent suspension, the filtrate was placed in the electrophoresis cell for zeta potential measurements. The measurements were repeated at least five times.

Scanning electron microscopy (SEM) analysis was carried out on the extracted cellulose to study its surface texture. The micrograph was recorded on a ZEISS-ULTRA55 SEM device, operating at 10 kV. Prior to the analysis by SEM, the surface was coated with a gold/palladium layer.

Several methods were used to characterize the extracted cellulose, before and after dye adsorption. The FT-IR spectra were recorded in a bio-Rad spectrophotometer. The analysis was applied on different samples to determine the surface functional groups. The acquisition conditions were 64 scans and 4 cm^{-1} resolution. The wavenumber scanning was in the range of 500-4000 cm^{-1}.

The X-ray photoelectron spectroscopy (XPS) experiments were performed with a XR3E2 apparatus (Vacuum Generators, UK) equipped with a monochromated MgKα X-ray source (1253.6 eV) and operating at 15 kV under a current of 20 mA. Samples were placed in an ultrahigh vacuum chamber (10^{-8} mbar) with electron collection by a hemispherical analyzer at an angle of 90°. The signal decomposition was realised using Spectrum NT, and the CAH signal was used as a reference peak at 285.0 eV. XPS analysis was performed in extracted cellulose before and after dye adsorption.

Dyes and analysis

AB (molecular formula = C$_{20}$H$_{14}$N$_2$O$_5$S, λ$_{max}$ = 602 nm) and BB (molecular formula = C$_{16}$H$_{18}$N$_3$SCl, λ$_{max}$ = 655 nm) were supplied by Reactifs–RAL and used without further purification. The characteristics and structure of these dyes is presented in Figure 1. The test solutions were prepared by diluting stock solutions to the desired concentrations. The UV–vis spectra of aqueous dye solutions were measured by a double beam UV/vis spectrophotometer (Cecil instrument, Model CE2021-2000).

Batch sorption studies

A fixed amount of extracted cellulose (0.25 g) was added to a set of bottles (Teflon) containing 50 mL of known concentration of AB and BB dyes in single solution at pH around 6.2 (without addition of acid or base). The bottles were agitated at 150 rpm (Heidolph Vibramaxe 100 shaker) and at temperature of 21°C until equilibrium was reached. Then, samples were centrifuged at 2000 rpm for 2 min, and the supernatant was analyzed for the residual AB and BB concentration (by measuring the absorbance at 602 and 665 nm for AB and BB, respectively) using a double beam UV/VIS spectrophotometer (Cecil instrument, Model CE2021-2000).

On the other hand, bottles' series of AB solutions (50 mL) at different initial concentrations (0-1 g.L^{-1}) were placed in contact with 0.25 g of extracted cellulose saturated by the basic dye BB. The flasks were agitated at 150 rpm (at temperature of 21°C) until equilibrium was reached and centrifuged at 2000 rpm for 2 min. The residual dye solutions after adsorption were scanned over the whole visible range. In this work, the classical equations Freundlich and Langmuir were used to model experimental curves and the equilibrium models were fitted employing the non-linear fitting method, using the non-linear fitting facilities of the software Microcal Origin 6.0.

Results and Discussion

Characterization of extracted cellulose

Figure 2 shows the SEM micrograph of the extracted cellulose sample. The extracted cellulose has considerable numbers of heterogeneous layers and pores which may be very interesting for dye adsorption.

Table 1 summarizes the results of Boehm titration of extracted cellulose and shows that most of acidic functional groups are carboxylic, followed by phenolic and then lactonic.

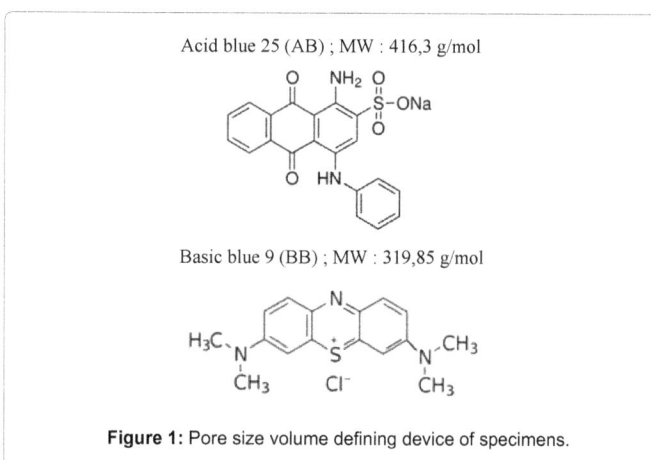

Acid blue 25 (AB) ; MW : 416,3 g/mol

Basic blue 9 (BB) ; MW : 319,85 g/mol

Figure 1: Pore size volume defining device of specimens.

The wide survey scan spectra are shown in Figure 3. The spectrum of extracted cellulose shows two strong symbol peaks for C1s and O1s at 285 and 531 eV respectively. Three quite weak ones for N 1s, Ca 2p3/2 and Cl 2p3/2 can be observed at, 395.7, 345.7 and 200 eV, respectively [17]. Those observations suggest the presence of abundant C, O and minor N, Ca and Cl on the surface of the extracted cellulose. The elemental surface composition of extracted cellulose obtained by XPS is given in Table 2, which shows that the surface of the extracted fiber did not consist of pure cellulose. The O/C ratio was much lower than in pure cellulose (0.83 in theory). Some nitrogen, calcium and chlorine were observed, which should not be present in pure cellulose. The calcium atom has been often detected on natural celluloses [18] and the chlorine atom could be present from bleaching operation. Those results confirm that the extracted cellulose yet contains lignin and hemicelluloses.

The FT-IR spectrum of extracted cellulose shows a strong absorption band at 3430 cm⁻¹ due to stretching of O-H groups and

that one at 2891 cm⁻¹ to the C-H stretching. The band at 1651 cm⁻¹ corresponds to the bending mode of the absorbed water [19] but can also be assigned to the absorption of carboxylic groups.

Dye adsorption in single-dye solutions

Effect of the solution pH on dye adsorption: The influence of equilibrium pH on the BB and AB sorption was investigated between 3 and 11 and the results are shown in Figure 4. Dyes sorption onto extracted cellulose appeared to be strongly pH-dependent. Increasing the pH value from 3 to 4 resulted in a considerable increase in BB sorption and higher decrease in AB one, whereas further increasing the pH value from 4 to 10 yielded a slight change in sorption, and this for both dyes. This tendency correlated well with the variation of surface charge of the extracted cellulose (the pH of point zero charge pH_{ZPC} measured by electrophoresis is 3.5). Consequently, for pH values lower than pH_{ZPC} the sorbent exhibits a positive surface charge. The higher amount of anionic dye (AB) sorbed at pH values lower than 3.5 can be explained by the electrostatic interactions between the positively charged surface of the sorbents and of the negatively charged AB dye. For pH higher than 3.5 the extracted cellulose surface is negatively charged. This further indicates that cationic dye (BB) sorption was also driven primarily via electrostatic interaction.

Equilibrium sorption isotherms: The sorption studies of AB and BB dyes in single dye solutions onto extracted cellulose were carried out in a batch reactor by varying the initial concentrations of the studied dyes at an initial pH of 6.2 ± 0.4. Nonlinear regression was used to fit the dye sorption isotherm to the Freundlich and the Langmuir (Table 3). Based on the R^2 values the best isotherm model fitted for extracted cellulose sorbents is the Langmuir model (Figure 5). Adsorption capacities of BB and AB on the extracted cellulose were determined to be 0.955 and 0.370 mmol/g, respectively. The successful application of this model to the present data supports that all adsorption sites are energetically and sterically independent of the adsorbed amount [20]. The solid is assumed to have a limited adsorption capacity, which defines sorbent's total capacity.

Dye-cellulose interaction characterization

FTIR study: In order to illustrate interaction between, firstly BB and extracted cellulose and secondly AB and extracted cellulose, to suggest responsible sites for adsorption, analysis FTIR of this sorbent after adsorption were investigated. Figure 6 shows the FTIR spectra of BB, AB and extracted cellulose before and after dye adsorption. Figure 6a shows the extracted cellulose spectra. For BB (Figure 6b), stretching vibration of tertiary amine groups is observed at 3422 cm⁻¹. Bands corresponding to the C-H stretching vibration are at 2919 and 2849 cm⁻¹. The stretching vibration of nitrile groups is at 2700 cm⁻¹. Vibrations of the aromatic ring are found at 1597 and 1538 cm⁻¹. The C–N stretching vibration is at 1337 cm⁻¹. The C–H 'in plane' vibrations are found between 1253 and 1037 cm⁻¹ and 'out of plane' vibration is at 945 cm⁻¹ [21].

Similarly, for AB Figure 6c reveals the presence of several peaks of AB's characteristic functional groups. The stretching vibrations observed at 3410 and 1185 cm⁻¹ mark the presence of amine groups, band corresponding to the CH stretching vibration is at 2920 cm⁻¹, the functional groups of the aromatic ring such as C=C and C=O are located at around 1560 cm⁻¹, the CN stretching vibration is at 1360 cm⁻¹, CH vibration is at 1262 cm⁻¹ and the band at 1017 cm⁻¹ can be assigned to the S = O bond according to Auta and Hameed [22].

It is noted that the shoulder observed between 1100 and 1000

Figure 2: Scanning electron microscope image of extracted cellulose.

pH$_{PZC}$	Acidic groups (mmol.g⁻¹)			
	Carboxylic	Lactonic	phenolic	total
3.5	0.875	0.015	0.275	1.165

Table 1: pH$_{PZC}$ value and Boehm titration results of extracted cellulose.

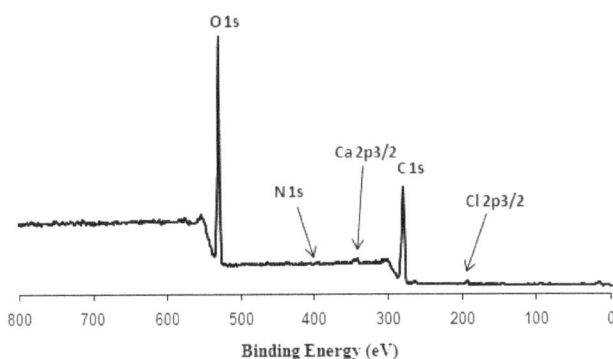

Figure 3: XPS wide survey scans spectra of the extracted cellulose.

	Surface composition (at percentage)							
	O	C	N	Ca	Cl	S	O/C	S/C
Extracted cellulose	35.62	61.78	0.68	0.41	0.67	-	0.57	-
Extracted cellulose after AB adsorption	36.17	62.90	0.91	0.09	0.13	0.04	-	6,3.10⁻⁴
Extracted cellulose after BB adsorption	28.20	67.93	2.33	0.17	0.71	0.08	0.42	-

Table 2: The elemental surface composition from XPS wide survey spectra.

cm^{-1} in cellulose spectrum, which corresponds to the deformation vibrations in the plane of hydroxyl groups of alcohols, is modified after incorporation of the dyes. The peaks that appear in (d) and (e) to 1341 cm^{-1} correspond to the CN stretching vibration at each dye (Figures 6b and 6c). In addition, after adsorption of AB (Figure 6d) the band at 1020 cm^{-1}, which corresponds in the case of raw material (a) to the CO stretching vibration has changed appearance at the cellulose adsorbed AB (d) marking the presence of S=O bond at 1017 cm^{-1}. Two other peaks appear at the shoulder at 1128 and 1080 cm^{-1} which probably correspond to the involvement of the carbon group C2 and C6 of the representative unit of the cellulose after dyes adsorption.

Besides peaks' shift positions, the intensity and sharpness of the bands observed for the raw material were also affected after dyes adsorption. The broadband at 3420 cm^{-1} of extracted cellulose (Figure 6a) becomes slightly narrow after dyes adsorption (Figures 6d and 6e), indicating the possibility of the involvement of hydroxyl groups of the extracted cellulose in the adsorption process.

Thus, it can be concluded that the OH group of the material interacts with the nitrogen atom on the BB or AB with hydrogen

bonding [21]. Since the N atom in BB is linked to the aromatic rings or to the terminal methyl groups, or in AB is linked to the primary or secondary amine, the vibrational frequencies of the aromatic ring and other groups will be affected when nitrogen is involved in a hydrogen bonding interaction.

X-ray photoelectron analysis (XPS): After dyes adsorption atomic composition of the surface varied considerably. After BB sorption, increasing of C, N and Cl amounts and decreasing O and Ca one (Table 2), and the appearance of S 2p3/2 at 164 eV binding energy was observed. The increase in C, S, N and Cl atomic composition come from the presence of BB species at the surface. While, increasing of C,O and N amounts and decreasing Cl and Ca one (Table 2), and the appearance of S was observed after AB adsorption.

It can be seen from BB molecule formula, that this latter present no oxygen atom, thus, it was interesting to compare O/C ratio of extracted cellulose and the resulting material after adsorption (Table 2). Reciprocally, S/C ratio of AB and extracted cellulose adsorbed the anionic dye was compared. From these values, it is easy to calculate the adsorption capacity of each dye and it could be measured as 2.05 and 0.34 mmol.g^{-1} for BB and AB, respectively. The adsorption capacity (from adsorption isotherm study) of samples analyzed with XPS was measured as 0.955 and 0.0465 mmol.g^{-1} for BB and AB, respectively, which is lower than the calculated one from XPS results. The AB and BB species might exit preferentially at the surface of the fiber rather than in its porosity.

AB adsorption on extracted cellulose loaded with MB dye

Highlighting dye-dye interactions: The visible absorption spectra of AB and BB were used for the analysis of dyes in single dye solutions. The single solutions of AB and BB containing 20 and 10 mg L^{-1}, respectively of dyes were prepared and the visible spectra of these solutions were recorded between 400 and 800 nm (Figure 7). As it can be seen from Figure 7, maximum wavelength (λ_{max}) of AB and BB were determined as 602 nm and 665 nm, respectively. The calibration curves were prepared at λ_{max} of each dye. On the other hand, this figure shows the addition of absorbance spectra of single solutions of AB and BB (superposition spectra) which is not in agreement with the absorbance spectra of binary solution made by 20 mg L^{-1} of AB and 10 mg L^{-1} of BB. This shows that the Beer–Lambert rule is not applicable due to possible interaction of dyes. So that, derivative spectrophotometry method [23], which is based on the analysis of derivative spectra, cannot therefore be applied in this case to determine the composition of a binary mixture of dyes. A series of binary solutions was prepared to cover the wide

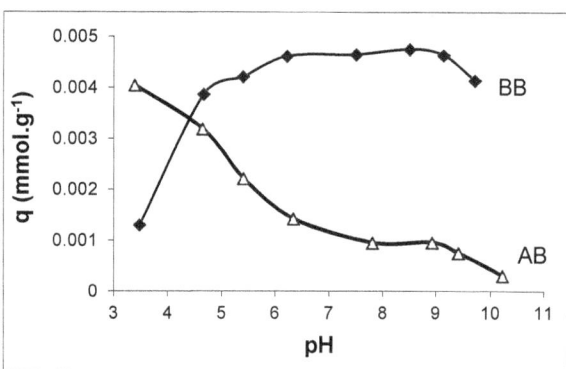

Figure 4: Effect of pH on AB and BB adsorption onto extracted cellulose (initial dye concentration: 10 mg.L^{-1}, temperature: 22°C).

Dye	Langmuir			Freundlich		
	q_m (mmol.g^{-1})	K_L (L mmol^{-1})	R^2	K_F (mmol$^{n-1/n}$ L$^{1/n}$ g^{-1})	n	R^2
BB	0.955	11.64	**0.99**	2.22	1.56	0.98
AB	0.370	0.95	**0.95**	0.17	1.49	0.91

Table 3. Isotherm parameters for AB and BB sorption on extracted cellulose.

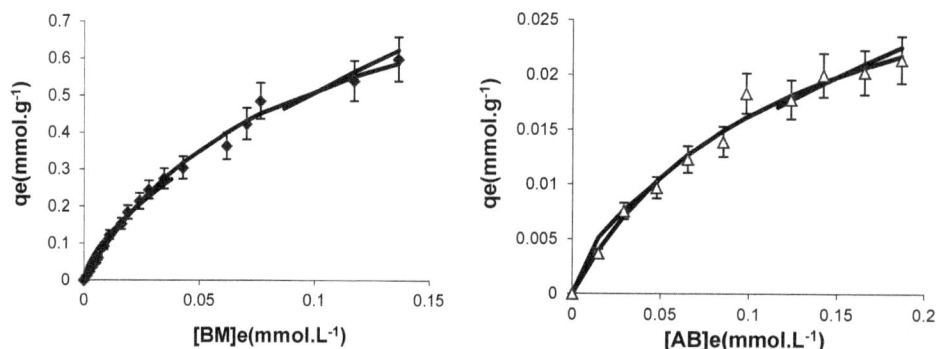

Figure 5: Non-linear sorption isotherms of AB and BB in single dye solution (solid lines indicate the Langmuir isotherm and dashed one the Freundlich isotherm).

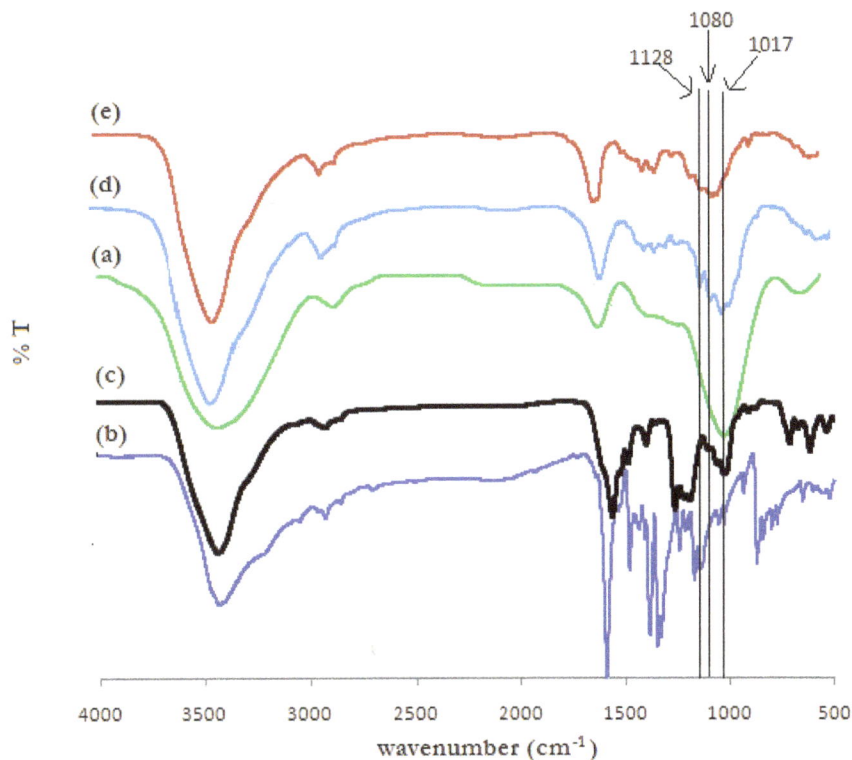

Figure 6: FT-IR spectra of (a) extracted cellulose, (b) BB, (c) AB, (d) extracted cellulose after AB adsorption and (e) extracted cellulose after BB adsorption.

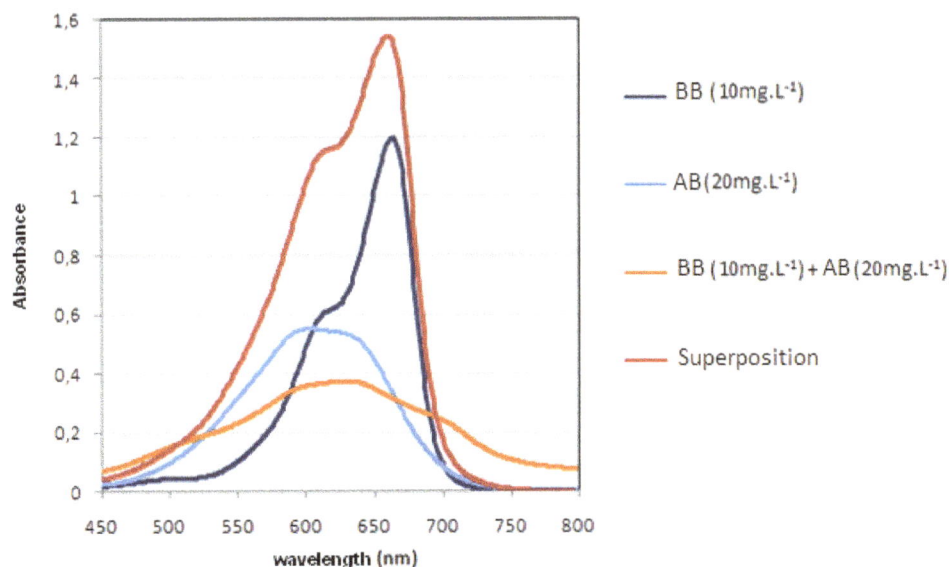

Figure 7: Visible absorption spectra of AB and BB in single and binary solutions.

range of concentrations which will allow, for identification of spectra as fingerprints to determine the composition of the mixtures in residual solutions after dye adsorption. Figures 8-10 show the spectra obtained for a wide range of binary composition mixtures.

The data spectra allow for some findings:

- If [AB] < [BB] the mixture spectrum has a similar shape to that of BB spectrum (Figure 8). It is thus found that for a constant concentration of BM (eg. [BB] = 0.626 mmol.L⁻¹), the absorbance at 665 nm decreases as AB concentration increases (provided that [AB] remains below [BB]) and the shape of the spectrum is closer to that of BM spectrum.

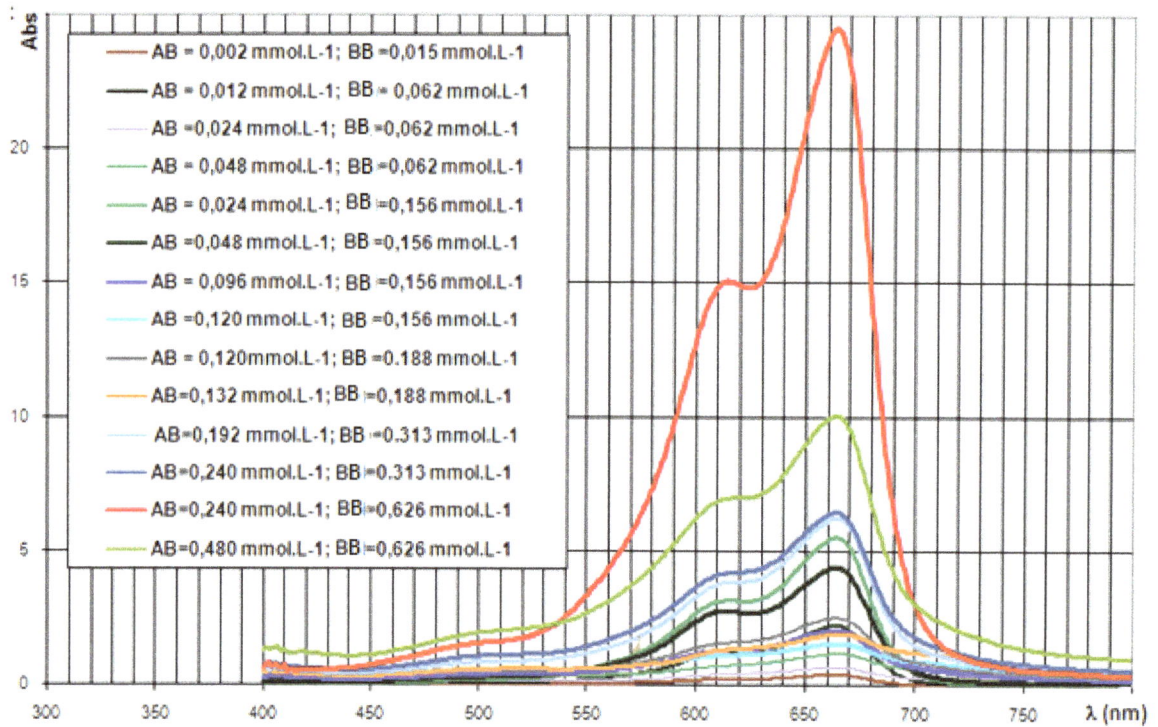

Figure 8: Spectral scanning binary mixtures for which [AB] < [BB].

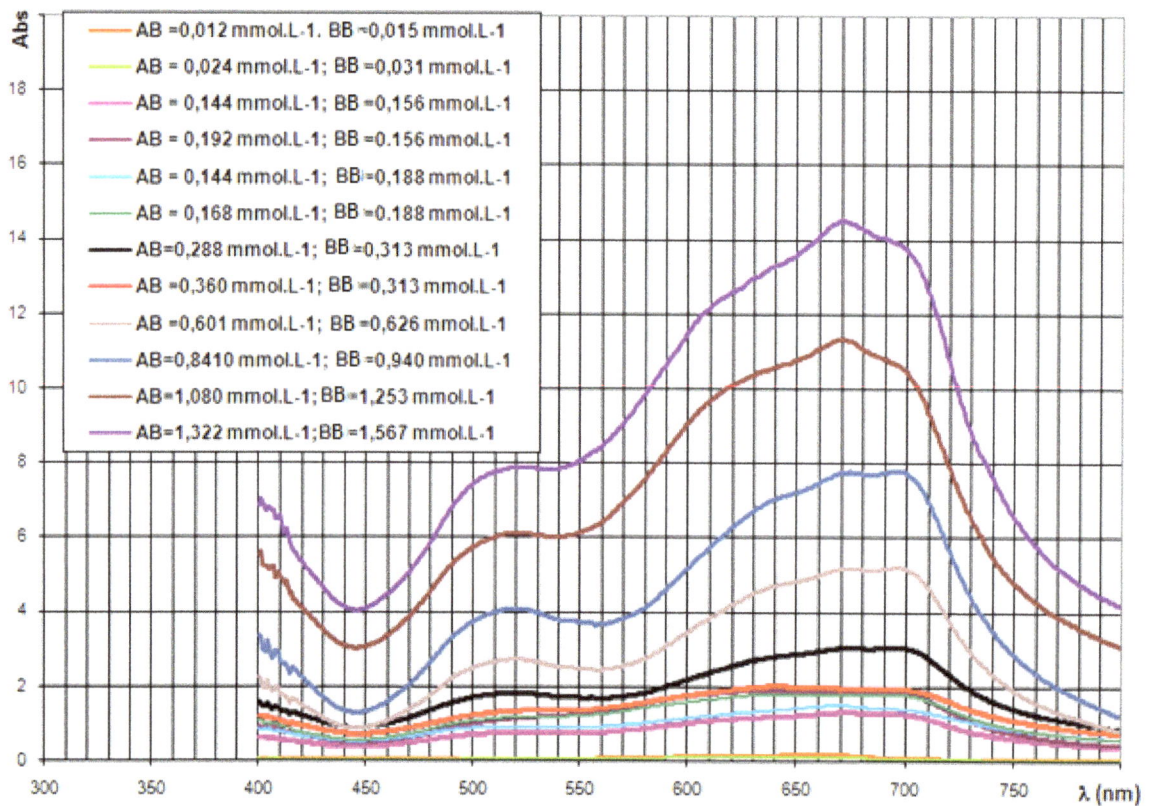

Figure 9: Spectral scanning binary mixtures for which [AB] = [BB].

- If [AB] > [BB] the mixture spectrum has rather the shape of AB spectrum (Figure 10).

- If [AB] ≈ [BB] the mixture spectrum has a third different shape from that of the two dyes, with the appearance of shoulders at 705 and 510 nm in addition to the maximum peak wavelength of the two dyes (Figure 9). It should be noted that in this case the absorbance of the shoulder at 705 nm is proportional to both concentration of BB and AB. This suggests an intermediate form takes place when both dyes are present in equimolar amounts. This form, which may be a molecular complex, is detected at 705 and at 510 nm.

Adsorption study: In this study, adsorption of the anionic dye (AB) was carried out after prior adsorption of the cationic dye (BB). The aim is to change the charge state of the adsorbent surface by the binding of the cationic dye and thus enable better sorption of anionic dye. The equilibrium adsorption of AB on the raw extracted cellulose and previously saturated one by BB dye was presented in Figure 11. On the practical side, the amounts of adsorbed AB dye in the adsorption isotherm study onto extracted cellulose saturated by BB dye (**O**) were determined following a direct reading of the absorbance at 602 nm (λ_{max} of AB) of residual solutions without taking account of the BB interference. Compared to adsorption isotherm of AB obtained on the

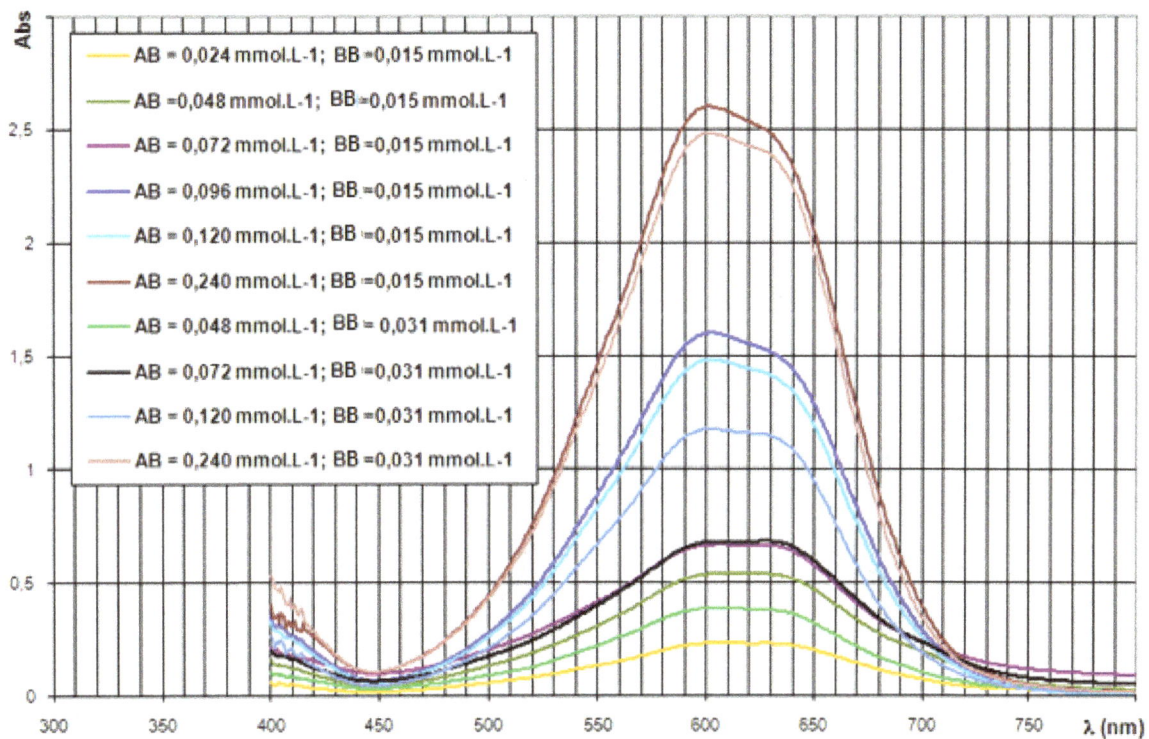

Figure 10: Spectral scanning binary mixtures for which [AB] □ [BB].

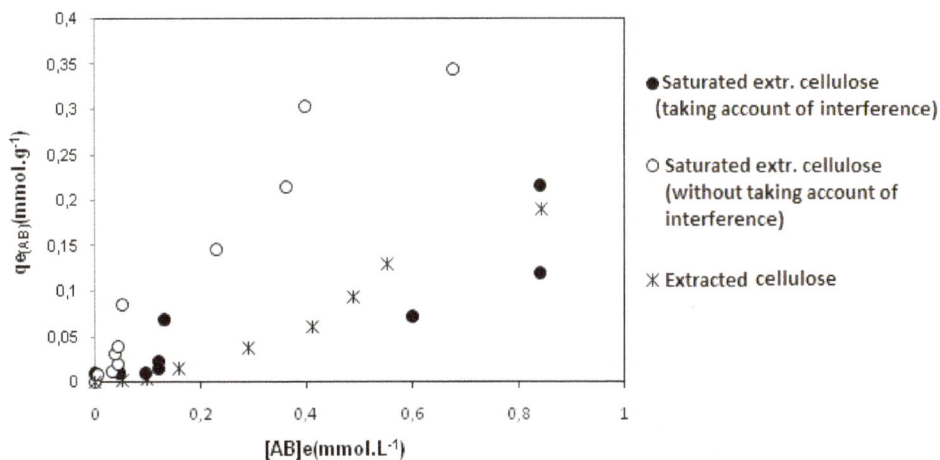

Figure 11: Comparison of adsorption isotherms of AB.

raw extracted cellulose (Ж), it is noted that the quantities of adsorbed AB onto extracted cellulose previously saturated with BB dye increased significantly. Thus, we can easily conclude that cationic dye (BB) helped to modify the surface state of the extracted cellulose by increasing the number of active sites for the anionic dye and thereby improving the adsorption. However, Figures 8-10 have demonstrated the effect of the presence of a dye on the spectrophotometric analysis of the other dye. Besides the probable BB release in the adsorption case of AB affects the analysis of the latter and will be the cause of wrong results interpretation. So that, the residual dyes solutions after adsorption were scanned over the entire visible range and the results are shown through Figure 12. The recorded spectra exhibit an absorption maximum at 665 nm and a shoulder at 610 nm (Figure 12a). The shape of these spectra corresponds to BB one, which proves the presence of the latter which is released during AB adsorption.

Moreover, Figure 12b shows that greater the initial AB concentration increases (which corresponds to the increase of the residual concentration) more the shape of the spectra is transformed with the appearance of a shoulder at 520 nm which appears on the dye mixture spectra (Figure 7). Thus, the identification of the solutions composition after adsorption was carried out by superposing the spectra obtained after adsorption (Figure 12) to those presented through Figures 8-10 and which correspond to the solutions at known composition. Consequently, it will be possible to determine AB adsorbed capacity and draw again the adsorption isotherm of the AB dye by taking into account the interfering presence of BB dye (Figure 11(●)).

Thus, Figure 11 highlights the error made by a direct reading of the AB absorbance neglecting the presence of BB and shows that adsorption was overstated. Indeed, comparison of AB isotherms onto raw extracted

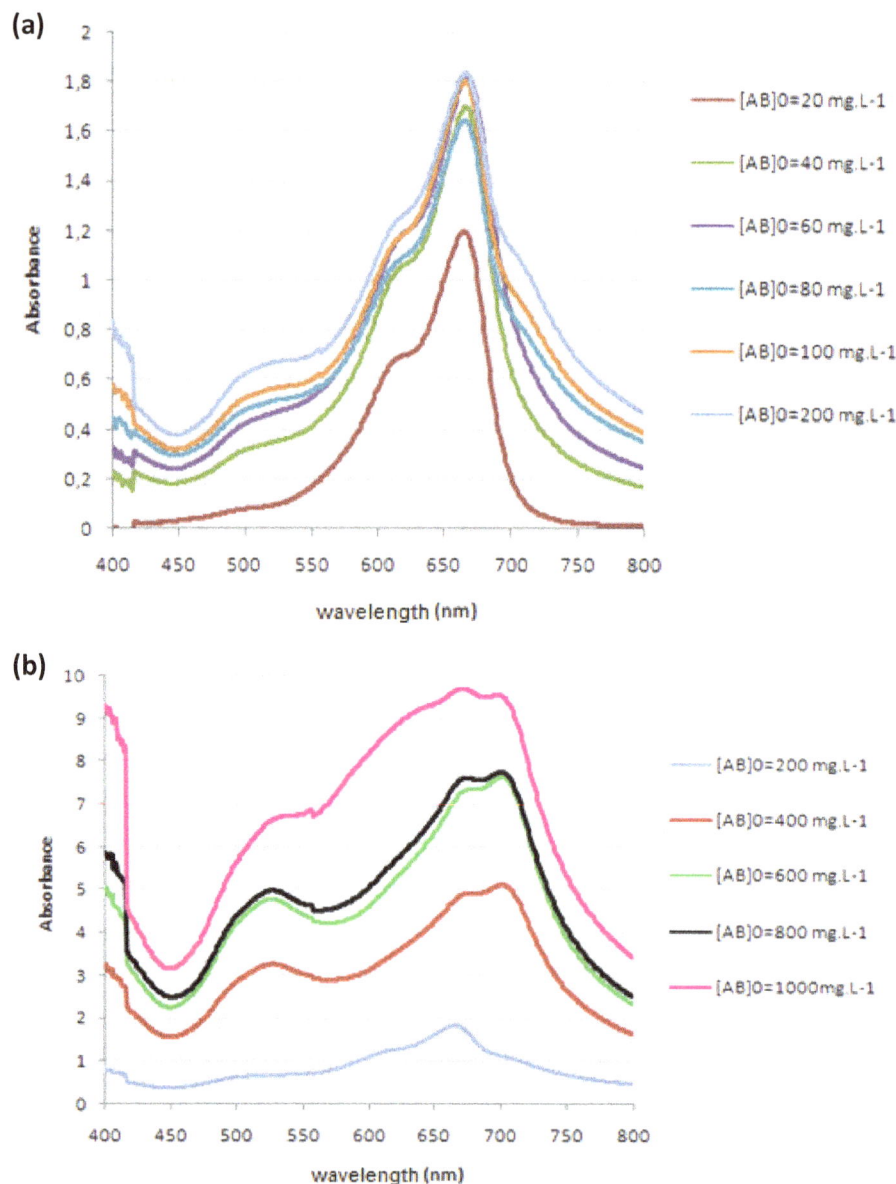

(a) 20 mg.L^{-1} < [AB]$_0$ < 200 mg.L^{-1} ; (b) 200 mg.L^{-1} < [AB]$_0$ < 1000 mg.L^{-1}

Figure 12: Spectral scanning of residual solutions after adsorption of AB onto the extracted cellulose saturated with BB dye.

cellulose and saturated one with BB dye by taking into account of the cationic dye release, shows quantitatively that adsorption of anionic dye has not enhanced as much as if each dye adsorbed on its own site. However, the shape of the isotherms appears slightly modified because the saturated surface state is different from the raw one.

A previous study [24], which investigated competitive adsorption of anionic and cationic dyes onto multiwalled carbon nanotubes showed that, in binary dye system, synergistic effect was observed. So cationic dye presented favorite adsorption on the sorbent due to hydrophobic effect and π–π bonds and authors demonstrate that the surface adsorbed cationic dye has a positive charge, which would serve as the sites for electrostatic attraction to negatively charged of the anionic dye. This result is concluded from macroscopic adsorption study, which shows an improvement in the adsorption of anionic dye in binary system compared to single one. However it should be noted that the author did not take into account the dyes mutual interaction in mixture, which leads to drawn wrong conclusions for the adsorption process.

Conclusion

The extracted cellulose from *P. oceanica*, locally available and low cost material, presented as good alternative adsorbents for Basic blue 9 (BB) and Acid blue 25 (AB) removal from aqueous solutions. Equilibrium data agreed well with Langmuir isotherm model with adsorption capacity of 0.955 and 0.370 mmol.g^{-1} for BB and AB, respectively. Extracted cellulose before and after dye adsorption was characterized by FTIR spectroscopy and XPS analysis indicating that adsorption of both dyes could be achieved mainly on external fibre surface with different adsorption interactions. It seems that electrostatic and hydrogen bonding involving respectively charged groups and OH groups takes place in the interaction between dye species and the fibre surface.

On the other hand, the adsorption of AB dye onto extracted cellulose previously saturated by the BB was investigated and showed that the determination of residual AB concentrations cannot be carried directly from the maximum wavelength of the dye, due to its interaction in solution with BB. Indeed, the formation of molecular complex detected at wavelengths 510 and 705 nm when the anionic (AB) and the cationic (BB) dyes were taken in equimolar proportions, has been revealed and has made it more difficult to determine the binary mixture composition after adsorption. The AB adsorption isotherm in this case was investigated and showed that the presence of BB on the surface affects the shape of the adsorption isotherm of the anionic dye but not the adsorbed amounts.

References

1. Kuo CY, Wu CH, Wu JY (2008) Adsorption of direct dyes from aqueous solutions by carbon nanotubes: determination of equilibrium, kinetics and thermodynamics parameters. J Colloid Interface Sci 327: 308-315.

2. Robinson T, McMullan G, Marchant R, Nigam P (2001) Remediation of dyes in textile effluent: a critical review on current treatment technologies with a proposed alternative. Bioresour Technol 77: 247-255.

3. Barragan BE, Costa C, Carmen Marquez M (2007) Biodegradation of azo dyes by bacteria inoculated on solid media. Dyes Pigments 75: 73-81

4. Zhou Y, Liang Z, Wang Y (2008) Decolorization and COD removal of secondary yeast wastewater effluents by coagulation using aluminum sulfate. Desalination 225: 301-311.

5. Gupta VK, Jain R, Varshney S (2007) Electrochemical removal of the hazardous dye Reactofix Red 3 BFN from industrial effluents. J Colloid Interface Sci 312: 292-296.

6. Hage R, Lienke A (2006) Applications of transition-metal catalysts to textile and wood-pulp bleaching. Angew Chem 45: 206-222.

7. Wang S (2008) A Comparative study of Fenton and Fenton-like reaction kinetics in decolourisation of wastewater. Dyes Pigments 76-714-720.

8. El-Qada EN, Allen SJ, Walker GM (2008) Adsorption of basic dyes from aqueous solution onto activated carbons. Chem Eng J 135: 174-184.

9. Hameed BH, Daud FBM (2008) Adsorption studies of basic dye on activated carbon derived from agricultural waste: Hevea brasiliensis seed coat. Chem Eng J 139: 48-55.

10. Aguir C, Khalfaoui M, Laribi N, Mhenni MF (2009) Preparation and characterization of new succinic anhydride grafted Posidonia for the removal of organic and inorganic pollutants. J Hazard Mater 172: 1579-1590.

11. Dridi-Dhaouadi S, Ben Douissa-Lazreg N, Mhenni MF (2011) Removal of lead and Yellow 44 acid dye in single and binary component systems by raw Posidonia oceanica and the cellulose extracted from the raw biomass. Environ Technol 32: 325-340

12. Guezguez I, Dridi-Dhaouadi S, Mhenni MF (2009) Sorption of Yellow 59 on Posidonia oceanica, a nonconventional biosorbent: comparison with activated carbons. Ind Crop Prod 29: 197-204.

13. Ncibi MC, Mahjoub B, Seffen M (2007) Kinetic and equilibrium studies of methylene blue biosorption by Posidonia oceanica (L.) fibres. J Hazard Mater B 139: 280-285.

14. Khiari R, Mhenni M F, Belgacem MN, Mauret E (2010) Chemical composition and pulping of date palm rachis and Posidonia oceanica - A comparison with other wood and non-wood fibre sources. Bioresour Technol 101: 775-780.

15. Aguir C, Mhenni FM (2006) Experimental study on carboxymethylation of cellulose extracted from Posidonia oceanica. J Appl Polym Sci 99: 1808-1816.

16. Boehm HP (1994) Some aspects of the surface chemistry of carbon blacks and other carbons. Carbon 32: 759-769.

17. Akdima O, Chamouna R, Demircia ub, Zaatarb Y, Khouryb A, et al. (2011) Anchored cobalt film as stable supported catalyst for hydrolysis of sodium borohydride for chemical hydrogen storage. International Journal of Hydrogen Energy 36: 14527-14533.

18. Fras L, Johansson LS, Stenius P, Laine J, Stana-Kleinschek K, Ribitsch V (2005) Analysis of the oxidation of cellulose fibres by titration and XPS, Colloids Surf. A: Physicochem Eng Aspects 260: 101-108.

19. Sun RC, Sun XF, Tomkinson J (2004) Hemicelluloses and their derivatives. ACS Symposium Series 864: 2-22.

20. Langmuir I (1918) Adsorption of gases on plain surfaces of glass mica platinum. J Am Chem Soc 40: 1361-1403.

21. Somani PR, Marimuthu R (2003) Thermal degradation properties of solid polymer electrolyte (poly (vinyl alcohol) +phosphoric acid)/methylene blue composites. Polym Degrad Stabil 79: 77-83.

22. Auta M, Hameed BH (2011) Preparation of waste tea activated carbon using potassium acetate as an activating agent for adsorption of Acid Blue 25 dye. Chem Eng J 171: 502- 509

23. Ben Douissa N, Dridi-Dhaouadi S, Mhenni MF (2014) Study of antagonistic effect in the simultaneous removal of two textile dyes onto cellulose extracted from Posidonia oceanica using derivative spectrophotometric method. J Water Process Eng 2: 1-9

24. Wang Sh, Wei Ng C, Wang W, Li Q, Hao Z (2012) Synergistic and competitive adsorption of organic dyes on multiwalled carbon nanotubes. Chem Eng J 197: 34-40.

15

Sustainability and Biotechnology – Natural or Bio Dyes Resources in Textiles

Carvalho C* and Santos G*

CIAUD, University of Lisbon, Portugal

Abstract

Local is global and global is local. Globalization changed the way we view society for the past decades and it presents advantages and disadvantages. Although to be global means an increased cultural intertwining or a higher flow of information and a social tolerance as well as the existence of a world market that enables productivity and accessibility they are also disadvantages related to the loss of cultural identity of certain cultures or sustainable issues that must be addressed.

On the other hand, this new century's challenges and issues are often strongly related to the usage of non-renewable resources and production procedures putting at risk the environment and people's health.

Regarding textiles, or its resources, the growing concern over environmental quality and users health has led to a gradual interest of the reintroduction of natural dyes (and preservation of biodiversity) into the fashion and textile design industries as opposed to the current production processes.

This study analyses the evolution of natural dyes and colour throughout the centuries focusing in the sustainability of textile industry and the conservation of biodiversity, local production and ancient knowledge on dyeing techniques. It also reveals that a revival of natural dyes (and ancient/local know how) in addiction to new cutting edge technologies (such as biotechnology) allows for an industrial feasibility. Results also indicate significant reduced environmental impact and new strategies for sustainable development regarding colours for textiles..

Keywords: Sustainability; Biotechnology; Textile dyeing processes; Natural dyes; Bio dyes

Introduction

Textiles and sustainability

Globalisation surely has many advantages - increased cultural intertwining (thus more acceptance and social tolerance), higher flow of information, vast market enabling production and so on. However, it has also created a set of challenges that must be addressed urgently [1].

People and lifestyle transform rapidly, and so their expectations. The past decades represent worried generations, increasingly aware of the significance of biodiversity's protection and the importance of ecological footprint's reduction [2].

The demand for natural products is highly observable in the market; there's an increase on organic goods' consumption, recycled and recyclable materials, non-animal tested products, etc. It is evident that the lifecycle of products, as well as their production processes, became a concern to exigent consumers, observant of sustainability issues that are presented to and from the industries [2,3].

Industries, in the other hand, are aware of the level of difficulty that represents attending to these necessary demands. Products integrating, simultaneously, culture, communities, environment, economy, green technologies and sustainable materials are a real challenge.

Textile industry is a vast world, with many materials and techniques employed. Involving one of the longest and most complex and difficult chains in manufacturing, it is one of the most pollutants sectors in the world. It contributes a great deal to poor labour conditions, non-renewable energy and water waste, contamination and environmental impact [1,4]. Some of its most problematic facets are the finishing processes such as dyeing [5]. Common dyeing processes involve the usage of fossil generated energy and heavy amounts of water. In addition, their effluents are a massive environmental concern when not conveniently treated before being released into natural waters [6].

Dyeing processes and challenges

Dyes represent a massive industry. Aware of the colour's influence over consumers, the textile and fashion industry explore this aspect in great detail, spending massive amounts of energy and money in the pursuit for the perfect colour. This is crucial so the product projects an intended message to increase sales [7,8]. Dyes' commercial availability is enormous and their process is one of the most fundamental aspects of textile industry commercial success; these must also be economical, available in high quantities and in diverse shades of colour. The higher demand for these compounds originated the synthesis of many millions of dyes in the past century [9].

Besides the pattern/printing, consumers demand for basic characteristics in textiles: high level of colourfastness regarding light, washings and perspiration. The colour must be uniform and of a solid shade throughout the substrate [9]. To guarantee these properties, the substances conferring colour to fibres must present high affinity with the substrates. These are factors depending on the substrate texture or composition as well as on treatments applied previously or after the dyeing process [6,9].

*Corresponding author: Carvalho C, CIAUD, Faculty of Architecture, University of Lisbon, Portugal, E-mail: cristina@fa.ulisboa.ptG. Santos CIAUD, Faculty of Architecture, University of Lisbon, Portugal E-mail: reinodeprata@gmail.com

Dyes do possess proprieties that render them susceptible to be manipulated or altered by an infinite number of chemical agents (mordants) Additionally, these properties are vital to help creating permanent bonding with fibres. These dyeing substances also differ from each other; they must be applied in different ways, through varied distinctive methods to produce colours in certain substrates [6,10]. Regarding dyeing application procedures, exhaust dyeing (batch), continuous (padding) and printing are amongst the most common used ones [9].

Dyeing matter possesses peculiar chemical properties that make it very distinct from other materials. This is the backbone of the dyeing process and the reason why colouring substrates is such a complex subject. There are, currently, thousands of different types of dyes, in the textile industry alone - a justified amount, as each fibre to be coloured requires dyes with specific features [11,12]. However, the extent usage of chemicals and water waste by the textile industry is an emergent ecological concern [6].

Regarding the usage of water, there are, normally, two types of wasted water during dyeing. One is the dye bath, which contains the remaining dye as well as other complex compounds that helped the bonding between colouring substance and fibre [9]. These residues and the amount of dye lost vary, depending on the type of dye used and variations of pattern and colour combinations. The other type is the wash/rinse water, a procedure to remove any excess dye present in the substrate. In addition to this, water is also needed to clean all the manufacturing components involved in the process [6].

These effluents are generally thrown into pure clean waters, representing one of the most critical environmental challenges. These contain hazardous substances that are easily able to reach reservoirs and water treatment stations. Some of the chemicals are harmful, toxic, carcinogenic, mutagenic, corrosive and irritant [6,13]. some are known hormone disruptors whilst others can affect the reproductive system [14]. Many of these do not break down in the environment, but instead build up in the body of animals and humans, creating mutations [6,13,15].

Sustainable Design

Bioresources – textile natural dyes

Designer and consumer's expectations towards sustainability in the textile field and their awareness of the issues surrounding current production processes triggered some consideration on matters such as climate, environment and health [3]. The preference for organic materials has, thus, increased. This has led to the reintroduction of ancient dyes in the market, natural dyeing colorants known for their biodegradable nature and less toxic features [16].

These ancient dyes are obtained through biological resources, usually plants or animals, and were used as far back as two thousand years ago. A dye is a coloured compound extracted only through physical-chemical (dissolution, precipitation, amongst others) or biochemical (fermentation) processes. This coloured substance must be soluble in an aqueous solution (dyebath) in which the material to be dyed is soaked in [9]. Natural dyes are perceived as safer due to their higher level of affinity with the environment, thus, causing less impact. Some can provide for high quality dyeing, great colourfastness properties and bright colour shades (applied with our without mordants).

Some of the known natural dyes also contain many properties appealing to the most of us, consumers. They are eco-friendly compounds, possessing medicinal advantages such as antimicrobial [17,18] and anti-inflammatory properties [19]. Besides, clothes dyed with natural dyes revealed a higher level of protection against UVR than the ones dyed with synthetic ones [20,21].

Natural colorants implementation

However, whilst nature is teeming with coloured compounds, not all that nature provides can be used in dyeing. Only a small percentage of these natural substances are applied to textiles - mostly deliver dull, uneven hues and poor fastness, when washed and exposed to light or perspiration. Most natural dyes still unspecified in terms of fastness properties and methods of extraction, production and application involved. As varied studies indicate, there's great need for research to overcome issues related to the implementation of natural dyes in modern dye houses, particularly regarding efficient dyeing recipes and their variations [22].

Additionally, the amount of dyestuff and colour shades provided is very limited. Dyeing with natural dyes is, consequently, highly costly. This class of colorants involve rather complex processes - long and difficult extraction methods and a high level of difficulty to produce and apply in order to obtain a quality dyeing. To better deal these aspects, varied studies suggest technologies such as ultrasound methods of extraction [23] and application [24], or less water usage procedures to reduce the ecological footprint of finishing processes [25].

Although all the natural dyeing substances' benefits and the significant interest of their re-introduction in the market, the challenges around scaling such compounds are complicated. Besides, transferring traditional natural dyeing methods to a modern dye houses require intricate experimentations or redesigning already implemented systems [26].

As mentioned, dyeing is a complex process involving chemical and physical occurrences. By the time of synthetics dyes' introduction to the market dyeing with natural dyes became an obsolete practise and most knowledge on techniques and procedures were lost. In addition, there is an extensive amount of data that has yet to be recorded; one still knows very little about what can biodiversity provide or which biological resources possess dyeing potential material. Recent studies prioritise the search for new biological resources, identifying new species of fauna and flora producing substances with dyeing potential [27]. The significance of such data lies on the hypothesis of certain isolated coloured compounds to be tested in textile fibres, to better evaluate their dyeing efficiency, hence applied to the industry. Again, further research on the subject is imperative and will be determinant to overcoming sustainability issues in the textiles sector.

These are a few reasons why natural dyes' usage is such a challenging subject for the industry. Although presenting many beneficial properties, its viability at a massive industrial scale is, so far, not easily attained. These demands call for finding alternatives in sustainable dyeing and radically new ways of creating and manufacturing materials as well as deeper scientific research.

Biotechnology Textile – Sustainable Materials

Alternatives in product design and manufacturing

Textiles' sustainability issues are gradually being exposed and debated. Legislation is emerging as more demanding and pollution control boards are progressively restricting guidelines for the textile industry. General targets are synthetic dyes'production, application and related effluents; their usage represents serious toxicological issues and a threat to the ecosystems and human health [28,29].

Until fairly recently, manufacturers focused attention on industry elements that would enable quick profit (maintaining final product's cost low or efficiency in production). Emerging design strategies in sustainable and responsible design question the current systems of manufacturing considering aspects like environment, consumers and technologies.

Some fashion designers and brands have adopted a conscious, attentive and critical position towards sustainable issues, integrating human well-being and green philosophies at the core of their corporate identity [15]. These companies, generally part of the slow fashion movement, provide design prepared to sustain communities, encourage and support local employment while, simultaneously, aiming towards environmental protection. They represent local production, raw materials such as natural dyes or fibres and traditional know how by producing and resourcing locally [2].

Environmentally, is of great significance that designers and manufacturers understand, in greater depth, the lifecycle and real capital of products being created. It is mandatory that designers/ brands generate strategic and tactical approaches to design process and question how can goods be projected and manufactured to better serve and suit consumers, and the planet. Aspects such sustainable improvement, environment, consumer's values, wishes and needs must be taken into account as well as the search for radical new production alternatives, to efficiently meet the requirements of this new century [30].

It is crucial to examine the possibilities of new substance or materials emerging as well as new manufacturing methods and potential impact on the world. It is no longer only about resources exploitation but also about providing, through products, wellbeing and social improvement. Furthermore, is about providing consumers with highly smart and eco-friendly products and materials; creating goods or textiles which lifecycle enables to adapt accordingly to the consumer's characteristics and inevitable changes - age, shape, taste, needs, etc. [31].

The scope of science currently allows for a completely new radical way of rethink materials or for innovative methods of producing. With increasingly more technological innovation and breakthroughs as well as collaborations between scientists and designers the world has been noticing the intensification of biotechnologies as the foundation (green strategies) of many sustainable design projects [32].

Bio-cooperation - biotechnology impact on the world

Biotechnology application in textiles dates back over two thousand years - from fibres to natural dyes [33]. Kandra et al. [34] more recently, the management of residual waste, by microbes Novotny et al. [35]. The leap from classic to modern biotechnology involved only the ideal innovative tools to discover different uses and functions of bio resources, thus allowing for the improvement of this new field [36]. These sophisticated technologies' evolution is accentuated through significant advances in genetic engineering and synthetic biology, where the use of living organisms are seen as a biological process that may replace industrial or mechanical systems [37].

Synthetic Biology is, in broader terms, the engineering of biology. Its purpose is to make biology available to the requirements of everyday life, considered as highly effective to overcome environmental issues and tackle pollution [38].

Bio resourced technologies' increase is rapidly transforming the world with its infinite number of applications and possibilities; biotechnological processes play a key role in increasing and promoting sustainable production. In the Design or Textiles fields, biotechnologies' practices are normally associated with sustainable development and green manufacturing processes - pollution control and prevention, resources conservation, cost reduction. This approach enables for a new radical way of rethinking materials [38] as it consists in building new biological functions or systems or re-designing existing natural ones (Synthetic biology org). The production of biomaterials is a significant impact of this technology [38].

Textile and fashion designers are looking to science as part of the creative process, producing unique and often surprising results [31]. Cut edge and complex technologies serve better apparel industry in terms of quality innovative garments. Although most technology is considered underexplored, the recent trend of extensive use of biomaterials in product and fashion garments is growing exponentially, given their possibilities.

Bioresourced colorants – natural and bio dyes

As mentioned, natural dyes possess many benefits, particularly for consumer's health and the planet. However, their extraction, production, application and implementation are rather complex, not to mention the lack of knowledge surrounding techniques and bioresources. Nevertheless, some issues regarding their usage might be attenuated through cut edge, sophisticated, technologies such as modern biotechnology. Additional sustainable alternatives to toxic synthetic dyes consist in the use of fibres possessing natural colour (classic biotechnology) or modified to do so (modern biotechnology).

We are already familiarised with natural colorants to dye (from fauna and flora) or naturally dyed cotton fibres (pale shades of brown, beige, green, red), both used since ancient times [39]. There's also reference in the literature to silkworms producing naturally wide-ranging coloured silk fibres, when manipulating their diet or environment [40]. Another example can be analysed in the Madagascar's origin Orb Weaver Spider producing silk fibres that are naturally golden dyed [41].

Such choices are effective to decrease energy and water waste, saving fibres from most of textile finishing procedures. Other naturally dyed fibres consist in genetic engineered experiments [33], with varied coloured silk fibres being produced.

Following the biological technologies approach, there's also the production of coloured compounds through a rather unusual mediums – microorganisms or microalgae [42]. This is a phenomenon that occurs naturally in nature, often found in different plants' microsystems (rhizosphere) or glaciers [43] Manipulating at times their environment, different colour shades can be produced [31,44].

This mechanism, using living bacteria to produce substances, might be considered as a sustainable strategy for dyes' mass manufacture, creating less-environmentally damaging materials [45].

Its effectiveness finds in Synthetic biology a way to enable the production of specific dyes designed to meet the industry's demands and consumer's expectations. Microorganisms can, not only grow rapidly but also, be programmed to provide varied dyes with different properties and different colours, being simultaneously cost effective.

Some pigments created through bio machines were already isolated in lab. "E-chromi" is one of such projects, born from the collaboration of scientists and designers who genetically engineered bacteria to secrete a variety of coloured pigments, visible to the naked eye. Standardised sequences of DNA (biobricks) were designed and inserted *into E.coli* bacteria enabling for the production of colours such as red, yellow, green, blue or violet, possibly allowing for its application on textiles [45].

The usage of sophisticated technologies, such as Synthetic Biology, contributes to profound transformations in the sector and to the concept of Design itself. Its impact will, surely, invite new scenarios and generate vital debating [32]. As we are slowly capable of manipulating nature and build things with biology, we are increasingly witnessing a new era in Design, manufacturing products and producing materials that empowers for a more sustainable future [45].

Conclusion

It is a challenging period for the textile industry, with the economic downturn threatening sales and a growing awareness of real social and environmental challenges, such as climate change, wars over resources and increasing consumer's expectations of brands.

One of the most complex aspects of the industry is the dyeing process. It is one of the most pollutant components of textiles with a heavy ecological footprint. Toxicity, water waste and contamination, non-renewable generated energy consumption, health hazards for humans and ecosystem in general, etc.

Awareness on environmental issues has led to the interest of natural dyes reintroduction in the market. Although having many benefits, the reality is that their implementation on a massive industrial scale is rather incompatible to the needs of manufacturing. Industry demands for a great variety of colour shades as well as high quality colourfastness as well as economical dyes - vital requirements for commercial success. Additionally, natural colorant dyeing techniques, and bio-resources involved, lack deeper research. Nevertheless, with the significant improvements on biotechnologies witnessed during the last decade, some fundamental alternatives might be considered.

The appropriation of biotechnologies exerts increasing influence in our daily lives. Technological innovation and breakthroughs in textiles are establish to meet a variety of objectives such as improvement of varied species of plants used in the manufacture of fibres or their properties, production of new types of fibres, different types of dyes, effluents' management, amongst others. Environmental biotechnology (white and brown biotechnology) and specially Synthetic Biology, plays a key role in increasing and promoting sustainable production.

Textile and fashion designers are looking to science as part of the creative process, producing unique and often surprising results [31] (En Vie, 2013; Grow your own, 2013). Cut edge and complex technologies serve better apparel industry in terms of quality innovative garments; the recent trend of extensive use of biomaterials in product and fashion garments is growing exponentially, given their possibilities. Their contribution is vital to sustainable development or green manufacturing processes – prevent pollution, reduce costs and resource's conservation.

Sometimes introducing visionary and radical strategies for improving the performance of objects around, biotechnology multiple applications are focused and designed to deal with sustainable development and core industrial issues. The possibilities of such technologies suggest different approaches on responsive design assimilating science and technology, the environment, sustainable strategy, wellbeing and social innovation.

Acknowledgments

The authors would like to thank the following institutions: FCT (Foundation of Technology and Science-Portugal)CIAUD (Research centre for Architecture, Urbanism and Design)University of Lisbon, PT-Faculty of ArchitectureUniversity of Leeds, UK-School of Design.\

References

1. Klein N (2000) No logo: Taking Aim at the Brand Bullies. Canada.

2. Awamaki lab (2012) Consultado em Maio.

3. Stoddar R (2014) Who are these purpose-driven consumers.

4. El-Hagar S (2010) Sustainable Industrial Design and Waste Management: Cradle-to-Cradle for Sustainable Development. Elsevier Academic Press, USA.

5. Fletcher K (2008) Sustainable fashion and textiles. Routledge Publication, London.

6. Malik A, Ghromann E, Akhtar R (2014) Environmental deteriorism and human health: natural and anthropogenic determinants. Springer.

7. Davis F (1994) Fashion, culture and identity. University of Chigaco Press, London.

8. Scully K, Cobb DJ (2012) Colour Forecasting for Fashion. Laurence King Publishing.

9. Clark M (2011) Handbook of textile and industrial dyeing: Principles, processes and types. Woodhead Publishing, limited. Cambridge, UK.

10. Bancroft E (2008) Experimental researches concerning the philosophy of permanent colours: and the best means of producing them, by dyeing, calico printing, etc. (1st edtn) T, Dobson, Universidade de Harvard.

11. Zollinger H (1991) Color Chemistry. (3rd edtn), VCH Publishing, Newyork, USA.

12. Zollinger H (2003) Color Chemistry: Synthesis, Properties and Applications of Organic Dyes and Pigments. (3rd edtn) Weinheim: Wiley-VHCA, Germany.

13. Christie RM (2007) Environmental aspects of textile dyeing. Woodhead Publishing, England.

14. (2011) Unravelling the corporate connections to toxic water pollution in China.

15. Clarke EA, Anliker R (1980) Handbook of Environmental Chemistry Chemical Safety.

16. Glover B (1995) Are natural colorants good for your health? Are synthetic ones better? Textile Chemistry Colorist 27: 17-20.

17. Prusty AK, Das T, Nayak A, Das NB (2010) Colourimetric analysis and antimicrobial study of natural dyes and dyed silk. Journal of Cleaner Production 18: 1750 -1756.

18. Singh R, Jain A, Panwar S, Gupta D, khare SK (2005) Antimicrobial activity of some natural dyes. Dyes and Pigments 66: 99-102.

19. Hamburger M (2002) Isatis Tinctoria- From the discovery of an ancient medicinal plant towards a novel anti-inflammatory phytopharmaceutical. Phytochemistry Reviews 1: 333-344.

20. Kozlowski R, Zaikov GE, Pudel F (2006) Renewable resources and plant biotechnology. Nova Publishers, Newyork, USA.

21. Hustvedt G, Crews PC (2005) Textile technology - the Ultraviolet Protection Factor of Naturally-pigmented Cotton. Journal of Cotton Science 9: 47-55

22. Bechtold T, Turcanu A, Ganglberger E, Geissler S (2003) Natural dyes in modern textile dyehouses-how to combine experiences of two centuries to meet the demands of the future? Journal of Cleaner Production 11: 499-509.

23. Sivakumar V, Vijaeeswarri J, Anna JL (2011) Effective natural dye extraction from different plant materials using ultrasound. Industrial Crops and Products 33: 116-122.

24. Vankar PS, Shanker R, Dixit S, Mahanta D, Tiwari SC (2008) Sonicator dyeing of modified cotton, wool and silk with Mahonia napaulensis DC and identification of the colorant in Mahonia. Industrial crops and products 27: 371-379.

25. http://www.dyecoo.com/

26. Leitner P, Fitz-Binder C, Mahmud-Ali A, Bechtold T (2012) Production of a concentrated natural dye from Canadian Goldenrod (Solidago Canadensis) extracts. Dyes and Pigments 93: 1416-1421.

27. Jasmin Malik Chua (2013) Silkworms Fed on "Green" Dyed-Leaf Diet Spin Naturally Colored Silk.

28. Clarke EA, Steinle DJ (1995) Dyes Color. J Soc.

29. Vandevivere PC, Bianchi R, Verstraete W (1998) Treatment and reuse of wastewater from the textile wet-processing industry: Emerging Technologies. Journal Chemistry Technology Biotechnology 72: 289-302.

30. Niinimaki K, Hassi L (2011) Emerging design strategies in sustainable production and consumption of textiles and clothing. Journal of Cleaner Production 19: 1876-1883.

31. http://thisisalive.com

32. Myers W (2012) Bio-Design. Thames and Hudson Publishing, London.

33. http://www.biocouture.co.uk

34. Kandra P, Challa MM, Jyothi HKP (2012) efficient use of shrimp waste: present and future trends. Applied Microbiology Biotechnolgy 93: 17-29.

35. Novotny C, Svobodová K, Benada O, Kofronová O, Heissenberger A et al., (2011) Potential of combined fungal and bacterial treatment for color removal in textile wastewater. Bioresource Technology 102: 879-888.

36. Church MG, Regis E (2014) Regenesis: how Synthetic biology will reinvent nature and ourselves. Published by Basic Book.

37. http://www.synbioproject.org/topics/synbio101/definition

38. Schmidt M (2012) Synthetic Biology: Industrial and Environmental Applications. John Wiley and Sons, London UK.

39. Vreeland JM (1999) the Revival of Colored Cotton. Scientific American 280: 112-118.

40. Bridgette Meinhold (2011) New Silkworm Diet Naturally Dyes Silk, Reduces Water Consumption. Eco-Textiles.

41. Josephine Moulds (2015) Can big brands catch up on sustainable fashion?

42. Lu Y, Wang L, Xue Y, Zhang C, Xing XH, et al. (2009) Production of violet pigment by a newly isolated psychrotrophic bacterium from a glacier in Xinjiang. Biochemical engineering Journal 43: 135-141.

43. Chua JM (2015) London Designer Dyes Silk Scarves With Living Soil Bacteria. Eco-Textiles.

44. Zhao H (2013) Synthetic biology tools and applications. Elsevier/Academic Press, Boston.

45. Ginsberg AD (2015) Design Pour La Sixième Extinction.

16

Novel Photo-Fenton Oxidation with Sand and Carbon Filtration of High Concentration Reactive Dyes both with and without Biodegradation

Jablonski MR[1]*, Ranicke HB[2], Qureshi A[3], Purohit H[3], Reisel JR[2] and Satyanarayana KG[4]

[1]Department of Civil Engineering and Mechanics, University of Wisconsin, Milwaukee, 3200 N. Cramer St. Milwaukee, WI 53211, USA

[2]Department of Mechanical Engineering, University of Wisconsin, Milwaukee, 3200 N. Cramer St., Milwaukee, WI 53211, USA

[3]Environmental Genomics Division (EGD), National Environmental Engineering Research Institute (NEERI), Nagpur India, Nehru Marg, Nagpur, 440020, India

[4]Honorary Professor, Poornaprajna Institute of Scientific Research (PPISR), Sy. No. 167, Poornaprajnapura, Bidalur Post, Devanahalli, Bangalore-562 110, Karnataka, India

Abstract

There is an increasing need to provide rural textile dye operations in developing nations with an effective and low-cost method to clean dye wastewater. Such operations often have no choice in the location of their wastewater disposal due to their lack of funds and influence in the industry, resulting in wastewater disposal that is detrimental to environmental safety. Photo-Fenton oxidation, an advanced oxidation process used to degrade low-concentration textile dye wastewater, has shown promise using expensive chemicals in laboratory-scale projects. Aerobic biodegradation, a common biological treatment method used in large-scale low-concentration textile industrial applications, generates large amounts of hazardous biological waste. This paper presents successful decolorization of high-concentration reactive dye wastewater using a wide range of temperatures and solar irradiances in two locations for the first time. To fully degrade dye wastewater, full oxidation times combined with sand and carbon filtration rank more important than different iron surface areas. UV-visible spectrometry, GC/MS, and ICP techniques along with measured COD levels were used to support these findings. This study is expected to provide a low-cost method to clean high concentration dye effluent as it deals with testing sustainable decolorization of textile dye wastewater using photo-Fenton oxidation and sand and carbon filtration in a reactor and the filter that is constructed of recycled rusty metal and locally available sand.

Keywords: Reactive dye; Photo-fenton oxidation; Biodegradation; Filtration; Scrap iron

Introduction

Clothing is an essential part of daily human life worldwide, but rarely any thought is given to how clothing is actually made. Before fabric is sewn into clothing, it must be dyed. Much of the cloth manufacturing process is often outsourced to developing countries, where it is frequently located in rural communities. Rural dyeing communities exist in many countries throughout the world, including India, Morocco, Mexico, and Turkey. In these areas, the heavy pollution caused from the industry is not readily seen, but still is damaging to the environment. For example, India is the second-largest global producer of silk, employing over one million families, where the majority of silk dyeing is accomplished in small villages, in which a typical community contains small-scale silk dyeing businesses owned and operated by local families [1]. In central Bangalore, there are approximately 150 silk dyers that are currently using no wastewater treatment as well as many other small communities of cottage-scale silk dyers surrounding the city. Most of the operation of dyeing silk saris there occurs in small communities that are then saddled with environmental problems due to the high levels of chemicals in the dyes. The fundamental problem is that once the fabrics are dyed, the dye waste is discharged back into the community's water source or thrown directly onto the ground, where it can seep into the groundwater that feeds the local water systems. Many of the residents in these communities are very poor and are forced to use the contaminated water for their daily needs.

Reactive dyes have a very high stability that sustains their color once the textile is washed. As a result, they are used frequently in the silk dyeing industry and account for 32% of all dye operations. In these operations, only 70% of the dye is absorbed during a dye job, leaving 30% to enter the waste stream [2]. Considering that the homespun silk

industry in places such as rural Bangalore currently use no treatment for their dye wastewater prior to discharging it into streams or onto land, the typical textile wastewater entering the groundwater of the area contains a pH ranging between 7-9, Biological Oxygen Demand (BOD) 80-6000 mg/l, Chemical Oxygen Demand (COD) 150-12000 mg/l, Total Suspended Solids (TSS) 15-8000 mg/l, Total Dissolved Solids (TDS) 2900-3100 mg/l, chloride levels between 1000-1600 mg/l, total Kjeldahl Nitrogen from 70-80 mg/l, and color from 50-2500 color units [3].

Dye effluent has very complex molecular composition that when discharged to the natural environment can cause irreparable harm to ecosystems. The total discharge of dye effluent is more than 0.01 metric tonnes per annum because of the high production rate of the industry. This pollution, if left untreated, thereby creates a huge impact on flora and fauna. Reactive red dyes are types of azo dyes that continue to be used in some countries in great quantities although they have been banned in many developed countries of the world. These dyes have one or more double nitrogen bonds in their chemical structure bonded to aromatic rings; this is one of the most difficult synthetic linkages

*Corresponding author: Jablonski MR, Department of Civil Engineering and Mechanics, University of Wisconsin, Milwaukee, 3200 N. Cramer St., Milwaukee, WI 53211, USA, E-mail: jablons5@uwm.edu

to break [4,5]. It is also reported by these researchers that Fenton's reagent, discovered a century ago, shows great promise in breaking these environmentally-destructive bonds through using hydroxyl radicals. However, the use of iron as a catalyst in the reaction tends to create additional unforeseeable pollutants in solution. It is also reported that removal of azo dyes from solution is difficult, but the reaction that forms hydroxyl radicals removes 85% of COD and complete color removal in 20-40 minutes [6].

Residents of rural areas would certainly benefit from reductions in the contamination by the dye effluents generated in their localities. For example, families in rural Bangalore expressed great interest in finding a system that would allow them to economically treat the dye wastewater [Personal Communications]. It is logical to conclude that there is need to develop a suitable methodology to meet the aspirations of these rural families throughout the world. Accordingly, to help achieve this goal, the authors took up this task in several stages (objectives), first with the proposal to test two methods to clean contaminated dye wastewater using very affordable materials that can be easily found in and around the homes of the dyers.

Large and medium dyers who produce 40,000-3,000,000 liters per day of dye effluent can afford treatment to include settling tanks, biodegradation, drying beds, sand and carbon filtration, reverse osmosis systems, etc. that costs between 0.003-0.005 USD per liter [Personal Communications], not including exorbitant infrastructure and equipment costs (including reverse osmosis) that exceed 30,000 USD. Small cottage-scale industries do not have the luxury or community support to get such a discount on their treatment. The aim of this study however, is still to create a low-cost treatment option that includes reasonably priced infrastructure. Table 1 displays a comparison of infrastructure and treatment costs of large and medium dyers compared with the sustainable photo-Fenton oxidation and filtration method for small-scale dyers. Preliminary work has been done to develop a system for a small-scale dyer. The costs determined in that project are reflected there.

In the present study, results of the treatment of Reactive Red 120, Reactive Yellow 81 and Reactive Blue 4 in laboratory scale experiments are presented. The two forms of treatment used were sustainable photo-Fenton oxidation with sand and carbon filtration and sustainable photo-Fenton oxidation with sand and carbon filtration coupled with biodegradation. Both of these techniques have been shown to be affordable, efficient methods of decolorizing dye wastewater using acid, hydrogen peroxide, iron scrap, and sunlight [7-11]. This work is the first of its kind to use (i) a 5 g/l dye concentration equivalent to effluent from cottage-scale textile dye industries, and (ii) sand filtration to raise the pH back to neutral and to separate the sludge

formed during the oxidation process from the aqueous solution. To meet the objective to help poor communities in developing countries through the development of an economically viable and easily adoptable process, a system was designed and constructed from readily available, inexpensive (or scrap) materials. Accordingly, the laboratory study used sheet metal as its iron source, to create a rusty sheet metal container for a future demonstration project in rural places of not only India, but in other countries of the world.

With the success of this new technology and proper education, rural populations will be able to clean their dye effluent before discarding it to the environment, thereby reducing soil and ground water pollution. Alternatively, water could be reused in future dye batches thus decreasing their overall water consumption. It is expected that this laboratory-scale project will be successfully extended to a demonstration-scale project.

Experimental Methodology

The laboratory-scale treatment process encompassing eight stages was carried out both in India (National Environmental Engineering Research Institute (NEERI), Nagpur - hereafter referred to as 'India') and in the USA (University of Wisconsin-Milwaukee (UWM), Milwaukee, WI, USA). The Reactive Dye wastewater is (i) created in a 1-liter pot, (ii) acidified to pH 2.5-3.0 with hydrochloric acid, (iii) stirred and infused with pharmacy-grade 6% hydrogen peroxide (quantity listed in Tables 2 and 3), (iv) poured into the mobile photo-Fenton oxidation reactor that contains multiple pieces of rusty metal on its base, (v) left open to the sun for partial/full oxidation time (time listed in Table 4), (vi) poured through a sand filter to separate the iron/dye sludge from solution, (vii) filtered through activated carbon to absorb any remaining organics, and (viii) tested for color change, temperature, acidity (pH), concentrations of COD, any other organic material present, metals and ions, and biodegradability. It should also be noted that six main variables were used throughout the study to test their effect on treated water quality: (a) decolorization of different colors of dye (red, yellow, and blue) at a concentration of 5 g/L, (b) angle of incoming natural UV sunlight (India or USA), (c) volume of hydrogen peroxide (optimized according to color), (d) quantity of rusty iron (high iron of 645 cm^2, medium iron of 323 cm^2, low iron of 161 cm^2), (e) time of oxidation (partial or full), and (vi) filtration (none, sand, sand and carbon).

Materials employed

Dyes: Following literature indicating that reactive dyes have been the leading silk dyes used by industry [2], this study considered three such dyes, Red 120, Yellow 81 (also known as Procion Yellow H-E3G),

Size of Dyer	Type of Treatment	Initial Infrastructure Cost (USD)	Cost per Liter Treatment (USD)	Liters per Day to be Treated in System	Daily Cost (USD)	Effluent Content	Treatment Level (Complete or Pretreatment)
Large-scale [Personal Discussions]	2 settling tanks, and 2 reverse osmosis units	NA	0.005	3,000,000	15,000	Zero-discharge, easily reuseable for new dye batches	Complete
Medium-scale [Personal Discussions]	Settling tanks, biodegradation, sand/carbon filtration, drying beds	30,000	0.003-0.005	40,000	133-200	Permissible levels of organics and inorganics for industrial disposal	Complete
Small-scale [Preliminary Small-Scale System]	Sustainable photo-Fenton Oxidation Rusty container and Cotton/Sand/Carbon Filtration Unit	50	0.026	1,000	26	High conductivity, safer for dumping compared with untreated but requires reverse osmosis for complete treatment	Pretreatment

Table 1: Comparison of infrastructure and treatment costs of large and medium dyers compared with the sustainable photo-Fenton oxidation and filtration method for small-scale dyers.

Dye Type	Dye Color	Iron Quantity	6% Hydrogen Peroxide Volume (mL)
Reactive	Red	Low	250
Reactive	Red	Medium	150
Reactive	Red	High	100
Reactive	Yellow	Low	250
Reactive	Yellow	Medium	150
Reactive	Yellow	High	100
Reactive	Blue	Low	500
Reactive	Blue	Medium	300
Reactive	Blue	High	200

Table 2: Amount of 6% hydrogen peroxide used with various dye colors, types, and amounts of rusty iron.

Oxidation Time (Full or Partial)	Filtration Type	Iron Surface Area (High, Medium, Low)	Abbreviation
Full	None	High	R1H
Full	Sand	High	R2H
Full	Sand and Carbon	High	R3H
Partial	None	High	R4H
Partial	Sand	High	R5H
Partial	Sand and Carbon	High	R6H
Full	None	Medium	R1M
Full	Sand	Medium	R2M
Full	Sand and Carbon	Medium	R3M
Partial	None	Medium	R4M
Partial	Sand	Medium	R5M
Partial	Sand and Carbon	Medium	R6M
Full	None	Low	R1L
Full	Sand	Low	R2L
Full	Sand and Carbon	Low	R3L
Partial	None	Low	R4L
Partial	Sand	Low	R5L
Partial	Sand and Carbon	Low	R6L

Note: If iron surface area is not taken into account, the notation R1*-R6* will be used.

Table 3a: Abbreviations of sample names for experiments.

Oxidation Time (t_0 or Full)	Added Chemicals (None, HCl, HCl and H_2O_2)	Abbreviation
t_0	None	C1
t_0	HCl	C2
t_0	HCl and H_2O_2	C3
t_F Full	HCl	C4
t_F Full	HCl and H_2O_2	C5

Note: Control samples used no filtration.

Table 3b: Abbreviations of control sample names.

and Blue 4 for analysis of the performance of the treatment process. Other factors in favor of their use include ease of procurement and the ability to compare with other dye oxidation studies. Molecular masses of these dyes are 1469.98 g/gmol, 1632.18 g/gmol and 637.43 g/gmol, respectively, while their CAS numbers were 61951-82-4, 59112-78-6 and 13324-20-4, respectively. Normal tested or documented wavelengths of these dyes are at 535 nm for Red 120 [12], documented maximum wavelength of Yellow 81 were 270 nm and second at 356 nm [13] and finally, that of Blue 4 was at 595 nm [14].

The Linear chemical formulae of these dyes are $C_{44}H_{24}Cl_2N_{14}O_{20}S_6Na_6$, $C_{52}H_{34}Cl_2N_{18}Na_6O_{20}S_6$, and $C_{23}H_{14}Cl_2N_6O_8S_2$, respectively, while their chemical structures are shown schematically in Schemes 1a-1c.

Water, acid, and oxidant: Tap water was used to dissolve all dyes. Since water quality varies throughout the world and the difference between the water quality in the USA and India and within India (irrespective of a rural community or a city) alters dye samples with enough variability, one could expect large variations in the results obtained.

Hydrochloric acid (HCl) of approximately 1M was chosen as the acidifier to bring the dye solution to a pH of 2.5-3.0. HCl was used due to its ready availability throughout India. It is well known that HCl can react with the oxidation chemicals and the dye intermediates thus producing some unwanted byproducts such as chlorides. Sodium levels are often high in treated dye effluence because of the strong attraction between chloride and sodium ions, creating sodium chloride. Sodium chloride is very soluble in water and will remain in solution even after filtration. Although chloride levels from HCl are high, possible byproducts resulting from acidification using other acids are much more obtrusive.

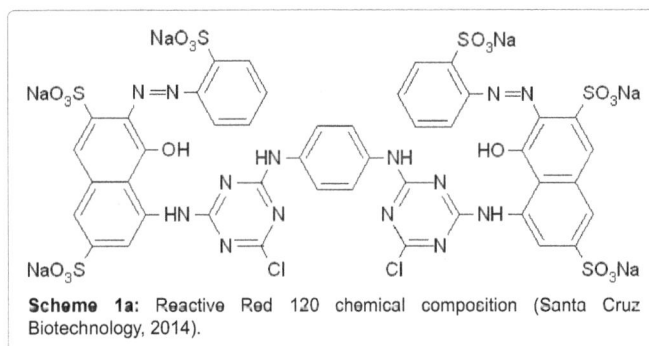

Scheme 1a: Reactive Red 120 chemical composition (Santa Cruz Biotechnology, 2014).

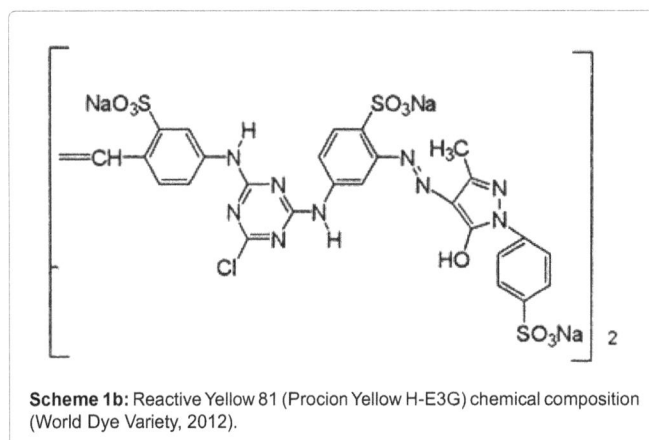

Scheme 1b: Reactive Yellow 81 (Procion Yellow H-E3G) chemical composition (World Dye Variety, 2012).

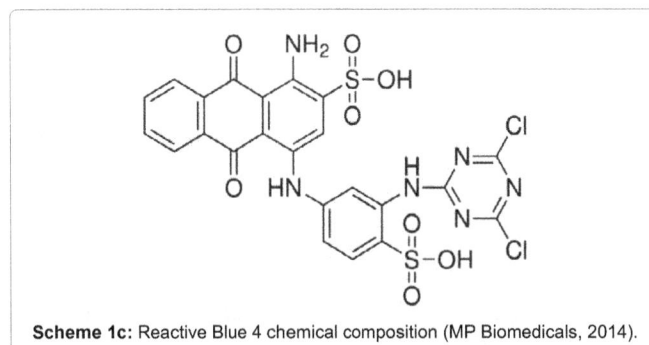

Scheme 1c: Reactive Blue 4 chemical composition (MP Biomedicals, 2014).

Hydrogen peroxide was used as an oxidizer in the photo-Fenton process in the present study. A 6% solution, which is 1.76 M, was used due to its low cost and availability to small-scale dyers. Table 2 shows the amount of 6% hydrogen peroxide used with various dye colors and amounts of rusty iron.

Sand and carbon filters: A sand filter for the 1-liter experiments carried out in the present study was made of readily-available construction sand from the area and required approximately 0.5 m depth of sand with a diameter of 40 cm. In the case of the laboratory experiments in the USA, the fully oxidized samples were first filtered, followed by the partially oxidized sample; after this, the sand was changed. The reason for this is because un-oxidized dye color permanently dyes the sand in the filter and cannot be removed without using excessive amounts of water.

Further, activated carbon was used after sand filtration with the expectation that it will adsorb any residual color or chemical intermediate in solution after oxidation and sand filtration. The carbon filter was rinsed twice between filtration of different samples.

Following one of the philosophies of the present study that the process developed for the treatment of effluent of dyes to be economical, activated carbon was initially used in the laboratory experiments, although one may think of using wood charcoal instead of more expensive activated carbon in future studies for the field experiments.

Methods

By mimicking the cottage-scale silk dyers' wastewater in laboratory-scale studies, the intent was to augment dyers' processes with a low-cost treatment option. For this reason, the following experimental methodology was chosen. (1) Reactive Dyes at 5 g/L concentration were mixed with household hydrochloric acid (Lion Brand toilet bowl cleaner cost 0.80 USD per 1-L). (2) 6% Pharmacy-grade hydrogen peroxide (0.37 USD per 100 mL) was poured into the solution in containers holding rusty metal from old yogurt containers (free as they were found in dump). (3) Natural sunlight at ambient temperature was employed. (4) Following treatment, sand and carbon filtration of the dye effluent was used (container cost minimal may be found in dump, sand found in local environment). Cost is limited to chemical consumption and ranges per liter dye effluent from 0.37-1.84 USD for peroxide and is minimal for acid at approximately 0.01 USD for 10 mL/L, totaling 0.38-1.85 USD per liter. Although this is significantly higher than the cost of effluent treatment for a large-scale dyer mentioned previously and detailed in Table 1 (0.005 USD per L), the infrastructure for the sustainable photo-Fenton oxidation and filtration method here uses recycled scrap materials with minimal costs as compared to the expensive reverse osmosis and biological and chemical settling tanks at the large-scale operation.

Photo-Fenton oxidation studies: It is known that variables that affect the photo-Fenton oxidation process are the (i) UV light source (sunlight angle in India or the USA), (ii) color of dye (red, yellow, or blue), (iii) water source for mixing dyes, (iv) acid used to bring pH to 2.5-3.0 (here hydrochloric acid is used) (a pH increase of 1 generally occurs during oxidation, all samples require filtration to separate sludge from liquid, and sand raises pH further as is noted later), (v) hydrogen peroxide volume added as the main oxidizing agent, (vi) surface area of rusty iron pieces as the chemical catalyst (high, medium, low), (vii) time (partial or full oxidation), and (viii) filtration (none, sand, or sand and carbon). Accordingly, photo-Fenton research was carried out using direct sunlight during the summer/fall months in

USA and summer/start of rainy season in India. Solar radiation was measured using CEM's DT-1307 Solar Power Meter measuring up to 1999 W/m^2 or 634 BTU/ft^{2*}hr (CEM Instruments, 2012). Milwaukee also has a website that reports measured solar irradiance [15].

Reactor: Considering the outcome of this study was intended to function in small-scale cottage industry silk dyers, a reactor was fitted with rusty metal sheets cut from the sides of old yogurt containers found in an open dumping site in India, where 50% of the studies were performed. As this work was intended to function for small-scale cottage industry silk dyers, the materials (the rusty metal and sand) used were as found in the field.

In order to optimize the time to stop the photo-Fenton oxidation reaction, experiments were carried out by taking a 5 mL sample every 5 minutes during oxidation, adding 1 drop of 0.1M sodium hydroxide to raise the pH above neutral (9 for Reactive Red 120, 9.37 for Reactive Yellow 81, and 8.86 for Reactive Blue 4). The undiluted samples were then analyzed using UV-visible spectrometry to find the time when the color became partially degraded by watching the peak height of the color and any organics left in solution decrease with increased oxidation time. Full oxidation time was chosen when all peaks on the spectral analysis decreased to zero. Partial oxidation time was chosen when the peaks on the spectral analysis were visibly separated from each other and could each be identified as an individual chemical peak.

All the samples listed in Table 3a were then tested for the three dye colors (red, yellow, and blue) after experiencing three filtration methods (none, sand, sand and carbon), Table 3b lists the control samples.

Measurements: UV-visible spectrometry was used as an analysis method to determine the optimal time for partial and full oxidation for the reactive dyes using photo-Fenton oxidation after optimizing hydrogen peroxide quantity much like other studies [4,5,7,16,17]. In addition, with a view to monitoring the color changes and also peaks of color intensity, an HP 8453 Agilent UV-visible scanning spectrophotometer was used with scans from 190-1100 nm wavelengths in 10 mm quartz cuvettes. Experiments on each dye yielded peak absorbance at particular wavelengths.

Broad organic chemical levels present in the water samples were carried out through COD as an indirect measurement of the amount of organic compounds present in a solution similar to other studies [18]. This was measured colorimetrically using a Hach Meter 890. Samples were digested at 148°C for 120 minutes with one sample of de-ionized water as a control. The digested samples were then measured at 600 nm colorimetrically with the control as zero to measure the color change relative to water.

GC/MS analysis was used to measure (qualitative analysis) and label organic chemical intermediates found in solution just as other studies [17,19,20] before and after oxidation as well as after filtration. For this purpose, Solid-phase Micro Extraction Coupled with Gas Chromatography/Mass Spectrometry (SPME-GC/MS) was used. First the samples were extracted using Solid Phase Micro-Extraction (SPME) with a 65-µm Polydimethylsiloxane/Divinylbenzene (PDMS/DVB), Stableflex 24Ga fiber assembly (Supelco) controlled by a manual Micro-Extractor holder immersed in each sample for a minimum of 50 minutes. A 65-µm PDMS/DVB partially polar fiber is typically used for amines and nitro-aromatic compounds with molecular weights between 50-300 g·mol^{-1}. This fiber was chosen since the oxidation of dyes typically results in the formation of amines. After the extraction was complete, the fiber was injected into a GC System Hewlett Packard,

HP 6890 Series, MS Agilent Technologies Mass Selective Detector 5973 Network and column HP-5MS (Crosslinked 5% PH ME Siloxane) with 30 m x 0.25 mm, 0.25 μm film thickness (Hewlett Packard Special Performance Capillary Column). Maximum injector volume was 1.0 μl with an inlet heat of 250°C and a pressure of about 52 kPa; total flow was 103 mL/min and split ratio of 100:1 with flow of 99.5 mL/min and an average velocity of 36 cm/s. Helium was used as the carrier gas at a flow-rate of 103 mL/min at a pressure of 52 kPa. The oven was set for an initial hold at 50°C for 5 min, followed by a ramp-up of 10°C/min until a temperature of 160°C was reached. Then, there was a hold for 2 min, followed by a ramp-up of 5°C/min until 235°C was reached. This was followed by a hold for 17 min for a total run-time of 50 minutes. The maximum temperature of the column was 300°C. The detector heat was set at 150°C, hydrogen gas flow set at 40 mL per minute and air at 450 mL per min. Analysis of samples was completed by the attached mass spectrometer connected with NBS75K.L library.

With a view to qualify the chemical peaks detected by the GC, they must reach a minimum of 10,000 abundance (based on the electrical signal passed through the sample) and match the MS library with 80% or more accuracy. In order to verify the quantification of abundance of organics found in solution, a 95% pure sample of Tolylene, 2,4-diisocyanate (also called Benzene, 2,4-diisocyanato-1-methyl, one of the aromatic and cyanate chemicals found in solution after treatment) was tested using the same GC/MS methods listed above.

Identification of metals and their quantity left in solution was analyzed using an Inductively Coupled Plasma Atomic Emission Spectroscopy (ICP-AES) as this technique has been normally used for the determination of metals and ions in order to quantify and qualify the levels of metal ions left in solution after the following tests: (1) photo-Fenton oxidation, (2) photo-Fenton oxidation followed by sand filtration, and (3) photo-Fenton oxidation followed by sand and carbon filtration (both fully and partially oxidized).

The test methods followed in this study were similar to previous reports [21-23]. Under the standard methods for the examination of water and wastewater [24] all results were determined in filtered and acidified samples and the flowing stream of argon gas was ionized by applied radio frequency field ~27.1 MHz. The Instrument Parameters of ICP used were 1.00 kW power, 15 L/min of plasma, 200 kPa of Nebulizer pressure, argon gas flow of 1.5 L/min with viewing height of 10 mm, pump rate of 15 rpm and Spectrometer Atomic Emission Spectroscopic studies having a photodiode array with dwell time of 30 sec.

Finally, scrap sheet iron used was characterized according to surface area as it is believed that the iron that acts as a catalyst in the photo-Fenton oxidation reaction acts at the interface between UV light and iron. For 1-liter solutions, low iron was 165 cm², medium iron was 323 cm², and high iron was 645 cm². This measurement differs from other studies that use scrap metal that measured iron quantity by volume or weight [19,25-27].

Sludge that is produced during the photo-Fenton oxidation process included both organic dye and inorganic iron as described in review reports [28]. ICP-AES measured quantities of iron in control samples and treated samples. For the Indian samples, iron and dye sludge was quantified volumetrically. In brief, one-liter solution was poured into a volumetric cylinder and allowed to settle overnight. A measurement was then taken of the foam that rose to the top as well as the sludge that settled to the bottom of the solution.

Respirometric analysis: To determine the biodegradable quality

of the residual solution, respirometric analysis was performed as in a previous publication [29]. Biodegradability of solution is an important factor to test as can be seen in previous reports [7]. For this, the sludge from a local Common Effluent Treatment Plant (CETP) in India was used as biomass for this analysis. This allowed for the determination of whether the sludge from a CETP can uptake oxygen in the presence of the contaminants left in the photo-Fenton oxidized samples. Oxygen released due to consumption of the pollutant given to the bacteria was measured in the digital oxygen system (Model 10, Rank Brothers, Bottisham UK) instrument.

Scanning electron microscopy (SEM): To determine whether there were any other impurities in solution affecting the oxidation process and to ensure that the ferrous ions in the rust were in fact working as catalyst for the photo-Fenton oxidation process, surface content of iron pieces was measured using SEM (Hitachi S-4800) much like other works using SEM for photo-Fenton oxidation studies [4,5,22,30]. This also had an attachment of 15 mm Energy Dispersive X-ray Spectrometer (EDX) with a detector distance of 1.5 mm for elements analysis used to map various elements (Fe, Cl, S, Al, Fe, O, C).

Statistical analysis: Considering the number of variables used in this study, to determine the main effects and the interaction effects due to three factors: color (red, yellow, or blue), quantity of iron (low, medium, or high), and time (t_0, partial oxidation time, full oxidation time), the biological variable (biodegradability based on bacteria's oxygen uptake) was treated as dependent and was analyzed using three-way analysis of variance [31]. ANOVA was performed with a full model separately for all the three variables similar to that reported earlier [32]. For each combination of factors, the dependent variable data was available in triplicate, resulting in a balanced factorial design.

Results and Discussions

Scanning electron microscopy study

Figure 1 shows scanning electron micrograph of the iron rust from the scrap iron material at 1.5K, found to be composed of Fe, S, Al, and C as determined by EDAX (not shown). It can be seen from Figure 1 that the surface morphology of the iron rust is an uneven surface with different sizes (submicron) of rust spread all over the surface. Similar observations of surfaces of different materials used in photo-Fenton studies have been observed by other researchers [4,5,22,30,33].

UV-visible spectrometry results

Peak absorbance yielded on each dye at particular wavelengths is listed in Table 4. It may be noted that the present values found have previously been reported as phenol and benzene in other studies. For example phenol values reported are 209 nm, 269 nm [34]; while those

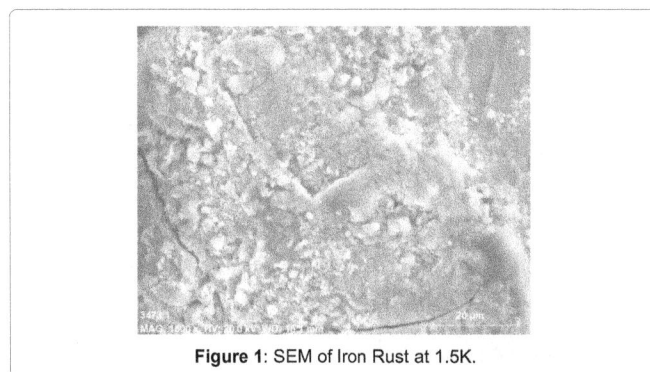

Figure 1: SEM of Iron Rust at 1.5K.

Dye	Phenol (nm)	Benzene (nm)	Color Wavelengths (nm)	Unknown Peaks Found (nm)
Reactive Red 120	206	237	536	291, 371, 512
Reactive Yellow 81	206	229, 243	268, 352	410, 520
Reactive Blue 4	206	230, 258	597	369, 520

Table 4: Peak absorbance wavelengths for the tested dyes and their likely responsible chemicals.

of benzene reported are 235 nm [35], 226 nm [18], 230 nm [36], and 260 nm [16]. However, naphthalene was not found in the present treated samples contrary to those reported in other studies such as at 315 nm [36], and at 318 nm [18]. On the other hand, color wavelength has been confirmed by the present study while unknown peaks reported in the present study have not been reported by others.

Rate reaction: Spectra obtained by UV-visible spectrometry for Reactive Dyes (Undiluted C1, R1H, and R4H spectra of Reactive Red 120, Reactive Yellow 81, and Reactive Blue 4) using photo-Fenton oxidation after optimizing hydrogen peroxide quantity are shown in Figure 2 displaying the unmodified spectra peak heights for t_0, partial, and full oxidation samples C1, R1L, and R4L spectra of the three Reactive Dye colors for low iron (high and medium iron trend is similar). Generally, UV and visible peaks for t_0 are beyond the 4.0 maximum thus denoting the need for decolorization and degradation. It can also be seen that visible peaks of partial oxidation have decreased to within the 4.0 scale and display clear peaks that can be described as individual organic chemicals. On the other hand, full oxidation samples displayed peaks at a lower UV range and no peaks in the visible range meaning that full oxidation time was chosen when complete decolorization had occurred. The actual control peak heights C1-C5 are calculated from dilutions for Reactive Red 120, Reactive Yellow 81, and Reactive Blue 4 respectively. Table 5 lists the partial and full oxidation times chosen from Figure 2 for the three Reactive Dye colors using different iron surface areas.

Partial and full oxidation results: Figure 3 compares spectra of Reactive Red 120 dye (a) R1H-R3H, (b) R4H-R6H, (c) R1M-R3M, (d) R4M-R6M, (e) R1L-R3L, (f) R4L-R6L with that of the diluted control peak C1. Similar spectra were obtained for Reactive Yellow 81 and Reactive Blue 4 dyes. These are not included as there will be too many figures in the paper.

In Figure 3, peaks found at 206 nm were only present after acid and hydrogen peroxide were added to solution (C2-C5 not shown). Peaks at 237 nm, 291 nm, 371 nm, 512 nm, and 536 nm were also visible in C1. In general and despite treatment, the peak at 206 nm stayed much higher than the other pollutants in solution. The contaminants displaying UV peaks at 206 nm, 237 nm, 291 nm, 371 nm were reduced most by full oxidation. Sand and carbon filtration following oxidation greatly decreased the height of the peaks by physically removing the organics out of solution. The greatest impact of iron on peak degradation occurred for medium iron. The contaminants displaying visible peaks at 512 nm and 536 nm degraded more by partial oxidation than by full oxidation consistently for all iron levels. However, sand and carbon filtration following full oxidation was more effective at removing the chemicals responsible for peaks at 512 nm and 536 nm (thus making full oxidation followed by sand alone or sand and carbon the most effective treatment). The medium iron level was still the most effective for removal of these organics. As with the UV wavelength peaks, generally sand filtration accomplished more than the carbon filter due to physical capturing of the solids created by the photo-Fenton oxidation process and physically removing them from solution, although carbon filtration did no harm to the process.

Dye	Iron Quantity	t_0 (min.)	$t_{partial}$ (min.)	t_{Full} (min.)
Reactive Red 120	High	0	40	60
Reactive Red 120	Medium	0	55	75
Reactive Red 120	Low	0	75	135
Reactive Yellow 81	High	0	35	70
Reactive Yellow 81	Medium	0	85	155
Reactive Yellow 81	Low	0	75	135
Reactive Blue 4	High	0	100	125
Reactive Blue 4	Medium	0	155	180
Reactive Blue 4	Low	0	180	220

Table 5: Partial and full oxidation times for Reactive Dyes.

In the case of Reactive Yellow 81 peaks were found at 206 nm, 229 nm, 243 nm, and 520 nm in C2-C5 only after acid and hydrogen peroxide were added to solution. Peaks at 268 nm, 352 nm, and 410 nm were also found in C1. Just as with Reactive Red 120, the peak at 206 nm remained after treatment. In addition, the peak at 229 nm stayed much higher than the other peaks despite treatment. The contaminants that displayed UV peaks at 206 nm, 229 nm, 243 nm, 268 nm were reduced the most from full oxidation compared to partial oxidation. Sand and carbon filtration following oxidation greatly decreased the heights of the peaks by physically removing the organics out of solution. In general, medium iron had the greatest impact on the degradation of these peaks.

The contaminants displaying peaks at 352 nm, and 410 nm exhibited behavior similar to both the UV and visible peaks of Reactive Red 120. Partial oxidation decreased peak height to below full oxidation samples; however, sand and carbon filtration raised the levels again. Full oxidation followed by sand and carbon decreased the levels to near negligible. Medium iron gave the lowest numbers, however the same samples using low and high iron were similar.

The contaminants displaying a visible peak at 520 nm was reduced most from full oxidation followed by sand and carbon filtration. Unlike all other dyes tested, low iron levels removed pollutants from the treated Reactive Yellow 81 solution the most, closely followed by medium and then high iron.

In the case of Reactive Blue 4 dye, peaks were found at 206 nm, and 520 nm only after acid and hydrogen peroxide were added to solution (C2-C5), while peaks at 230 nm, 258 nm, 369 nm, and 597 nm were also found in C1. Just as with Reactive Red 120 and Reactive Yellow 81, peaks of the UV spectra of Reactive blue4 dye at 206 nm, 230 nm, and 258 nm remained much higher than the visible peaks despite treatment. Degradation of the contaminants displaying these UV peaks occurred most after full oxidation using high iron followed by sand and carbon filtration.

The contaminants displaying peaks at 369 nm, 520 nm, and 597 nm presented a mixture of results. The greatest degradation was still with high iron, followed closely by low iron and medium iron.

It may be noted that all the results obtained in UV-visible spectrometry can be compared with published data [16,18,34-36] to describe which organics were left after treatment. Further, as will be

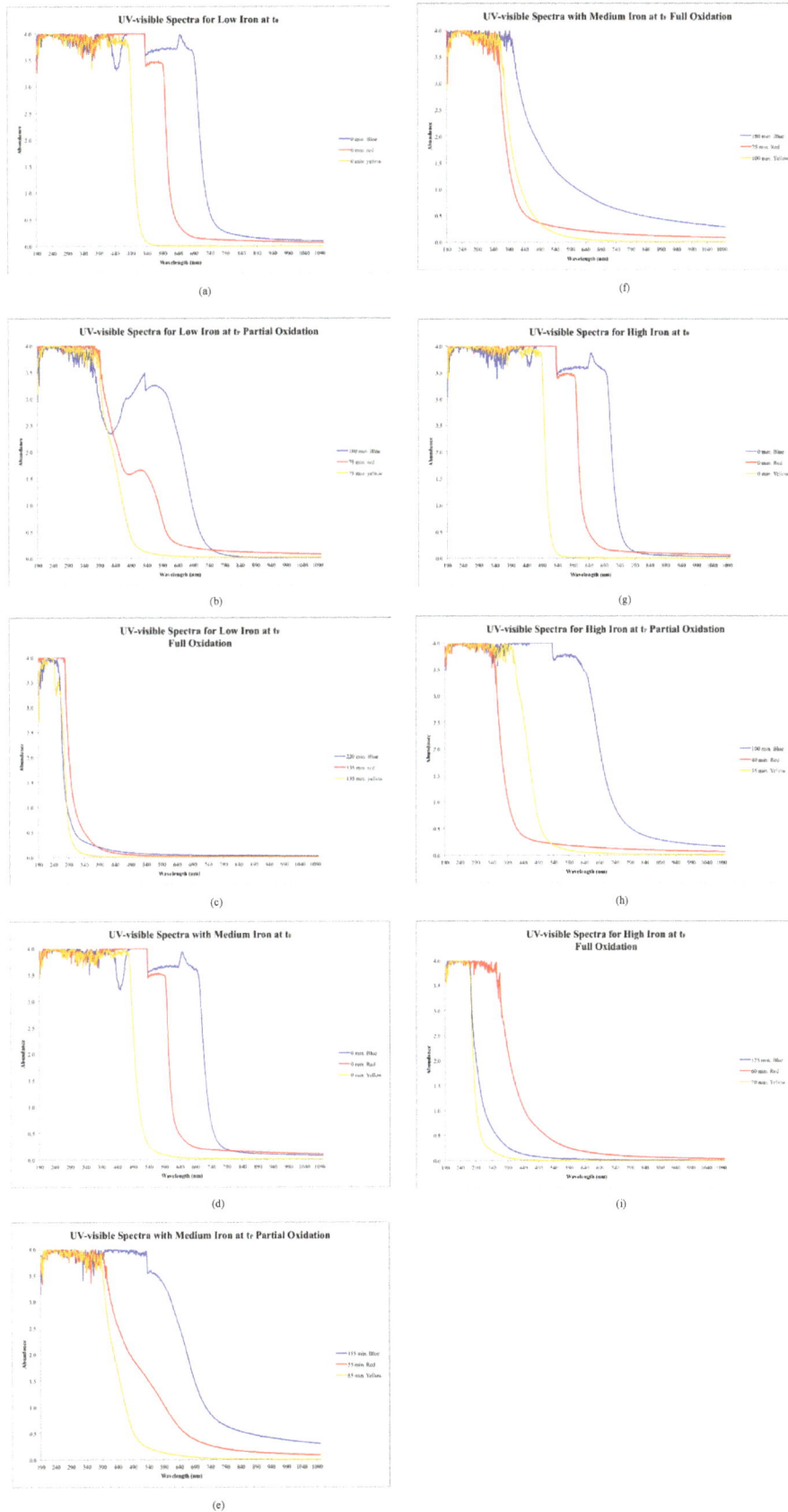

Figure 2: UV-visible spectra for t_0, partial oxidation t_p, and full oxidation t_F of the three Reactive Dye colors using low, medium, and high iron.

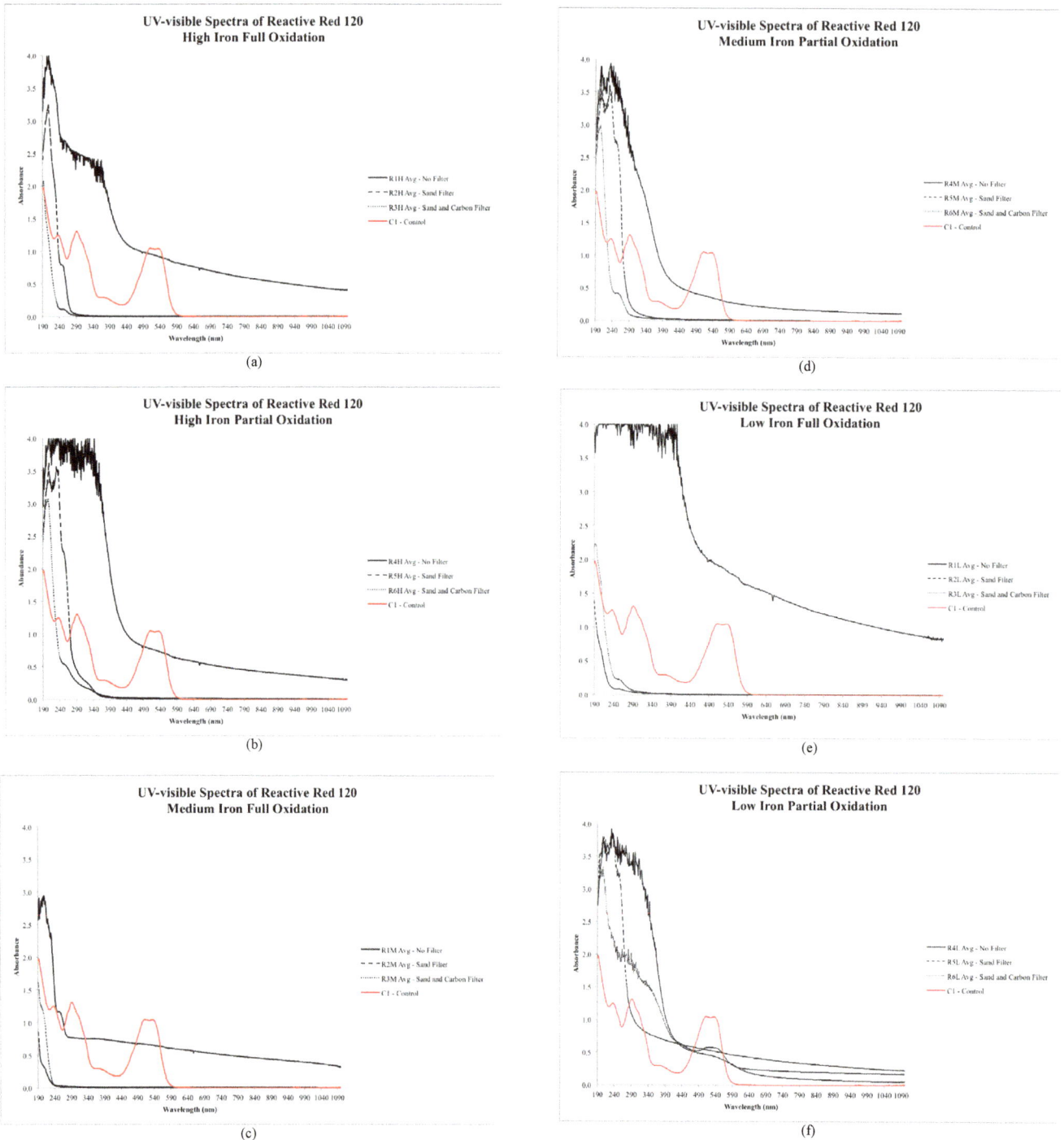

Figure 3: Reactive Red 120 UV-visible spectra of the diluted control C1 compared with those of (a) R1H-R6H full oxidation, (b) R1H-R6H partial oxidation, (c) R1M-R6M full oxidation, (d) R1M-R6M partial oxidation, (e) R1L-R6L full oxidation, (f) R1L-R6L partial oxidation.

seen later, observed UV-visible spectrometry results are comparable with those observed in the COD studies in respect to total organic load remaining. Generally, full oxidation followed by sand and carbon filtration was most effective for red, yellow, and blue dye removal; while optimal iron level cannot be determined as peaks lowered differently with changes in iron surface area.

COD results

COD levels in the present study showed great variability, even when comparing similar samples. This may be due to the fact that the high level of dye used in this study has an exceptional level of organic material. Photo-Fenton oxidation also forms sludge in solution that is comprised of dye and rust from the iron catalyst. For this reason COD levels that dropped below the Karnataka State Pollution Control Board's (KSPCB) allowable 250 ppm are particularly to be noted. In the analysis of the data, it is important to also consider ambient temperature, sunlight, and pH during the experiment in order to capture the effects of these variables on the COD reduction caused by the photo-Fenton oxidation process.

COD results related to the three reactive dyes with respect to temperature, pH, iron, and sunlight were compiled. Figure 4a-4c depicts the COD results comparing C1 to R1H-R6H, R1M-R6M, and R1L-R6L in both India and the USA for Reactive Yellow 81 (trend for Reactive Red 120 and Reactive Blue 4 similar). In general, a decrease of COD from C1 to R4*, R1*, R5*, R6*, R2*, R3* for the respective surface areas of iron in solution is observed.

The control samples for Reactive Red 120 had COD levels that ranged between 2392 and 7320 ppm. Full oxidation using medium iron followed by sand filtration lowered COD levels to a greater extent and this agrees with the UV-visible spectrometry results. COD decreased well below the permissible effluent levels using full oxidation followed by filtration despite iron surface area. From highest to lowest, the COD value trends, regardless of iron differences, are R4*, R1*, R5*, R6*, R2*, R3*. The fully oxidized samples produced more sludge than the partially oxidized samples. Sludge kept the COD higher than permissible effluent levels; however, after physical removal of the sludge using sand and carbon filtration, the COD was lowered close to negligible. Since partial oxidation does not fully precipitate the organics out of solution (and thus creates less sludge), the sand and carbon filtration cannot fully separate them from the water base which results in a higher COD. The low iron sample in India did not exactly follow the same trend since full oxidation brought the COD to permissible effluent levels and much lower than the partially oxidized sand and carbon filtered samples.

COD levels ranged between 2070 and 9760 ppm, in the case of Reactive Yellow 81, while the control samples had ranged between 2262 and 7240 ppm. As in the case of Reactive Red 120 dye, in the case of these two dyes also, full oxidation using medium iron followed by sand filtration lowered COD levels to a greater extent. This agreed with the UV-visible spectrometry results for both these dyes. Also, COD values decreased well below the allowable effluent levels using full oxidation followed by filtration, independent of the iron surface area in both the cases of dyes. Similarly, from highest to lowest, the COD value trends were the same as Reactive Red 120. The trend is independent of iron level and the location of the study. However, it may be noted that while the trends were consistent, the individual numbers could differ greatly between repeated experiment samples.

The control samples for Reactive Blue 4 had COD levels that ranged between 2262 and 7240 ppm. Full oxidation using medium iron

followed by sand filtration lowered COD levels the most. This agrees with the UV-visible spectrometry results. COD decreased well below the allowable effluent levels using full oxidation followed by filtration, independent of the iron surface area.

The fully oxidized samples produced sludge that kept the COD higher than permissible effluent levels. However, after physical removal of these organics, the sand and carbon filtered samples brought COD close to negligible. Since partial oxidation does not fully precipitate the organics out of solution, the sand and carbon filtration cannot fully separate them from the water base, these resulted in a higher COD values. The high iron sample in India did not follow this same trend, but the fully oxidized and filtered samples had higher COD than the partially oxidized and filtered samples. High COD levels were found in control samples due to the high concentration of dye in solution.

Sludge production

The percentage sludge production using sustainable photo-Fenton oxidation was expected and discussed in previous reports [28], and was found in this study to be low. Sludge production was notable during the photo-Fenton oxidation process with more sludge being produced for longer duration experiments. Sludge volume was quantified during the India studies and percent volume sludge per 1-liter laboratory tests for each dye color after full oxidation are as follows.

Reactive Red 120 dye gave about 0.97% with high iron, 6.7% with medium iron and about 6.5% for low iron. Values of these for Yellow 81 and Reactive Blue 4 dyes are about, 0.96% and 3.3% (high iron), about 7.3% and 4.3% (medium iron) and about 3.3% and 5.4% respectively. It can be seen from the above values that high iron surface area produced the least amount of sludge, followed by low iron, and then medium iron. Since sludge was removed from solution using filtration, presence of sludge in non-filtered samples could greatly impact the overall COD level. It was learned that sludge production created from biodegradation (customary method of dye effluent treatment in medium and large-scale textile plants) of their wastewater, was approximately 15% of total solution [Personal Communications].

Effect of solar irradiance and temperature

Generally, the solar irradiance in the USA tests was higher (383-720 W/m^2) than solar irradiance tests in India (81-485 W/m^2), since studies in India were completed at the beginning of the rainy season. While the USA tests were completed at lower ambient temperatures (0.41-23.9°C), India tests were performed during the period when the temperature was in the range of 25.6-28.9°C. Decolorization and degradation occurred for both test sites, indicating that the process can work for a wide range of temperatures and solar irradiances. COD levels varied more as a result of time and filtration rather than solar irradiance and temperature. The COD levels that resulted from the extreme ends of solar irradiance or temperature were mid-range COD results. It can therefore be concluded that while UV irradiance and temperature are important parts of the chemical reaction, without which the reaction could not take place, they are not the key factors in the success of the study. The tests were successful despite the lack of direct sunlight in overcast or partly sunny conditions as long as there was daylight. This is important for eventual field implementation of the process, as both solar irradiance and temperature are variables that cannot be easily controlled in practice.

GC/MS results

The GC/MS graph of the control is shown in Figure 5. This 95%

Reactive Yellow 81 High Iron COD

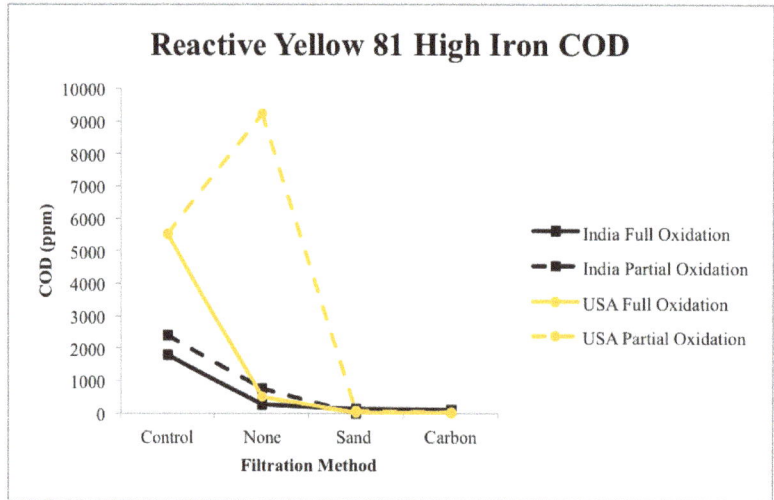

(a)

Reactive Yellow 81 Medium Iron COD

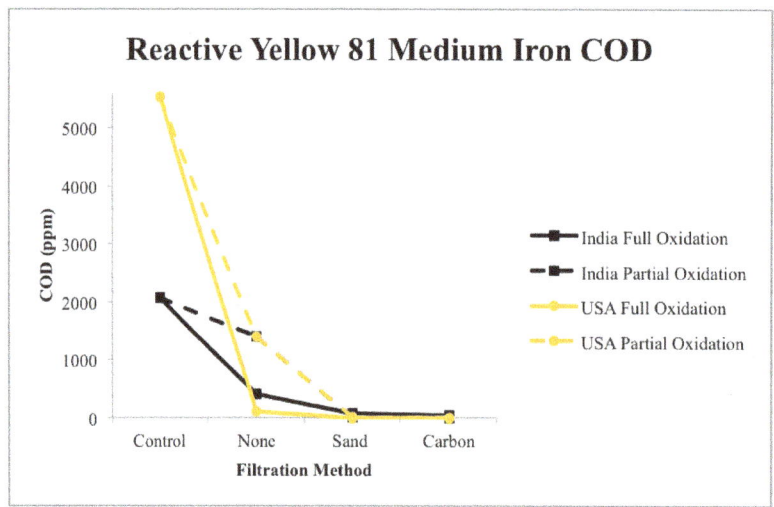

(b)

Reactive Yellow 81 Low Iron COD

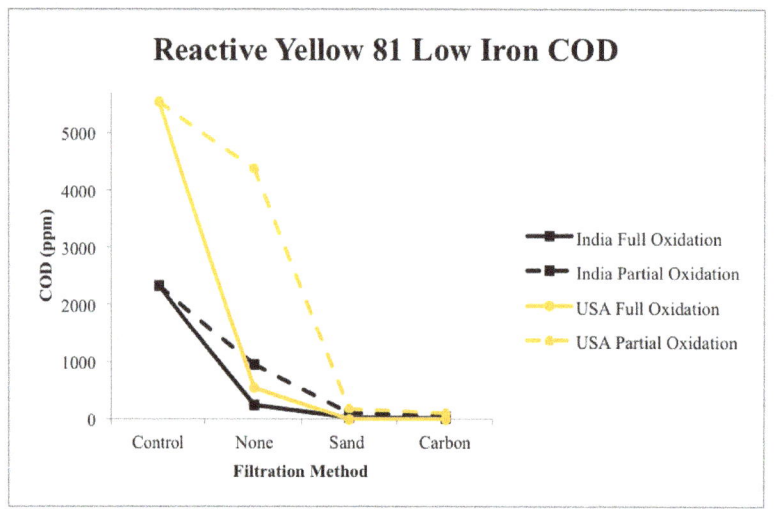

(c)

Figure 4: COD results comparing Reactive Yellow USA and India samples C1 to (a) R1H-R6H, (b) R1M-R6M, (c) R1L-R6L (trends similar for Reactive Red 120 and Reactive Blue 4).

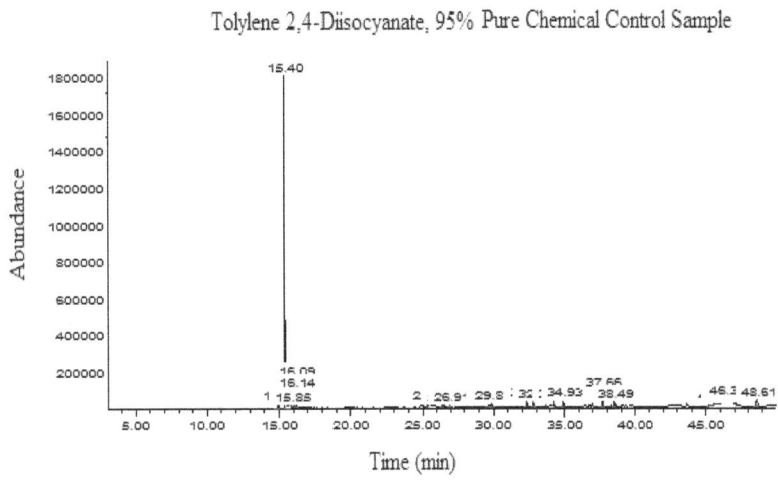

Figure 5:Total ion GC-MS chromatogram of 5 g/L 95% Tolylene, 2,4-diisocyanate (Benzene, 2,4-diisocyanato-1-methyl).

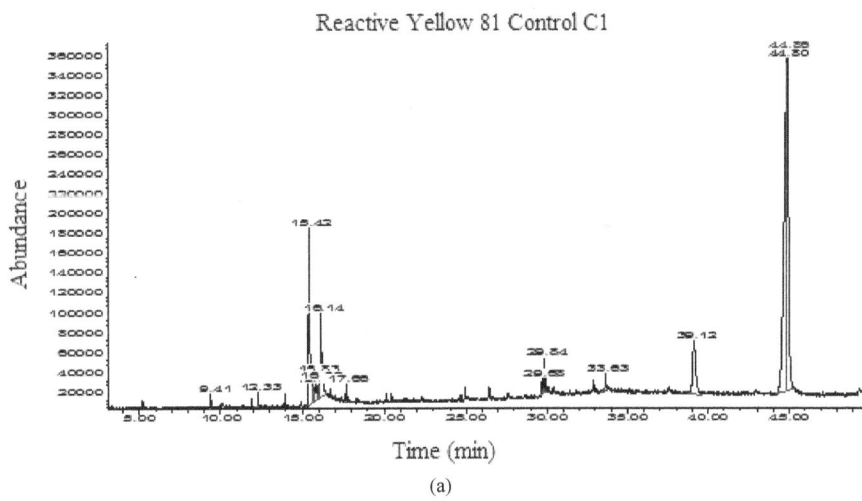

(a)

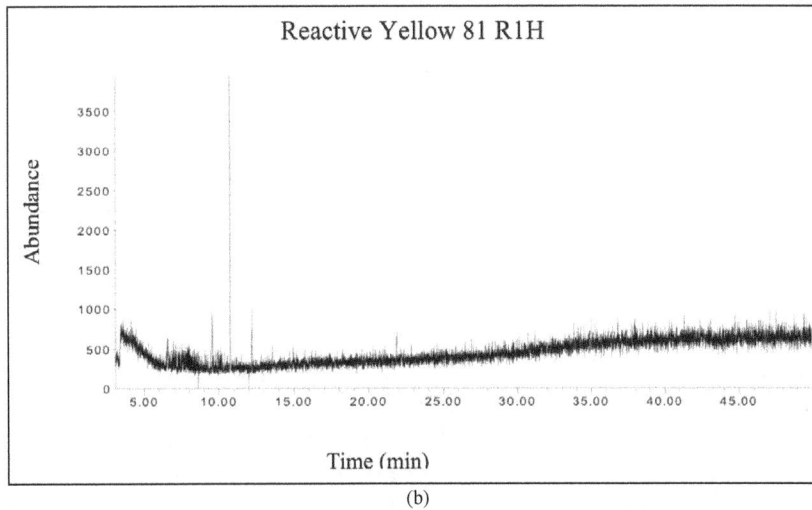

(b)

Figure 6: GC/MS graphs for Reactive Yellow 81 (a) C1 (b) R1H.

pure sample absorbed onto the Solid Phase Micro-Extraction (SPME) fiber was found to have an abundance of Benzene, 2,4-diisocyanato-1-methyl at 1,800,000, as well as Dodecanoic Acid, Ethyl Ester at 28,000; 2H-Benzimidazol-2-one, 1,3-dihydro-5-methyl at 100,000; 2-Propenoic acid, 3-(4-methoxyphenyl)-,2-ethylhexyl ester at 40,000; Bis(2-ethylhexyl) phthalate at 95,000 and squalene at 50,000. This verifies the lesser relative abundance found in samples tested using SPME and GC/MS.

Results obtained for Reactive Red 120 can be found in Supplementary Table S.1, which gives an overview of GC/MS results of all chemicals found in C1-C5, R1H-R6H, R1M-R6M, and R1L-R6L for Reactive Red 120 and what type of compound each result is and their responsible functional groups including when complete degradation had occurred (which resulted in no noticeable peaks). Similar results were obtained in the case of the other two dyes (Reactive Yellow 81 and Reactive Blue 4). From Supplementary Table S.1, it becomes evident that various organic compounds were left in solution after treatment, but with decreasing abundance as was observed for all dyes.

In the case of Reactive Red 120 dye, aromatics in control samples measured at very high levels while combined compounds were lower. Due to chemical alteration during oxidation, treated samples showed higher abundance of only combined compounds. In the case of Reactive Red 120, mostly aromatic and alcohols, and aromatic and cyanate combined compounds with less aromatic and ether and aromatic and aldehyde combined compounds were present. Percent remaining in treated samples was taken from total compounds remaining in USA C1 compared with treated samples. USA sample R1L decreased to 4.3% of original while R4L to 2.9%, and R5H to 3.6%. India samples R2H decreased to 3.7% of original, R2M to 0.9%, R2L to 2.6%, R3H to 3.5%, R4H to 1.0%, R5H to 3.7%, and R6H to 5.9%. All other samples showed no remaining compounds in solution using GC/MS after treatment. Similarly for Reactive Yellow 81 most samples decreased completely while few remained that included mostly aromatic and alcohol combined compounds, aromatic and cyanate combined compounds, with less aromatic and aldehyde combined compounds, and amine combined compounds, aromatic, azine, and chloride combined compounds, and finally aromatic, chloride, and imino combined compounds with slightly higher percentages for India samples R2M and R3H.

In the case of Reactive Blue 4, C1 had undetectable levels of organics in solution possibly due to its different dye chemical composition. C3 however due to its lower pH and its peroxide level was found to contain aromatics, ketones, azines, amines, and chlorides and therefore was used for the control comparison. As with Reactive Red 120 and Reactive Yellow 81, the majority of treated samples were found to contain no detectable organics. However for Reactive Blue 4, R1H in the USA and R4H in India were found to contain higher levels of organics after treatment. Other samples such as R5H in India, and R4H and R4M in the USA were found to decrease only 24-47% of C3.

From the GC/MS graphs obtained it became evident that control samples had the most detectable organic chemicals found in solution in the case of all the dyes. For example, Figure 6a displays Reactive Yellow 81 C1 while Figure 6b shows Reactive Yellow 81 R1H with much lower peaks. Similar graphs were obtained for Reactive Red 120 and Reactive Blue 4 dyes controls compared with treated samples. In the case of Reactive Red 120 dye treated samples, fully oxidized samples degraded more than partially oxidized samples. Medium iron samples had the least amount of organics as found from GC/MS analysis. India samples had many more organics in solution than USA samples. Full

oxidation using medium iron followed by sand filtration in the USA revealed the least amount of chemicals (carbon samples were not tested if sand filtered samples found no chemicals).

Although C1 had the most detectable organic chemicals found in solution in the case of Reactive Yellow 81 dyes, treated samples, partially oxidized samples degraded more than fully oxidized samples. Low iron samples had the least amount of organics as indicated by GC/MS measurements. India samples had many more organics in solution than USA samples. Medium iron oxidized samples in the USA returned no organics in solution.

Unlike Reactive Red 120 and Reactive Yellow 81, C4 for Reactive Blue 4 had the most detectable organic chemicals found in solution. This means that after acidifying the solution and leaving it to react with neither oxidant nor iron catalyst in the sun, more organic compounds were found in solution. Just as with Reactive Yellow 81 samples, low iron samples had slightly less chemicals in solution than medium iron. USA samples had many more organics than India samples. Medium iron oxidized samples in the USA or India returned no organics in solution.

ICP results

Figure 7 shows ICP results testing for iron, calcium, and magnesium from Reactive Red 120 dye samples both in India and USA. It can be seen from Figure 7 that iron always decreased below permissible levels after filtration; calcium and magnesium were always below the permissible limits. In the case of USA samples, sand influenced calcium levels, while sodium was never below the permissible limits in neither USA nor India samples.

Although 18 metals were measured, only four were found regularly at high levels: calcium, iron, magnesium, and sodium. Their levels were measured for all samples from C1-C5, R1H-R6H, R1M-R6M, and R1L-R6L to track when levels decreased below permissible limits. Arsenic, cadmium, lead, and manganese were occasionally found above permissible limits and it could be noticed when this occurred. Data were taken for all the samples, including from the tests carried out both in India and USA. India's available permissible effluent standards for wastewater in Dye and Dye Intermediate Industries for disposal in surface waters do not list calcium, iron, magnesium, and sodium in their effluent regulations [37]. Therefore, levels of these observed metals were compared to global permissible drinking water standards [37-39]. Permissible calcium levels are limited to 75-200 mg/L because calcium causes water hardness between 40-100 mg/L, while the permissible magnesium levels are limited to 30-150 mg/L in view of the taste of the water being affected at 500 mg/L. Similarly, in the case of iron, permissible levels are limited to 0.1-1.0 mg/L because of the taste and aesthetics of the water being affected at 0.3 mg/L. On the other hand, permissible sodium levels are limited to 20 mg/L because, (i), sodium levels in natural water is between 5-50 mg/L and (ii) at 250 mg/L the taste and aesthetics of natural water is affected and can alter natural biological activity. Similarly, permissible manganese levels are limited to 0.1-0.5 mg/L, but at 0.05 mg/L the taste and aesthetics of the water are affected. In the case of metals such as arsenic, cadmium and lead, which affect human health, permissible levels are 0.01-0.05 mg/L, 0.005-0.01 mg/L and 0.05-0.1 mg/L respectively. This is because, at 0.05 mg/L arsenic affects human health; cadmium at 0.005 mg/L is toxic to humans, while lead causes irreversible neurological damage to humans, especially children [40,41].

The general trend of these ions in obtained solution in this study

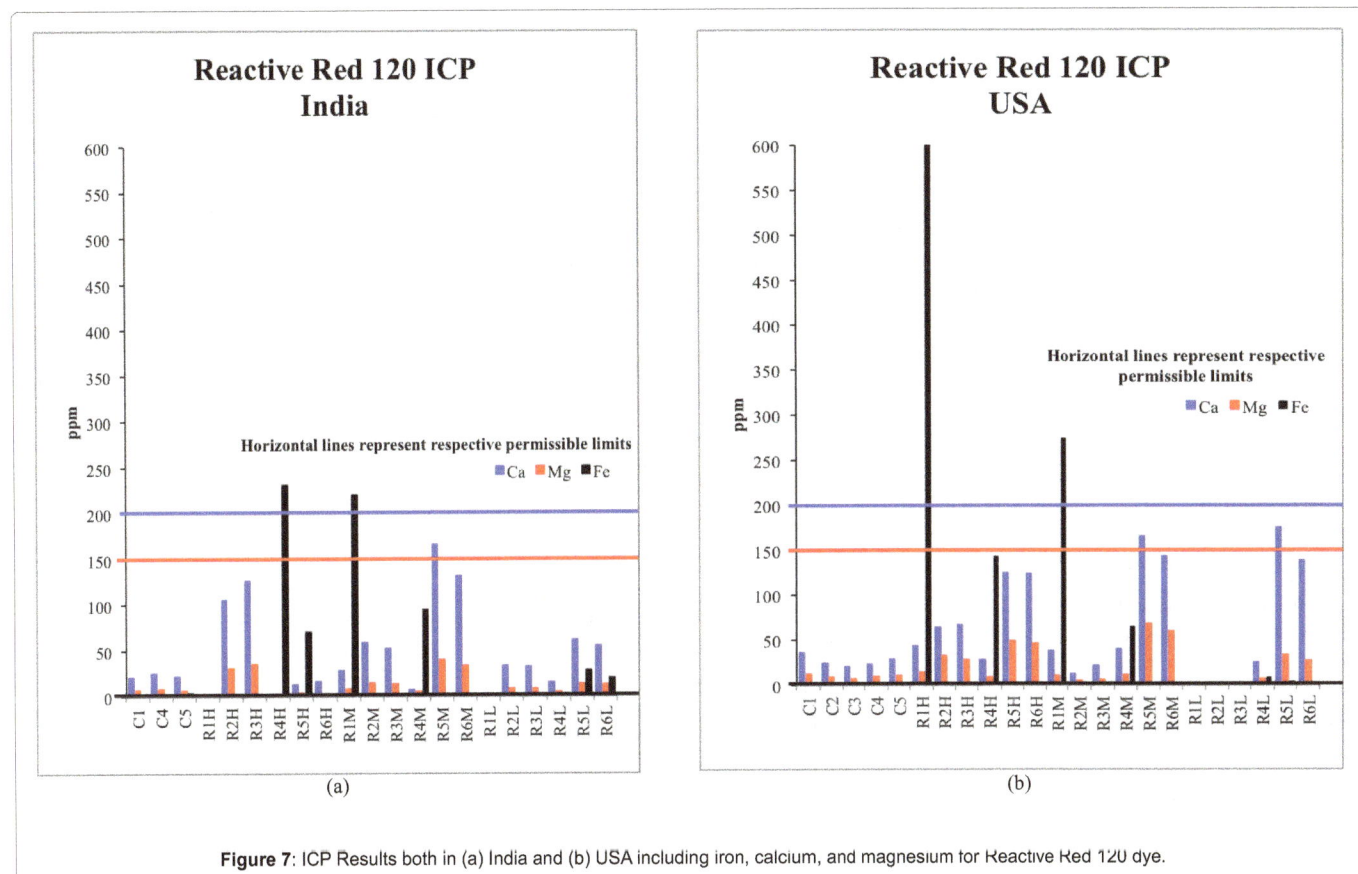

Figure 7: ICP Results both in (a) India and (b) USA including iron, calcium, and magnesium for Reactive Red 120 dye.

(a)

Sample	Amount of Fe	No Filter/Filter	O$_2$ Uptake (nmol/min/mg)
R4H	High	No Filter	3.84
R5H	High	Sand Filter	0.00
R4M	Medium	No Filter	40.6
R5M	Medium	Sand Filter	75.4
R4L	Low	No Filter	0.00
R5L	Low	Sand Filter	0.00

(b)

Sample	Amount of Fe	No Filter/Filter	O$_2$ Uptake (nmol/min/mg)
R4H	High	No Filter	1.58
R5H	High	Sand Filter	0.00
R4M	Medium	No Filter	365
R5M	Medium	Sand Filter	2.33
R4L	Low	No Filter	0.00
R5L	Low	Sand Filter	101

(c)

Sample	Amount of Fe	No Filter/Filter	O$_2$ Uptake (nmol/min/mg)
R4H	High	No Filter	0.00
R5H	High	Sand Filter	0.00
R4M	Medium	No Filter	7.36
R5M	Medium	Sand Filter	620
R4L	Low	No Filter	0.00
R5L	Low	Sand Filter	0.00

Table 6: Calculated oxygen uptake according to low, medium, or high iron and non-filtered or sand filtered sample results for (a) Reactive Red 120; (b) Reactive Yellow 81; (c) Reactive Blue 4.

was an increase of calcium and magnesium after sand filtration. Iron had the opposite trend, as it increased during the oxidation process and decreased to an almost negligible amount after sand filtration. This was a result of the physical capturing of the iron/dye sludge formed in solution during the oxidation process. Sodium increased or remained stable during oxidation and decreased after sand and carbon filtration. It may be noted that the levels of increase or decrease do not fit any mathematical trend, because of the variability in chemical composition of the wastewater at such a high dye concentration. In addition, there was little control over the contents of Indian sand, since it was used for construction and had been stored outside and open to nature, prior to use. Because of the difference in quantitative results between India and the USA, only the trends are described. These trends often matched despite these numerical differences. These trends were often comparable despite varying iron surface areas described in this study as well differing locations of testing (India and USA). Considering that similar results were found for the three dyes, for the sake of brevity, only Reactive Red 120 results are described here.

For example, in the case of Reactive Red 120 dyes, it was observed that all ICP results were within the drinking water permissible limits for calcium (below 200 mg/L) and magnesium (below 150 mg/L). Further, other than calcium, magnesium, iron, and sodium, manganese was also present occasionally in the India samples, but only after sand or carbon filtration was performed. Most India filtered samples showed a level of manganese in solution (except R2H, R2L-R3L), along with the presence of arsenic in R3H. On the other hand, USA samples (particularly R1H, R4H, R1M, R1L, R3L, and R5L) showed at times, low but varying levels of manganese, lead, and arsenic than those of Indian samples, and typically after no filtration.

Calcium controls ranged from 20 to 37 ppm with the highest level in C1, magnesium ranged from 7 to 14 ppm with the highest level in C1, iron from 0.1 to 4 ppm with the highest level in C4-C5, and sodium from 580 to 1100 ppm with the highest level in C4. Similarly, calcium and magnesium content in solution changed as iron surface area and location of test changed. For high iron, fully oxidized Indian samples had more metals than partially oxidized samples, while those of USA samples had fewer metals than partially oxidized samples. However, all of these samples increased with filtration.

On the other hand, for medium iron, partially oxidized samples had more metals than fully oxidized samples in the case of both Indian and USA samples. However, iron in all of these samples increased with filtration except for USA samples, opposite results were observed for R2M-R3M samples. Finally, for low iron, fully oxidized samples had more metals than partially oxidized samples. Here also, all the samples increased with filtration except USA samples R2M-R3M.

Similarly, iron concentration in solution changed as iron surface area and location of test changed. Longer oxidation time resulted in higher iron metal content for non-filtered samples, while filtration resulted in lower iron content especially after full oxidation (except India samples R5L-R6L). Medium, low, and high iron surface area resulted in less metal in solution respectively for all samples. One exception is for India samples R5L-R6L having a larger metal content than R4L. Samples R2H, R3H, R2M, R3M, R2L, and R3L showed iron content below the permissible limits. However, samples R6H, and R6M also showed below the permissible limits, although 3-10 times higher than the fully oxidized equivalents.

For sodium, none of the samples decreased to below permissible limits except R6H in India, and R2M and R3M in the USA. On the other hand, samples R2H-R3H, R2M-R3M, and R2L-R3L (except R2H and R3H in India) showed lower level than R5H-R6H, R5M-R6M, and R5L-R6L. Generally, it was observed that more sodium was removed from solution with more filtration. Low, medium, and high iron surface area resulted in less metal in solution respectively for all samples.

Respirometric results

Since COD values of fully oxidized samples were too low to biodegrade further, respirometric analysis was carried out only for partially oxidized samples of India of differing colors (red, yellow, blue), with different amounts of iron (low, medium, high), and that used different filtration methods (no filter, sand filter). Table 6a-6c show the results of the calculated oxygen uptake based on the obtained data of endogenous and exogenous activities describing the oxygen uptake of R4H, R4M, R4L or R5H, R5M, R5L in the case of three dyes.

In the case of Reactive Red 120 dye, it was observed that for the samples with low iron surface area, oxygen uptake was found to be 0.00 nMol min^{-1} mg^{-1} for both R4L and R5L. This suggests that filtration had little effect on the sample, but partial oxidation for 40 minutes using an iron surface area of 161 cm^2 was not enough to leave the sample biodegradable for CETP sludge.

For the medium iron group, R4M had an oxygen uptake of 40.6 nMol min^{-1} mg^{-1} and 75.4 nMol min^{-1} mg^{-1} for R5M. Therefore oxidation for the same 40 minutes was enough for the partially oxidized sample to create biodegradability by CETP sludge, although almost double the

amount for the filtered sample. The group with high amount of iron was found to have zero theoretical biodegradability by CETP sludge for R5H and very low oxygen uptake of 3.84 nMol min^{-1} mg^{-1} for R4H.

Similarly, for Reactive Yellow 81 with low iron surface area, oxygen uptake was found to be 0.00 nMol min^{-1} mg^{-1} for R4L, and 101 nMol min^{-1} mg^{-1} for R5L. This suggests that filtration had a more substantial effect on the sample, and partial oxidation after 25 minutes using low surface area iron followed by sand filtration was enough to leave the sample biodegradable for CETP sludge. The medium iron group R4M had an oxygen uptake of 365 nMol min^{-1} mg^{-1} and 2.33 nMol min^{-1} mg^{-1} for R5M. Being partially oxidized for the same 25 minutes was enough for the sample to create biodegradability by CETP sludge, whereas the filtered sample had an extremely low oxygen uptake of 2.33 nMol min^{-1} mg^{-1}, due to lack of nutrient availability for consumption. The group with a high amount of iron was found to have very low theoretical biodegradability by CETP sludge with or without filtration at 1.58 nMol min^{-1} mg^{-1} for R4H and 0.00 nMol min^{-1} mg^{-1} for R5H.

In the case of Reactive Blue 4 with low iron surface area, oxygen uptake was found to be 0.00 nMol min^{-1} mg^{-1} for R4L, and 0.00 nMol min^{-1} mg^{-1} for R5L. This suggests that filtration had little effect on the sample, but partial oxidation for 100 minutes using low surface area iron was not enough to leave the sample biodegradable for CETP sludge. For the medium iron group R4M had an oxygen uptake of 7.36 nMol min^{-1} mg^{-1} and 620 nMol min^{-1} mg^{-1} for R5M. Being oxidized for the same 100 minutes was enough for the partially oxidized sample to barely create biodegradability by CETP sludge, whereas the filtered sample had an extremely high oxygen uptake almost 100 times that of non-filtered. The group with high amount of iron was found to have no theoretical biodegradability by CETP sludge with or without filtration. Overall oxygen uptake results for reactive dyes show the greatest promise for biodegradability created after medium iron usage with or without filtration.

Results of three-way ANOVA used to statistically analyze the interaction between three different variables (color, iron surface area, filtration method) for the reactive dye experiments and how they affected respirometric analysis of the partially oxidized samples show that none of the three factors have a statistically significant effect on oxygen uptake (P>0.05). Also, the interaction effect was insignificant. It may be noted while the numbers of tests performed were too low to be statistically significant with regards to variables like iron surface area, the results do indicate that the process is beneficial, and there are indications that certain combinations are potentially better than others. For instance, medium iron tends to lead to oxygen uptake, indicating that pollutants are being consumed, while that is seen less with other iron surface areas.

In brief, it may be noted that the purpose of the studied photo-Fenton oxidation procedure was to completely degrade organic compounds from a reactive dye solution and to bring inorganics to the permissible drinking water level. The level of degradation achieved through various combinations of experimental variables described above and finally, the combined results obtained can be used to identify an experimental procedure that will achieve the most effective treatment of the dye effluent in a cost-effective and sustainable manner.

Since the work was not completed using laboratory-grade supplies, some contamination was expected to be found in the above results in the form of outliers in the analysis. Organic and inorganic chemicals found before (in the controls), during (after partial oxidation and followed by filtration), and after (after full oxidation, and followed by filtration) the sustainable photo-Fenton oxidation process, agreed with the previous literature studies [16,18,34-36]. The authors feel, since this study has a practical application because of the need for low-cost decolorization and dye wastewater treatment, it is necessary to prioritize the importance of certain variables over others in order to create a viable treatment unit to be used by small-scale silk dyers worldwide. The overall intention is to make sure that the color degrades and the chemical composition oxidizes to a simpler form before reusing the treated water for a second batch of dye, or before disposing of it as is normally done.

Conclusions

The following conclusions can be drawn from this study.

- This work has demonstrated that the photo-Fenton oxidation process is able to clean dye wastewater from cottage-scale textile industries using higher concentrations of dye in effluent that are not treatable by other means.

- There are many variables in the photo-Fenton oxidation process that impact the success of dye degradation. These include chemical composition of dyes, concentration of dye, acidified pH of solution, quantity of hydrogen peroxide used, solar irradiation, oxidation time, surface area of the scrap iron catalyst, and filtration method.

- Laboratory-scale studies using sustainable photo-Fenton oxidation to degrade 5 g/L dyes using pharmacy-grade 6% hydrogen peroxide, household hydrochloric acid, scrap iron sheets, and sunlight is best done for Reactive dyes using an intermediate amount of iron surface area (323 cm^2) as the iron catalyst, while allowing for full oxidation time followed by sand and carbon filtration to remove sludge.

- The sustainable photo-Fenton oxidation process using scrap metal as the iron catalyst coupled with sand and carbon filtration shows great promise for the decolorization and chemical degradation of red, yellow, and blue colors in Reactive dyes. The process may not need biodegradation after complete oxidation and filtration, as little additional benefit was seen with biodegradation.

- The low-cost and simplicity of the sustainable photo-Fenton oxidation reactors may in the future offer a treatment option to cottage industries to implement this system quickly and easily in their own communities.

- Once the process grows past a pilot scale built for small-scale dye units, there is a great potential to scale up the system for larger units and commercial applications.

- Life cycle assessment (including O&M and labor costs) of low-cost photo-Fenton oxidation with filtration compared with biodegradation with reverse osmosis treatment is proposed for future work.

Acknowledgements

The authors sincerely thank the Chemical Technology Department of Milwaukee Area Technical College (MATC), USA, and the Director, National Environmental Engineering Research Institute (NEERI), Nagpur of Council of Scientific and Industrial Research (CSIR), New Delhi for providing all the necessary facilities during the course of this study.

References

1. Datta RK, Nanavaty M (2005) Global Silk Industry: A Complete Source Book. Boca Raton, FL, USA: Universal Publishers. ISBN: 1-58112-493-7.

2. Aplin R, Waite TD (2000) Comparison of three advanced oxidation processes for degradation of textile dyes. Water Sci Technol 42: 345-354.

3. Al-Kdasi A, Idris A, Saed K, Guan CT (2004) Treatment of Textile Wastewater by Advanced Oxidation Processes-A Review. Global Nest J 6: 222-230.

4. Daud NK, Hameed BH (2010) Decolorization of Acid Red 1 by Fenton-like process using rice husk ash-based catalyst. J Hazard Mater 176: 938-944.

5. Daud NK, Ahmad M, Hameed BH (2010) Decolorization of Acid Red 1 dye solution by Fenton-like process using Fe-Montmorillonite K10 catalyst. Chem Eng J 165: 111-116.

6. Hsing HJ, Chiang PC, Chang EE, Chen MY (2007) The decolorization and mineralization of acid orange 6 azo dye in aqueous solution by advanced oxidation processes: a comparative study. J Hazard Mater 141: 8-16.

7. Tantak NP, Chaudhari S (2006) Degradation of azo dyes by sequential Fenton's oxidation and aerobic biological treatment. J Hazard Mater 136: 698-705.

8. Lodha B, Chaudhari S (2007) Optimization of Fenton-biological treatment scheme for the treatment of aqueous dye solutions. J Hazard Mater 148: 459-466.

9. Chamarro E, Marco A, Esplugas S (2001) Use of Fenton reagent to improve organic chemical biodegradability. Water Res 35: 1047-1051.

10. Rodriguez M, Sarria V, Esplgas S, Pulgarin C (2002) Photo-Fenton treatment of a biorecalcitrant wastewater generated in textile activities: biodegradability of the photo-treated solution. J Photochem 151: 129-135.

11. dos Santos AB, Cervantes FJ, van Lier JB (2007) Review paper on current technologies for decolourisation of textile wastewaters: Perspectives for anaerobic biotechnology. Bioresour Technol 98: 2369-2385.

12. Santa Cruz Biotechnology (2014) Reactive Red 120.

13. World Dye Variety (2012) Reactive Yellow 81.

14. MP Biomedicals (2014) Reactive Blue 4.

15. Curtronics (2013) Curt Blank's Milwaukee Weather Report September - October 2013 Archives: Solar Radiation Average per Hour (24 Hours).

16. Jozwiak WK, Mitros M, Kaluzna-Czaplinska J, Tosik R (2007) Oxidative decomposition of Acid Brown 159 dye in aqueous solution by H_2O_2/Fe^{2+} and ozono with GC/MS analysis. Dyes Pigm 74: 9-16.

17. Dutta K, Bhattacharjee S, Chaudhuri B, Mukhopadhyay S (2003) Oxidative Degradation of Malachite Green by Fenton Generated Hydroxyl Radicals in Aqueous Acidic Media. J Environ Sci 4: 754-760.

18. Sun JH, Sun SP, Wang GL, Qiao LP (2007) Degradation of azo dye Amido black 10B in aqueous solution by Fenton oxidation process. Dyes Pigm 74: 647-652.

19. Lin JJ, Zhao XS, Liu D, Yu ZG, Zhang Y, et al. (2008) The decoloration and mineralization of azo dye C.I. Acid Red 14 by sonochemical process: rate improvement via Fenton's reactions. J Hazard Mater 157: 541-546.

20. Jadhav SB, Phugare SS, Patil PS, Jadhav JP (2011) Biochemical degradation pathway of textile dye Remazol red and subsequent toxicological evaluation by cytotoxicity genotoxicity and oxidative stress studies. Int Biodeterior Biodegrad 65: 733-743.

21. Bobu M, Yediler A, Siminiceanu I, Shulte-Hostede S (2008) Degradation studies of ciprofloxacin on a pillared iron catalyst. Appl Catal B 83: 15-23.

22. Muthuvel I, Swaminatham M (2008) Highly solar active Fe (III) immobilized alumina for the degradation of Acid Violet 7. Sol Energy Mater Sol Cells 92: 857-863.

23. Dash D, Venkateswarlu G, Thangavel S, Rao SV, Chaurasia SC (2011) Ultraviolet photolysis assisted mineralization and determination of trace levels of Cr, Cd, Cu, Sn, and Pb in isosulfan blue by ICP-MS. Microchem J 98: 312-316.

24. Eaton AD, Clesceri LS, Greenberg AE, Franson MAH (1995) (19th edtn) Inductively Coupled Plasma: In Standard methods for the examination of water and wastewater. New York: American Public Health Association.

25. Ganesan R, Thanasekaran K (2011) Decolourisation of textile dyeing wastewater by modified solar photo-Fenton oxidation. Int J Environ Sci 1: 1168-1176.

26. Ganesan R, Latha A, Thanasekaram K (2014) Treatment of Textile Dyeing Wastewater by Modified UV Photo-Fenton Process Using a New Composite Steel Scrap/H_2O_2. Int J Emerg Technol and Adv Eng 4: 108-113.

27. Zhang H, Zhang J, Zhang C, Liu F, Zhang D (2009) Degradation of C.I. Acid Orange 7 by the advanced Fenton process in combination with ultrasonic irradiation. Ultrason Sonochem 16: 325-330.

28. Pignatello JJ, Oliveros E, MacKay A (2006) Advanced oxidation processes for organic contaminant destruction based on the Fenton reaction and related chemistry. Crit Rev Environ Sci Technol 36: 1.

29. Jablonski MR, Shaligram S, Qureshi A, Purohit H, Reisel JR (2013) Degradation kinetics of resorcinol by Enterobacter cloacae isolate. Afri J Microbiol Res 7: 3632-3640.

30. Parra S, Guasaquillo I, Enea O, Mielczarski E, Mielczarki J, et al. (2003) Abatement of an Azo Dye on Structured C-Nafion/Fe-Ion Surfaces by Photo-Fenton Reactions Leading to Carboxylate Intermediates with a Remarkable Biodegradability Increase of the Treated Solution. J Phys Chem B 107: 7026-7035.

31. Andonian K, Hierro JL, Khetsuriani L, Becerra P, Janoyan G, et al. (2011) Range-expanding populations of a globally introduced weed experience negative plant-soil feedbacks. PLoS One 6: e20117.

32. Narde GK, Kapley A, Purohit HJ (2004) Isolation and characterization of Citrobacter strain HPC255 for broad-range substrate specificity for chlorophenols. Curr Microbiol 48: 419-423.

33. Rodrıguez A, Ovejero G, Sotelo JL, Mestanza, M, Garcıa J (2010) Heterogeneous Fenton catalyst supports screening for mono azo dye degradation in contaminated wastewaters. Ind Eng Chem Res 49: 498-505.

34. He F, Lei LC (2004) Degradation kinetics and mechanisms of phenol in photo-Fenton process. J Zhejiang Univ Sci 5: 198-205.

35. Bilgi S, Demir C (2005) Identification of photooxidation degradation products of CI Reactive Orange 16 dye by gas chromatography-mass spectrometry. Dyes Pigm 66: 69-76.

36. Feng W, Nansheng D, Helin H (2000) Degradation mechanism of azo dye C. I. reactive red 2 by iron powder reduction and photooxidation in aqueous solutions. Chemosphere 41: 1233-1238.

37. Hasan R (2014) Ministry of Environment and Forests Notification. The Gazette of India.

38. Kumar M, Puri A (2012) A review of permissible limits of drinking water. J Occup Environ Med 16: 40-44.

39. Kumar M, Puri A (2012) A review of permissible limits of drinking water. Indian J Occup Environ Med 16: 40-44.

40. Paul S, Chaven S, Khambe S (2012) Studies on Characterization of Textile Industrial Waste Water in Solapur City. Int J Chem Scie 10: 635-642.

41. Sengupta B, Verma NK, Basu DD, Ansari PM, Kumar P, et al. (2008) Status of Water Treatment Plants in India.

17

Reduction of the Weaving Process Set-up Time through Multi-Objective Self-Optimization

Saggiomo M*, Gloy YS and Gries T

Institut für Textiltechnik (ITA) der RWTH Aachen University, Aachen, Germany

Abstract

Real (physical) objects melt together with information-processing (virtual) objects. These blends are called Cyber-Physical Production Systems (CPPS). The German government identifies this technological revolution as the fourth step of industrialization (Industry 4.0). Through embedding of intelligent, self-optimizing CPPS in process chains, productivity of manufacturing companies and quality of goods can be increased. Textile producers especially in high-wage countries have to cope with the trend towards smaller lot sizes in combination with the demand for increasing product variations. One possibility to cope with these changing market trends consists in manufacturing with CPPS and cognitive machinery. This paper focuses on woven fabric production and presents a method for multi-objective self-optimization of the weaving process. Multi-objective self-optimization assists the operator in setting weaving machine parameters according to the objective functions warp tension, energy consumption and fabric quality. Individual preferences of customers and plant management are integrated into the optimization routine. The implementation of desirability functions together with Nelder/Mead algorithm in a software-based Programmable Logic Controller (soft-PLC) is presented. The self-optimization routine enables a weaving machine to calculate the optimal parameter settings autonomously. Set-up time is reduced by 75% and objective functions are improved by at least 14% compared to manual machine settings.

Keywords: Weaving; Industry 4.0; Optimization; Cognition; Set-up time

Introduction

Weaving is the most common as well as the oldest process for fabric manufacturing. Until today a fabric is created by crossing warp and weft threads in a right angle, like it was done since approximately 4000 B.C. Today's applications are for example:

- Apparel (jeans, lining fabric, etc.),

- Geotextiles (erosion protection, soil reinforcement, etc.) and

- Technical textiles (filters, fireproof fabric, airbag fabric, reinforcements for fiber composites, etc.).

Because of the low production costs, the textile production has been relocated to the Asian countries, whereas the production of high-quality and technical textiles is progressively shifted to Europe. The textile industry in high-wage countries like Germany is facing numerous challenges today. For example, the tendency to small lot sizes requires shorter cycle times and aggravates the economical production of goods [1,2].

Small lot sizes in the fabric production often involve a change of the fabric. A weaving machine with about 200 parameters has to be reconfigured after each change of the fabric to fulfill the expectations of the customer. In order to find the optimal configuration for the machine, the operator of the weaving machine has to conduct weaving trials. These time-consuming and wasteful trials require - depending on the experience of the operator - the weaving of up to 120 m of fabric until the optimal parameters are found [3,4]. This research paper presents an algorithm for multi-objective self-optimization of the weaving process and the integration of this algorithm into the machine control of the weaving machine. A weaving machine is upgraded to a cognitive unit (CPPS) on the shop floor. The algorithm for self-optimization identifies a combination of machine parameters, with the result that the essential objective functions can be adjusted to the individual preferences. Figure 1 visualizes the principle of a weaving machine.

Concept and Implementation of Multi-Objective Self-Optimization of the Weaving Process

Self-optimization systems apply adoptions of their inner state or structure in case of changes in input conditions or disturbances. Target values for self-optimization can be e.g. capacity, lot size, quality or energy consumption [1]. According to [5] self-optimization systems are characterized by the following continuous steps:

- Analysis of actual situation.

- Determination of targets.

- Adaption of system behaviour in order to reach the targets.

The presented concept of self-optimizing production systems will now be applied to the weaving process. The following objective functions are considered by the multi-objective self-optimization (MOSO) of the weaving process:

- Warp tension,

- Energy consumption of the weaving machine (air- and active power consumption),

- Quality of the fabric,

The objective functions are optimized according to the following parameters:

---

*Corresponding author: Marco Saggiomo, Institut für Textiltechnik der RWTH Aachen University, Aachen, Germany
E-mail: Marco.Saggiomo@ita.rwth-aachen.de

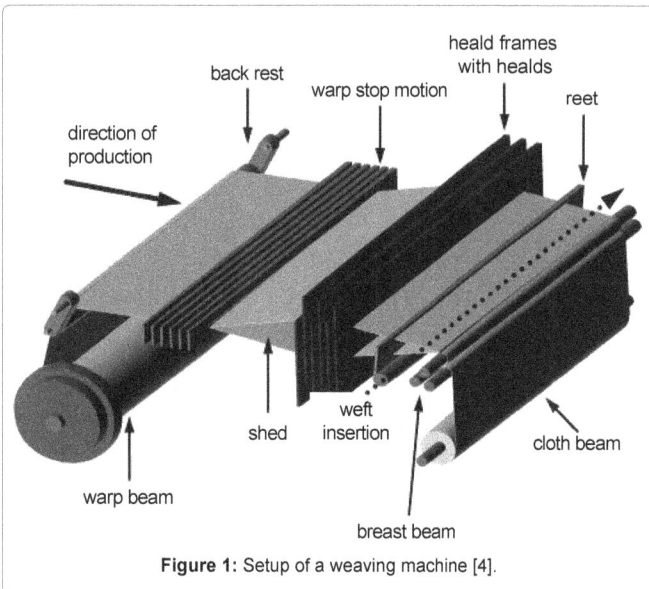

Figure 1: Setup of a weaving machine [4].

- Basic warp tension (bwt),
- Revolutions per minute (n),
- Vertical warp stop motion position (wsm_y),

With the MOSO of the weaving process, a weaving machine is enabled to automatically find an optimal configuration. A program for self-optimization is implemented in a programmable logic controller (PLC). Figure 2 provides an overview of the required hard- and software infrastructure.

Signal processing and control system

For the signal processing and the execution of the self-optimisation routine the ibaPADU-S Module system by iba AG, Fürth, Germany is used. The System consists of the following modules:

- ibaMS16xAI-20 mA: Analog input module for current signals in the range of (0…20) mA
- ibaMS16xAl-10 V: Analog input module for voltage signals in the range of (-10…10) V
- ibaPADU-S-IT-16: Central Processing Unit (CPU) for the modular system

The analog input modules ibaMS16xAI-20 mA and ibaMS16xAl-10 V collect and process the signals from the sensor system. Both analog input modules are connected to the base unit ibaPADU-S-IT-16 using a back panel bus. The base unit receives the data from the analog input modules through the back panel bus.

The central unit ibaPADU-S-IT-16 is connected to a computer using the Transmission Control Protocol/Internet Protocol (TCP/IP) interface. On the computer the software ibaLogicV4 from iba AG is installed. IbaLogicV4 is a programming environment which forms a software-based programmable logic controller (soft-PLC) together with the introduced modular system. An ibaLogicV4 program is created on the computer and transmitted to the central unit using TCP/IP. The central unit provides a runtime platform for the ibaLogicV4 program (runtime system). In the ibaLogicV4 program the data from the analog input modules are collected and processed. The program for MOSO is developed within the environment of ibaLogicV4 and uses

the ibaPADU-S-IT-16 module as runtime system. The program for MOSO is presented in section II.D.

Measurement technology for warp tension

For measuring the warp tension, the yarn tension sensor TS44/A250 by BTSR International S.p.A. Partita, Olgiate Olona, Italy is used. The yarn tension sensor generates a voltage signal in the range of (0…10) V which is proportional to the present yarn tension. The yarn tension sensor is placed in the middle of the weaving machine, between the back rest and the warp stop motion. The data connector of the yarn tension sensor is connected to the soft-PLC using an analog/digital converter. For additional information on the yarn tension sensor.

Measurement technology for energy consumption

Air consumption measurement: The air consumption of the weaving machine is measured using the flow sensor SD8000 by ifm Electronic GmbH, Essen, Germany. The flow sensor generates a signal, which is proportional to the compressed air consumption in the range of (4…20) mA. The output data from the flow senor are wirelessly transferred to the soft-PLC.

Active power measurement: The power measurement module collects characteristic values of the three-phase supply and saves the values into the process image. To access the measurement values from the power measurement module, the power measurement module is connected to a Fieldbus controller using a terminal bus.

The process image of the power measurement module is provided to the fieldbus controller via the terminal bus. The fieldbus controller is connected to the soft-PLC using the Transmission Control Protocol/Internet Protocol (TCP/IP) Interface. The communication between fieldbus controller and soft-PLC is carried out in the Modbus-Protocol format. The soft-PLC sends out a specific request (Request) in the Modbus-Protocol formal to the fieldbus controller and receives the requested value from the process image (Response).

As soon as the response has been received by the soft-PLC, the requested data are available for the signal processing. Both, Modbus-Request and Modbus-Response, consist of binary codes and are organised as bytes. The runtime platform with soft-PLC is the Modbus-

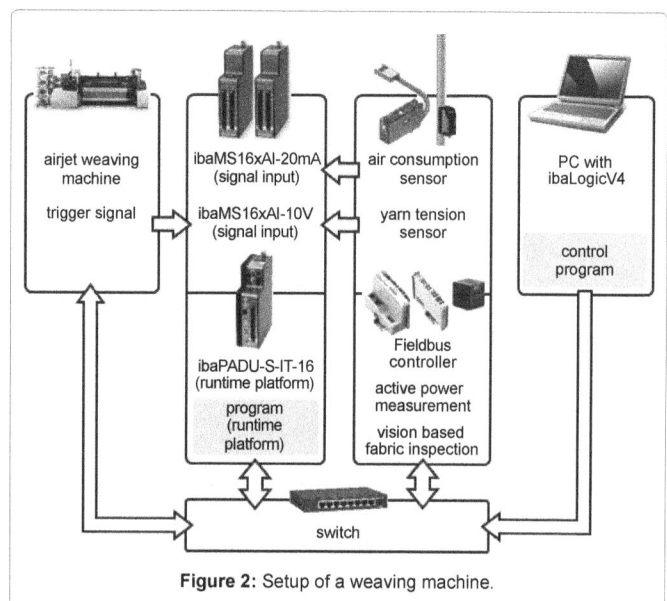

Figure 2: Setup of a weaving machine.

Master and sends the request to the fieldbus controller, which is the Modbus-Slave. The request contains information regarding the requested value from the process image of the power measurement module. In response to the request the fieldbus controller identifies the requested value and stores it into the response. The response is send via TCP/IP interface to the Modbus-Master. In the scope of this paper the active power is requested from the process image of the power measurement module.

Measurement technology for fabric quality: At Institut für Textiltechnik der RWTH Aachen University (ITA), Aachen, Germany a measuring system for online error detection during fabric production was developed [6]. A camera takes pictures of the fabric. Subsequently the pictures are checked for defects in the fabric, using digital image processing. The software for digital image processing runs on a separate computer. The camera system is installed over the section of the weaving machine where the fabric is created, as shown in Figure 3.

The camera system is able to detect defects immediately after the fabric is produced. The digital image processing software is calibrated using a flawless piece of fabric. The digital image processing classifies deviations from the calibrated condition as a defect [6]. Depending on the share of incorrect pixels in the pictures, the fabric is assigned to a quality category.

The examination for defects is carried out in real time during the weaving process. The computer running the digital image processing is connected to the soft-PLC via TCP/IP interface. Depending on the status of the fabric, the number of the quality category (0 to 4) is continuously transmitted via TCP/IP. A quality category of 0 is achieved in case the fabric quality is accurate. A quality category of 4 stands for a destroyed fabric.

Program steps of multi-objective self-optimization

The program for MOSO consists of the steps shown in Figure 4. Though continuous communication between weaving machine and soft-PLC, the weaving machine is enabled to run the entire program autonomously.

In the first step an experimental design is calculated automatically. Within this design, the three setting parameters static warp tension, vertical position of warp stop motion and revolutions per minute are varied. The user sets the parameter spaces to ensure that the algorithm acts within a feasible range.

During the second step, the test procedure, the weaving machine sets-up every test point. Sensor data describing the objective functions are recorded for the respective parameter setting.

In the third step, the obtained data are used to calculate three regression models (one model per each objective function) which describe the objective functions in dependence of the setting parameters.

In the last step, an optimized set-up of the weaving machine based on predefined quality criteria is calculated by application of desirability functions and a numerical optimization algorithm. Before execution of the optimisation procedure, user-defined preferences regarding the objective functions (warp tension, energy consumption and fabric quality) can be integrated through target weights. The preference scale for each objective function is divided into three sections (low, middle, high).

The program for MOSO is implemented within the ibaLogicV4

programming environment and runs on a central processing unit as depicted in section II.A.

The next chapter illuminates desirability functions and the optimization algorithm used for MOSO.

Desirability Functions and Nelder/Mead Algorithm

Desirability functions

The origin of the application of desirability functions in the multi-dimensional optimization goes back to Derringer and Suich [7]. The aim of using desirability functions is to summarize the objective functions which need to be optimized into one common function. The aggregation of the objective functions is conducted using the so-called desirability. For each objective function, one desirability function is developed. The desirability function assigns a desirability to each value of the objective functions. The desirability function has a value range of (0;1). If the value of one objective function reaches a desirability of zero, the result is invalid within the optimization routine. In case the desirability reaches the value one, the value of the objective function is optimal. In Figure 5 the exemplified shape of a desirability function is shown. The desirability w_z is plotted over the normalized objective function Z(X). Desirability functions can be constructed in three different ways, as shown in Figure 5. If the goal is to achieve the highest possible value for one objective function, the desirability f unction for aximizing has to be used. The desirability increases when the objective function value increases, etc.

The aim of the utilization of desirability functions is to aggregate the target functions into one common function, the so-called total desirability d_{tot}. d_{tot} is calculated by using the geometric mean of the individual desirabilities:

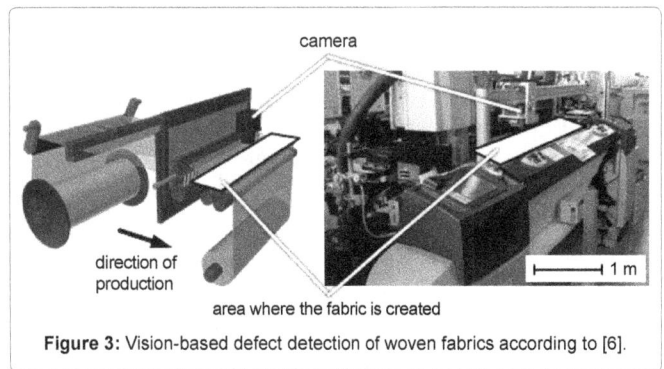

Figure 3: Vision-based defect detection of woven fabrics according to [6].

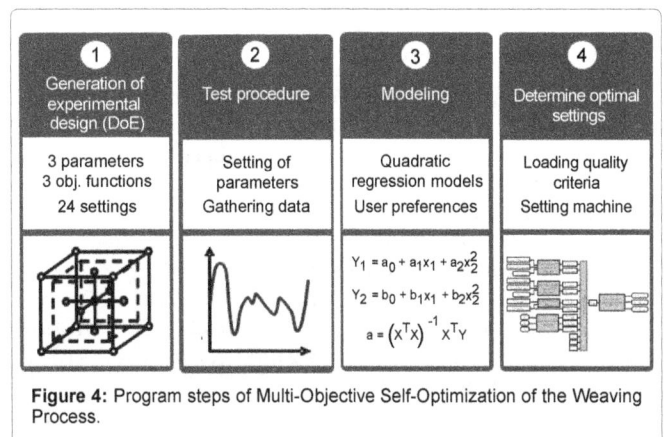

Figure 4: Program steps of Multi-Objective Self-Optimization of the Weaving Process.

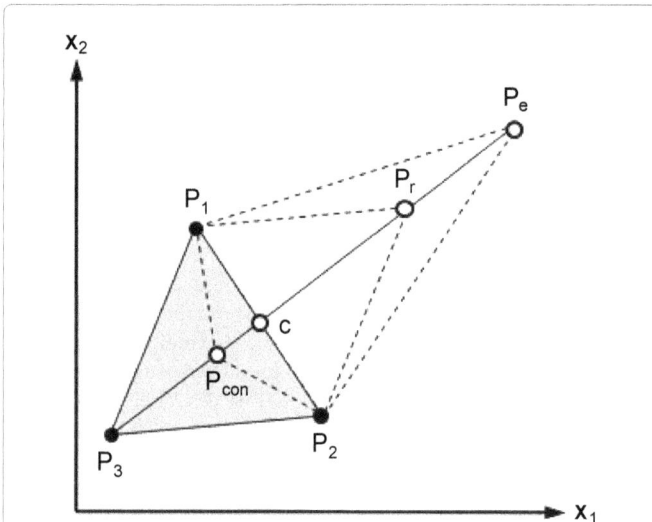

Figure 5: Basic operations of Nelder/Mead algorithm by the example of two parameters x_1 and x_2 (P_1, P_2, P_3: Calculated objective function values by algorithm; c: focal point; P_r: reflection; P_e: expansion; P_{con}: contraction.

$$d_{tot}=(w_1 \cdot w_2 \cdot \ldots \cdot w_n)^{1/n} \tag{1}$$

Whereas w_1, w_2, ..., w_n are the desirabilities of n objective functions.

The total desirability reveals how close the individual desirabilities are to the optimal range. Because of the multiplication of the individual desirabilities, d_{tot} is in the range of (0;1). A total desirability of one is reached, when all target functions are in the optimal range. In case only one target function has an invalid value, the total desirability equals to zero.

The combination of process parameters which maximizes d_{tot}, represents the optimal operating point for the weaving process.

Numeric algorithms are suitable for maximizing the total desirability. The application of numeric algorithms is more efficient than for example grid search methods [8]. It is advised in several references to utilize Nelder/Mead algorithm [9] to maximize the total desirability, see e.g. [10,11].

Nelder/mead algorithm

The Nelder/Mead algorithm is a numeric optimization procedure [9]. To find a subjective optimal operating point of the weaving machine, d_{tot} is maximized. The Nelder/Mead algorithm searches for a combination of the three parameters basic warp tension, revolutions per minute and vertical warp stop motion position that maximizes d_{tot}. The basic operations of Nelder/Mead algorithm are shown in Figure 6.

Setting the start values for the considered parameters leads to the starting point for the algorithm. The start values are set before the first iteration and are moved towards the optimal values during the utilization of the algorithm. Starting from a minimization problem with m parameters, the algorithm considers m+1 parameter combinations $(P_1, P_2, ..., P_{m+1})P_1$. The values of the objective functions functions are calculated in the m+1 points and sorted ascendingly. Figure 6 shows the minimization problem for a function of two parameters. The next step is the examination of the three points P_1, P_2 and P_3 in the parameter space. At each of the three points the algorithm calculates the value ob the objective functions $F(x_1,x_2)$:

$$F_i=F(P_i), i=1, 2, 3 \tag{2}$$

Afterwards the function values are sorted:

$$F_1 \leq F_2 \leq F_3 \tag{3}$$

Considering this example, F_3 is the worst (highest) and F_1 is the best (lowest) value in the context of the optimization. The minimization of the target function is achieved by applying several iterations of the algorithm. In each iteration one new point in the parameter space is created, which replaces the point P_{m+1} with the biggest value of the function to be minimized. In the present case the point P3 results is the worst value of the target function and is therefore replaced in the next iteration. The replacement of the worst point is achieved through the basic operations of the Nelder/Mead algorithm which are visualized in Figure 6.

Experimental Results

In this chapter, the ibaLogicV4 program for MOSO of the weaving process is validated during a long-term test in the laboratory of ITA. To establish industrial conditions, the duration of the long-term test is eight hours, like usual shift duration. A long-term test is carried out using the MOSO against not using the optimization procedure respectively, to examine the influence of MOSO on production figures. The long-term test is conducted with an air-jet weaving machine OmniPlus 800 by Picanol n.v., Ieper, Belgium. During the long-term test, polyester filament yarn with 330 dtex for warp and weft was used (binding: twill 3/1). The configuration of MOSO used for the long-term test are listed in Table 1.

After program execution of MOSO (Figure 4), the algorithm calculates the following optimal parameter settings: bwt=3,71 kN; n=522 RPM; wsm_y=20 mm.

The following settings are used as reference settings coming from an

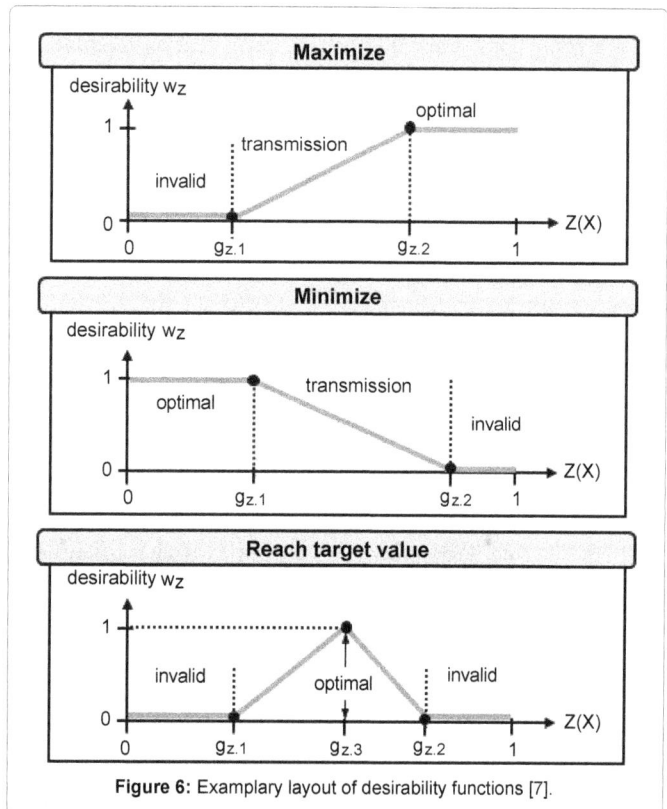

Figure 6: Examplary layout of desirability functions [7].

Setting	Value
Lower/upper limit bwt	2 kN/4 kN
Lower/upper limit n	400 RPM/900 RPM
Lower/upper limit wsm_y	0 mm/20 mm
Target weight warp tension	LOW
Target weight energy consumption	LOW
Target weight quality	HIGH
Algorithm start point bwt/n/wsm_y	3,5 kN/750 RPM/15 mm

Table 1: Configuration of MOSO used for long-term test.

Recorded data	Results	
	MOSO	Ref. settings
Efficiency (prod. time/total time)	98,6%	97,2%
Produced fabric	8,16 m	15,41 m
Weft insertions	125157	215982
Weft defects	2	6
Warp breakages	0	0
Average warp tension	1,27 N	1,49 N
Average air consumption	134,23 m³/h i. N.	155,26 m³/h i. N.
Average active power usage	2,49 kW	4,62 kW
Average quality category	0,93	1,55
Set-up time	30 Min.	120 Min.

Table 2: Results of long-term test.

industrial weaving mill that processes the same material as mentioned above: bwt=4 kN; n=900 RPM; wsm_y=0 mm. During the long-term test the following parameters of the weaving process are recorded:

- efficiency of the weaving machine
- produced amount of fabric
- amount of weft insertions
- warp/weft defects and breakages

Additionally, data of the objective functions are recorded. The results of the long-term test using MOSO and reference settings are shown in Table 2.

During the long-term test, sensor data regarding the objective functions are recorded using the software ibaPDA from iba AG, Fürth, Germany. The measured data is illustrated in Figures 7-10. Data are plotted over the main shaft position of the weaving machine which is the rotating angle of the machine's main drive.

The program for self-optimization enables the weaving machine to autonomously find an operating point, which improves all objective functions compared to conventional (reference) machine settings.

The target weight of fabric quality was set high and correlates with the course of the quality category displayed in Figure 10. MOSO is able to find an operating point of the weaving machine, in which the quality category reaches the optimal value of zero almost constantly.

The efficiency presents the relation of production time of the weaving machine to the total time. The efficiency of the weaving machine is higher using the optimal setting than in case of using the reference settings, see Table 2. Higher efficiency is mainly achieved by reduced machine downtime.

Using the optimal machine settings, two weft defects caused be the collision of the weft threads with sagging warp threads occurred.

In contrast, using the suboptimal settings, six weft defects occurred. During the long-term test it was observed, that the machine runs more stable with less RPM. The higher amount of weft defects can be explained by a disadvantageous machine speed of 900 RPM. Weft defects result from the faulty transport of weft threads across the width of the weaving machine.

Without MOSO a machine operator needs around 120 min. for the configuration of the weaving machine and to find appropriate settings for the process. The program for self-optimization is concluded in 30 minutes and successfully reduces the set-up time by 75%.

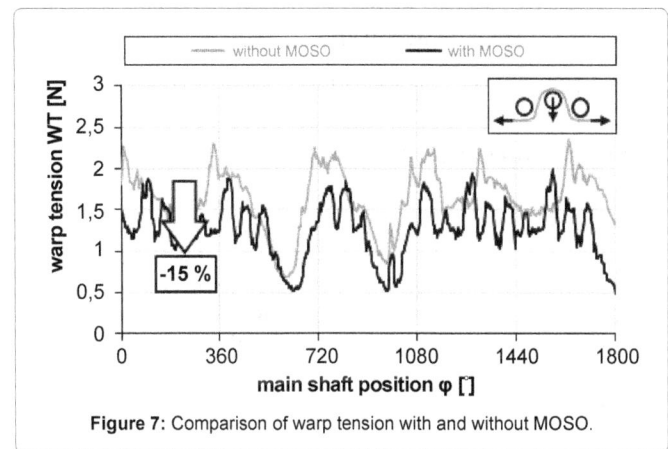

Figure 7: Comparison of warp tension with and without MOSO.

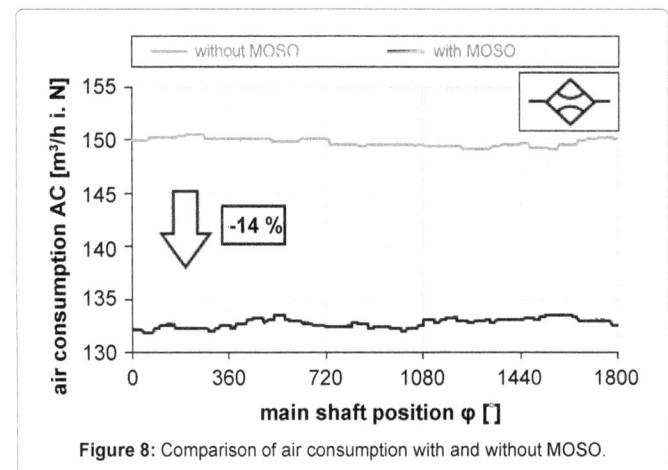

Figure 8: Comparison of air consumption with and without MOSO.

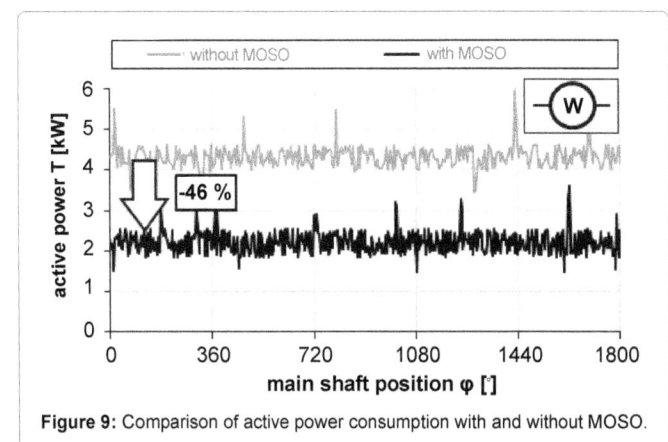

Figure 9: Comparison of active power consumption with and without MOSO.

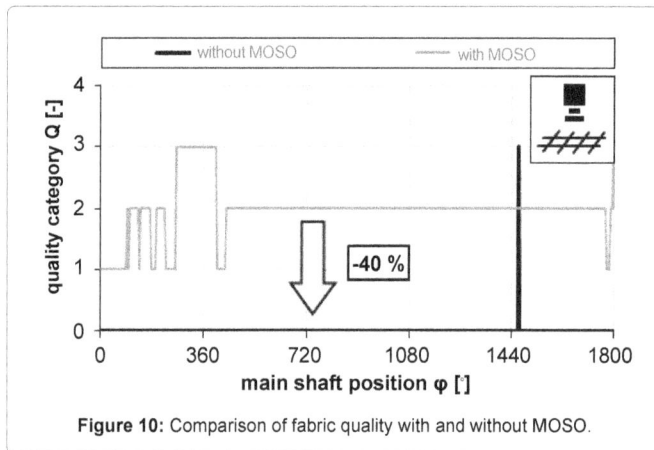

Figure 10: Comparison of fabric quality with and without MOSO.

Summary and Outlook

This paper presented a concept and implementation of multi-objective self-optimization of the weaving process. An optimization routine which is implemented into a software-based programmable logic controller enables the weaving machine to calculate optimal parameter settings autonomously. Individual preferences of operators or plant management are integrated into the calculations. By assistance of the resulting cognitive weaving machine, the set-up time was reduced by 75%.

Further research will focus the development of mobile applications to form an assistance system for operators basing on the presented results. Moreover, the cognitive weaving machine will be embedded into an intelligent textile process chain in the sense of Industrie 4.0 where all production units are interconnected to each other.

Acknowledgments

The authors would like to thank the German Research Foundation DFG for the kind support within the Cluster of Excellence "Integrative Production Technology for High-Wage Countries".

References

1. Brecher C (2011) Integrative Production Technology for High-Wage Countries. Springer-Verlag, Berlin/Heidelberg 1: 747-1057.

2. Osthus T (1996) Process optimization and changeover time reduction for weaving through automatical adjustment of backrest and warp stop motion. Dissertation, Rheinisch Westfälische Technische Hochschule Aachen, Aachen, Germany. Original title: "Prozessoptimierung und Rüstzeitverkürzung in der Weberei durch automatische Einstellung von Streichbaum und Kettwächterkorb"

3. Chen M (1996) Computergestützte Optimierung des Webprozesses bezüglich Kettfadenbeanspruchung und Kettlaufverhalten. Dissertation, Universität Stuttgart, Stuttgart, Germany.

4. Adanur S (2001) Handbook of Weaving. CRC press, Taylor and Francis, Boca Raton, London, New York.

5. Gausemeier J, Rammig FJ, Schäfer W, Josef F (2009) Design Methodology for Intelligent Technical Systems: Develop Intelligent Technical Systems of the Future. Springer Science and Business Media.

6. Schneider D, Gloy YS, Merhof D (2015) Vision-Based On-Loom Measurement of Yarn Densities in Woven Fabrics. Instrumentation and Measurement, IEEE Transactions 64: 1063-1074.

7. Derringer G, Suich R (1980) Simultaneous Optimization of Several Response Variables. Journal of Quality Technology 4: 214-219.

8. Blobel V, Lohrmann E (1998) Statistische und numerische Methoden der Datenanalyse.

9. Nelder JA, Mead R (1965) A Simplex Method for Function Minimization. The Computer Journal 4: 308-313.

10. Bera S, Mukherjee I (2010) Performance Analysis of Nelder-Mead and A Hybrid Simulated Annealing for Multiple Response Quality Characteristic Optimization. Proceedings of the International MultiConference 3: 1728-1732

11. Gloy YS, Sandjaja F, Gries T (2015) Model-based self-optimization of the weaving process. Journal of Manufacturing Science and Technology (CIRP) 9: 88-96.

18

Direct Dyeing of Jute: Effect of Cationic Treatments on Color Fastness

Sarwar Z*, Azeem A, Munir U and Abid S

Department of Postgraduate Studies, National Textile University, Faisalabad, Pakistan

Abstract

Direct dyes represent one of the cheapest and the simplest dyeing systems usually require only an electrolyte for their application. They are widely used in the textile industry because they are cheap and the only problem is their fastness properties that were solved in this research. Different techniques have been developed to enhance their fastness properties, one of them being the use of a cationic dye fixing agents. Bleached jute fabrics were dyed with direct dye before and after the treatment of cationizing agents and their properties are compared. It was found that in terms of fabric rubbing fastness, washing fastness and K/S value, cationization after dyeing is superior to cationization before dyeing.

Keywords: Jute; Cationizing agent; Direct dye; K/S value; Rubbing fastness; Washing fastness

Introduction

Jute fiber consists of strands i.e., bast bundle fiber assemble in parallel manner with overlapping to produce filaments throughout the length of the stalk. It is physically coarse, meshy, harsh, and irregular in length and diameter. Jute contains bundle of fibers which are joined together with natural cementing materials as lignin and hemi-cellulose [1]. The eco-friendly and bio-degradable nature of jute fiber along with its tenacity and long staple length has made this fiber popular in the field of textile and other decorative end uses products (Table 1) [2].

Direct dyes are sodium salt of sulphonic acid and most of them contain an azo group as the main chromophore. Direct dye is a class of dyestuffs that are applied directly to the substrate in a neutral or alkaline bath. Direct dyes give bright shades but exhibit poor wash-fastness. Various after-treatments are used to improve the wash-fastness of direct dyes, and such dyes are referred to as "After-treated Direct Colors". Direct dyes are molecules that adhere to the fabric molecules without help from other chemicals. Direct dyes are defined as anionic dyes with substantivity for cellulosic fibres, normally applied from an aqueous dyebath containing an electrolyte, either sodium chloride (NaCl) or sodium sulfate (Na_2SO_4) [3]. During dyeing process, about 20% dye hydrolyzed with water and drained out with water which is not eco-friendly [4]. Direct dyes thus pollute the environment by discharging highly colored species and higher electrolyte concentration. Higher electrolyte concentrations in effluents cause worse effects such as impairing the delicate biochemistry of aquatic organisms, destructive attack on concrete pipes is sodiumsulphate is used as electrolyte due to the formation of alumino-sulphato complexes which swell and crack concrete with considerable alumina content. Evolution of hydrogen sulphide gas under anaerobic conditions when sodium sulphate is used as electrolyte. Dissolution of such sulphide and subsequent bacterial oxidation to harmful sulphuric acid [5]. To overcome these problems and improve the dye ability of fabric the surface modification of fabric is done. This can be done by treating the fabric with strong cationic reactive agent which react with hydroxyl groups of fabric and produce positive charge on the surface of the fabric. By introducing cationic group, the fabric become cantonized and has columbic attraction between cationic fabric surface and anionic dyestuff. This cantonized fabric can be dyed without the use of electrolyte (Table 2) [6].

Cationization is the chemical modification of cellulose to produce cationic (positively charged) dyeing sites in place of existing hydroxyl (-OH) sites at which negatively charged dye can attach. Dyeing cationic-treated fabric results in greater use of dye and higher color values (Table 3). In addition, the strong dye-fiber interactions resulting from cationizing allow dyeing with no added electrolytes and minimal rinsing and after washing. Cationized fabric shows increase in the uptake of direct dyes, acid dyes and reactive dyes [7].

S. No.	Types of cationizing agents	Concentration (%)	Dye (%)
1	Ultrafix WS Conc.	1	3
2	Ultrafix WS Conc.	2	3
3	Cyclanon fast HWF	1	3
4	Cyclanon fast HWF	2	3
5	(3-chloro-2-hydroxypropyl) trimethyl ammonium chloride	1	3
6	(3-chloro-2-hydroxypropyl) trimethyl ammonium chloride	2	3

Table 2: Application of cationizing agent before dyeing.

Direct dye (RED)	Concentration (%)	Washing fastness	K/S value	Rubbing fastness	
				Dry	Wet
Simple dye	3	01-Feb	20.56	03-Apr	2

Table 3: Properties of controlled sample.

S. No.	Types of cationizing agents	Dye (%)	Concentration (%)
1	Ultrafix WS Conc.	3	1
2	Ultrafix WS Conc.	3	2
3	Cyclanon fast HWF	3	1
4	Cyclanon fast HWF	3	2
5	(3-chloro-2-hydroxypropyl) trimethyl ammonium chloride	3	1
6	(3-chloro-2-hydroxypropyl) trimethyl ammonium chloride	3	2
7	Simple dye	3	

Table 1: Application of cationizing agent after dyeing.

*Corresponding author: Sarwar Z, Department of Postgraduate Studies, National Textile University, Faisalabad, Pakistan, E-mail: zahidsarwar38@yahoo.com

The literature review exposed that a lot of study has been done on the dyeing of jute. Jute is mostly dyed with natural and reactive dyes. These dyes give good fastness properties and these dyes are costly so the dyeing process become costly, but dyed jute is mostly used in packing, fashion and apparel industry where fastness is not a priority. So it is imperative that we dye the jute with that type of dyes that are not costly and their fastness properties are good [1,2,8-14]. In the present work, an attempt was made to dye the jute fabric with direct dye and check the effect of different cationizing agents on fastness properties (Table 4).

Materials and Methods

Substrate

Grey jute fabric was taken from Nishat Textile Bikhi.

Chemicals

The following chemicals of analytical grade were used in the experiment: hydrogen peroxide, sodium hydroxide, wetting agent, sequestering agent, stabilizer, salt and acetic acid.

Cationizing agent

Three cationizing agents were used.

I.   Ultrafix WS Conc.

II.   Cyclanon fast HWF

III.   (3-Chloro-2-hydroxypropyl) trimethyl ammonium chloride.

Methods

Bleaching

Bleaching was done on lab-scale jigger machine at 95°C for 45 minutes by using 40 g/L hydrogen peroxide, 24 g/L caustic soda, 4 g/L wetting agent, 4 g/L sequestering agent and 14 g/L stabilizer followed by rinsing with tap water.

Cationizing treatment

Three types of cationizing agents with two concentrations of each were used. On six samples, cationizing agents were applied before dyeing and six samples were cationized after dyeing.

Dyeing

Jute was dyed by using red direct dye. Seven samples were dyed by using 1 g/L dye, 2 g/L alkali, 1 g/L salt, 2 g/L wetting agent and 2 g/L sequestering agent. Six samples on which cationizing agent was applied before dyeing, were dyed without alkali and salt.

Experimental Design

Testing

Dry and wet rubbing fastness values of the dyed samples were evaluated according to AATCC TM-08. Washing fastness values of dyes samples were evaluated according to AATCC test method 61-2010. K/S values of dyed samples were evaluated according to the AATCC 6-2008 (Table 5).

Results and Discussion

Rubbing fastness

A comparison of rubbing fastness of cationizing agent apply before dying and after dyeing shows in Figure 1. It can be noticed that both dry and wet rubbing fastness of cationizing agent apply before dyeing are poor in comparison to cationizing agent apply after dyeing.

Washing fastness

A comparison of washing fastness of cationizing agent apply before dying and after dyeing shows in Figure 2. It can be noticed that washing fastness of cationizing agent apply before dyeing is poor in comparison to cationizing agent apply after dyeing.

K/S value

A comparison of K/S value of cationizing agent apply before dying and after dyeing. It can be noticed that K/S Value of cationizing agent apply before dyeing is poor in comparison to cationizing agent apply after dyeing.

Conclusion

It is concluded from above research that application of cationization agent after dyeing give better results as compared to application

Cationizing agent	Concentration (%)	Washing fastness	K/S value	Rubbing fastness	
				Dry	Wet
Ultrafix WS	1	03-Apr	23.76	4	03-Apr
Ultrafix WS	2	03-Apr	30.68	4	3
Cyclanon Fast HWF	1	03-Apr	18.24	03-Apr	01-Feb
Cyclanon Fast HWF	2	03-Apr	19.85	03-Apr	01-Feb
CHTAC	1	03-Apr	7.02	4	03-Apr
CHTAC	2	03-Apr	18.78	4	03-Apr

Table 4: Properties of samples treated with cationizing agent after dying.

Cationizing agent	Concentration (%)	Washing fastness	K/S value	Rubbing fastness	
				Dry	Wet
Ultrafix WS	1	3	12.32	4	3
Ultrafix WS	2	3-4	9.11	3-4	1-2
Cyclanon Fast HWF	1	3	11.91	3-4	1-2
Cyclanon Fast HWF	2	1-2	7.99	4	3
CHTAC	1	1-2	15.85	3-4	1-2
CHTAC	2	1-2	7.56	4	3

Table 5: Properties of samples treated with cationizing agent before dying.

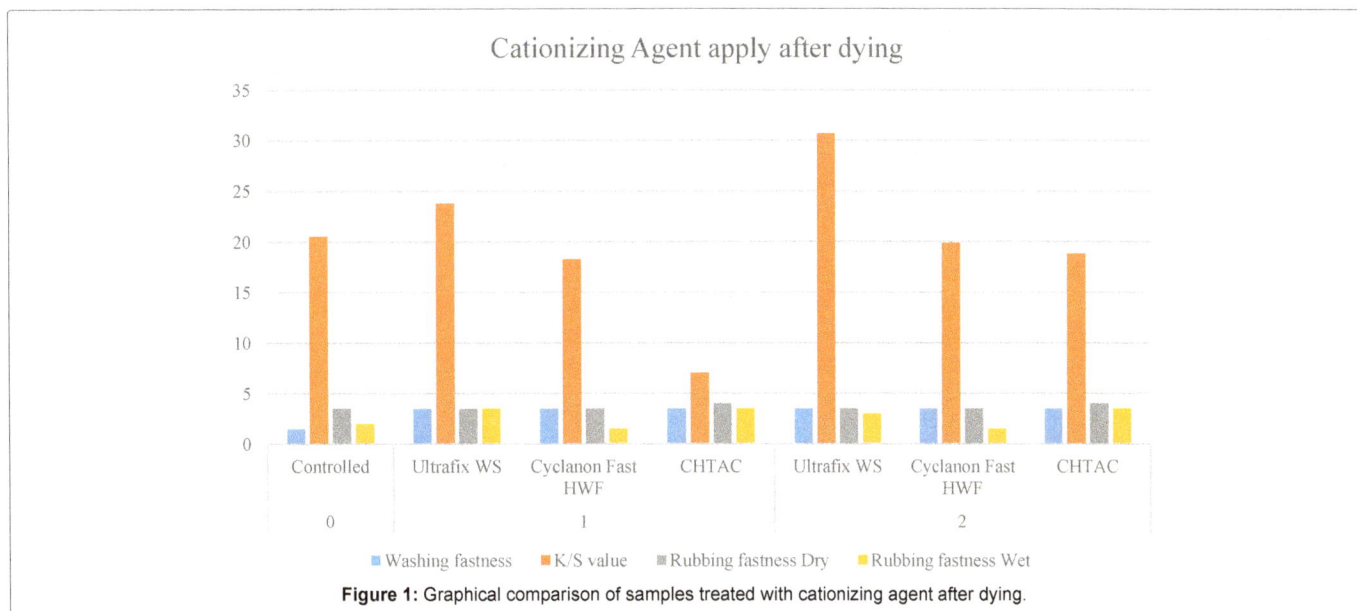

Figure 1: Graphical comparison of samples treated with cationizing agent after dying.

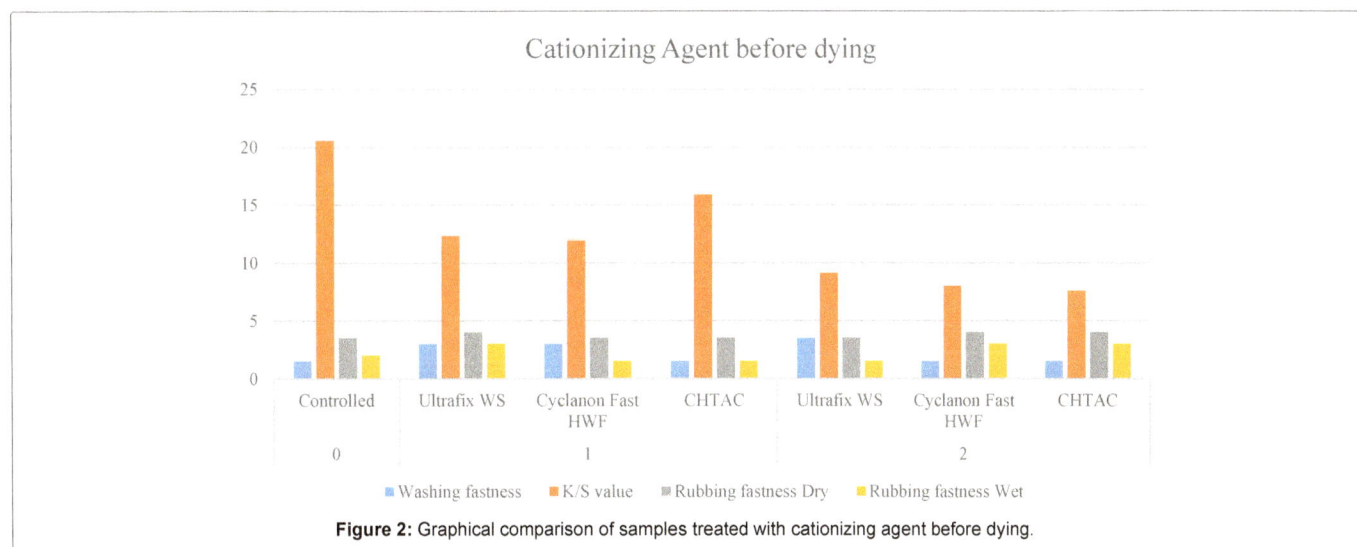

Figure 2: Graphical comparison of samples treated with cationizing agent before dying.

before dyeing. The use of cationizing agent increase the dye fixation by chemical reaction of positively charged substrate and negatively charged dye molecule. This method of dyeing is ecofriendly causes it is free of salt and percentage of dye in effluent is reduced.

Future Work

- Development of such cationic fixer which is cheapest easily available in market.

- Development of such cationic fixer which easily attached with fabric and strongly bonded.

- Development of such cationic fixer which effective at less concentration.

- To measure the effluent of the dye bath of both cationized and un-cationized jute using different cationizing agents.

References

1. Reza S, Haque AKMM, Yousuf A, Anwarul M, Hoque A (2014) Color Build up on Jute Fabric with Reactive Dye after Bleaching and Mercerizing. IJAERS1: 58-62.

2. Tec F, Park R (2002) Effect of Pretreatments on Ambient Temperature Bleaching and Reactive Dyeing of Jute. IJFTR 27: 417-421.

3. Esche SP (2004) Direct dyes-Their application and uses. AATCC 4: 14-17.

4. Bhatt P, Rani A (2013) Textile dyeing and printing industry: An environmental hazard. Asian Dye10: 51-54.

5. Ramasamy M, Kandasaamy PV (2005) Effect of cationization of cotton on its dye ability. Indian Journal of Fibre and Textile Research 30: 315-323.

6. Montazer RMA, Malek, Rahimi A (2007) Salt free reactive dyeing of cationized cotton. Fibers Polymer 8: 608-612.

7. Choudhury BAKR (2014) Coloration of Cationized Cellulosic Fibers-A Review 1.

8. Bhuiyan MAR, Shaid A, Bashar MM, Haque P, Hannan MA (2013) A Novel Approach of Dyeing Jute Fiber with Reactive Dye after Treating with Chitosan: 87-91.

9. Rashid MA, Hossain D, Islam A, Engineering T (2013) Evaluation of effective jute dyeing process with basic dye and reactive dye: 2319-2322.

10. Mondal IH, Islam K (2014) Effect of pH on the dye absorption of jute fibre dyed with Direct Dyes. An International Research Journal of Pure and Applied chemistry 30: 1-5.

11. Chattopadhyay SN, Pan NC, Day (2006) Reuse of reactive dyes for dyeing of jute fabric. Bioresour Technol 97: 77-83.

12. Chattopadhyay SN, Pan NC, Day (2002) Ambient-temperature bleaching and reactive dyeing of jute: the effects of pre-treatment, bleaching, and dyeing methods. Journal of Textile Institute 93: 306-315.

13. Ahmed Z (2009) Effect of Different Pretreatment on Various Properties of Undyed and Dyed Jute Fabrics 6: 1-9.

14. Imtiazuddin BSM, Tiki S, Chemicals AVM (2012) What is the Jute fiber ? Bleaching and Dyeing Processes of Jute Fiber: 36-37.

19

Effect of Bleaching Agents on Colour Depth of Jute Fabric Dyed with Natural Dyes

Patel S[1], Sharan M[1] and Chattopadhyay DP[2*]

[1]Faculty of Family and Community Sciences, Department of Clothing and Textiles, The Maharaja Sayajirao University of Baroda, Vadodara, India
[2]Faculty of Technology and Engineering, Department of Textile Chemistry, The Maharaja Sayajirao University of Baroda, Vadodara, India

Abstract

Raw jute fabric was subjected to two bleaching agents namely hydrogen peroxide and peracetic acid. Bleached jute fabrics were pre mordanted with alum, copper sulphate and ferrous sulphate and then dyed with four natural dyes: madder, turmeric, eucalyptus leaves and Indian almond leaves. The effect of bleaching agents on various physical properties like weight loss, tensile strength, whiteness, yellowness and brightness indices were studied. The effect of bleaching on colour development using different natural dyes was examined. Peracetic acid bleached samples were found to have high degree of whiteness with very less damage compared to hydrogen peroxide bleached samples. Peracetic acid bleached samples were found to be relatively darker.

Keywords: Jute; Bleaching; Fabric; Natural dyes

Introduction

During the past decade, increasing environmental awareness, new global agreements and international governmental regulations have been driving forces behind the renewed interest in the natural fibres. The attractiveness of a plant-based fibre comes from its high-specific strength and stiffness, natural availability and environmental friendliness [1,2].

Jute is a natural biodegradable fiber with advantages such as high tensile strength, excellent thermal conductivity, coolness, ventilation function etc. Besides its traditional usages like sackings, hessian, carpet backing, etc. jute is now-a-days being used to produce various fancy and household products. Jute is a natural fibre, its environmental friendliness remains intact where dyed with natural dyes and colour effect on this fibre is enhanced by proper bleaching [3,4].

Bleaching holds the key for successful production of diversified jute products. The primary object of jute bleaching is to improve its whiteness and to have better look after dyeing [5].

Substantial research has been carried out on the effects of different bleaching agents on jute and dyeing of jute with natural dyes but the correlation between effects of bleaching agents on dyeing performance of jute using different natural dyes is hardly available in the literature. In the present investigation, therefore an attempt has been made to study the effect of various bleaching agents on the physical properties of jute fabric as well as its effect on the dyeing properties when dyed with natural dyes.

Materials and Methods

Material

Fabric: Locally available jute fabric with the following specifications was selected for this study (Table 1).

Dyes and chemicals: Natural dyes for the study were Madder, Turmeric, Eucalyptus Leaves and Indian almond leaves. The mordants for the study were used Alum, Copper Sulphate and Ferrous Sulphate. Sodium carbonate and detergent were used for the scouring of jute fabric. Glacial Acetic acid and Hydrogen Peroxide (30%) were taken for the preparation of Peracetic acid. All the chemicals used for the study were LR grade.

Methods

Scouring: The fabric was scoured using 2 g/L of soda ash and 2 g/L of detergent keeping material to liquor ratio 1:40 at 85°C for 30 minutes. The samples were then rinsed in water to remove traces of soap and dried in shade. The scoured fabric was subjected to two bleaching systems namely hydrogen peroxide and peracetic acid bleaching.

Bleaching systems

Hydrogen peroxide: The scoured fabric was bleached with 1% (v/v) of hydrogen peroxide using 2 g/l of sodium silicate as stabilizer maintaining the material liquor ratio 1:40. The fabric was treated in this solution for 30 minutes at 50°C. The samples were then neutralized using dilute acetic acid.

Peracetic acid: Peracetic acid is prepared by reacting hydrogen peroxide with acetic anhydride or acetic acid. The most widely used method for the preparation of peracetic acid is direct, acid catalyzed reaction of 30-98% hydrogen peroxide with acetic acid. For this study, peracetic acid was prepared by reacting hydrogen peroxide (30%) with glacial acetic acid in 1:2 molar ratio at room temperature for 24 hours using sulphuric acid as catalyst. The scoured jute fabric samples were treated with freshly prepared peracetic acid for 1 hr at room temperature.

Extraction of dyes: As a natural dye source, four dyes namely Madder, Turmeric, Eucalyptus leaves and Indian Almond leaves were collected. Madder and Turmeric were taken in powdered form. Eucalyptus Leaves and Indian Almond leaves were dried in shade and then powdered. In order to extract the dye, 3% (wt/vol) of the natural source of the dye was immersed in water and was boiled for 30 minutes and then filtered. The dye extract thus obtained was directly used for dyeing of the bleached samples.

*Corresponding author: Chattopadhyay DP, Faculty of Technology and Engineering, Dept. of Textile Chemistry, The Maharaja Sayajirao University of Baroda, Vadodara, India, E-mail: dpchat6@gmail.com

Mordanting: The mordants used for the study were alum (10% owf), copper sulphate (4% owf) and ferrous sulphate (4% owf). The fabric samples were treated in the aqueous solution of the mordant for 30 minutes keeping material to liquior ratio 1:40. After treatment samples were squeezed and directly taken for dyeing.

Dyeing: The dyebath was prepared with the requisite amount of dye. The pre mordanted samples were dyed using exhaust method at room temperature for 10 minutes maintaining material to liquor ratio 1:40. The temperature of the dyebath was then gradually increased to boil and the dyeing was continued for 1 hour. After dyeing, the samples were thoroughly washed, rinsed and dried.

Determination of K/S: The K/S values of the dyed jute fabric samples were determined by computer colour matching system using Spectrascan 500, Premier Colour Scan, India.

Results and Discussion

Effect of bleaching agents on physical properties

After scouring the jute fabric samples were bleached with two bleaching agents namely Hydrogen Peroxide and Peracetic acid. The effect of bleaching on various physical properties like tensile strength, weight loss, whiteness, yellowness and brightness indices were investigated, the results of which are presented in Table 2. It is clear from the results that whiteness and brightness of the sample bleached using peracetic acid was better compared with the sample bleached using hydrogen peroxide. The result of whiteness is also supported by the yellowness indices which show a reversed effect. The higher weight loss of Hydrogen peroxide bleached sample caused higher loss in tensile strength compared to peracetic acid bleached sample. Similar effect was also observed by Chattopadhyay et al. [3].

The higher weight loss in case of hydrogen peroxide is because of alkaline bleaching condition. The alkaline condition causes partial removal of hemicelluloses which is a cementing material for the ultimate cells of jute fibre. The loss in hemicelluloses weakens the fibres

Fabric	Fiber content	Weave	Fabric count (yards/sq. cm)		Weight per unit area (gms/sq.mt)
			Ends	Picks	
Jute	100% jute	Plain	25	27	875.5

Table 1: Specification of the fabrics.

Figure 1: Scanning electron micro photograph of sample bleached with Hydrogen peroxide.

and causes reduction in tensile strength as a consequence leads to loss in weight. Peracetic acid bleaching was conducted in neutral condition; hence there was no major damage to the fibre which was reflected in tensile strength also after bleaching.

The damage to the fibre in hydrogen peroxide bleaching compared to peracetic acid bleaching can be clearly seen from the SEM photographs for these two bleached sample represented in Figures 1 and 2. The surface of peracetic acid bleached sample shows lesser physical damage compared to hydrogen peroxide bleached sample.

Effect of bleaching agents on the colour depth

The bleached samples were pre- mordanted with alum, copper sulphate and ferrous sulphate and then dyed with various natural dyes like madder, turmeric, Eucalyptus leaves and Indian almond leaves. The K/S values of the dyed samples were examined after dyeing. The results of this investigation are presented in Tables 3-5.

The values in the parenthesis indicate per cent improvement in K/S values compared to hydrogen peroxide bleaching: Peracetic acid bleached samples were darker compared to hydrogen peroxide bleached counter parts for all the natural dyes used in this study which can be attributed to better whiteness achieved in case of peracetic acid bleaching. The colour depth of Indian almond leaves was found to be much darker compared to Turmeric, Madder and Eucalyptus leaves.

The values in the parenthesis indicate per cent improvement in K/S values compared to hydrogen peroxide bleaching: The results of colour depth obtained for all the dyes using copper sulphate as a mordant are shown in Table 4. Here also the colour depths of peracetic acid bleached samples were higher for all the dyes. Samples dyed with Indian almond leaves were found to be the darkest samples among all the four dyes.

The values in the parenthesis indicate per cent improvement in K/S values compared to hydrogen peroxide bleaching: Table 5 indicates the effect of ferrous sulphate as a mordant on K/S for different natural dyes. Peracetic bleaching in general was resulted higher K/S for all the dyes. The per cent improvement in K/S values of turmeric was much higher followed by madder, Indian almond leaves and eucalyptus leaves. Like copper sulphate, in case of ferrous sulphate mordant also, the sample dyed with Indian almond leaves generated comparatively darker shade.

Figure 2: Scanning electron micro photograph of sample bleached with Peracetic acid bleached sample.

Sr. No.	Bleaching Agents	Physical Properties					
		Loss in weight (%)	Loss in Breaking Load (kg) (%)	Breaking Extension (mm)	Whiteness Index	Yellowness Index	Brightness Index
1.	Hydrogen Peroxide	4.5	15.1	18.4	63.22	31.92	33.79
2.	Peracetic Acid	1.7	6.5	10.7	64.15	30.87	34.81

Table 2: Effect of bleaching agents on physical properties of jute fabric.

Sr. No.	Bleaching Agents	Natural Dye			
		Madder	Turmeric	Eucalyptus Leaves	Indian Almond Leaves
1.	Hydrogen Peroxide	10.38	12.63	6.21	25.63
2.	Peracetic Acid	11.35 (9.3%)	13.43 (6.3%)	7.32 (17.8%)	27.65 (7.9%)

Table 3: Effect of bleaching agents on K/S values of Jute fabric dyed with different natural dyes for samples pre mordanted with alum.

Sr. No.	Bleaching Agents	Natural Dye			
		Madder	Turmeric	Eucalyptus Leaves	Indian Almond Leaves
1.	Hydrogen Peroxide	7.66	7.80	7.66	18.66
2.	Peracetic Acid	8.25 (7.7%)	7.84 (0.5%)	7.75 (1.2%)	18.69 (0.2%)

Table 4: Effect of bleaching agents on the K/S values of Jute fabric dyed with different natural dyes for samples pre mordanted with Copper Sulphate.

Sr. No.	Bleaching agents	Natural Dye			
		Madder	Turmeric	Eucalyptus leaves	Indian almond leaves
1.	Hydrogen Peroxide	7.22	7.65	17.33	23.75
2.	Peracetic acid	7.56 (4.7%)	8.75 (14.4%)	17.47 (0.8%)	24.56 (3.4%)

Table 5: Effect of bleaching agents on the K/S values of Jute fabric dyed with different natural dyes pre mordanted with Ferrous Sulphate.

Sr. No.	Bleaching Agents	Natural Dye										
		Madder			Turmeric			Eucalyptus Leaves			Indian Almond Leaves		
		M1	M2	M3	M1	M2	M3	M1	M2	M3	M1	M2	M3
1.	Hydrogen Peroxide	10.3	7.6	7.2	12.6	7.8	7.6	6.2	7.6	17.3	25.6	18.6	23.7
2.	Peracetic Acid	11.3	8.2	7.8	13.4	7.8	8.7	7.3	7.7	17.4	27.6	18.6	24.5

M1: Alum; M2: Copper Sulphate and M3: Ferrous Sulphate.

Table 6: Effect of the bleaching agents on K/S values of Jute fabric dyed with natural dyes using different mordants.

Table 6 summaries the effect of different mordants on color depth of all the natural dyes used as well as the effect of bleaching agents. Different dye sources exhibited different effects for the three types of mordants. For madder, turmeric and Indian almond leaves alum was found to be a better choice so far as colour depth is concerned. Ferrous sulphate mordanting exhibited much enhanced depth for the eucalyptus leaves and was found to be quite ahead compared to the other two mordants.

When the samples dyed with natural dyes were assessed visually it was found that the sample mordanted with alum and dyed with madder was brighter and redder in colour while the sample mordanted with ferrous sulphate was duller and was maroon in colour. Alum mordanted samples for turmeric dye were brighter and yellower compared to the rest of the mordants. The ferrous mordanted sample showed a greenish tint when compared to the alum mordanted sample. The samples dyed with Indian almond leaves were of greener tint.

Conclusion

Bleaching holds the key for successful production of diversified jute products. The primary object of jute bleaching is to improve its whiteness and to have better look after dyeing. This study was aimed at investigating the effect of bleaching agents on dyeing performance of jute using natural dyes.

- The whiteness and brightness of the sample bleached using peracetic acid was found to be better compared with the sample bleached using hydrogen peroxide.

- The higher loss in weight for hydrogen peroxide bleaching may be attributed to the alkaline bleaching condition which causes partial removal of hemicelluloses. However, peracetic acid bleaching was conducted in neutral condition; hence there was no major damage to the fibre which was also reflected in tensile strength.

- The relatively higher damage to the fibre in hydrogen peroxide was also manifested by the SEM microphotograph.

- For all the natural dyes used peracetic acid bleached samples were found to develop darker shades and were found to be a better choice over hydrogen peroxide.

- Different mordants exhibited different results for all the four natural dyes used. Ferrous sulphate mordanting resulted much enhanced depth for the eucalyptus leaves and was found to be quite ahead compared to alum and copper sulphate mordants, whereas for madder, turmeric and Indian almond leaves, alum was found to lead the other mordants used, so far as K/S value is concerned.

References

1. Bhattacharya N, Doshi BA, Shasrabudhe AS (1998) Dyeing jute fibers with natural dyes. American Dyestuff Reporter 87: 26-29.

2. Chattopadhyay DP, Sharma JK, Chavan RB (2003) In-situ peracetic acid bleaching of jute. Indian Journal of Fibre and Textile Research 28: 456-461.

3. Chattopadhyay DP, Sharma JK, Chavan RB (1999) Sequential bleaching of Jute with Eco-friendly peracetic acid and Hydrogen peroxide. Indian Journal of Fibre and Textile Research 24: 120-125.

4. Wang W, Cai Z, Yu J (2008) Electrospun Fibrinogen-Polydioxanone Composite Matrix: Potential for In Situ Urologic Tissue Engineering. Journal of Engineered Fibers and Fabrics 3: 12-21.

5. Sinha E, Rout S (2008) Influence of fibre-surface treatment on structural, thermal and mechanical properties of jute. Journal of Mater Science 43: 2590-2601.

20

Engineering Fiber Volume Fraction of Natural Fiber Staple - Spun Yarn Reinforced Composite

El Messiry M* and EL Deeb R

Textile Engineering Department, Faculty of Engineering, Alexandria University, Egypt

Abstract

There is a large demand for high fiber volume fraction natural fiber reinforcement polymer composites. Among many parameters affecting the mechanical properties of a composite, the fiber volume fraction is the most decisive factor. The effect of fiber volume fraction on the physical and tensile properties of aligned natural fiber staple spun yarn composites has been intensively investigated. However, there have been no direct studies on determining the relation between composite fiber volume fraction and the yarn fiber volume fraction for fiber staple-spun yarn composite. With the intention to improve the utilization of most fibers of the yarn cross section in the composite, the variation of the yarn diameter in staple-spun yarn reinforced composite is investigated in this study. The analysis of the yarn diameter indicates that it has a high variability that necessitates an increase in the composite diameter to envelope all the yarn body leading to reduction in its fiber volume fraction. A model for the determination of composite fiber volume fraction was driven, grounded on the analysis of the yarn diameter variability along its length. An equation has also been developed to calculate the diameter of the composite to cover majority of yarn cross section.

Keywords: Fiber composites; fiber volume fraction; Thermoplastic polymer; Composite fiber volume fraction

Introduction

Renewable bio-based composite materials provide a stimulating prospect to develop sustainable materials. Natural fibers in particular are an attractive source of reinforcement for fiber reinforced plastics. The low density, low cost of raw material, high specific properties and ecological profile of plant fibers has portrayed them as a prospective replacement for E-glass in traditional fiber reinforced polymer [1].

One of the many advantages of composite materials, in general is the possibility of tailoring material properties to meet different requirements. It is well-known that the macroscopic behavior of heterogeneous fiber reinforced polymer depends on many factors; including the (volumetric) composition, the stress–strain behavior of each component, the geometrical arrangement of the phases and the interface properties [1,2]. Composites can be made with fibers as mats and as aligned assemblies impregnated with matrix polymer [3,4].

Aligned fibers composites are generally stronger and stiffer in the fiber direction than composites with randomly orientated fibers. Consequently, the aligned yarns composites have higher strength than woven and mat one [1]. The fabrication route of the aligned plant fiber yarn composites is aligned by filament-winding [3,5].

The staple fibers in a conventional twisted yarn are held together by the fiber-to-fiber friction and produced on the different spinning systems. The flax yarn fibers in the present study show very good reinforcement efficiency with a tensile strength and modulus [6].

The fiber volume fraction of fiber reinforced composite directly correlated with the mechanical properties of the composite. The ability of composites reinforced with short fibers to support loads depending on the presence of the matrix as the load-transfer an intervening substance, and the efficiency of this load transfer is directly related to the ratio of the fiber/matrix. Given the fiber volume fraction, the theoretical elastic properties of a composite can be determined. The elastic modules of a composite can be expressed as:

$$E = (1 - V_f) E_m + V_f E_f$$

Where:

V_f is the fiber volume fraction.

E_m, E_m is the elastic modulus of matrix and fiber, respectively.

The value of the volume fraction varied depending on the structure of reinforcement: unidirectional, woven, or random mat. Therefore, there is an optimal space between fibers that will fully exploit the uniform load transfer between fibers [1]. Minimum and critical fiber volume fraction, if there are very few fibers present ($0 < V_f < V_{f,min}$), the stress on a composite may be high enough to break the fibers. The broken fibers, which carry no load, can be then regarded as an array of aligned holes ($0 < Vf < Vf,min$) [7].

The reinforcing action of the fibers is only observed once the fiber volume fraction exceeds the critical fiber volume fraction ($V_f > V_{f,crit}$). Which implicitly illustrates the minimum and critical fiber volume fractions for short banana fiber reinforced vinyl-ester composites to be $V_{f,min} \cong 15\%$ and $V_{f,crit} \cong 25\%$ [8]. As the theoretical maximum fiber volume fraction of flax and jute composites is known, the maximum theoretical tensile modulus can be determined. This is found to be 17.3 GPa for flax–polyester (at V_f=33.1%) [9], for jute–polyester (at V_f=46.8%). A high $V_{f,crit}$ (of the order of 10%) and low $V_{f,max}$ (of the order of 45%) implies that the range of useful fiber volume fractions for vacuum infused PFCs containing staple fiber twisted yarns is only 35%.

As the theoretical maximum fiber volume fraction of flax and jute composites is known, the maximum theoretical tensile strength can be determined [10].

*Corresponding author: Magdy El Messiry, Textile Engineering Department, Faculty of Engineering, Alexandria University, Egypt
E-mail: mmessiry@yahoo.com

Many researchers analyzed the value of the yarn fiber volume fraction in the spun yarns [11-13], either theoretically or experimentally. The fiber distribution in the yarn cross section was found to be varied in the radial direction. The structure of the spun yarn the fibers distribution in its cross section is different [2].

The theoretical distribution of the fibers in the yarn cross section may be assumed hexagonal or square, giving different values of packing density. The yarn packing density is also a function of the yarn twist level [13]. Semi-empirical relationship between twist level T (turns per meter) and fiber packing density of staple fiber yarns indicates that its value increased with the increase of twist level. The effect of fiber twist on the tensile properties of single sisal yarns was investigated. It was shown that lower twist level led to higher mechanical properties of its reinforced composite because the polymer matrix provides enough bonding force between sisal fibers in the impregnated yarn, the low twist sisal yarn with a low twist angle would possess better tensile properties [13].

In practical cases, there is a variation in fiber diameter and irregular fiber packing increases the difficulty in calculating the $V_{f.comp}$ without apprehension the different variables existing along the yarn length.

The objectives of this work are to give a method of calculating the fiber volume fraction of composite taking into consideration the actual variation of the yarn diameter to accomplish suitable structural performance for a composite material. The fiber volume ratio plays a critical role.

Material and Methods

Material

Sets of different yarns were spun on ring spinning with specifications given in Table 1.

Yarn diameter measurement

Most of the researchers consider the yarn diameter is constant, which is true only in the case monofilament or multi-filament yarns. Several ring spun yarns were produced and the diameters were measured on Quick Quality management QQM3 instrument in this study, Faculty of Textiles lab, The Technical University of Liberec, CZ. The QQM has 2 Optical sensors of 2mm width, equipped with infra diodes and transistors positioned in the direction of yarn delivery, the value of yarn diameter, 2 mm apart, is recorded at the sampling rate 300 m/min [4]. Quick Quality management QQM3 instrument makes it possible to get the individual data of yarn diameter. Also, the yarns were tested on Uster 4 for characterization of yarn evenness and yarn diameter. Table 2 gives the results of the tested yarns.

Results and Discussion

Fiber volume fraction

In the case of continuous filament yarns, the impregnation is defined here in order to saturate every single reinforcement filament by the molten thermoplastic polymer. The molten thermoplastic flows through the capillaries between the reinforcement filaments [13].

In this case the fiber volume fraction:

Yarn ID	Yarn count tex	tpm	Spinning system	Material	Yarn ID	Yarn count tex	tpm	Spinning system	Material
6394	24.6	744.64	Carded	Cotton	6402	29.5	661.88	Carded	Cotton
6395	24.6	705.45	Carded	Cotton	6403	29.5	643.99	Carded	Cotton
6396	10.9	1146.36	Combed	Cotton	6404	19.7	766.81	Carded	Cotton
6397	10.9	1175.76	Combed	Cotton	6405	19.7	722.99	Carded	Cotton
6398	19.7	876.36	Combed	Cotton	6406	19.7	701.08	Combed	Cotton
6399	19.7	898.26	Combed	Cotton	6407	24.6	764.24	Carded	Cotton
6400	14.8	885.44	Combed	Cotton	6408	24.6	666.26	Carded	Cotton
6401	14.8	885.44	Combed	Cotton	6409	24.6	685.86	Carded	Cotton

Table 1: Yarn specifications.

Yarn ID	CV [%]	CV (1 m) [%]	CV (3 m) [%]	φ 2D [mm]	CV1D 0,3 mm [%]	CV2D 0,3 mm [%]	CV2D 8 mm [%]	s2D mm	H	sH	Neps +140% [1/ km]
6394	13.55	3.83	3.16	0.277	16.88	15.26	11.89	0.031	6.43	1.39	310
6395	12.86	3.56	2.94	0.274	16.5	14.44	11.23	0.031	6.37	1.47	170
6396	13.27	3.34	2.38	0.15	17.19	15.9	11.4	0.018	4.74	1.17	630
6397	13.11	3.77	2.78	0.151	16.76	15.44	11.16	0.017	4.65	1.17	490
6398	14.91	4.31	3.71	0.218	19.82	18.68	13.4	0.027	6.21	1.55	1620
6399	14.72	4.73	3.69	0.213	19.07	17.43	12.88	0.027	6.01	1.56	1005
6400	15.79	3.97	2.96	0.182	19.3	17.77	13.55	0.025	5.05	1.28	1365
6401	16.07	4.56	3.38	0.184	19.73	13.84	18.01	0.024	5	1.3	1220
6402	14.45	4.47	2.97	0.282	17.85	16.23	12.5	0.033	6.45	1.5	410
6403	14.72	4.86	3.73	0.28	18.43	16.45	12.74	0.035	6.42	1.56	555
6404	13.99	3.35	2.37	0.249	17.61	15.47	12.3	0.031	5.51	1.5	560
6405	14.64	4.27	3.51	0.26	18.76	16.69	12.94	0.033	7.21	1.65	805
6406	15.12	4.71	3.93	0.264	19.36	16.81	13.17	0.033	7.12	1.69	640
6407	12.72	3.09	2.2	0.253	15.3	13.44	10.7	0.026	4.89	1.22	180
6408	13.33	4.39	3.81	0.251	15.74	13.45	10.8	0.028	4.82	1.25	225
6409	13.39	3.88	2.92	0.275	16.41	14.25	11.26	0.03	5.95	1.41	210

Table 2: The analysis of the yarn evenness, yarn diameter and hairiness of the samples.

$$V_{f\,comp} = \frac{V_f}{(V_c)}.$$ (1)

Where V_f and V_m are the volumes of the fibers and matrix, respectively.

The fiber volume can be calculated for a circular shape of the fibers and composite, as illustrated in Figure 1, by:

$$V_f = 0.785\, n_f D_f^{\,2}$$ (2)

$$V_c = 0.785\, D_c^{\,2}$$ (3)

$$V_{f\,comp} = \frac{n_f D_f^{\,2}}{D_c^{\,2}}.$$ (4)

Where; n_f is the number of reinforcement fibers, $D_f^{\,2}$ and $D_c^{\,2}$ are the fiber diameter and composite diameter (Figure 1).

Spun yarn diameter

Actually, in the structure of the spun yarn the fiber's distribution in its cross section is different (Figure 2). Besides to that, the variation of the yarn diameter along the yarn length which add another variability to the value of V_f (Figure 2).

The yarn diameter of natural fiber spun yarn was varied along its length due to the variation of the number of fibers in the yarn cross section, especially in the case of the bast fibers which are widely used in the manufacturing of nature fiber polymer composites. The variation of yarn diameter along the yarn length is illustrated in Figure 3.

The histogram of the yarn diameter is shown in Figure 4, indicating that in most of the cases the frequency curve is skewed to the right with the high value of kurtoses (Figure 4).

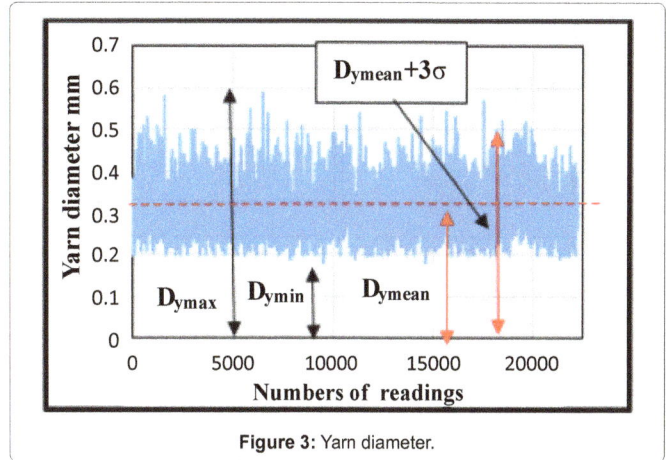

Figure 1: Sketch of composite cross section.

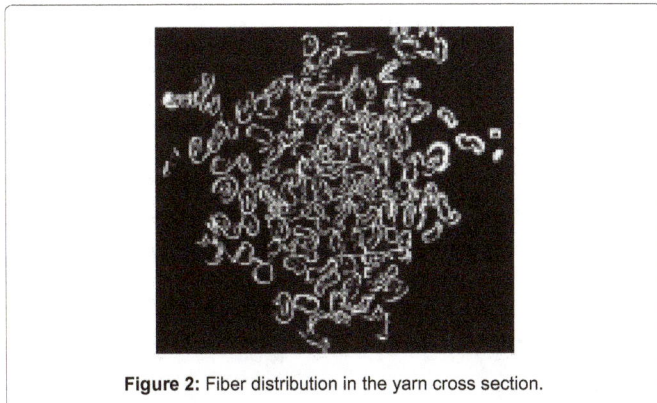

Figure 2: Fiber distribution in the yarn cross section.

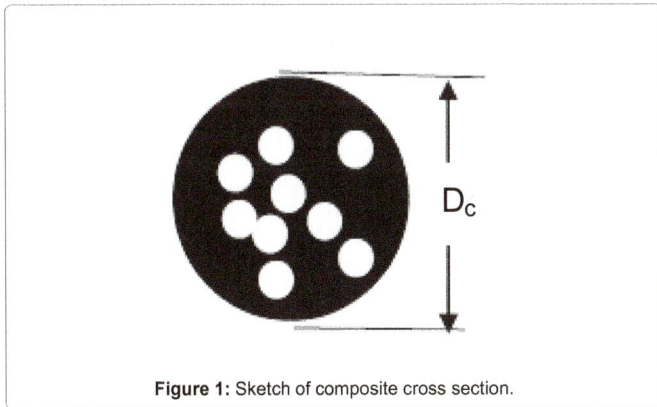

Figure 3: Yarn diameter.

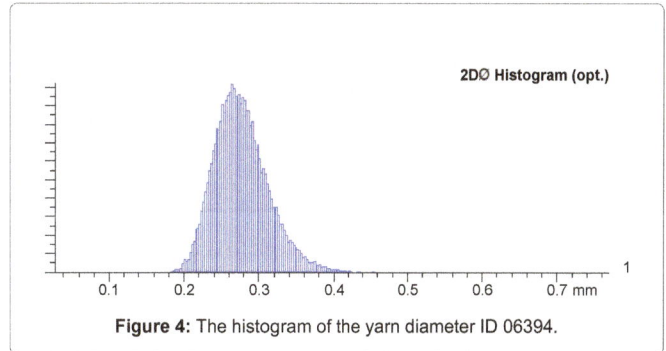

Figure 4: The histogram of the yarn diameter ID 06394.

Moreover, the value of the yarn fiber volume fraction V_f is found to vary across the yarn cross section [2]. So, the composite fiber volume will also vary, even the composite diameter is constant. The yarn volume fraction depends on the spinning system, twist level (turns per meter), yarn count, and fiber count [2] (Figure 5).

The yarn diameter data measured on Quick Quality management QQM3 instrument was analyzed and given in Figure 5. Which indicates the mean diameter of the yarn cannot be used for determined final composite diameter since large number of fibers will be outside the surface limit. Further, the number of the sections which has large diameter values varied from one yarn to the other. Consequently, the diameter of the composite that required to cover most of the yarn diameters along its length should be predetermined.

Data analysis

In order to estimate the value of the composite fiber volume fraction, the data of the yarn diameter measured for samples, each 2 mm, along the yarn length. Sample sections 22230 are considered. The total volume of the yarn is calculated V_y. Assuming the yarn fiber volume fraction α_y, then the composite fiber volume fraction will be:

$$V_{fcomp} = \alpha_y (V_y / V_{comp})$$ (5)

Where:

α_y is fiber volume fraction,

$$V_y = \sum_0^n \frac{\pi}{4} D_i^{\,2} \Delta l$$

$$V_{fcomp} = \frac{\pi}{4} \left(D_c \right)^2 \sum_0^n \Delta l,$$

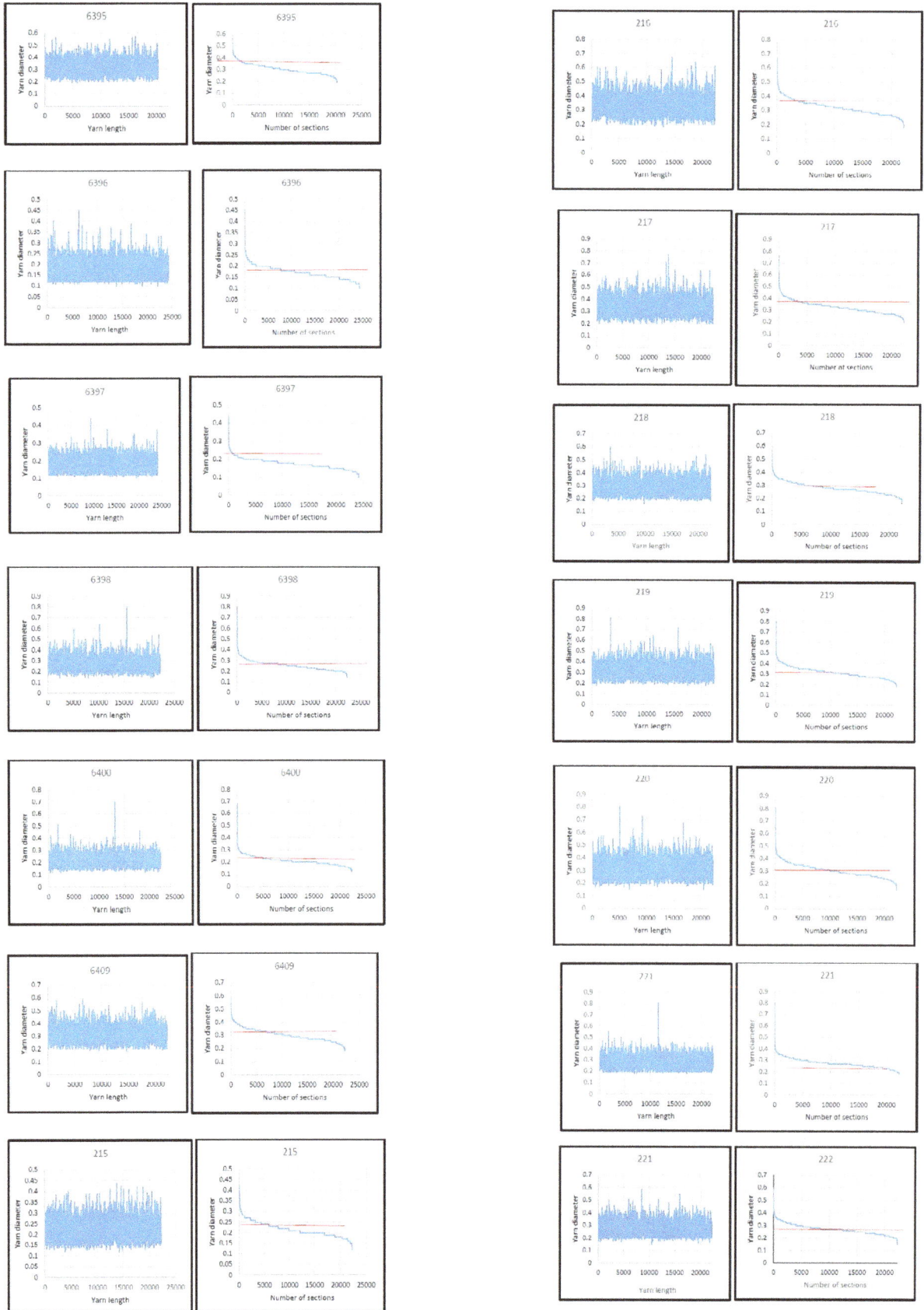

Figure 5: Analysis of the yarn diameter of tested yarns.

n is number of the measured sections.

The choice of the composite diameter where consider two cases:

Case I : take $D_{comp} = D_{y\ mean} + 3\sigma$

Case II: take $D_{comp} = D_{ymax}$

Where: D_{ymax}, is maximum diameter

D_{ymean} is mean yarn diameter

σ is the standard deviation of the yarn diameter.

Case study I: The variation of the yarn diameter (Figure 6) indicates that in order to cover the surface of the yarn by the matrix for the benefit of the utilizing all the fibers in the yarn cross section, the diameter should be 0.22 mm greater than the yarn mean diameter, in this case is 0.328mm. Which leads to a reduction of the composite fiber volume fraction value? As the diameter of the composite is reduced, a certain percentage of the fibers will not share the load applied to the composite. It is clear that the percentage of the yarn cross section containing the large value of the yarn diameters is very limited. Accordingly, when taking $D_{comp} = (D_{y.\ mean} + 3\sigma)$ will increase the value of composite fiber volume fraction V_{fcomp} (Figure 6).

The analysis of the yarn diameters of the different measured samples are given in Table 3.

The value of composite fiber volume fraction, assuming $D_{comp.} = (D_{y.\ mean} + 3\sigma)$, versus the yarn fiber volume fraction is illustrated in Figure 7, indicating that in all cases the yarns with the different characteristics result in the lower composite fiber volume fraction, even in the case of when the yarn fiber volume fraction is equal to one. The lower value of composite volume fraction is due to the variation in the yarn diameters at the different cross sections.

As it indicated in Figure 8, the increase of the coefficient of variation of the yarn diameter has an impact on the composite fiber volume fraction (Figures 7 and 8).

Thus, the relation between V_{fcomp} and V_{fyarn} depends on the variability of the yarn diameter in the different cross sections of the yarn. The distribution shape of the yarn diameter is expressed by its kurtosis, which is a measure that describes the shape of a distribution's tails in relation to its overall shape and reflects on the value of V_{fcomp} as illustrated in Figure 9.

In order to reduce the value of kurtosis of the yarn, it's recommended to adjust the setting on the winding machine to remove thick places and long thin places for yarn to be used in composite applications.

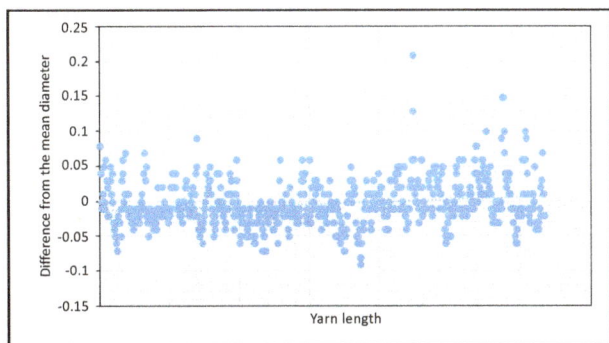

Figure 6: Distribution of the yarn diameter along the yarn length.

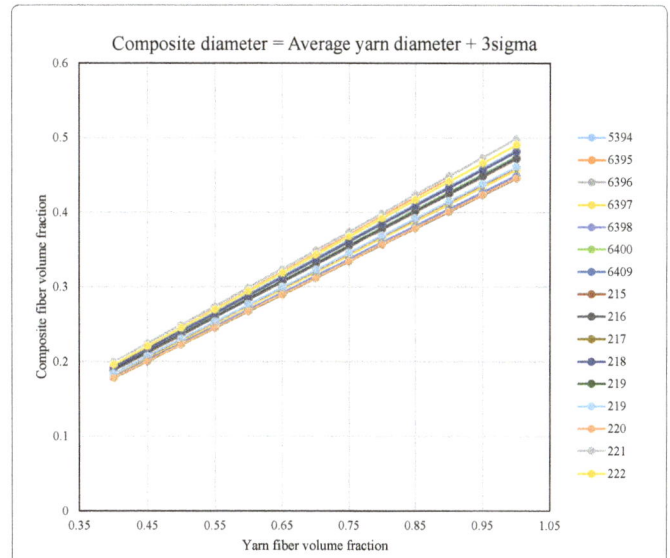

Figure 7: Composite fiber volume fraction versus the yarn fiber volume fraction.

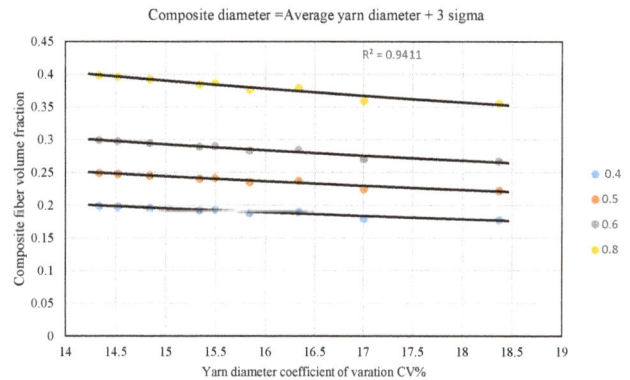

Figure 8: The composite fiber volume fraction versus yarn diameter coefficient of variation.

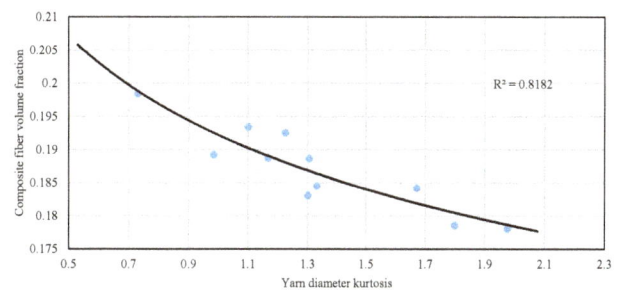

Figure 9: The composite fiber volume fraction versus yarn diameter kurtosis.

Case study II: Assume the composite diameter D_{comp} was chosen to cover all the yarn diameters along the yarn length which is its maximum diameter $D_{y.max}$ of the yarn as illustrated in Figure 4. In that case the ratio of matrix to fiber weight will increased, resulting in a lower value of V_{fcomp}. As indicated in Figure 10, the comparison of two cases specified the higher value of V_{fcomp} when applying conditions given in Case I.

Yarn diameter data analysis are given in Table 3. demostrating that

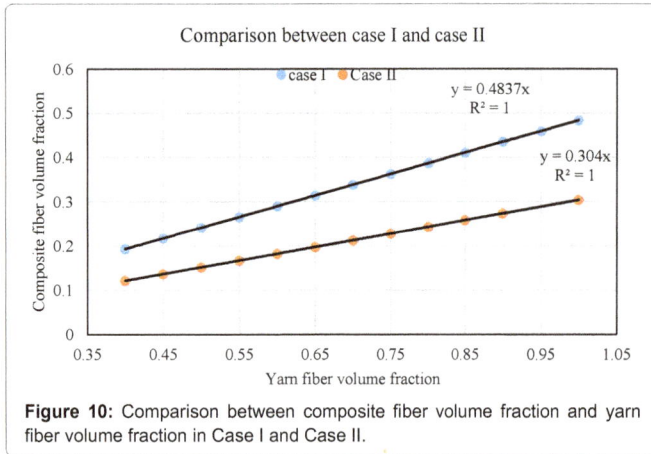

Figure 10: Comparison between composite fiber volume fraction and yarn fiber volume fraction in Case I and Case II.

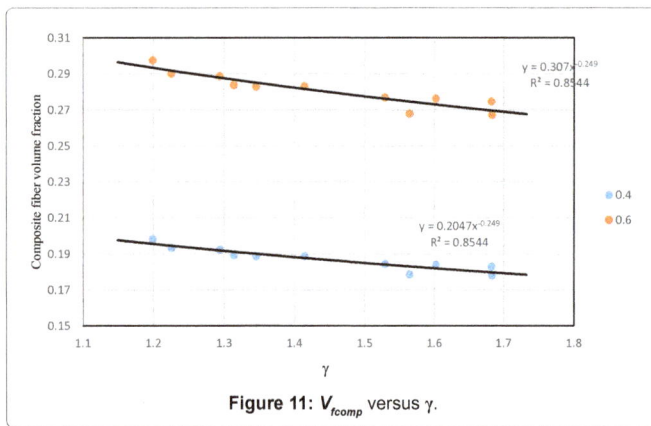

Figure 11: V_{fcomp} versus γ.

Sample ID	$D_{y.\,mean}$	CV%	σ	$D_{y.\,max}$ mm	$\gamma=D_{y.\,max}/D_c$	$D_c=X+3\sigma$
5394	0.328526	15.49357	0.0509	0.59	1.226	0.481
6395	0.305499	14.51941	0.044357	0.57	1.300	0.439
6396	0.172352	18.36763	0.031657	0.45	1.683	0.267
6397	0.176745	15.97463	0.028234	0.44	1.683	0.261
6398	0.176745	15.97463	0.028234	0.64	2.448	0.261
6400	0.211491	16.33498	0.034547	0.66	2.094	0.315
6409	0.305134	15.71713	0.047958	0.59	1.314	0.449
215	0.216537	15.84439	0.034309	0.43	1.346	0.319
216	0.321795	15.72429	0.0506	0.67	1.415	0.474
217	0.326379	15.75656	0.051426	0.77	1.602	0.481
218	0.280466	15.33761	0.043017	0.53	1.294	0.410
219	0.314984	16.49429	0.051954	0.72	1.529	0.471
220	0.305321	17.62183	0.053803	0.73	1.564	0.467
221	0.27606	14.33111	0.039562	0.71	1.799	0.395
222	0.271455	14.84114	0.040287	0.59	1.504	0.392

Table 3: Analysis of the yarn diameter for different samples.

ratio $\gamma=(D_{y.\,max}/D_{y.\,mean})$ related to the values of V_{fcomp}. As illustrated in Figure 11, it is proportional to the yarn diameter distribution kurtoses. Accordingly, it can be taken as a single variable to express the value V_{fcomp} as given by Equation (6) for investigated data.

$$V_{fcomp} = c^{-a}$$

$$V_{fcomp} = c\,\alpha\gamma^{-a} \,. \tag{6}$$

Where: a, c are constants depending on the variability of the yarn diameter cross section.

$$V_{fcomp} = 0.5135\,\alpha\gamma^{-0.249} \,. \tag{7}$$

The value of γ can be obtained directly from the data of measuring the yarn diameter by using Uster 4 tester which gives $D_{y.max}$, $D_{y.mean}$ and $CV_y\%$. For continuous filament yarn, the values of constants a, c and $\gamma=1$ as given by Equation (4).

Conclusion

The fiber volume fraction plays an important role in the behavior of a composite under different loading. The quantity of fiber in a fiber-reinforced composite directly relates to the mechanical properties of the composite. In the case of natural fiber polymer composites, the relation between V_{fcomp} and V_{fyarn} should be predetermined based on the above analysis and the equation giving to an engineer the required value of V_{fcomp}. It is recommended to use the yarns of low coefficient of variation and high V_{fyarn} which can be attained through an increase in the twist factor or use of the compact yarns. Another technique is to apply the pultrusion to the yarns that increases the composite fiber volume fraction.

References

1. Madsen B (2004) Properties of plant fiber yarn polymer composites: An experimental study. Technical university of Denmark.

2. Magdi El Messiry (2013) Theoretical analysis of natural fiber volume fraction of reinforced composites. Alexandria Engineering Journal 52: 301-306.

3. Thygesen A (2006) Properties of hemp fibre polymer composites. Ph.D. theses, The Royal Agricultural and Veterinary University of Denmark.

4. Onal L, Karaduman Y (2009) Mechanical Characterization of Carpet Waste Natural Fiber-reinforced Polymer Composites. Journal of Composite Materials August 43: 1751-1768.

5. Lehtiniemi P, Dufva K, Berg T, Skrifvars M, Jarvela P (2011) Natural fiber-based reinforcements in epoxy composites processed by filament winding. Journal of Reinforced Plastics and Composites 30: 1947-1955.

6. Mehmood S, Madsen B (2012) Properties and performance of flax yarn/thermoplastic polyester composites. Journal of Reinforced Plastics and Composites 31: 1746-1757.

7. Shah DU, Schubel PJ, Licence P, Clifford MJ (2012) Determining the minimum, critical and maximum fiber content for twisted yarn reinforced plant fiber composites. Composites Science and Technology 72: 1909-1917.

8. Ghosh R, Reena G, Krishna AR, Raju BHL (2011) Effect of fiber volume fraction on the tensile strength of Banana fibre reinforced vinyl ester resin composites. Int J Adv Eng Sci Technol 4: 89-91.

9. Shah DU, Schubel PJ, Clifford MJ, Licence P (2011) Mechanical characterization of vacuum infused thermoset matrix composites reinforced with aligned hydroxyethylcellulose sized plant bast fibre yarns. In: 4th International conference on sustainable materials, polymers and composites. Composites: Part A. Birmingham, UK.

10. Pan N (1993) Theoretical Determination of the Optimal Fiber Volume Fraction and Fiber-Matrix Property Compatibility of Short Fiber Composites. Polymer composites 14: 85-93.

11. Golzar M, Brunig H, Mader E (2007) Commingled Hybrid Yarn Diameter Ratio in Continuous Fiber-reinforced Thermoplastic. Journal of thermoplastic composite materials 20: 17-26.

12. Yilmaz D, Göktepe F, Göktepe O, Kremenakova D (2007) Packing density of compact yarns. Text Res J 77: 661-667.

13. Hao Ma, Yan Li, Di Wang (2014) Investigations of fiber twist on the mechanical properties of sisal fiber yarns and their composites. Journal of Reinforced Plastics and Composites 33: 687-696.

21

Functional Modification on Adhesive Bandage Using Natural Herbs

Sumithra M[1]* and Amutha R[2]

[1]Department of Textile and Apparel Design, Bharathiar University, Coimbatore, India
[2]Department of Costume Design and Fashion, PSG college of Arts and science, Coimbatore, India

Abstract

A Bandage is a standard of biomaterial used on wound to protect from infections and also to cure the wound. An adhesive bandage, also called a sticking plaster (and also known by genericized trademarks Band-aid or Elastoplasts) is a small dressing used for injuries not serious enough to require a full-size bandage. The adhesive bandage protects the cut from friction, bacteria, damage and dirt. In this present study of 50%:50% Bamboo Cotton web was selected for the construction of bandage functional part. In functional part of web was finished with eco-friendly natural leaves of *Galinsoga parviflora* and *Azadirachta indica*. For the finished web the antibacterial assessment EN ISO 20645 and anti-allergy assessment of the finished fabric contact allergy test (in house method) was carried out. From the test, it was concluded as 50%:50% Bamboo cotton finished with *Galinsoga parviflora* and *Azadirachta indica* has excellent wound curing property when compared to 50%:50% Bamboo cotton finished with *Galinsoga parviflora*. This study used to prevent the skin allergy, protect from the Bacteria and also to cure the skin diseases.

Keywords: Adhesive bandage; Anti-allergy; Antibacterial; Azadirachta indica; *Galinsoga parviflora*

Introduction

Textiles are an integral part of everyone's life associated with him from cradle to grave. It is used to cover human body, thus encompassing and protecting it from dust, sunlight, wind and other foreign matter present in the external environment that may be harmful to him. Textiles in apparel have retained an important place in human life, starting now into developing of newer high technology and interdisciplinary products describes [1].

Technical textiles are one of the fastest growing sectors of the global textile industry, reveals [2]. The term technical textiles was coined in the 1980s to describe the growing variety of product and manufacturing techniques being developed primarily for their technical properties and performance rather their appearance or other as aesthetic characteristics, remark [3]. Medical textiles constitute one of the most dynamic research field's characteristic of technical textiles and its range of applications suggested by [4].

Adhesive bandage, also known as sticking plaster, is a wound care dressing product that is utilized as small dressing. Adhesive bandages are applied on the patients who have not undergone serious accident but have minor abrasion (scratches) and cut on their body describe [5].

Consumers' attitude towards hygiene and active lifestyle has created a rapidly increasing market for antimicrobial textiles suggested [6]. The term antibacterial finishes indicates controlling or limiting the growth of bacterial colonies and their extinction, defines [7]. The antibacterial finish protects wearers of the textile product for against bacterial, dermatophytic fungi, yeasts, viruses and other deleterious microorganisms, states [8]. Antibacterial control, destroy or suppress the growth of microorganisms and their negative effects of odour, staining and deterioration.

Nature has been a source of medicinal agents since times immemorial. The importance of herbs in the management of human ailments cannot be over emphasized. It is clear that the plant kingdom harbors an inexhaustible source of active ingredients invaluable in the management of many intractable diseases. Ayurveda is ancient health care system and is practiced widely in India, Srilanka and other country expresses [9] Ayurveda system of medicine use plants to cure the ailments and diseases. Despite the availability of different approaches to that of its isolated and pure active components denoted [10].

Galinsoga parviflora Cav., comes from the Andes region. The chemical composition, activity and use are similar for both species. *Galinsoga* species are used in folk medicine as anti-inflammatory agents and accelerators for wound healing describe [11].

Azadirachta indica have been known to possess a wide range of pharmacological properties, especially as antibacterial, antifungal, antiulcer, antifeedant, repellent, pesticide, inhibitor and sterilant and is thus commercially exploitable, and hence, traditionally used to treat large number of diseases. The internal medicinal uses of Neem include malaria, tuberculosis, rheumatism, arthritis, jaundice and intestinal worms as well as skin says [12].

Cotton fibres are particularly suitable for manufacturing textiles for sports, non-implantable medical products, and health care/hygienic product. However the ability of cotton fibres to absorb large amount of moisture makes them more prone to microbial attack under certain conditions of humidity and temperature. Cotton may acts as a nutrient, becoming suitable medium for bacterial and fungal growth. Therefore, cotton fibres are treated with numerous chemicals to get better antimicrobial cotton textiles. Among the various antimicrobial agents, silver nanoparticles (AgNPs) have shown strong inhibitory and antibacterial effect says [13].

Bamboo fibre is a cellulose fibre extracted or fabricated from natural bamboo, and possibly other additives, and is made from (or in the case of material fabrication, is) the pulp of bamboo plants. It is

---

*Corresponding author: Sumithra M, Department of Textile and Apparel Design, Bharathiar University, Coimbatore, India
E-mail: mithrasumi6@rediffmail.com

usually not made from the fibres of the plant, but is a synthetic viscose made from bamboo cellulose presented by [14].

Hence, an attempt finished adhesive bandage has been made to prevent skin allergy and skin diseases by using eco-friendly herbs.

Methodology

Selection of fibre

50% bamboo and 50% cotton,

Bamboo and cotton fiber has good absorbency and more over bamboo has anti-bacterial character in nature.

Selection of herbs

The particulars of the medicinal herbs and natural materials used for the development of health care product are furnished in Table 1.

Galinsoga parviflora, Azadirachta indica were herbs which has got it is antibacterial character in nature which are abundantly available and hence it has been chosen for the study.

Collection of herbs: *Galinsoga parviflora, Azadirachta indica* were collected around the area of Erode Figures 1 and 2.

Sample No	Common name for the medicinal herbs used	Botanical name for the medicinal herbs used	Parts used
1.	Galinsoga	*Galinsoga parviflora*	leaves
2.	Neem, nimdi, miracle	*Azadirachta indica*	leaves

Table 1: Herb particulars used for the development of health care product.

Figure 1: *Galinsoga parviflora.*

Figure 2: *Azadirachta indica.*

Selection of fibre web formation

Fibre taken – Cotton, Bamboo,

Blending – Cotton and Bamboo,

Cotton – 50%.

Sample-1: Cotton+Bamboo fibre weight - 42 grams,

Final web weight - 38 grams,

Opening of the fibre was done manually and taken for the carding process. Carding was done with computerized carding machine in Kumarguru College of technology, Coimbatore.

Application of antibacterial finish on samples

Extraction and coating of natural herb: Solvent extraction of collected herbs

• Leaves were collected in a selected medical plants and it was dried into shadow dry method. Then it was converted into powder format.

• Extraction was carried out by dissolving 6 grams of the powder in 100 ml of 80% methanol.

• The mixture was kept overnight under shaking condition. The extract was filtered using Whatmann no.1 filter paper.

• The filtrate was collected and evaporated at room temperature. The concentrated extract was stored at 4°C and used for further studies by Prashith kekuda (2014) was shown in Figures 3-8.

Evaluation of antibacterial finish

The antibacterial activity of the finished fabrics was tested according to EN ISO 20645 against *Staphylococcus aureus* and *Escherichia coli*. Nutrient agar plates were prepared and 0.1% inoculums was swabbed uniformly and allowed to dry for 5 minutes. The finished fabric with the diameter of 2.0 ± 0.1 cm was placed on the surface of medium and

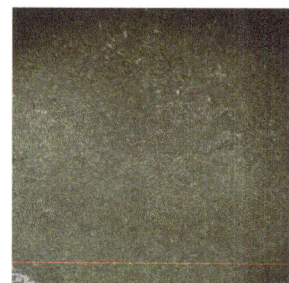

Figure 3: *Galinsoga parviflora* powder.

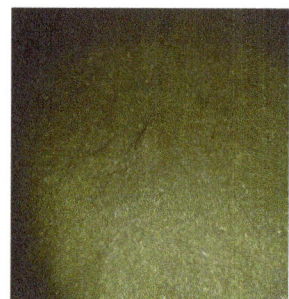

Figure 4: *Azadirachta indica* powder.

Figure 5: Methonol+aqua+herb powder.

Figure 6: Mixed herbal solution.

Figure 7: In testing condition.

the plates were kept for incubation at 37ºC for 24 hours. At the end of incubation, the zone of inhibition formed around the fabric was measured in millimetres and recorded.

Evaluation of anti-allergy finish on treated samples: Antiallergy assessment of the finished fabric contact allergy test (in house method)

Procedure: The fabrics patched on the normal skin were observed for the specified period of time for the development of the symptoms related to contact dermatitis allergy. Non hairy part of the skin of the

subjects was selected. The surface of the skin was cleaned with moistered sterile cotton swabs. The patches of the fabrics sample were made and plastered on the surface of the cleaned skin. The site of patching was observed for any immediate allergic response. Observations were made up to 24 hours for the symptoms such as Skin rashes, redness and irritations. (Erythema and edema). The time of observation may be extended for another 24 hours to confirm the effect.

Evaluation: After the contact time, the fabric patches were removed and observed for the following reactions:

(NIR) - No irritant reaction,

(IR) - Irritant reaction.

Mechanical testing

Fibre strength: Strength factor is identified using "Stelo meter" instrument.

Fibre fineness: Fineness properties is identified using "Sheffield Micronaire" instrument.

Development of adhesive bandages

Hence the product (ADHESIVE BANDAGES) is developed using bamboo and cotton web formation

Nomenclature

The Nomenclature of the control web sample, finished web sample were finished with antibacterial herbs are given below Table 2.

Result

Evaluation of anti bacterial finish (bamboo cotton)

The antibacterial test result of the developed both bamboo cotton samples are given below in the Table 3.

In that result:

<25 – result failed (negative),

>25 – possitive result.

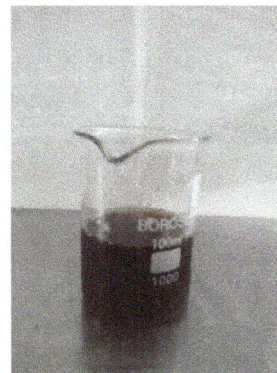
Figure 8: After filtered solution.

Sample No	Samples	Samples abbreviations
1	CWS	Control Web Sample
2	FWS1	Finished Web Sample [*Galinsoga parviflora* (12 grams)]
3	FWS2	Finished Web Sample [*Galinsoga parviflora* (6grams), *Azadirachta indica* (6 grams)]

Table 2: Nomenclature.

From the Table 3 had shown FWS2 have good antibacterial capacity when compare to CWS and FWS1.

From the Figures 9-11 it has been proved the FWS1 and FWS2 has good antibacterial activity, and the CWS has zone inhibition. Also it is surrounded by colonies bacteria.

From the Figures 10 and 11 it has been proved FWS2 has good antibacterial activity when compare to FWS1.

The above tables and figures the Anti-bacterial activity by test for the bacteria's *Escherichia coli* and Staphylococcus aureus. From the

S.No	Sample No	Zone of Bacteriostasis (mm)	
		Staphylococcus aureus	*Escherichia coli*
1	CWS	0	0
2	FWS1	35	38
3	FWS2	40	39

Table 3: Evaluation of antibacterial finish.

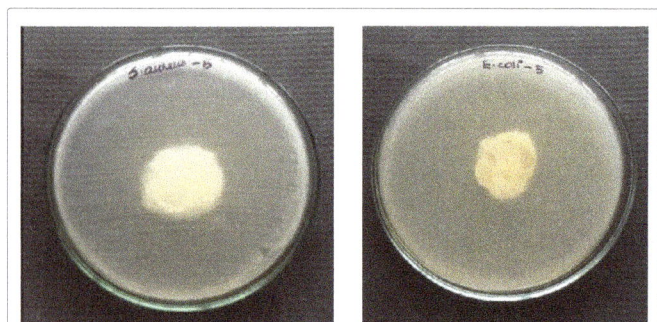

Figure 9: Antibacterial assessment – *Staphylococcus aureus* and – *Escherichia coli* bacteria (CWS).

Figure 10: Antibacterial assessment – *Escherichia coli* bacteria (FWS1 & FWS2).

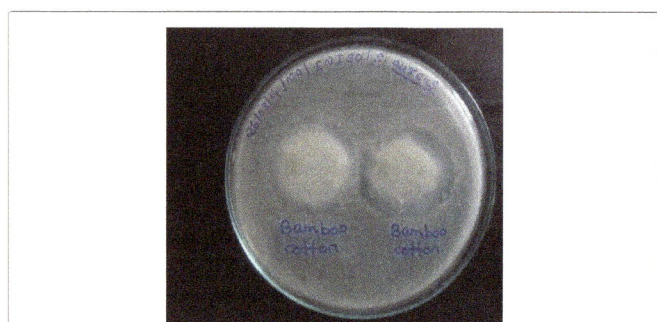

Figure 11: Antibacterial Assessments *Staphylococcus aureus* bacteria (FWS1 & FWS2).

test results, that is best comparative study against the bacterial attack. Hence the anti – bacterial finish on the both cotton finished material found to be effective compared to the other samples.

Evaluation of anti allergy finish

Galinsoga parviflora **(12 grams) finished bamboo cotton web:** From the Table 4 had shown, 17, 22 and 24 years subjects have no irritant reaction.

Sample 2

Galinsoga parviflora **(6 grams),** *Azadirachta indica* **(6 grams) finished bamboo cotton web:** From the Table 5 had shown, 17, 22 and 24 years subjects have no irritant reaction.

Skin-before and after testing (anti allergy) test shown in Figures 12-14.

Mechanical testing

Fibre strength

The bundle fibre strength of cotton in m Kg at 3 mm gauge length=3.86,

Sample No	Subject	Fabric sample (FWS1)
1	Subject 01 (Female/17 yrs)	No irritant reaction
2	Subject 02 (Male/22 yrs)	No irritant reaction
3	Subject 03 (Female/24 yrs)	No irritant reaction

Table 4: The anti-allergy test result of the developed FWS1 samples.

Sample. No	Subject	Fabric sample (FWS2)
1	Subject 01 (Female/17 yrs)	No irritant reaction
2	Subject 02 (Male/22 yrs)	No irritant reaction
3	Subject 03 (Female/24 yrs)	No irritant reaction

Table 5: The anti-allergy test result of the developed FWS2 samples.

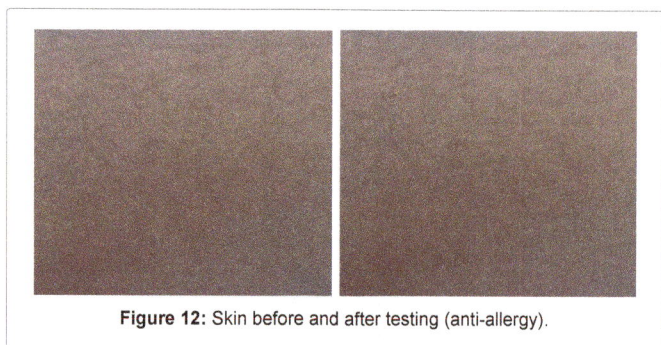

Figure 12: Skin before and after testing (anti-allergy).

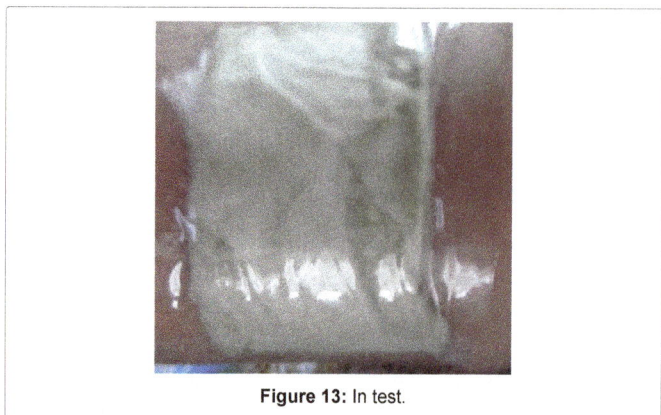

Figure 13: In test.

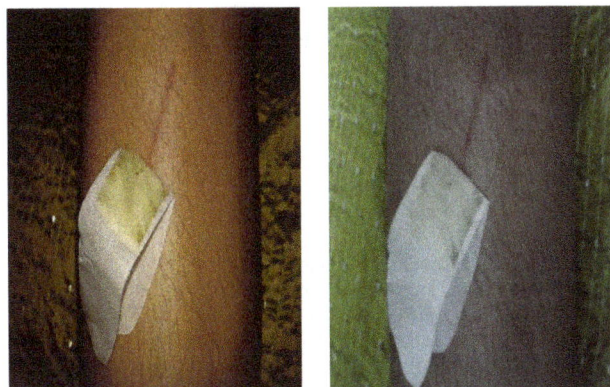

Figure 14: In test.

The bundle fibre strength of cotton in m Kg at 0 mm gauge length=2.5,

The bundle fibre strength of bamboo in m Kg at 3 mm gauge length=43.36%,

The bundle fibre strength of bamboo in m Kg at 0 mm gauge length=38.08%,

Cotton fibres Tenacity in grams/tex=0 mm=2.98,

3 mm=7.16,

Cotton fibres Elongation percentage=0 mm=15.4,

3 mm=8.2.

Fibre Fineness

Cotton fibre fineness=3.7 Micrograms/inch,

Bamboo fibre fineness=1632 nm.

Conclusion

The use of plant and plant products by the human beings are innumerable. Herbs are used very popular all over the world for medicinal purpose. The herbal extracts from these plants were applied on the bamboo cotton web and spun laid nonwoven material. The finished webs and spun laid non-woven materials were subjected to antibacterial test, anti-allergy, mechanical test and wear ability to identify the best samples. After the evaluation it is concluded that the best finished sanitary napkin have all the expected character like comfort, size, color, cost and protection from skin diseases and itchiness.

References

1. Malarvizhi G (2015) Development Of Herbal Finished Baby Diapers With Bamboo Fiber. International Journal of Humanities, Arts, Medicine and Sciences 3: 41-46.

2. Gulrajani ML (2008) Finishing techniques for medical textiles. Asian Technical Textiles 2: 29.

3. Nadiger GS, Ghosh PS (2008) An important technical textile segment: Medical textiles. Asian Technical Textiles 2: 35.

4. Parveenbanu K, Pradeepa P (2014) Non Implantable Materials in Medical Textiles. International Journal of Current Research 6: 61-20.

5. Adhesive Bandages market (2013) Global industry analysis, size, share, growth. Trends and forecast.

6. Gao Y, Cranston R (2008) Recent Advances in Antimicrobial Treatments of Textiles. Textile Research Journal 78: 60-725.

7. Achwal (2003) Antibacterial finishes and their modification. Colourage 50: 58.

8. Jasuja (2004) Antibacterial and anti-insect finishes. New Cloth Market 18: 13.

9. Chopra A, Doiphode VV (2002) Ayurvedic medicine: core concept, therapeutic principles and current relevance. Medical Clinics of North America 86: 75-89.

10. Shariff ZU (2001) Modern Herbal Therapy for common Ailments Nature Pharmacy Series. Spectrum Books LTD, Ibadan, Nigeria in Association with Safari Books Ltd. UK 1: 9-84.

11. Bazylko A, Boruc K, Borzym J (March 2015) Aqueous and ethanolic extracts of Galinsoga parviflora and Galinsoga ciliata. Investigations of caffeic acid derivatives and flavonoids by HPTLC and HPLC-DAD-MS methods 11: 394-398.

12. Manisha SY, Sachin NA, Amrita KA (2014) World Journal of Pharmacy and Pharmaceutical Sciences 3: 590-598.

13. Ravindra S, Mohann YM, Reddy NK, Raju M (2010) Fabrication of antibacterial cotton fibres loaded with silver nano particles via Green Approach 367: 31-40.

14. Federal Trade Commission (2010) FTC Warns 78 Retailers, Including Wal-Mart, Target, and Kmart, to Stop Labeling and Advertising Rayon Textile Products as "Bamboo".

22

Effect of PVAmHCl Pre-treatment on the Properties of Modal Fabric Dyed with Reactive Dyes: An Approach for Salt Free Dyeing

Periyasamy AP*

Department of Material Engineering, Technical University of Liberec, Studentska, 46117, Liberec, Czech Republic

Abstract

In this research, polyvinylamine chloride (PVAmHCl) was used as a physical modification agent on regenerated cellulosic fabric such as modal by a pad-batch process. The modified modal samples were dyed with different reactive dyes containing various reactive groups. The dyeability of the modified modal samples with reactive dyes without electrolyte was significantly improved due to an increase in the ionic attraction between the reactive dyes and modified modal fabrics. It has been confirmed through zeta potential analysis, as well as the result of various fastness properties such as light, wash and rubbing fastness of polyvinylamine chloride pre-treated modal with different reactive dyes are similar to those of untreated modal fabric. Also, the tensile strength, flexural rigidity and crease recovery angle of pretreated sample were determined, out of that crease recovery angle and flexural rigidity of pretreated sample showing significant improvement.

Keywords: Dyeing; Modal; Pre-treatment; PVAmHCl; Reactive dyes; Salt free reactive dyeing

Introduction

Modal (CMD) is a regenerated cellulosic fiber from Lenzing and it is extracted from Beechwood trees [1]. Modal has all the benefits of being a regenerated cellulosic fiber, and also it is fully biodegradable, higher soft in nature, excellent moisture absorbent and outstanding in handling properties; and possible to blend with all types of fibers [2]. It has a relatively high strength, which allows for the production of finer yarns and lighter fabrics as compared to another regenerated fiber such as viscose rayon. As a result, fabrics produced from modal are breathable and moisture absorbent and have high dimensional stability [3].

Cotton and regenerated cellulosic fibers can be dyed with various classes of dyes [4]. Most prominently, reactive dyes are widely employed for dyeing cellulose fibers because of their brilliance, wide shade gamut, and excellent fastness properties [5]. However, dyeing of cellulose fibers with reactive dyes still suffers from two major disadvantages; one is poor dye uptake and other one is unsatisfactory dye fixation, both were leads to environmental pollution [6]. Poor dye uptake is related to the existence of the charge barrier effect between the negatively charged fiber surface and anionic reactive dyes. This problem can be solved by adding a large amount of electrolytes, to the dye bath in order to suppress the negative charge on the fiber surface and then allowing reactive dye molecules to diffuse inside the fiber [7]. In the case of poor dye fixation, the contributing factor is the presence of inactive hydrolyzed dye [8]. Since fixation of reactive dyes onto cellulose fibers requires alkaline dyeing conditions in order to activate the hydroxyl group of cellulose to be able to react with the dye, some of the reactive dye can inevitably undergo the competing hydrolysis reaction with hydroxide nucleophiles [9], resulting in a reduction in the efficiency of the reaction with the cellulose substrate [10].

Increasing concern on environmental impacts has prompted regulators to enact rigorous environmental legislation to mandate dye users to minimize color in the dye-house effluent [5,11]. To meet such requirements, the dyeing industry has to adopt a more efficient dyeing process as well as selecting dyes which have a high dye fixation value [12]. Reactive dyes, the newest addition of existing dyes are the center of attraction in dyestuff research. Several new reactive systems have been introduced from time to time, hot brand reactive dyes have been widely considered due to their higher fixation yield on various fibers

[13]. Theoretically, Poly functional reactive dyes contain more than one reactive group in each molecule [5,14]. In practice, these additional reactive groups can have an impact on important physical properties such as solubility, aggregation, substantivity, and migration, which offers a very high level of fixation and it leads to low color usage to achieve a given depth of shade and a lower unfixed color load in effluent [15]. These multifunctional reactive dyes were especially designed for cellulosic fibers, and are claimed to save energy and time and reduce water consumption; a special auxiliary to assist the washing-off of unfixed dyes [16]. PVAmHCl has been used as a physical modifying agent for cellulosic materials [17]. It has been used for modification of cotton as well as other regenerated fibers for salt free dyeing as been previously reported [16,17]. Interest in PVAmHCl arises from the presence of a large number of cationic sites $NH_3^+Cl^-$ which illustrated in Figure 1 [18] conclude in their research, dyeability for the cotton fabric significantly improved after pretreatment with PVAmHCl without addition of electrolyte as exhausting agent.

In this study, various reactive dyes with two different reactive groups which including bi-functional and polyfunctional were applied onto the PVAmHCl pre-treated (modified) modal fabrics. The dyeing was performed with and without electrolyte (Na Cl). To determine the modification of modal fabric samples was done by zeta potential, retained

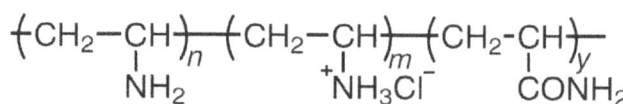

Figure 1: Chemical structures of PVAmHCl (Ma et al. [18]).

Corresponding author: Aravin Prince Periyasamy, Department of Material Engineering, Technical University of Liberec, Studentska, 46117, Liberec, Czech Republic, E-mail: aravin.prince.periyasamy@tul.cz

water quantity, TDS, color strength, fastness properties and various physical properties like tensile strength, flexural rigidity, fabric crease recovery angle were also determined to examine the effect of PVAmHCl.

Materials and Methods

Materials

100% Modal (bleached and desized) plain woven fabric was used for this study, which is having Arial density of 102 g/m² and it has been obtained from Texpert, Mumbai, India and the fabric characteristics are listed in the Table 1. The reactive dyes used with different reactive groups, including bi-functional and polyfunctional which are listed in Table 2, and both were procured from Huntsman, India. PVAmHCl (Laboratory reagent grade) were purchased from Triveni chemicals, Chennai, India and all other chemicals such as Sodium carbonate, Sodium hydroxide, Potassium di hydrogen phosphate, Sodium chloride, acetic acid were supplied from color chemicals, Tirupur, India, and above these reagents were used in laboratory reagent grade.

Pre-treatment and dyeing of modal fabric

Pretreatment: Various concentrations of PVAmHCl (2.5, 5.0, 10, 15 and 20 g/L) solution was applied to the modal fabric by padding technique. The wet pick up was 100% and the room temperature (30-35 °C) was maintained during the padding. The pH of pretreatment liquor was maintained at 7 by using of buffer comprising potassium dihydrogen phosphate (7 g/L) and sodium hydroxide (1.4 g/L). The padded fabric was put in the polyethylene sheet for 6 hours to prevent chemical migration and water evaporation. After 6 hours, the padded fabric was dried at room temperature and baked at 102 °C for 10 min in rapid baker. After baking the pre-treated fabric was washed with dilute acetic acid and followed by water till neutralization, finally the fabric was dried at room temperature. Table 3 shows sample labels and treatment parameters.

Dyeing: Dyeing of both unmodified and modified (pre-treated) modal samples was carried out using an Infrared dyeing machine at the liquor ratio of 20:1. Dyeing with both unmodified and modified modal fabrics with reactive dyes was performed according to the procedure which offered by the dye manufacturer. 1% o.w.f dyes was added for both bi-functional and polyfunctional reactive dyes. The procedure for poly functional reactive groups shown in the Figure 2a and 2b, 30 g/L NaCl, 15 g/L of Na_2CO_3 (Figure 2a) was used for unmodified modal fabrics and for the no salt dyeing procedure, 15 g/L of Na_2CO_3 (Figure 2b) was used for the modified modal samples. Similarly the procedure for bi-functional reactive groups shown in the Figure 2c and 2d (Figure 2c, 40 g/L NaCl, 15 g/L of Na_2CO_3) was used for unmodified modal fabrics and for the no salt dyeing procedure (Figure 2d, 15 g/L of Na_2CO_3) was used for the modified modal samples. After dyeing both unmodified and modified modal fabrics was removed from the bath and rinsed consecutively in cold, hot and cold water. The dyed fabric was boiled in a 2 g/L solution of anionic detergent ladiquest 1097 liq (Pure chem, India.) for 15 min until the complete surface or undiffused was removed, and then rinsed, and then the fabric was allowed to air dry in room temperature.

Testing

Dye Exhaustion, color strength and Fixation: The dye uptake for the modal fabric was determined with the help of UV-Visible light spectrophotometer and the dye-bath absorbance was measured at the wavelength of maximum dye absorption (λ_{max}). The exhaustion of the dyebath can be calculated by the Eq. (1).

Warp density, ends /inch	Weft density, picks/inch	Arial Density (g/m²)	Warp yarn count (Ne)	Weft yarn count (Ne)
106	96	102	40	40

Table 1: Characteristics of used modal fabrics.

Commercial name	C.I. Number	Reactive groups	Amount (% o.w.f)
Novacron Blue FN-G	C.I Reactive Blue 268	Bi-functional	1
Novacron Red FN-R	C.I Reactive Red 238		1
Novacron Yellow FN-2R	C.I Reactive Yellow 206		1
Avitera Blue SE	C.I Reactive Blue 281	Poly-functional	1
Avitera Red SE	C.I Reactive Red 286		1
Avitera Yellow SE	C.I Reactive Yellow 217		1

Table 2: Characteristics of different reactive dyes used for this research.

Designation	Treatment
CS	Conventional dyeing with salt
CW	Conventional dyeing without salt
P2.5	Pre-treatment of fabric with 2.5 g/L of (PVAmHCl) and followed by dyed.
P5.0	Pre-treatment of fabric with 5 g/L of (PVAmHCl) and followed by dyed.
P10.0	Pre-treatment of fabric with 10 g/L of (PVAmHCl) and followed by dyed.
P15.0	Pre-treatment of fabric with 15 g/L of (PVAmHCl) and followed by dyed.
P20.0	Pre-treatment of fabric with 20 g/L of (PVAmHCl) and followed by dyed.

Table 3: Sample labels and treatment parameters.

$$E\% = \frac{(1-A)}{A_0} \times 100 \tag{1}$$

Where A_0 and A are the absorption of the dye solution at the maximum wavelength before and after dyeing process [10]. After dyeing, reflectance spectra of each sample were measured with the help of Minolta spectrophotometer CM-3600A. A color depth of the dyed fabrics was analyzed by measuring the (K/S) color strength values of samples. It seems the higher value of color strength is nothing but higher the dye absorption by the fabric. Color strength values of the dyed samples were calculated using the following instrument settings (illuminant D65, 10 supplemental standard observers, specular included, UV included). The effective color, strength value was calculated by summation of the color strength values at 10 NM intervals from the wavelength of 360-750 nm. While testing, the sample was folded two times. Each sample was measured ten times by changing the measuring point randomly to calculate the average value [19]. Color measuring instrument (spectro-photometer) determines the color strength value of a given fabric through Kubelka-Munk Eq. (2).

$$\frac{K}{S} = \frac{(1-R)^2}{2R} \tag{2}$$

Where R=reflectance percentage, K=absorption and S=scattering of dyes.

The fixation of adsorbed dye (F) can be calculated by using of the Eq. (3) and the total fixation of the originally applied dye (T) can be

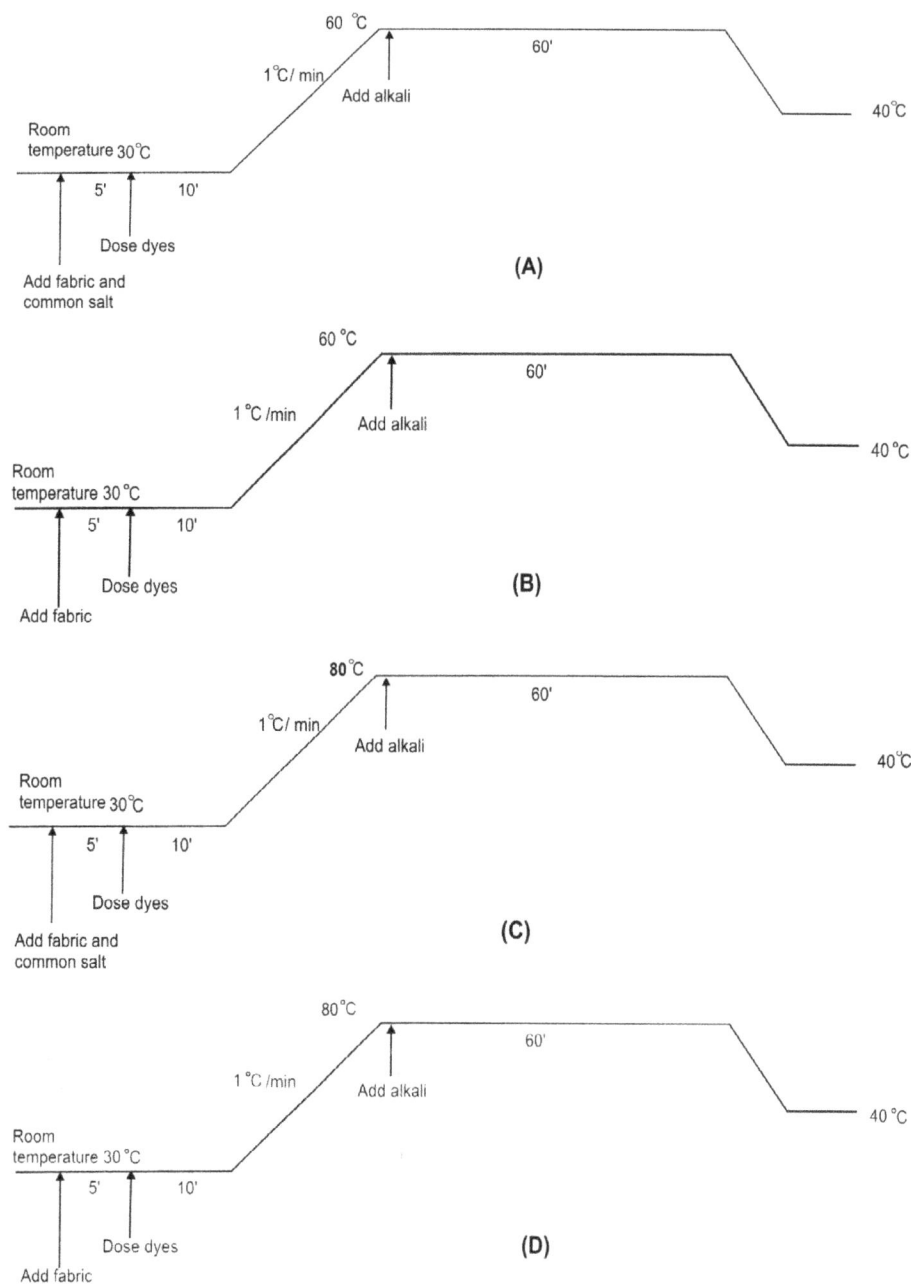

Figure 2: Dyeing procedure for control and treated modal fabrics. (a) Conventional dyeing procedure for unmodified modal fabrics with poly functional reactive dyes. (b) Conventional dyeing procedure for modified modal fabrics with poly functional reactive dyes. (c) Con-ventional dyeing procedure for unmodified modal fabrics with bi-functional reactive dyes. (d) Conventional dyeing procedure for modified modal fabrics with bi-functional reactive dyes.

$$F\% = \frac{(K/S)_1}{(K/S)_2} \times 100 \qquad (3)$$

$$T(\%) = E \times \frac{(K/S)_1}{(K/S)_2} \times 100 \qquad (4)$$

Where $(K/S)_1$ and $(K/S)_2$ indicates the before and after soaping respectively [10].

The percentage of dye intensity increases (I) on PVAmHCl pre-treated samples compared to untreated sample was obtained from the following Eq. (5).

$$I\% = \frac{(K/S)_a - (K/S)_b}{(K/S)_b} \times 100 \qquad (5)$$

Where $(K/S)_a$ refers to modified modal fabric samples and $(K/S)_b$ is to samples without pre-treatment and dyed with salt (CS).

Evaluation of zeta potential: The electrokinetic potential of pre-treated and untreated modal fibers were measured by the streaming potential (U_p) and streaming current (I_p) method using Brookhaven-Paar Electrokinetic Analyzer (EKA). The ζ-potential for the modal fabric was measured by streaming current method depending on pH values of 0.001 M KCl solution in a pH range of 2 to 10. The zeta potential was calculated according to the Helmholtz-Smoluchowsky Eq. (6).

$$\zeta = \frac{U_p \cdot \eta \cdot L}{\varepsilon \cdot \varepsilon_0 \cdot Q \cdot R \cdot \Delta p}$$

or   (6)

$$\zeta = \frac{I_p \cdot \eta \cdot L}{\varepsilon \cdot \varepsilon_0 \cdot Q \cdot R \cdot \Delta p}$$

where ζ: zeta potential (mV), U_p: streaming potential (mV), I_p: streaming current (mA), ε: permittivity electrolyte solution (Fm^{-1}) ($Kgm^{-3}S^4A^2$) η: dynamic viscosity of solution (Pas) ($Kgm^{-3}S^4A^{-1}$), ε_0: vacuum permittivity (Fm^{-1}) ($Kgm^{-3}S^4A^2$) $\Delta\rho$: pressure difference between capillary ends (Pa) (Nm^{-2}), L: capillary length (m), R: electrical resistance (Ω), Q: capillary cross-section (m^2) [17,20,21].

Evaluation of TDS: Total dissolved solids (TDS) were measured through a process of evaporation. During the first rinse of the dye bath, dye liquor samples were collected for measurement of TDS. The TDS for the dye liquor samples were determined by gravimetric method that involved evaporating the liquid solvent to leave a residue that can subsequently be weighed with a precision analytical balance (normally capable of 0.0001 gram accuracy). The TDS (mg/L) in the effluent was determined with the help of Eq. (7).

$$TDS\left(mg/L\right) = \frac{(A-B) \times 1000 \times 1000}{Sample\ volume}$$ (7)

Where, the final weight (A) is the weight of the dish plus the dried residue in gram and the initial weight (B) is the weight of the dish in gram.

Note: TDS were tested for the all concentration of PVAmHCl with respect to Reactive Blue 268 only.

Water retention value: To identify the water retention values, pre-treated and untreated modal fabric sample has to be immersed in distilled water for 24 hours and then centrifuged at 7800 min^{-1} for 2 minutes, then it should be dried for 105 °C. The water retention values can be calculated by this Eq. (8).

$$WRV\left(\%\right) = \frac{(W_{wet} - W_{dry})}{W_{wet}} \times 100$$ (8)

Where W_{dry}, W_{wet} refers the mass of dried and before dried the samples [2,3,17,20-24].

Evaluation of fastness properties: Washing fastness test for the dyed samples were performed as per the AATCC 61-2003 by using the Texcare Launderometer. A multi-fiber test fabric was attached to each sample to evaluate the staining and the test conditions were set according to the standard. Color fastness to rubbing was measured according to AATCC 08-2003 and both dry and wet crocking test were measured using the AATCC automated crockmeter. The washing and rubbing fastness results can be given grading with the help of appropriate gray scale. Perspiration fastness for the dyed samples was carried out according to the ISO 105-E04 test method. The light fastness was performed as per the ISO 105 B02:1994, standard method and the degree of color fading were assessed by using of SDC blue wool scales.

Evaluation of mechanical properties: The mechanical properties of fabrics, such as tensile strength were measured in Instron tester (3300 series) as per ASTM D5034-95; flexural rigidity was measured in

Taber fabric stiffness tester as per the BIS BS 3356-1991; crease recovery angle was measured in James Heal crease recovery tester as per AATCC 66-2003 standard. Five samples were tested for each test and calculated average values for the results.

Results and Discussion

Zeta potential

Generally, ζ-potential for textile fibers are negative, it is due to negative charge developed on the fiber surface. Modal is a cellulosic fiber and having the –COO^- groups which arise the ionization or dissociation reaction on the surface. The results of Zeta potential (ζ), Iso electric point (IEP), Point of Zero charge (PZC) and Specific surface charge (q) of modal fabrics were summarized in the Table 4, which specify the dissociation of functional groups such as hydroxyl (-OH), carboxyl acid (-COOH) groups of modal fabric and it introduces the negative charge. The result shows that untreated (CW) modal fabric surface has lowest ζ-potential (negative charge) in whole pH range. The ζ-potential for PVAmHCl pre-treated fabric was increased to positive charge at low pH range. This is probably because of the presence of PVAmHCl and it has a larger number of cationic sites (NH^+) and it gains the H^+ ions in the liquid phase which helps to change the positive charge of the fabric from:

ζ=-20.4 mV for CW to ζ=-14.7 mV for P2.5, ζ=-12.8 mV for P5, ζ=-11.7 mV for P10,

ζ=-9.4 mV for P15 and ζ=-7.2 mV for P20 at pH 10. It clearly shows that how PVAmHCl can enhance the adsorption of reactive dyes on modal fabric. Apart from the ζ- potential at pH 10, it is necessary to know the impact of Iso electric point (IEP) and Point of zero charge (PZC) on dyeing, because both are directly influenced on the dyeing process. The results of Iso electric point (IEP) and Point of zero charge (PZC) are shown in the Table 4 and Figure 3, results indicates that there is no IEP for untreated (CW) sample, because the ζ- potential for untreated sample have only negative values in the whole pH range. An IEP for modified modal samples moved towards the higher pH values when the concentration of PVAmHCl is increased, the range lies between 5.4 to 6.7. The point of zero-charge (PZC) was determined in the pH 10, because the fixation of reactive dyes can occur in this pH and also the ζ- potential can become higher and reaching the constant value. Results shows that PZC values are highest in the case of untreated samples (CW=69.76), and it decreases with increasing the concentration of PVAmHCl, for pre-treated samples having the PZC of 62.54 (P2.5) to 43.41 (P20), it seems that PVAmHCl have significant relationship with point of zero charge. The specific amount of surface charge can be determined with the help of back titration. The values were shown in the table and it concludes, the untreated fabric shows the negative values due to the higher negative surface charge. However, PVAmHCl pretreated samples were showing negative values in q=-1.0210, -0.8754 in case of P2.5 and P5 respectively. It seems that ionic modification for these (P2.5 and P5) samples have less as compared to higher concentrations of PVAmHCl. So the modified fabrics have

Sample	ζ at pH 10 (mV)	IEP (at pH)	PZC (µg/mL)	q (C/g)
CW	-20.4	-	69.76	-2.3472
P2.5	-14.7	5.4	62.57	-1.0210
P5.0	-12.3	6.1	57.43	-0.8754
P10.0	-11.6	6.3	53.94	0.1220
P15.0	-10.8	6.4	49.77	0.4527
P20.0	-10.2	6.7	43.41	0.9456

Table 4: ζ-potential for the control and treated samples.

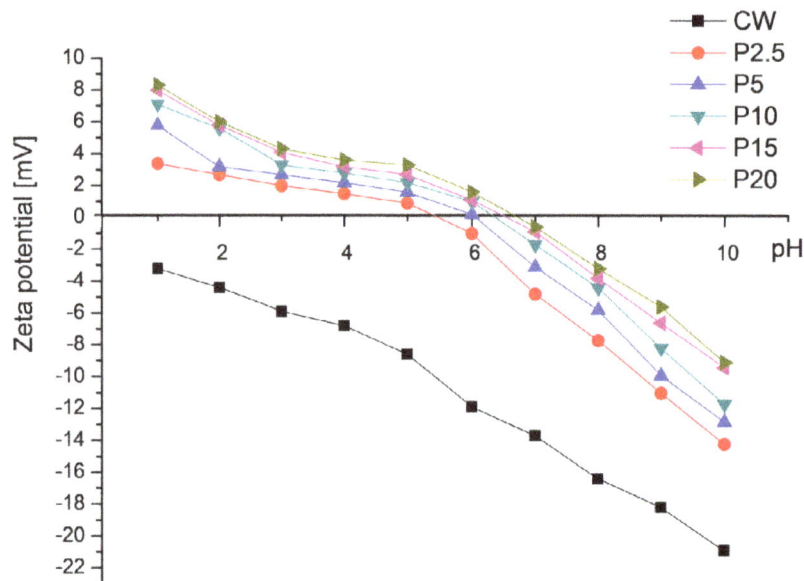

Figure 3: Zeta potential for modified and unmodified modal fabric with different concentration of PVAmHCl with respect to pH of electrolyte 0.001 M KCl.

significantly positive charge and therefore respective samples having good adsorption towards anionic dyestuffs without using of electrolyte (common salt) as well as surfactants.

The effect of PVAmHCl pretreatment on the dyeing performance

To determine whether the PVAmHCl modified modal fabrics would have enhanced dyeability with reactive dyes, the untreated modal fabric and PVAmHCl modified modal fabrics were dyed subsequently with six different reactive dyes for the concentration of 1% (o.w.f), the dyeing carried out with and without electrolyte, but in the presence of alkali, dyeing results are listed in Table 5. The results showed that all six dyes were given color strength (K/S) values for the PVAmHCl pretreated modal fabric was giving much higher value than the untreated samples. For untreated modal fabrics with salt (CS) shows higher K/S values (lowest K/S=2.5 and highest K/S=4.1 values are achieved) than the untreated modal fabric without salt (CW), In the case of untreated modal fabric without salt (CW) shows very low K/S values ((lowest K/S=0.7 and highest K/S=1.4 values are achieved). From the result, the K/S values for the treated modal fabric shows twice than the untreated samples. The K/S values were increased with the increasing concentration of PVAmHCl, but it started to reduce at 15 g/L and above the concentration of PVAmHCl. The decrease in color strength value may be due to, when excess of PVAmHCl was padded on the fabric, the bonding between the fiber and some cationic polymer become weak, and repulsion force also existed within the cationic sites of PVAmHCl. This would lead to presence of unbound polymer in the dye bath, thereby hindering the absorption of dye and possibly causing it to flocculate.

Low color strength values of the both untreated samples (CW and CS) are due to the negative zeta potential for the dyes and fiber and resulting the maximum dyes were never exhausted. However, unexhausted dyes go to pollution treatment plant and it also creates many issues while removing the color. It clearly shows that the importance of electrolyte in dye exhaustion and followed by color strength. Above observations indicate that the pretreatment of modal fabric increases dye uptake. This confirms the effectiveness of

pretreatment in enabling the fabric to be dyed without salt. It is also observed that there is a statistical significant difference in color strength of both dyes of which poly-functional reactive dyes has higher values than bi-functional reactive dyes. As compared to bi-functional reactive dyes, polyfunctional reactive dyes, having many advantages like higher color strength, good exhaustion as well as fixation.

It can be seen from Table 5 that dye reactivity on pretreated fabric was greater, due to presence of primary amino groups provided by the PVAmHCl, this is due to the maximum cationic sites on the surface of modified modal fabrics. However, the exhaustion values (%) may vary from dye to dye and concentration of PVAmHCl (highest exhaustion found 96% for Reactive Blue 281, similarly the lowest exhaustion was found in 88.5% in Reactive Red 238). Exhaustion values of C.I. Reactive blue 268 with respect to CS and CW shows 81%, 1168% less values than treated modal fabric. Generally, the treated fabrics have positive zeta potential at lower pH and it initiates the strong attraction between the anionic dyes on the cationic sites of modified modal fabric and the anionic dyes (reactive dyes) are highly exhausted without the presence of electrolyte in the dyebath. The results for color intensity on modified modal fabric show higher values than the untreated samples. Increased color intensity values are due to the reduction of zeta potential in the dye bath at lower pH.

The effect of PVAmHCl pretreatment on total dissolved solids

The presence of total dissolved solids (TDS) in water makes high hardness which disturbs the osmotic balance of native species. Disposal of the salt laden effluents into ground and surface water bodies cause pollution and render them unfit for any purpose namely agriculture, domestic. The quality of irrigation water mainly depends on its salt content and the proportion of sodium to other ions. Sodium chloride corrodes steel and sodium sulfate corrodes concrete. So it is necessary to determine the TDS values of the polluted water. The TDS was measured from the first rinse of the exhausted dye bath. The TDS (mg/L) values for the dyeing of modified and untreated modal fabrics are presented in Table 6. The results show that the TDS of the effluent obtained from the dyeing of PVAmHCl pretreated samples shows less compared to untreated samples. This is due to a PVAmHCl pre-

Dye (1% o.w.f)	Sample	E (%)	F (%)	T (%)	(K/S)	I (%)
Reactive Blue 268	CS	51.8	82.1	42.5	2.8	-
	CW	7.3	90.9	6.7	0.8	-
	P2.5	88.9	90.6	80.5	3.2	14
	P5.0	92.6	92.1	85.3	3.8	36
	P10.0	92.6	95.1	88.0	4.1	46
	P15.0	90.6	95.0	86.1	4	43
	P20.0	91.6	94.9	86.9	3.9	39
Reactive Red 238	CS	54.1	88.0	47.6	2.5	-
	CW	11.2	87.5	9.8	0.8	-
	P2.5	88.5	96.7	85.5	6	140
	P5.0	89.6	96.8	86.7	6.2	148
	P10.0	92.1	95.5	87.9	6.6	164
	P15.0	93.2	96.2	89.7	6.6	164
	P20.0	90.8	97.3	88.4	6.4	156
Reactive Yellow 206	CS	47.4	83.9	39.8	3.1	-
	CW	11.2	80.0	8.9	1.1	-
	P2.5	89.6	85.1	76.3	4.7	52
	P5.0	94.5	88.0	83.2	5	61
	P10.0	95.9	92.5	88.7	5.3	71
	P15.0	93.2	93.7	87.3	5.2	68
	P20.0	90.8	91.9	83.4	5.4	74
Reactive Blue 281	CS	61.3	87.9	53.9	3.3	-
	CW	19.4	85.7	16.6	1.4	-
	P2.5	88.9	93.2	82.8	5.9	79
	P5.0	92.9	92.2	85.6	6.4	94
	P10.0	96.0	96.2	92.3	7.8	136
	P15.0	93.9	94.9	89.1	7.8	136
	P20.0	93.4	94.7	88.5	7.6	130
Reactive Red 286	CS	62.2	87.8	54.7	4.1	-
	CW	21.0	75.0	15.8	1.2	-
	P2.5	89.8	92.1	82.8	7.6	85
	P5.0	91.6	90.2	82.7	8.2	100
	P10.0	95.1	90.8	86.4	8.7	112
	P15.0	93.3	91.0	84.9	8.9	117
	P20.0	92.9	92.0	85.5	8.8	115
Reactive Yellow 217	CS	63.4	91.3	57.9	2.3	-
	CW	20.5	85.7	17.6	0.7	-
	P2.5	89.6	93.2	83.5	5.9	157
	P5.0	92.4	93.7	86.5	6.3	174
	P10.0	95.4	93.8	89.4	6.4	178
	P15.0	92.2	92.4	85.3	6.6	187
	P20.0	91.3	88.7	81.0	6.2	170

Table 5: Color strength value of control and treated samples.

treated modal fabric does not require any electrolyte for exhaustion of reactive dyes. Utilization of electrolyte in the dyeing may increase the TDS values and it shows (for CS) 18550 mg/L, whereas the PVAmHCl treated sample shows the range of 6400 to 6980 mg/L. It is nothing but, approximately 3 times less TDS value as compared to the untreated and dyed (CS) sample. The fabrics with the lowest percentage of PVAmHCl pre-treatment have slightly lower the TDS values than the fabrics with the highest percentage of pre-treatment. This may be some complex interaction between dyes and the pre-treatment chemicals which, is outside the scope of this study. The higher TDS values (CS=18550 mg/L) are necessary to reduce and it requires many ETP processes which increase the overall process cost (power cost, energy cost, manpower and etc), whereas PVAmHCl pre-treated sample does not requires as much.

The effect of PVAmHCl pretreatment on water retention values

The water retention value (WRV) is an important parameter for the fibers, due to many of the chemical processing such as dyeing and finishing were done by using water or liquids as a medium. Wetting, repulsion and transportation is a complex phenomena which depend on many parameters, namely, fiber surface morphology, functional groups or chemical composition of the fiber and structure of the fiber porous can decide the hydrophilic or hydrophobic characteristics of the fiber. Functional groups or chemical composition is one of the parameters which decide the hydrophilic characteristics of modal fabrics, it is necessary to determine the WRV for PVAmHCl pre-treated fabrics. The WRV results were shown in the Figure 4, which clearly explain the WRV values for treated and untreated samples. The overall WRV has been decreased for the pre-treated samples as compared to untreated samples; this is due to, while treatment with PVAmHCl were blocking the hydroxyl groups of cellulose and probably the pore structure of the fabric was slightly changed which resulting decrease the WRV. But this change is very small (maximum 10%) and it never affects the dyeing or finishing process.

The effect of PVAmHCl pretreatment on colorfastness properties

While usage of dyed or printed fabrics was frequently subjected to various conditions such as washing, rubbing, light radiation and perspiration. It is necessary the respective dyestuff has to withstand with respective conditions, which generally called as good fastness property. Related to this research, it is mandatory to determine the impact of pre-treatment on the dye fastness properties with respect to washing, rubbing and light are summarized in Table 7. In overall, pre-treated fabrics showing excellent results as compared to conventional dyed and without salt (CW), whereas it shows almost the same results to the sample which dyed conv-entional with salt (CS). Generally, (CW) dyed fabric without salt shows poor to moderate (rating1-2) in some dyes; it is because of the poor exhaustion and followed by fixation of reactive dyes. In particularly, the washing fastness for pre-treated and dyed with poly-functional reactive dyed modal fabrics against staining of its adjacent fabrics (wool and cotton) was excellent in few cases (5), and the color change was also in the acceptable range which is (2-3). It seems that the results of washing fastness are excellent for the salt free dyeing and it is due to the effectiveness of dye fixing due to the pretreatment with PVAmHCl. Regarding to the rubbing fastness, dry rubbing results for the majority of the cases are showing 4 which namely called as a good fastness property. In case of wet rubbing the results were showing (3-4) which is called average and it is due to the residual unfixed dyes on the surface of the fabric even after soaping process. Owing to the solubility of reactive dyes the unfixed and the surface dyes will dissolve in water during wet rubbing and get transferred to the crocking cloth. As seen in the Table 7, the ratings for color fastness to light in terms of the degree of color change and color staining were good to very good (rating 5). The results for the perspiration fastness under acidic and alkaline conditions were shown in the Table 8. There is no obvious improvement in perspiration fastness properties in most cases with, related to the concentration of pre-treatment. The result of the color change is showing (2-3) under acidic and alkaline conditions, it seems, dyed fabric have sensitive towards pH, and there is no significant difference between pre-treatment and perspiration fastness. However, the staining towards to the acidic conditions is (4-5), whereas in alkali conditions, it shows (3-4) similarly the result for (CW) dyed in conventional method without salt shows moderate to good (rating 3

Dye (1% o.w.f)	Sample	TDS (mg/L)
Reactive Blue 268	CS	18550
	P2.5	6400
	P5.0	6550
	P10.0	6800
	P15.0	6900
	P20.0	6980

Table 6: TDS value of control and treated samples.

Dyes	Sample	Wash fastness			Rubbing fastness		Light fastness
		Evaluation of color change	Evaluation of staining		Evaluation of staining		
			Cotton	Wool	Dry	Wet	
Reactive Blue 268	CS	3	4-5	4-5	4	3-4	3
	CW	2-3	2	2	1-2	1-2	2
	P2.5	2-3	4	4	4	4	5
	P5.0	2-3	4	4	4	4	5
	P10.0	2-3	4-5	4-5	4	4	5
	P15.0	2-3	4-5	4-5	3-4	3-4	5
	P20.0	2-3	4-5	4-5	3-4	3-4	4
Reactive Red 238	CS	2-3	4-5	4-5	4	3-4	3
	CW	2-3	2	2	1-2	1-2	2
	P2.5	2-3	4-5	4-5	3-4	3-4	5
	P5.0	2-3	4-5	4-5	3-4	3-4	4
	P10.0	2-3	4-5	4-5	3-4	3-4	4
	P15.0	2-3	4-5	4-5	3-4	3-4	4
	P20.0	2-3	4-5	4-5	3-4	3-4	4
Reactive Yellow 206	CS	3	4-5	4-5	4	3-4	4
	CW	2-3	2	2	1-2	1-2	3
	P2.5	2-3	4-5	4-5	4	4	5
	P5.0	2-3	4-5	4-5	4	4	5
	P10.0	2-3	4-5	4-5	4	4	5
	P15.0	2-3	4-5	4-5	3-4	3-4	5
	P20.0	2-3	4-5	4-5	3-4	3-4	4
Reactive Blue 281	CS	2-3	4-5	4-5	4	3-4	3
	CW	2-3	2	2	1-2	1-2	2
	P2.5	2-3	4-5	4-5	4	4	5
	P5.0	2-3	5	4-5	4	4	5
	P10.0	2-3	5	4-5	4	4	5
	P15.0	2-3	5	5	4	4	5
	P20.0	2-3	5	5	4	4	4
Reactive Red 286	CS	2-3	4-5	4-5	4	4	3
	CW	1-2	2	2	2	2	2
	P2.5	2-3	4-5	4-5	3-4	4	4
	P5.0	2-3	5	5	4	4	4
	P10.0	2-3	5	5	4	4	4
	P15.0	2-3	5	5	4	4	4
	P20.0	2-3	5	4-5	4	3-4	4
Reactive Yellow 217	CS	3	4-5	4-5	4	3-4	4
	CW	1-2	2	2	1-2	1-2	3
	P2.5	2-3	5	5	4	4	5
	P5.0	2-3	4-5	4-5	4	4	5
	P10.0	2-3	5	5	4	4	5
	P15.0	2-3	5	5	4	4	5
	P20.0	2-3	5	4-5	4	3-4	5

Table 7: Wash, Rubbing and Light fastness for control and treated samples.

Dyes	Sample	Perspiration fastness					
		Evaluation of color change		Evaluation of staining			
				Acidic		Alkaline	
		Acidic	Alkaline	wool	cotton	wool	cotton
Reactive Blue 268	CS	2-3	2-3	4-5	4-5	4-5	2-3
	CW	2-3	2-3	4	4	4	3
	P2.5	2-3	2-3	4-5	4-5	4-5	3-4
	P5.0	2-3	2-3	4-5	4-5	4-5	3-4
	P10.0	2-3	2-3	4-5	4-5	4-5	3-4
	P15.0	2-3	2-3	4-5	4-5	4-5	3
	P20.0	2-3	2-3	4-5	4-5	4-5	3
Reactive Red 238	CS	2-3	2-3	4	4	3-4	3-4
	CW	2-3	2-3	4-5	4	3-4	3
	P2.5	2-3	2-3	4-5	4	4-5	3-4
	P5.0	2-3	2-3	4-5	4	4-5	3-4
	P10.0	2-3	2-3	4-5	4	4-5	3-4
	P15.0	2-3	2-3	4-5	4	4-5	3-4
	P20.0	2-3	2-3	4-5	4	4-5	3-4
Reactive Yellow 206	CS	2-3	2-3	4-5	4-5	4-5	3-4
	CW	2-3	2-3	4-5	4-5	4	3
	P2.5	2-3	2-3	4-5	4-5	4-5	3-4
	P5.0	2-3	2-3	4-5	4-5	4-5	3-4
	P10.0	2-3	2-3	4-5	4-5	4-5	3-4
	P15.0	2-3	2-3	4-5	4-5	4-5	3-4
	P20.0	2-3	2-3	4-5	4-5	4-5	3-4
Reactive Blue 281	CS	2-3	2-3	4-5	4-5	4-5	3-4
	CW	2-3	2-3	4-5	4-5	4-5	3-4
	P2.5	2-3	2-3	4-5	4-5	4-5	3-4
	P5.0	2-3	2-3	4-5	4-5	4-5	3-4
	P10.0	2-3	2-3	4-5	4-5	4-5	3-4
	P15.0	2-3	2-3	4-5	4-5	4-5	3-4
	P20.0	2-3	2-3	4-5	4-5	4	3
Reactive Red 286	CS	2-3	2-3	4-5	4-5	4-5	3-4
	CW	2-3	2-3	4-5	4-5	4-5	3-4
	P2.5	2-3	2-3	4-5	4-5	4-5	3-4
	P5.0	2-3	2-3	4-5	4-5	4-5	3-4
	P10.0	2-3	2-3	4-5	4-5	4-5	3-4
	P15.0	2-3	2-3	4-5	4-5	4-5	3-4
	P20.0	2-3	2-3	4-5	4-5	4-5	3
Reactive Yellow 217	CS	2-3	2-3	4-5	4-5	4-5	3-4
	CW	2-3	2-3	4-5	4-5	4-5	3-4
	P2.5	2-3	2-3	4-5	4-5	4-5	3-4
	P5.0	2-3	2-3	4-5	4-5	4-5	3-4
	P10.0	2-3	2-3	4-5	4-5	4-5	3-4
	P15.0	2-3	2-3	4-5	4-5	4-5	3-4
	P20.0	2-3	2-3	4-5	4-5	4-5	3

Table 8: Perspiration fastness for control and treated samples.

and 2-3). It is due to the stability of the bonds leads to alkali hydrolysis. The results for perspiration fastness were observed quite similar for all six kinds of reactive dyed fabric. Hence there is no impact on the pre-treatment.

The effect of PVAmHCl pretreatment on mechanical properties

The results of various physical properties of pre-treated samples were summarized in the Table 9. The crease recovery behavior of

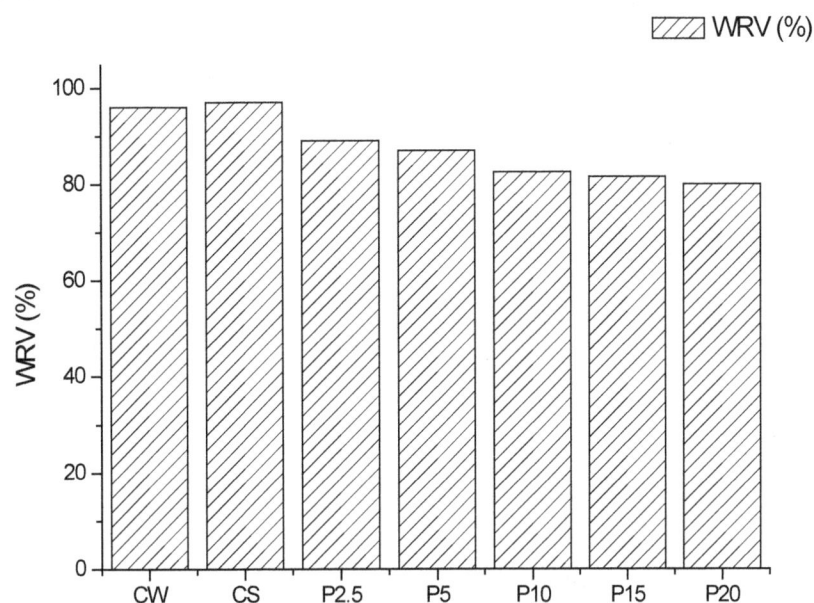

Figure 4: Water retention values (WRV) for control and PVAmHCl treated modal fabrics.

| Sample | Tensile Strength | | | | Flexural Rigidity mg.cm | | CRA degrees |
| | Warp way | | Weft way | | Warp way | Weft way | |
	Values, lbs	Loss, %	Values, lbs	Loss, %			
Raw fabric	75	--	60		108.6	87.2	132
CS	73	3	59	2	109.2	87.8	135
CW	72.5	3	59	2	109.5	88.1	134
P2.5	70	7	55	8	110	89.8	141
P5.0	69	8	54	10	111.4	91	148
P10.0	68	9	52	13	112.8	91.2	156
P15.0	66	12	51	15	114.3	95	155
P20.0	66	12	50	17	117.5	96.9	156

Table 9: Physical properties of control and treated samples.

all the PVAmHCl pre-treated samples were observed to be good, Crease recovery angle (CRA) has been increased with increasing the concentration of PVAmHCl. Maximum CRA was obtained for (P20) have 156 degrees; whereas raw fabric has 132 degrees only. Increasing the CRA is due to cross-linking of PVAmHCl between the cellulose molecules. These cross-links hinder the molecular and fibrillar slippage and stabilize the structure, thereby increasing the crease recovery angle. The effect of cross linking of PVAmHCl with cellulose generally results in excellent crease recovery behavior, but at the same time the strength loss also associated. The tenacity of pre-treated fabric is decreased with concentration of PVAmHCl is increased; the tenacity decreased 12% and 17% for warp way and weft way respectively at the 20 g/L concentrations of PVAmHCl (P20); it seems that the tenacity is directionally proportional to the PVAmHCl concentrations, it is due to the formation of intermolecular an interamolecular crosslink's reduce the possibilities of equalizing the stress distribution, it ceased to reduce the capacity to withstand towards the load. Generally Flexural rigidity is nothing but resistance to flexibility, flexural rigidity has increased in the range of 8% and 10% in war and weft ways. Increase in flexural rigidity shows that the fabric becomes slightly stiff because of cross linking of cellulose with PVAmHCl; it can be solved by adding a softener during the finishing treatment.

Conclusion

The main purpose of this work is to increase the dyeability of modified modal fabric with reactive dyes were carried out without utilization of electrolyte. Modification of modal fabric with different concentration of PVAmHCl which significantly decrease the negative surface of the pre-treated fabric, it has been confirmed during the zeta potential measurement. Generally the PVAmHCl can adsorb on the modal fiber surface and it increases the positive values of zeta potential. These results can lead good affinity towards to the reactive dyes (anionic) on modified fabric via good ionic interaction. All dyes have good exhaustion towards modifying fabrics, but out of that, polyfunctional reactive dyes have better exhaustion as compared to bi-functional reactive dyes, this is due to the number of reactive groups which help higher exhaustion. The color strength (K/S) values for the modified fabric gave better color strength than the comparable to dyeing without salt by conventional dyeing method (CS), almost the same results as compared to dyeing with salt by conventional dyeing method (CW). The water retention values (WRV) for the modified modal have gradually decreased when the concentration of PVAmHCl has been increased, this is because PVAmHCl can be blocked the hydroxyl groups which available on the surface of fabric, which lead to reduce the water retention capacity. After dyeing, modified modal fabric shows the very less total dissolved solids (TDS) values (approximately 4 times less than conventional dyed fabric with salt (CS), it seems, this method of dyeing can help to save environment from the pollution, as well as the saving the energy which can utilize for the pollution treatment plant. Wash, Rubbing, Light fastness of modified fabrics is showing approximately the same result like conventional dyed fabric with salt (CS). Fabric crease recovery and flexural rigidity increased because of the pretreatment. There is no change in the tensile strength of the fabric because of the pretreatment. Pretreatment with PVAmHCl on modal fabrics having the following advantages are observed.

- Elimination of salt as an electrolyte

- Maximum fixation of dye

- Improve the exhaustion and lead to reduce the hydrolysis of dye

- Significant saving in process cost (for after treatments and pollution treatments)
- Reduction of pollution load up to 80%.

It is considered that PVAmHCl is found to be effective for pretreatment in salt less dyeing of modal fabrics.

Acknowledgements

The author would like to thank Dr. Martina Vikova and Dr. Michal Vik, Associate professors, Technical University of Liberec, Czech Republic for providing the technical suggestion during this work; The author also thanks to Dr. P V Kadole, Principal, Mr. Anil Kumar Yadav, Mr. Pramod M. Gurave, DKTE's Textile Engineering Institute, India for providing the facilities also moral support to carry out this work.

References

1. Lenzing AG (2015) Modal from Lenzing. lenzing website.

2. Kreze T, Stana-Kleinschek K, Ribitsch V (2001) The sorption behaviour of cellulose fibres. Lenzinger Berichte 28-33.

3. Varga K, Kljun A, Noisternig MF, Ibbett RN, Gruber J, et al. (2009) Physiological investigation of resin-treated fabrics from tencel and other cellulosic fibres. Lenzinger Berichte 87: 135-141.

4. Broadbent AD (2001) Basic principles of textile coloration. Text Color 395-396.

5. Khatri A, Peerzada MH, Mohsin M, White M (2015) A review on developments in dyeing cotton fabrics with reactive dyes for reducing effluent pollution. J Clean Prod 87: 50-57.

6. Zhang Y, Zhang W (2014) Clean dyeing of cotton fiber using a novel nicotinic acid quaternary triazine cationic reactive dye: salt-free, alkali-free, and non-toxic by-product. Clean Technol Environ Policy 563-569.

7. Grancarić AM, Ristić N, Tarbuk A, Ristić I (2013) Electrokinetic phenomena of cationised cotton and its dyeability with reactive dyes. Fibres Text East Eur 21: 106-110.

8. Burkinshaw SM, Mignanelli M, Froehling PE, Bide MJ (2000) The use of dendrimers to modify the dyeing behavior of reactive dyes on cotton. Dye Pigment 47: 259-267.

9. Ma W, Meng M, Yan S, Zhang S (2015) Salt-free reactive dyeing of betaine-modified cationic cotton fabrics with enhanced dye fixation. Chinese J Chem Eng 24: 175-179.

10. Montazer M, Malek RMA, Rahimi A (2007) Salt free reactive dyeing of cationized cotton. Fibers Polym 8: 608-612.

11. Ibrahim NA, Abdel Moneim NM, Abdel Halim ES, Hosni MM (2008) Pollution prevention of cotton-cone reactive dyeing. J Clean Prod 16: 1321-1326.

12. Babu K M, Selvadass M (2012) Investigation on Ecological Parameters of Dyeing Organic Cotton Knitted Fabrics Abstract : Processes: Parameters of Dyeing. Univers J Environ Res Technol 2: 421-428.

13. Lewis DM (2014) Developments in the chemistry of reactive dyes and their application processes. Color Technol 130: 382-412.

14. Periyasamy AP (2012) Effect of alkali pretreatment and dyeing on fibrillation properties of lyocell fiber. In: RMUTP International Conference: Textiles and Fashion 2012.

15. Mousa AA (2007) Synthesis and application of a polyfunctional bis(monochlorotriazine/sulphatoethylsulphone) reactive dye. Dye Pigment 75: 747-752.

16. Periyasamy AP, Dhurai B (2011) Salt free dyeing: A new method of dyeing of Lyocell fabrics with reactive dyes. AUTEX Res J 11: 14-17.

17. Nebojsa Ristic IR (2012) Cationic Modification of Cotton Fabrics and Reactive Dyeing Characteristics. J Eng Fiber Fabr 7: 113-121.

18. Ma W, Zhang S, Tang B, Yang J (2005) Pretreatment of cotton with poly (vinylamine chloride) for salt-free dyeing with reactive dyes. Color Techonology 121: 193-197.

19. Wang L, Ma W, Zhang S, Xiaoxu T, Jinzong Y (2009) Preparation of cationic cotton with two-bath pad-bake process and its application in salt-free dyeing. Carbohydr Polym 78: 602-608.

20. Tarbuk A, Grancaric AM, Leskovac M (2014) Novel cotton cellulose by cationization during mercerization-part 2: The interface phenomena. Cellulose 21: 2089-2099.

21. Ramesh Kumar A, Teli MD (2007) Electrokinetic studies of modified cellulosic fibres. Colloids Surfaces A Physicochem Eng Asp 301: 462-468.

22. Christie R (2007) Environmental Aspects of Textile Dyeing. (1stedn) Woodhead Publishing Limited and CRC press, Cambridge.

23. Hu Z, Zhang S, Yang J, Chen Y (2003) Some Properties of Aqueous-Solutions of Poly (vinylamine chloride). Polymer (Guildf) 89: 3889-3893.

24. Shore J (1995) Cellulosics dyeing. (1stedn) Society of Dyers and Colourists, UK.

23

Processing, Structure and Properties of Melt Blown Polyetherimide

Kandagor V[1], Prather D[1], Fogle J[1], Bhave R[2] and Bhat G[3]*

[1]*The University of Tennessee, Knoxville, Tennessee, USA*
[2]*University of Georgia, Athens, Georgia, USA*
[3]*Oak Ridge National Laboratory, Oak Ridge, Tennessee, USA*

Abstract

Polyetherimide (PEI), an engineering plastic with very high glass transition temperature and excellent chemical and thermal stability has been processed into membranes of varying pore size, performance, and surface characteristics. A special grade of the polyetherimide was processed by melt blowing to produce microfiber nonwovens suitable as filter media. The resulting microfiber webs were characterized to evaluate their structure and properties. The fiber webs were further modified by hot pressing, a post processing technique, which reduces the pore size in order to improve the barrier properties of the resulting membranes. This ongoing research has shown that polyetherimide can be a good candidate for filter media requiring high temperature and chemical resistance with good mechanical properties.

Keywords: Meltblowing; Microfibers; Polyetherimide; Polymer membranes; Filtration

Introduction

The need for better filter media for applications in liquid and gas filtration has increased in recent years. In particular, filters for biomedical applications, water filtration, oil and gas separation and other applications for reducing environmental pollution have become increasingly necessary. Due to its robust biocompatibility properties, PEI materials are finding applications in the biomedical industry. Recent research has shown that PEI does not exert any considerable level of cytotoxicity or hemolysis, permitting the growth of cell on the surface [1,2]. Membrane filtrations for biological applications have to face two different environments: blood tissues and cell. Consequently, filters for biological and other separation applications must have surface characteristics that are compatible with specific environments that the membranes are intended to be used [1]. With some separation process requiring vigorous procedures which may include high temperature, high pressure, and in some cases high chemical environment, it has become necessary to design filters that can withstand these tough conditions without compromising the effectiveness, quality, and efficiency of the separation process. Nonwoven filters have particularly become attractive because of the simplicity in the fabrication process which does not requires use of solvents that can present challenges during processing as well as in the end product. The PEI resin is a copolymer with the ether molecules between imide groups. The Ultem resin combines the high performance associated with the exotic specialty polymer together with the excellent processability characteristics of engineering polymers, and finds specific applications in the aero, auto, and insulation industries, where performance at high temperatures is a stringent requirement [3].

The melt blowing has become a commercially successful process in producing nonwovens because of its ability to produce fibers of desired characteristics including fiber diameter ranging from 2-4 microns, and desired permeability characteristics, which is achieved by manipulation of the processing conditions. It involves application of hot air jet to an extruding polymer melt, which is then drawn into micro and nano size fibres [4].

The melt blowing technology, which was originally commercialized by the Exxon Chemical Company, is currently widely used for production of fine fiber nonwovens. The extruder melts the polymer and the molten polymer is forced through the melt-blowing die which consists of a row of orifices or jets, resulting in the formation of small diameter fibers [5]. The fibers are then drawn by the high velocity hot air, quenched and collected on a continuous moving belt forming the continuous fiber web. The properties of the meltblown webs are affected by various production parameters including air temperature, polymer/die temperature, die to collector distance (DCD), collector speed, polymer throughput, air throughput, die hole size and air gap [6]. The melt blown fiber webs can be compacted further by calendaring to form membranes with desired surface and pore characteristics.

Several polymers have been successfully meltblown using the pilot line, and many of them are commercially practiced [7,8]. Although ultem has not been melt blown before, it has been converted into fibers and membranes [9]. In this work, we have fabricated flat module membranes from Ultem with high temperature, high chemical resistance, and high pressure operating range and evaluated them for their performance and physical properties [10]. The PEI has been processed by meltblowing and then hot pressed to reduce their pore size at different pressure and temperature for varying structures and properties [11,12] (Figure 1).

Experimental Section

Materials and processing

Commercially available polyetherimide, Ultem 1285, was purchased from Sabic Innovative Plastics, and was used without any further modification. The polymer was dried at a temperature of 130°C for 6 hours to ensure that the moisture content was reduced to the recommended 0.02% before melt blowing. This Ultem had a Melt Flow Rate (MFR) of 23 g/10 min at a temperature of 325°C.

***Corresponding author:** Bhat G, Oak Ridge National Laboratory, Oak Ridge, Tennessee, USA, E-mail: gbhat@uga.edu

Meltblowing was performed using the 15 cm wide meltblowing line (Figure 2) at the University of Tennessee's Nonwoven Research Laboratory (UTNRL). Seven heating zones with independent heaters, including three in the extruder, facilitate incremental heating of the polymer to allow complete melting [13]. The Exxon type die used had 10 holes per cm and each hole was 450 μm in diameter. The die temperature was maintained around 365°C and the air temperature slightly higher (~390°C) to help maintain the die at the desired temperature. Since air pressure is the most critical variable in controlling the fiber diameter, three different air pressures were investigated, keeping rest of the processing conditions same. The primary variable in the production of these fibers was the air pressure because a slight variation in the temperature results in significant change in the permeability of the membrane compared to the variation of the air temperature or collection speed. It has been previously observed that for fibers used in separation technology, the ideal air temperature for Ultem fibers is 390°C. This is the main reason for varying the air pressure (Table 1).

The meltblown Ultem 1285 fiber web was then hot pressed using carver hot press model 3895.4 NE1000 at different temperatures and pressures resulting in further consolidation of the fiber web so that the pore size was reduced to a much lower value that is determined

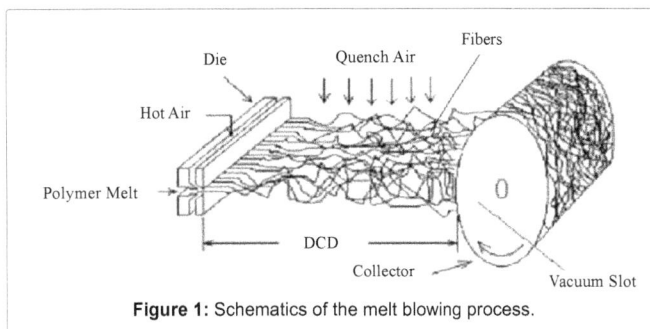

Figure 1: Schematics of the melt blowing process.

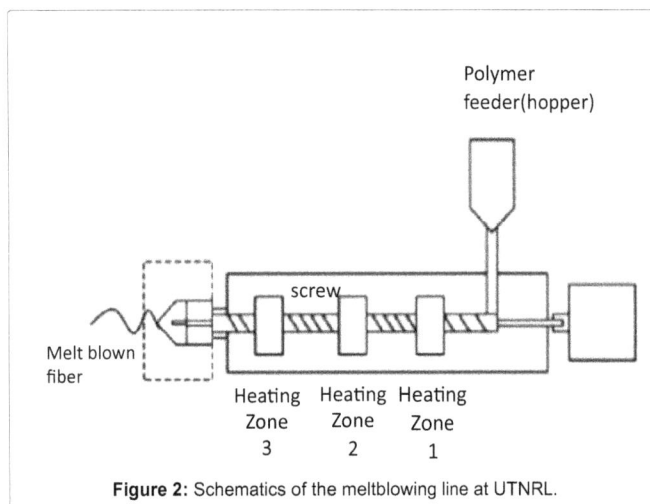

Figure 2: Schematics of the meltblowing line at UTNRL.

	Sample 1	Sample 2	Sample 3
Die temperature	365	365	365
Air Temperature (°C)	390	390	390
Air Pressure KPa	172	206	241
Collector Speed (m/min)	9.3	9.3	9.3
DCD (cm)	12	12	12

Table 1: Summary of the meltblowing conditions for the three samples that were prepared.

by the pressure and temperatures of the calender rolls. The resulting membrane is bulky and soft on the surface giving it the characteristics of selective permeability.

Characterization

The produced nonwovens were characterized to determine the pore size and pore size distribution. This was done using a Porous Materials Inc. capillary flow porometer model ASF-1100-AEX. The porometer measures the gas flow as a function of the applied pressure, and the curve that is determined for both the dry and wet measurements is used for the calculation of the pore size, mean flow pore size, the smallest pore size and the gas permeability of the resulting membrane. The Washburn Equation has been used to define the mathematical relationship between the applied pressure and the pore size by using the surface tension and the contact angle of the wetting fluid providing the porosity data [14].

SEM micrographs were obtained using an ETEC Auto-scan scanning electron microscope at 3 KeV after coating with a gold layer. The SEM images used in combination with a computer image processing software (imageJ NIST) helped determine the average fiber diameter and fiber diameter distribution of the samples. The tensile properties of the nonwovens were tested using a United SSEM-1-E-PC tensile tester. Five specimens of each nonwoven sample were cut and the resulting values from the tensile testing were averaged according to the ASTM D638 - 10. The air permeability was measured using TEXTEST FX3300 equipment according to ASTM standard D737-96. The Mettler Differential Scanning Calorimetry (DSC) model DSC821 was used to characterize the polymer meltblown fiber webs and the calendered membranes. The samples were heated at 10°C/min from room temperature to 380°C, held for 3 minutes at this temperature then cooled back to room temperature at 10°C/min [10].

The water flux measurement was performed using an in house water flux measurement set up shown below in Figure 3. The Ultem membrane is subjected to water under pressure and the permeate corresponding to different feed pressure measured. The feed pressure was varied between 13.8 and 68.9 KPa. For experimental accuracy, the time it takes to collect 100 Ml of the permeate was measured for the different feed pressures and the membrane cross flow permeability (LMHB) was calculated for each membrane [15].

Results and Discussions

Fiber diameter

The fiber diameter of melt blown Ultem has been characterized in

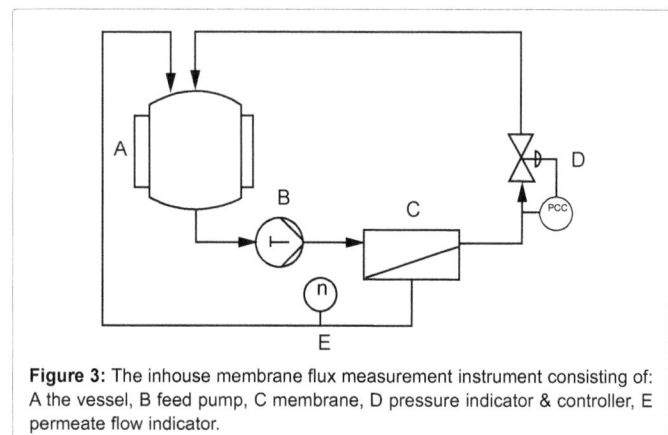

Figure 3: The inhouse membrane flux measurement instrument consisting of: A the vessel, B feed pump, C membrane, D pressure indicator & controller, E permeate flow indicator.

detail and the correlation between the fiber diameter and the processing air pressure reveals a relationship that is similar to those reported in literature. The lower the air pressure, the higher the resulting fiber diameter and the entangling of the fiberweb. Theoretically, the fiber diameter will increase as the airflow rate decreases [16,17].

Figures 4 and 5 show the fiber diameter and diameter distribution throughout the nonwoven meltblown material, and the data is consistent with the expected results. At the 241 µm, kPa at of 206 pressure, kPa, the average fiber diameter is about 9.5 fiber diameter and increase sat 172 kPa to 10 processing. 5 air pressure, the resulting fiber diameter was determined to be 12 µm. There is a linear relationship between processing air pressure and the resulting fiber diameter as shown in the graph below. The processing air has the most dominant effect on fiber attenuation. In fact, the drawdown of the fiber in the melt blowing process is due to the processing air and higher air pressure leads to acceleration of the molten polymer coming out of the die. That

is how the polymer coming out of the die at 450 micron goes down to few microns within a short distance. The higher air pressure means an increase in acceleration of the filament leading to higher velocity and effective draw ratio.

The fiber diameters are only slightly larger than that of the meltblown webs from typical polypropylenes. Considering the fact that the MFR of the ultem resin was very low and it was not designed to achieve fine fibers, the fiber diameters achieved are very good, especially since the webs were consistently uniform. In fact, the melt blown fibers from many other polymers as well as from earlier PP resins are in the same range. Only because current day commercial PP resins for melt blowing are of special high melt flow rate type, finer fibers in the range of 2-5 microns are possible.

The SEM micrograph of the fiber webs (Figure 6) shows smooth fiber morphology that would confirm the relative ease in the process ability of Ultem 1285 from the pellets by melt blowing, in spite of their high glass transition temperature and melt viscosity. There is no evidence of breaking up of the resulting fibers or variation in diameter along the length of the fiber, an indication of strong fiber web structural integrity. The fiber web formed at higher air pressure shows a lower fiber diameter but higher fiber volume per unit area, which results in a smaller effective pore size and separation characteristics.

Web structure and properties

The thickness of the hot pressed nonwoven fiberwebs decreased with increase in temperature of hot press, and also increasing pressing

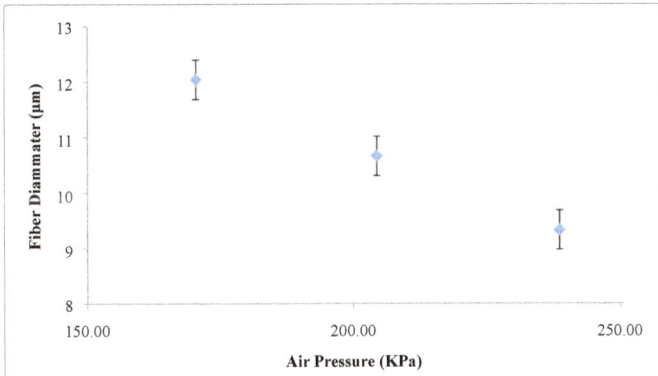

Figure 4: The relationship between the air pressure and the fiber diameter.

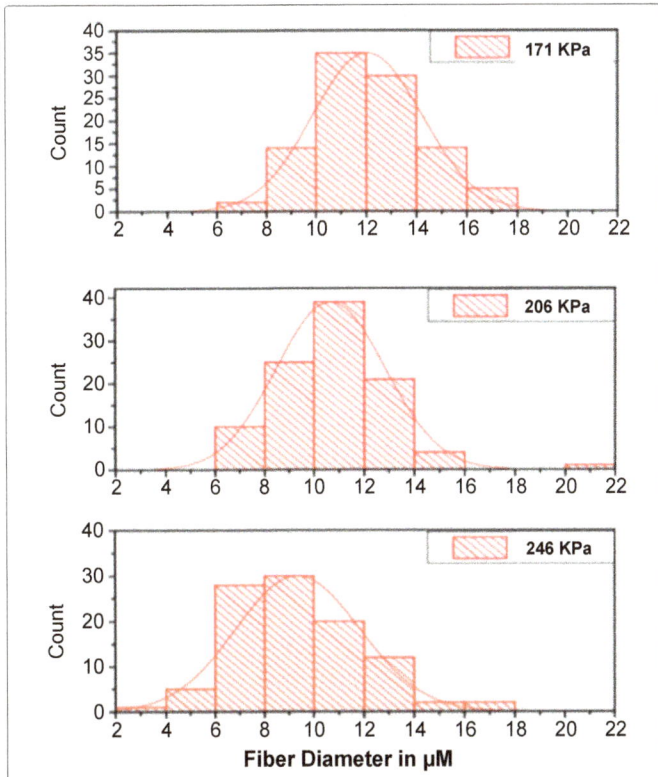

Figure 5: Fiber diameter and diameter distribution.

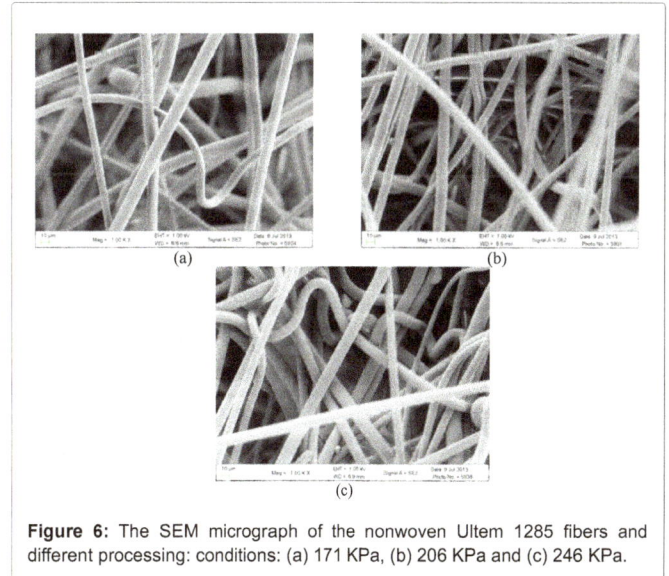

Figure 6: The SEM micrograph of the nonwoven Ultem 1285 fibers and different processing: conditions: (a) 171 KPa, (b) 206 KPa and (c) 246 KPa.

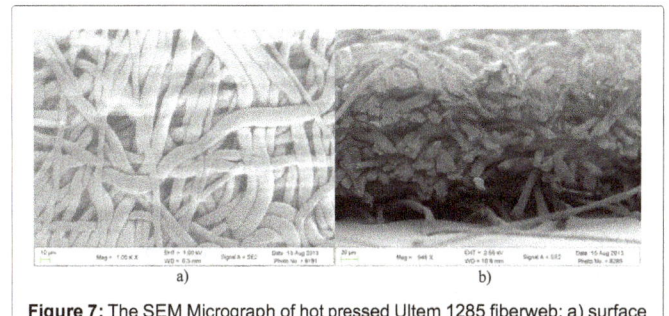

Figure 7: The SEM Micrograph of hot pressed Ultem 1285 fiberweb: a) surface and b) crossectional view.

time as shown below. The calendering process resulted in the bonding of the web as seen in the SEM micrograph (Figure 7). Comparable observations have been made earlier in structure analysis of calendered webs from various polymeric fibers. This calendering resulted in reduced pore size of the fiberwebs. A combination of shorter time of about 30 seconds and a temperature of 160°C (for the 206 KPa meltblowing condition) resulted in better Ultem membrane as will be seen in the results discussed in following sections [18].

The air permeability of the webs and membranes indicates that the smaller the thickness of the membranes produced after calendering the lower the permeability because of significantly reduced pore size. The membranes that were produced by calendering for longer time at higher temperatures (150°C for 30 sec) resulted in almost zero permeability compared to that which was calendered at lower temperatures for a shorter time as shown in Figure 8.

The change in average thicknesses on calendering. The low meltblowing pressure of 172.4 KPa had the highest thickness and at the pressure of 241 KPa the thickness was reduced membrane thickness, consequently reducing the pore size. This reduction in thickness due to the combined effect of heat and pressure, which increases the packing density of the fibers shown in Figure 9.

The relationship between the meltblowing pressure and the average mean flow pore diameter (MFPD) of the membrane is unique. The MFPD is that defined as value for which the flow in the membrane is reduced by half in a partial flow test in a capillary flow parameter instrument. Although at lower meltblowing pressure the fiber diameter is larger and increases with the increasing pressure as shown earlier,

it is the fiber density that will determine the MFPD. The graph in Figure 10 shows that at the meltblowing air pressure of 206 KPa, the fiber diameter is smaller than at 172 KPa air pressure, but larger than the 241 KPa processing pressure. However the pore size and pore size distribution are the results of fiber diameter as well as fiber packing. In general finer fiber diameter results in smaller pore size. The pore structure is further reduced when calendering is done at different temperatures due to more compact packing. Thus pore size can be manipulated by changing the temperature and pressure to achieve a desired pore structure.

Tensile properties

Reproducible data shows that the tensile strength of the nonwoven Ultem 1285 produced at the pressure of 241 KPa averaging 8 KN/m². The peak elongation for this sample was 11.1%. The tensile strength is a combined effect of fiber diameter, web consolidation as well as total mass per unit area of the fabric. The results for the fibers produced at 206 KPa and 172 KPa meltblowing pressure are relatively same, but the values are higher compared to that produced at 241 KPa air pressure as shown in Figure 11. This means that the strength of the webs is almost similar and as the meltblowing pressure decreases, the increase in the strength is not larger. The peak force for these fibers is about 12.7 KPa and an average elongation of about 13% as shown in Figure 11. The breaking load increased with increasing die air pressure that could be due to the increase in basis weight and thickness. The elongation decreased with increasing die air pressure in the production direction, due to increase in the breaking load. The breaking load increased due to stronger bounding of the fibres in the web as a result of increasing air pressure applied to the web by the vacuum [19,20].

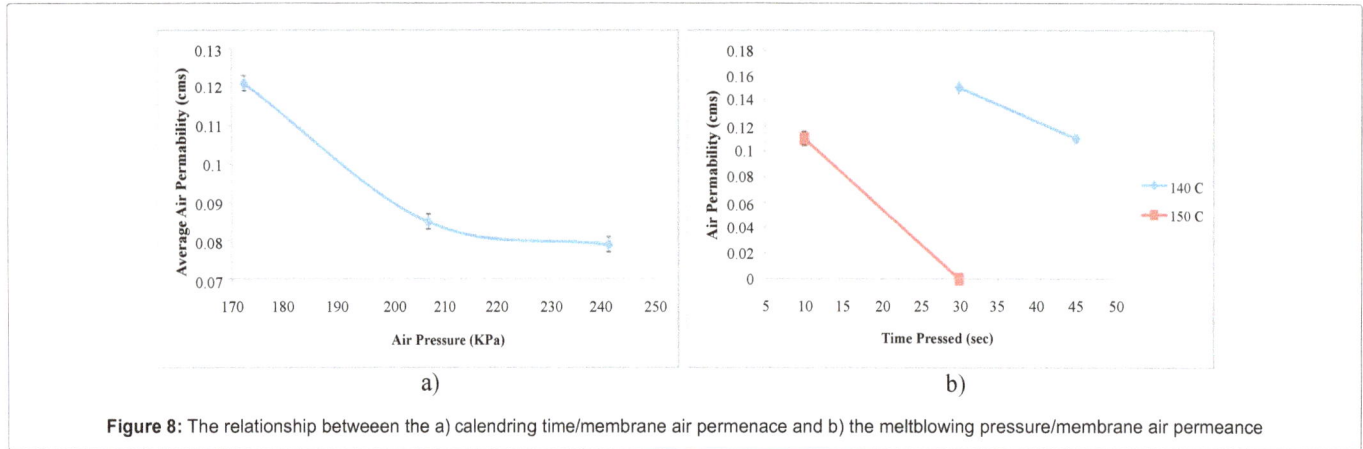

Figure 8: The relationship betweeen the a) calendring time/membrane air permenace and b) the meltblowing pressure/membrane air permeance

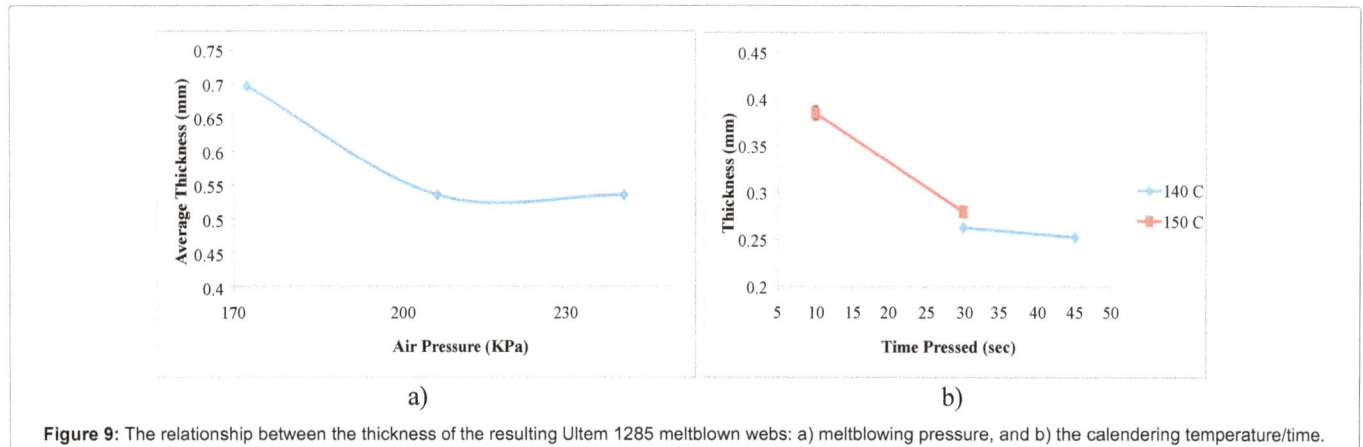

Figure 9: The relationship between the thickness of the resulting Ultem 1285 meltblown webs: a) meltblowing pressure, and b) the calendering temperature/time.

Thermal analysis

The glass transition temperature of the Ultem 1285 pellets and the meltblown webs obtained from different processing conditions was the same and measured at 177.62°C for heating and average of 185.2°C for the cooling process. No significant difference in morphology was observed for the different melt blown samples as shown in Figure 12. Ultem being an amorphous polymer does not show any melting peaks and only change in slope around the glass transition temperature. Accordingly, we observed the glass transition temperature in all the nonwoven samples at temperature close to that of the original polymer,

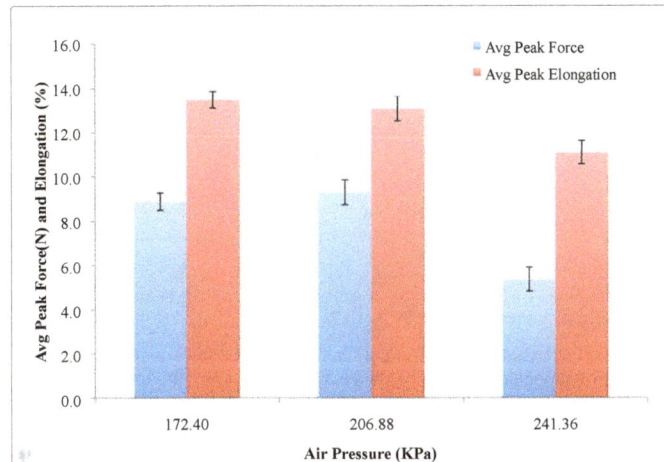

Figure 10: The graphical representation of the meltblowing pressure to the mean flow pore diameter.

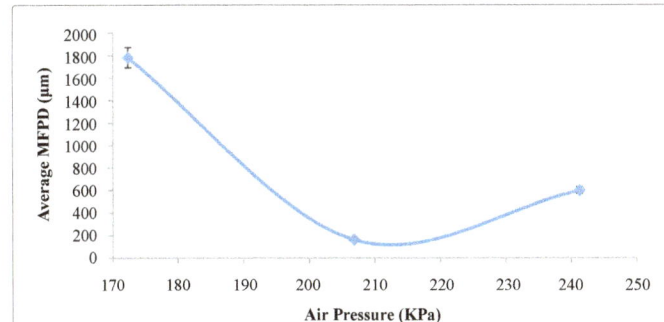

Figure 11: Tenisle strength and elongation of meltblown webs produced at different air pressures.

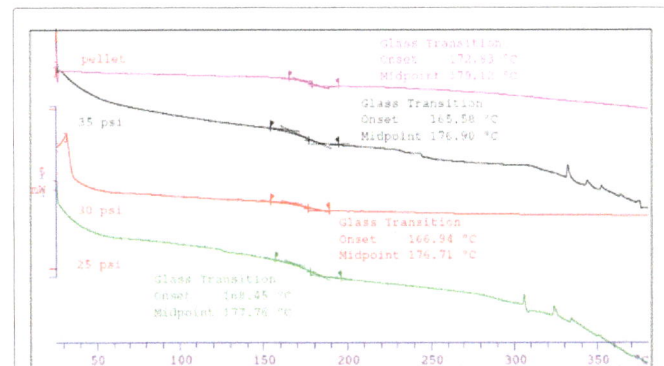

Figure 12: DSC heating curve for Ultem 1285 pellets and fiberweb produced under different processing.

only a few degrees lower due to the kinetic effect of the process as the finer fiber samples show better heat transfer due to higher surface area [21].

Liquid flux

All the membranes that were tested were produced at 241 KPa meltblowing pressure and hot pressed for 10 seconds at different temperature. The Ultem 1285 membrane calendered at 150°C reduced the LMHB by almost 50% compared and the membranes produced by calendering at 160°C and 175°C had a LMBH reduction of about 90%. This means that at higher calendering temperatures, the membrane pores are not reduced significantly and the permeation characteristics are almost similar. The ideal calendering condition is therefore 160°C for 10 seconds. The pore size can also be manipulated to a specified conditions depending on the desired separation characteristics by changing the calendering conditions (Figure 13) [15].

The separation characteristics of the membrane formed from meltblown Ultem 1285 can be used for different separation application at high temperature and high pressure. The increase in filtration efficiency of the membrane was mainly due to physical changes in porosity and permeability, as the surface of the membrane did not acquire any charge [22,23].

Conclusions

In this work, we have fabricated fibrous membranes by meltblowing Ultem, which is a high temperature stable, high performance polyetherimide, and investigated the structure and properties of the resulting meltblown web membranes. The microfibers were produced at different meltblowing air pressures and calendered in order to reduce the pore size of the membrane by varying the calendering temperature, pressure, and time. The fiber diameters varied by 8 to 13 microns depending on the air pressure used, and there were differences in air permeability as well as pore size of these webs as expected from differences in fiber diameters. It was established that the separation characteristics of the resulting membrane can be changed by varying the calendering parameters that determine the pore size of the membrane, and therefore its permeability. Membranes that were calendered between 165°C and 175°C, had the same pore size while the membrane formed by calendering the fiber web at 150°C had slightly larger pores. Their separation characteristics varied by about 20,000 liters of water per meters squared per hour per bar (LMHB). This is a significant improvement in the mechanical filtration quality of the membrane. It was clearly demonstrated that the filtration

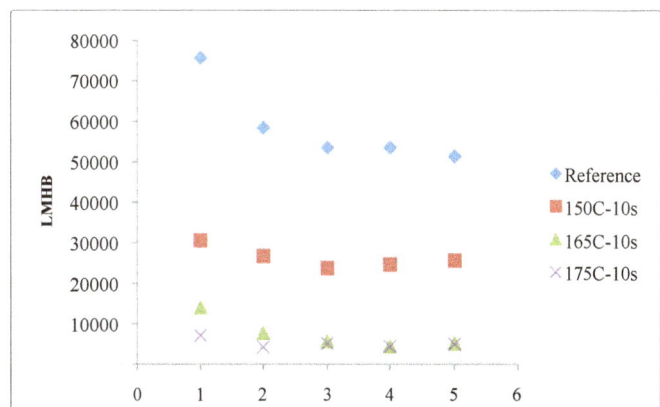

Figure 13: Calculated LMHB for each of the Ultem membrane fabricated at different condition ns.

characteristics of the membranes formed from this high temperature polymer can be tailored by manipulating different meltblowing and calendering parameters to produce membranes with desired pore size for a specific filtration quality.

Acknowledgments

The authors would like to thank Oak Ridge Associated Universities for funding a part of this research, and the University of Tennessee Center for Materials Processing for additional financial support.

References

1. Seifert B (2002) Polyetherimide: A new membrane-forming polymer for biomedical applications. Artificial Organs 26: 189-199.

2. Peluso G (1994) Polyetherimide as Biomaterial - Preliminary in-Vitro and in-Vivo Biocompatibility Testing. Journal of Materials Science-Materials in Medicine 5: 738-742.

3. Bhat GS, Wapner PG, Hoffman WP (2000) Processing of a high temperature imide copolymer into hollow fibers. Materials and Manufacturing Processes 15: 533-545.

4. Bhat GS, Malkan SR (2002) Extruded continuous filament nonwovens: Advances in scientific aspects. Journal of Applied Polymer Science 83: 572-585.

5. Yesil Y, Bhat G (2016) Porosity and Barrier Properties of Polyethylene Melt Blown Nonwovens. Journal of Textile Institute: 1-6.

6. Yesil Y, Bhat G (2016) Structure and Mechanical Properties of Polyethylene Melt Blown Nonwovens. International Journal of Clothing Science and Technology 28: 780-793.

7. Uppal R (2013) Meltblown Nanofiber Media for Enhanced Quality Factor. Fibers and Polymers 14: 660-668.

8. Hegde RR, Bhat GS, Deshpande B (2013) Crystallization kinetics and morphology of melt spun poly(ethylene terephthalate) nanocomposite fibers. Express Polymer Letters 7: 821-831.

9. Dai Y (2012) Ultem((R))/ZIF-8 mixed matrix hollow fiber membranes for CO2/N-2 separations. Journal of Membrane Science 401: 76-82.

10. Nanjundappa R, Bhat GS (2005) Effect of processing conditions on the structure and properties of polypropylene spun bond fabrics. Journal of Applied Polymer Science 98: 2355-2364.

11. Zhang J (2007) Effects of hot pressing conditions on the performances of MEAs for direct methanol fuel cells. Journal of Power Sources 165: 73-81.

12. Rong H, Leon RV, Bhat GS (2005) Statistical analysis of the effect of processing conditions on the strength of thermal point-bonded cotton-based nonwovens. Textile Research Journal 75: 35-38.

13. Han W, Bhat GS, Wang X (2013) Structure and Air Permeability of Melt Blown Nanofiber Webs. Journal Nanomaterials and Molecular Nanotechnology 2: 3.

14. Davies GM, Seaton NA, Vassiliadis VS (1999) Calculation of pore size distributions of activated carbons from adsorption isotherms. Langmuir 15: 8235-8245.

15. Kandagor CV (2014) Nanolayer Polymeric Coatings to Enhance the performance and service life of inorganic membranes for high temperature-high pressure biomass pre-treatment and other applications in the bredesen center, The University of Tennessee: Knoxville, TN: 169.

16. Lee Y, Wadsworth LC (1992) Effects of Melt-Blowing Process Conditions on Morphological and Mechanical-Properties of Polypropylene Webs. Polymer 33: 1200-1209.

17. Ting C, Wang XH, Huang XB (2005) Effects of processing parameters on the fiber diameter of melt blown nonwoven fabrics. Textile Research Journal 75: 76-80.

18. Chand S (2001) Structure and properties of polypropylene fibers during thermal bonding. Thermochimica Acta 367: 155-160.

19. Hassan MA (2013) Fabrication of nanofiber meltblown membranes and their filtration properties. Journal of Membrane Science 427: 336-344.

20. Lin T, Lukas D, Bhat GS (2013) Nanofiber Manufacture, Properties, and Applications. Journal of Nanomaterials.

21. Bhat GS, Schwanke R (1997) Thermal properties of a polyimide fiber. Journal of Thermal Analysis 49: 399-405.

22. Yoon K (2006) High flux ultrafiltration membranes based on electrospun nanofibrous PAN scaffolds and chitosan coating. Polymer 47: 2434-2441.

23. Chu KH (2014) Preparation and Characterization of Polypropylene Non-woven Fabrics Prepared by Melt-blown Spinning for Filtration Membranes. Bulletin of the Korean Chemical Society 35: 1901-1903.

24

TRIZ Inventive Principle in the Creative Process of Textile Products

da Silva de Santis SH[1]*, Franco B[1] and Marcicano CJPP[2]

[1]*University of Campinas, Campinas, Brazil*
[2]*Univesity of São Paulo, Brazil*

Abstract

The work was developed based on the concepts of the theory TRIZ (theory of Inventive Resolution) that was used to assist the creative process of the design of a technological fabric. It was evaluated the application of the concepts of sustainability for the design as an inventive principle. As a result, it was obtained various design alternatives working with the functions and requirements of the fabric.

Keywords: TRIZ; Process creative; Medology design

Introduction

The problems presented in this study joins the fabric development and the creative process ensuring the use of resources in an appropriate manner. Product development in textile industry and associates processes with creativity are part of the creative economy. The creative economy gathers a set of social entities focused on the development of creativity, culture, knowledge, media, arts, design and other linked to culture and creativity [1,2]. The growth of research in this area has provided an increase in technologies and incentives for creativity.

Industrial textile products which have a booming and very competitive market (market of fashion), these products need methods and instruments compatible with the necessary quality. The project methodology is concerned with providing an efficient scheme of production of goods and service that promotes the reduction of environmental impacts, increase internal controls and rapid realignment in the case of failures. The procedures for product design and management are important as an instrument in the production process.

Thus, the tools used to improve product design, increase of quality in processes and sustainability. In this respect, the evidence of validation for the premise is carried out by means of a case study in a textile manufacturing company.

Proposed Methodology

The proposed methodology for the product design begins with the use of mind mapping creation techniques after the theory TRIZ associated with design tools are utilized to create a model of stimulus to the creative process. The model establishes the analysis proposed in theory by means of idealization, contradiction, and resources for the creation of a technological fabric which can switch the composition to generate innovative products through functionality. In the first step develops the mental model from the object to the features, this stage is the description of the requirement.

The study demonstrates the existing relations and associations also met a group to identify the priorities of the characteristics of the fabric and the principles of TRIZ, showing what principles to assist in the construction of the technological knitted fabric. To do this, it was compiled a matrix that prioritizes the basic principles for creating a process that encourages creativity in the design of the textile product. In the second stage, the design project with the characteristics and the design sketch demonstrating the purpose. In the model, it was used tools to analyze the functions of the product, discover the main features of the object through the morphological framework of the knitted fabric technology.

TRIZ (Russian acronym for theory of inventive problem-solution)

According to Carvalho and Back [3], the narrow escape by Altshuller [4] had as object of study the patents of this period, aiming to seek creative solutions to problems of intuitive method. Mean-Shen [5] explains that the TRIZ was developed for application in inventive solution of problems because, although different methods in different areas of knowledge (Administration, advertising, arts), this methodology was born in engineering with the original purpose of developing a method to invent.

Carvalho and Back [3] and Mean-Shen [5] claim that the TRIZ promotes idea generation through structured tools, mainly by encouraging creativity. The concept works the ideality, desired and undesirable functions, contradictions between the requirements and the needs of the object, in addition, the system features (everything that is necessary for the performance of the product) and alternating the same for troubleshooting the system. The TRIZ is a structured procedure for innovation and creativity. With her, organizations will not need to hire "inventive geniuses" or be awarded solely on intuitive human processes to solve their problems.

Mean-Shen [5], Chai et al. [6] Rucht and Livotov [7] explain that this concept promotes analysis of the inventive principles by proposing the best use to produce through the idealization, contradiction, and resources. Consists of switch components of object seeking to encourage the creative process. The TRIZ is based on the 40 inventive principles that were used for modification and encourage the creative process, presents the principles through Table 1 adapted from Carvalho and Back [1]:

The principles translate the object property by encouraging the creative process, through amendment to encourage creativity. In this way, the principles should be used to contribute to the process

*Corresponding author: Sandra Helena da Silva de Santis, University of Campinas, Campinas, Brazil, E-mail: Brasil-s.h.santis@hotmail.com

Principles of TRIZ	Principles of TRIZ
1. Segmentation	21. Dispatch quickly
2. Extraction	22. Turn prejudice into profit
3. Quality in	23.Feedback
4. Asymmetry	24. Mediation
5. Union or Mixture	25. Self Service
6. Universalization	26. Copy
7. Alignment	27. Object use descartes
8. Balance	28. Replace the mechanical means
9. Prior Compensation	29. Use pneumatic or Hydraulic
10. Prior Action	30. Use of thin films or membranes
11. Cushioning or protection provided for	31. Use porous materials
12. Equipotential bonding	32. Color change
13. Reversal	33. Thermal voltage
14 Change shapes	34. Disposal and regeneration
15 Promotion	35. Change in physical state or chemical
16. Partial or Excessive action	36. Phase transition
17. Changing dimension	37. Thermal Expansion
18. Vibration	38. Use of strong oxidizers
19. Periodic Action	39. Using inert atmosphere
20. A useful action continuity	40. Use of composite materials

Table 1: 40 inventive principles.

of creation, so that the manufacture becomes more competitive and thus, promotes a series of methodologies based on the application of tools for innovation and creativity, supporting the idea that the creative process can be augmented through new methods and instruments. It also encourages the thinking in the use and disposal of the product in your development, promotes the sustainable development of the product [8].

Sustainability

The TRIZ establishes a set of measures that modify the object through Division of functions, coordination of actions, changing colors, functionality, weight and cost reduction. The analysis of the benefits of the product, component separation and verification of use and disposal. Smith [9] States that the characteristics of lean manufacturing can be defined as flexibility, reduction of constraints, such as the time of setup, or product, specialized tools, and techniques to improve production. Botero [10] comments that this philosophy aims to reduce costs and increase the company's results through the application of quality tools in the processes. Within this context, the Union of philosophy with the technique creates a template to give the necessary support to the manufacturing and creative incentive to avoid waste, cost, and non-conformities.

Porter [11] States that the increase in performance or activities change modifies the value to the buyer (customer), this includes the functionality to the product. Yet, according to the author the identification of worthwhile activities require the grouping criteria or dismemberment by categories. Thus, identification of the features of the product or process should consist of an analysis, a thorough examination of the system, process and procedures by determining its functions.

To manufacture becomes more competitive it is necessary to act upon the challenges faced by the market more and more demanding and competitive in a globalized world. The ecosystem is the environment in return for everything and the company. The system in this example is the company, your organization, its components, and its structure. The subsystem is the departments that are part of the company and that,

although each has its function, we have to work with the unification of ideas and thoughts. All join in a synthesis, it means that one must dialog the system since all components must engage each other for there to be unification. The management system must be prepared to respond appropriately, in this sense; the tools should give the necessary support to develop operational strategies of quick answers and simple.

It is important to outline goals involving contributions to organizational sustainability. One of the main goals of the globalized companies is to make the sustainable company through operational models. A manufacturer must promote the management of activities in order to achieve the operational capacity of its production the technique came about due to the shortage of raw materials during World War II, Csillag [12]. Researchers such as Juran [13], Campos Falconi [14] state that new technologies for products and manufacturing processes have created the need to invest in strategies to improve quality controls. In General, companies have problems to keep up with the market requirements, innovations and transformations. Many large how many small firms seek solutions [15].

Many companies are looking for solutions that can be tailored to your needs momentary. Manufacturing companies to carry out their activities require a management system targeted for improvement in production, performance, and reduction of losses [16,17]. The TRIZ offers alternatives to modify and change the creative process and proposes troubleshooting complexity through the inventive principles; the theory suggests significant changes by changing the composition, exploring the physical state, replacing forms and components. The variables in the creative process due to the complexity in creating require flexibility. The flexibility serves as a preposition as creativity has many decision-making process variability, uncertainty, complexity, and ambiguity because it is a choice that is based on the information. Therefore, the theory of inventive resolution brings a change to recreate, modifying the proposed object.

Creative process

The man has his creative manifestations through their individual aspirations, thoughts, and idealization. The need is a motivating factor that drives to the pursuit of knowledge, and satisfaction problem solution. To Lobach [18] the human being is also addressed by multiple and varied needs. The appearance of needs is not always logical, especially when other activities or processes have an occasional preference. The need to demand satisfaction, aspiration is the desire to get something spontaneous to prove the idea or the preview. Aspiration is the desire to get something that may or may not be achieved. The needs and follow the evolution of technology, the instruments of information and economic development.

Lobach [18] states that the design consists of a project, plan or systematic method that includes the solution of a problem incorporating ideas, innovation, creation of sketch, samples, templates to make the concrete solution. Over the centuries, the needs in their evolution have been accompanied by the development of tools, methods, and systems. The constant development through research and events show that the innovative creativity has had a key role. In this sense, the study of the methodology for the development of the creative process entered a logical and rational thinking in human evolution. The development of human creative process has also been marked by various frustrations, problems in creativity and innovation, these problems are reported consistently by various scholars.

A number of scholars and researchers [19-21] have already been attacked by creative inertia, the difficulty of exposing ideas, fears, lack of innovation or even problems that seemed intractable.

Product Development

Ostrower (20) States that the ability to understand, assimilate, configure and mean is the creative act.Create is a way of establishing a new relationship between the human mind and the object in order to understand the meaning or resign (give a new meaning, a new practice, ability to perceive an object for a different view). Already the creative process derives from the structure of cognition (knowledge of the facts), intelligence (human feature composed of logical thinking, communication, knowledge, wisdom, problem-solving, emotional control, etc.), ability to create (make sense of something or resign something) and innovation (create an unknown). To meet the new type from the consumer and social changes, the manufacturers seek to align existing requirements with the functionality and aesthetics creating values that can be applied to technological fabrics.

Barbará [8] style process as a set of ordered and integrated actions for a specific production order at the end of the cycle they generate products, services or information. In the manufacturing process with synthetic fibers began the 30, developed fibers become part of the manufacture of textiles and clothing. To give a small notion of what we call, I find it interesting the context of early history, remembering some key facts. In this sense, manufacture textile manufacturing uses the fibers to form the yarn and yarn weaved turns into the fabric. Manufacture textile production divided into three cores: the manufacture of yarn, fabric manufacturing, and clothing manufacturing.

According to Sanches [21-23], the fiber is the smallest element of the composition of the fabric in any natural or manufactured substance that has the appropriate features to allow its processing. Being the smallest component of hairy nature, that can be extracted or separated from a fabric. In the manufacture of yarn, the process of creation establishes the mixture of materials for processing. The processing consists of a rational part that modifies the form of a structure or system for the construction of a mixture, an irrational part that is gathering emotional, psychological aspects, innovative, creative and personal. This means that the transformation depends on the creative aspects to innovate in fiber mixture.

The procedure of creation promotes find strategies by encouraging the production of new ways of mixing the components, which can motivate, add capacity and add value to the basic functions and the secondary product or service to generate probabilities of more interactive information on the market. The set of productive or manufacturing operations must have as its main focus for improvement, the increase in productivity and quality. On an industrial scale (scale of manufacturing) in the contemporary good job of project methodology presents some techniques that promote encourage creative procedures, in short demand accommodating customer demand. The application processes of design methodology consist in the interaction of instruments, resources and converted into energy that carries out the connection between procedures and tasks.

The manufacture of textile manufacturing that is the object of study of this research produces knitted fabric, working in the field of circular knitting, among his articles produced are knitted fabric for fitness and beach. The knitting textile manufacturers that also serve as object for this research have a tradition in Brazilian economy, and considered as one of the major manufacturing sectors in Latin America, composed of several business units in the country, their most common products

made in fabric composed of combinations of polyamide, cotton and spandex (synthetic filament) in circular and rectilinear knitting machines. The study used the knitting on circular knitting machines, which is a reference to continue previous work by the author.

The Knitting is considered a reference in the textile area for development in more than 70 years of tradition, the sector considered of high growth in recent years, and this is due to the intensification of studies in the area. Offers cutting-edge and diverse resources in the development of the textile industry. The industry serves a reference for dealing with large organizations and major mark as Rhodia, Valiseré, Santos and others. Organizations have become increasingly flexible in its processes, mainly by using strategic factors that can motivate, add the ability to add value and basic functions and secondary goods production to provoke more interactive participation opportunities in the market. According to Agostinho [4] the search for competitiveness has several methodologies, scientific concepts, technologies for improving the organization.

This demand from the market has caused changes in organizations, so they are more flexible and dynamic in their structure, system and organizational model. The business system consists of a set of actions with its transformation to answer internal and external pressures to market, a logical reaction from the consumer needs, in short, an entity that by producing goods with or without profit. The actions constitute a form of planned processes that are linked in a virtual or physical structure. Processing (Figure 1) establishes a set of ordered processes in operations for modifying the resources in products to Agostinho [4].

The establishment of roadmaps and manufacturing processes secure manufacturing knowledge, or how to make, and is considered the cornerstone of fixing the knowledge of manufacturing. In sequence to the roadmaps and manufacturing processes, determines the time required for each operation of the script accordingly and set pieces that make up the product.

The manufacturing processes interrelate in a chain of interdependent functions that considers the independent variables (external environment) and the dependent variables (internal environment). This functional interrelation facilitates the systematization of production of goods and services. Each function has a sequential operation flow for the development of an operation from the resource entry to the exit of the goods or services. The set of actions in the supply chain shows the sequence of operations that is established for the construction of a product, a commodity or a service and this facilitates the understanding of the production system. Thus, it presents the results of the application of TRIZ.

Result and Discussion

Part of analysis of characteristics of the product in relation to the items noted by principles of TRIZ. In this sense, the creation of the design of the intelligent fabric product analyzes the potential required for the product, and it is used to prioritize needs factors correlating with the principles of TRIZ theory. In this way, review what the important criteria that are applied in the development of the product to stimulate creativity. We used the methodology of focus groups.

For prioritizing employed the brainstorming with the components of the Group (engineer, production engineer, Product Design, Teacher, Designer, etc.), a classification of the principles by the degree of importance of the factors and devised the correlation matrix. The brainstorming provides a more democratic climate doesn't just drive the freedom to the expression of ideas, but also allows greater opportunities

TRIZ principles / inventive factors	Ergonomics	Physical properties	Mechanical properties	Aesthetic and symbolic aspects	Tendencies	
3 Quality Location	5	5	5	3	1	19
5 Streamlining properties	5	5	5	3	1	19
27 Use and Disposal	5	5	3	3	1	17
4 Asymmetry	5	5	3	3	1	17
40 Use of composite materials	5	5	3	3	1	17
33 Thermal Stress	5	5	3	3	1	17
35 Mudança physical or chemical stat	5	5	3	3	1	17
1 Segmentation and fragmentation	5	3	3	3	1	15
9 Prior compensation	5	3	3	3	1	15
10 Prior Action	5	3	3	3	1	15
11 Prior Damping	3	3	3	3	1	13
12 Equipotentiality	3	3	3	3	1	13
13 Inversion	3	3	3	3	1	13
14 Esferoidicidade	3	3	3	3	1	13
15 Consolidation	3	3	3	3	1	13
16 partial or excessive action	3	3	3	3	1	13
17 Moving to a new dimension	3	3	3	3	1	13
18 Mechanical vibration	3	3	3	3	1	13
19 Periodic Action	3	3	3	3	1	13
20 Continuity of useful action	3	3	3	1	1	11
21 Acceleration	3	3	3	1	1	11
22 Loss transformation into profit	3	3	3	1	1	11
23 Feedback	3	3	3	1	1	11
24 Mediation	3	3	3	1	1	11
25 Self-service	3	3	3	1	1	11
26 Copy	3	3	3	1	1	11
37 Thermal Expansion	3	3	1	1	1	9
28 Substitution mechanical means	3	3	1	1	1	9
29 Pneumatic or hydraulic constructi	3	1	1	1	1	7
30 Use of thin films and flexible mem	3	1	1	1	1	7
31 Use of porous materials	3	1	1	1	1	7
32 Color change	3	1	1	1	1	7
8 Balance	3	1	1	1	1	7
34 Disposal and regeneration	3	1	1	1	1	7
7 Nesting	3	1	1	1	1	7
36 Phase Change	3	1	1	1	1	7
6 Universal	3	1	1	1	1	7
38 Use strong oxidizers	3	1	1	1	1	7
39 Use of inert atmospheres	3	1	1	1	1	7
2 Extraction	3	1	1	1	1	7
	140	110	96	78	40	

Source: Santis (2016)

Figure 1: Prioritization matrix.

the correlation matrix can prioritize necessary resources and establish important criteria for product development.

The array is equivalent to the discussion to check the items that should be prioritized in the creation of the product, what must be done to stimulate new ideas to technology innovation in fabric, properties that can differentiate and increase the value for the consumer. Under this section, an array created with the essential properties and parameters of the theory. In the first meetings discussed the importance of each factor in the development of intelligent fabric (Table 2):

The criteria were organized by assigning each note corresponds to a color (Figure 2):

- Note 5-red color corresponds to high priority;
- Note 3-yellow the average priority;
- Note 1-green for low priority.

With the use of the correlation matrix, it was possible to prioritize the features required by establishing the necessary resources. The first array was organized with many priorities because each element of the Group prioritized back to your experience and hasn't noticed what really was important to the creative process. In this model include the prioritization of activities that can encourage the creative and innovation process, starting from the modifications that can be made in the design of the fabric.

The array was accomplished through the methodology of focus groups consisting of a discussion group on the main properties and how it can change the fabric creating a differentiated customer value. The first attempts resulted in an array, but the priorities were many and this made the vision of what really should be done. In this way, a consensus was reached to make a new vote to narrow the criteria and approach a more decisive result of the priorities. After all, not everything can be a priority.

Prioritization matrix (Figure 1) defines the characteristics that can be changed to create customer value, technical, ergonomic factors, and properties. The value associated with the criteria of the TRIZ has for purpose to enter the best use taking account of the specifications required by the market. The alignment from this correlation has the purpose of promoting options for modification by creating value and functionality to the fabric. Understand the relationship between the characteristics of the TRIZ methodology and technological fabric promotes the understanding of solutions to problems and possibility of the redesign of the object. The functionality of the fabric can be changed, that promotes innovation in usability, the fabric may acquire new functions and benefits.

The TRIZ generates a comparison of the properties, characteristics, and needs, designing the possibilities of applying each principle identified by promoting a change of items.

The array defined as key features for application of TRIZ respectively: the quality location, streamlining of the properties, use, and disposal, asymmetry, the use of composite materials, thermal, voltage change of physical state, segmentation and fragmentation,

Correlation	Criterion
Strong	5
Average	3
Weak	1

Source: adapted from Agostinho (2012).

Table 2: Criteria for classification.

for the emergence of a cross-functional knowledge, preventing the members solve the problems so focused and isolated. With the use of

provided compensation and action envisaged. This characteristic of whisker, when applied to technological fabrics, encourage the creative process, is it possible to find solutions through modification of the properties of the knitted fabric.

The characteristics of the fabric may be crossword puzzle with the principles providing these materials changes can be: in thermal properties (absorb heat or turn up the heat, cool or heat), physical properties (have the stipulated format or preset), mechanical properties (conductive circuits of information, perform stimuli, activate electrically conducting elements, etc.), aesthetic aspects (change of colour, shape memory, outline the body , etc.) and fashion trends.

The graph you can see that only 10% of the correlations must be prioritized, the other correlations were considered a medium priority and low priority are practically representing only a difference of 6%, and the average priority appears with 48% and the low priority with 42%. Ergonomic characteristics, physical properties, and mechanical properties are more important than the aesthetic aspects and tendency as fabric because these characteristics are important for objects.

TRIZalready offers possibilities of modification of the properties, characteristics when applied provides the new combination. Each one of these principles is a possibility of modification of the fabric, which consists in a possibility of a change of the functionality and value of the product. These aspects are relevant in the creative process of technological fabric because they can provide options for product development. To demonstrate the possibilities of changes using the morphological framework applying the possibilities offered. The technological fabric analyzed features properties that should be highlighted in the project, providing greater visibility of the expected results and resources.

The creative process this case was encouraged by the systematic analysis of the project of construction and properties of the fabric. The same technological fabric was analyzed for morphological framework Figure 3, the same provides a different perspective to the creative process.

You can see that with this morphological framework can make various combinations and thus create different types of technological fabrics with treatment. The morphological framework provides an option to change the object, in this context, it is noted that the technological fabric can be: according to Dedini (2007), this method was developed by Fritz Zwicky consisting of decomposed the global problem in partial problems (or parameters). The Board assists in creative development analyzing the problem through the deconstruction of the object and thus encouraged to creative thinking.

Final Considerations

The work was developed based on the concepts of the theory

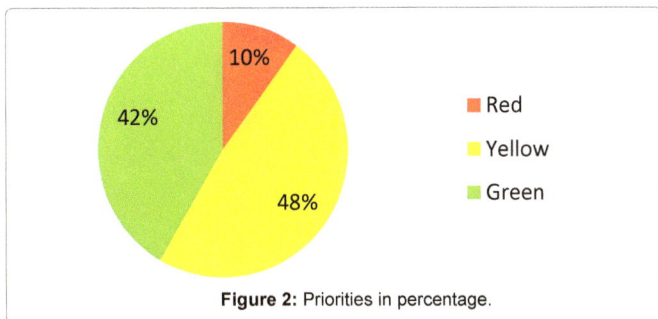

Source: Santis (2016)

Figure 3: Morphological Framework.

TRIZ (theory of Inventive Resolution) the proposal got an incentive to the creative process and evaluate the implementation of the project methodology and tools for the technological fabric designer. The study obtained various alternatives working with the functions and requirements of the fabric, we analyzed the functions through an array. Notice that the array provides build options for the development of new technological fabrics modifying features, functionalities and adding value to the object. This sustainability associated with the better use of resources, which must include the product disposal. One can notice that the tools used must provide an analysis of the properties and indicate possibilities to recreate changing the structure, functionality, and composition.

Resources are elements that make up the situation, or your environment, which can be mobilized to solve or contribute to solving a problem. They can be defined as any element of that helps to compose the system under examination or to the surroundings that were used to perform useful functions. Project methodologies make up part of the strategies to improve the creative process. The job must provide the solution for innovation and creativity, and also to problems of conflict between necessity and usability. The tools applied to the creation process must use the skills, research, and exploration of ideas by means of the methodology, thereby offering new alternatives to the process. Therefore, the use of these tools expands and organizes the creative process working with creativity and innovation.

References

1. Machado RM (2009) Da indústria cultural à economia criativa. Alceu Rio de Janeiro 9: 83-95.

2. Pires VS, Albagli S (2009) Estratégias empresariais, dinâmicas informacionais e identidade de marca na economia criativa.

3. Carvalho, Aurélio M, Nelson B (2001) Uso dos conceitos fundamentais da TRIZ e do método dos princípios inventivos no desenvolvimento de produtos. In: Congresso Brasileiro de gestão de desenvolvimento de produto.

4. Agostinho OL (2012) Sistemas de Manufatura. Apostilas de Curso. Universidade Estadual de Campinas.

5. Mean-shen I (2011) The study of green product design and development by applying triz innovation principles. African Journal of business management 5: 7740-7754.

Figure 2: Priorities in percentage.

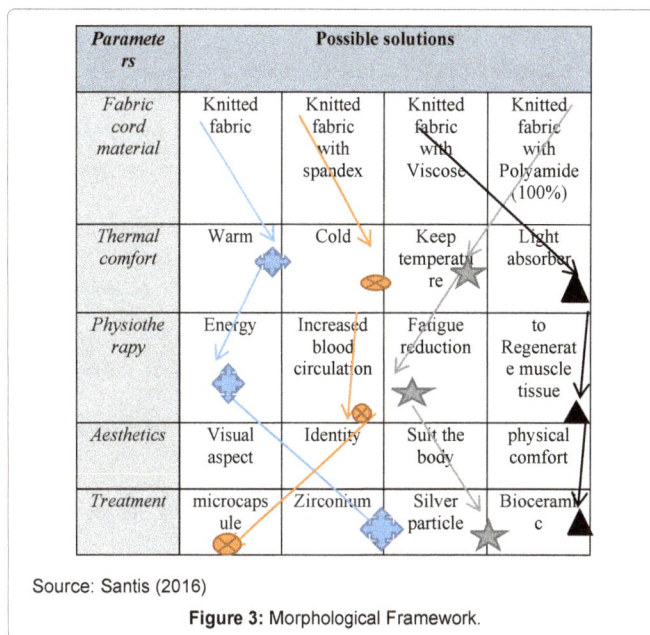

6. Chai KH, Zhang J, Tan KC (2005) A triz-based method for new service design. Journal of service research 8: 48-66.

7. Ruchti, B, Livotov P (2001) Triz-based innovation principles and a process for problem solving in business and management. The Triz journal 1: 677-687.

8. Barbará SE (2011) Gestão por Processos. Rio de Janeiro: Qualitymark.

9. Bezerra GM, Martins SB (2013) Equação da ergonomia no design de vestuário: espaço do corpo. modelagem e materiais.

10. Botero, Gómez PA (2010) Lean Manufacturing: flexibilidad, agilidad y productividad. Gestión and Sociedad 3: 88.

11. Porter ME (1996) Vantagem competitiva: criando e sustentando um desempenho superior. rio de janeiro: campus.

12. Csillag, Mario J (2012) Análise do valor: metodologia do valor: engenharia do valor, gerenciamento do valor, redução de custos, racionalização administrativa.

13. Juran, Joseph M (2004) A qualidade desde o projeto: novos passos para o planejamento da qualidade em produtos e serviços. são paulo: ed. pioneirathomson learning.

14. Campos, Falconi V (1992) Controle da qualidade total. Rio de Janeiro: Fundação Cristiano Ottoni.

15. Dalla CA, Souza-Santos ED (2011) Economia criativa: novas oportunidades baseadas no capital intelectual. economia and tecnologia Curitiba ano 7: 179-186.

16. Mesquita, Melissa (2001) Competências para melhoria contínua da produção: estudo de caso em empresas da indústria de autopeças. dissertação (mestrado em engenharia de produção) - Universidade Federal de São Carlos, São Carlos.

17. Dedini FG (2007) Projeto de sistemas mecânicos. apostila do curso em 964 da Faculdade de engenharia mecânica da Unicamp. Engenharia mecânica, universidade estadual de campinas, campinas.

18. Howkins J, Griesi A (2013) economia criativa. São Paulo.

19. Menegucci F, Martins E, Menezes M, Santos F, Abilio (2012) Experimentações têxteis e inovação no design de moda. in: 8º Coloquio de moda, Rio de Janeiro.

20. Ostrower F (1978) Criatividade e processos de criação.

21. Sanches, Regina AP (2010) tecnologia aplicada nos artigos de vestuário destinados à prática de esportes.

22. Santis SHS (2016) 6º Encontro Nacional de Pesquisa em Moda (ENPModa) and Fashion Colóquio São Paulo. Universidade de São Paulo (EACH).

23. Santis SHS (2016) Project Metodology Applyed to Smart Fabrics. Journal of Textile and Fashion Technology 2: 7-1.

25

Generation of Jute Fibre Length Distribution via Graphics and Computer Simulation with Gamma Distribution Function

Biswas SK*

Textile Physics Section, Department of Jute and Fibre Technology,Institute of Jute Technology, University of Calcutta 35, Ballygunge Circular Road, Calcutta, India

Abstract

Notwithstanding the complexity of production of jute fibre filaments a simple graphical presentation of jute fibre length distribution is made. The computer simulation of four types of frequency distribution of jute fibre length depending on method of sampling and testing is presented. These four distribution curves gives six points of intersections which merit technological significance. A study of fibre length-frequency distribution depending on methods of sampling and testing may lead to a general form of Gamma distribution function. The parameters in the mathematical formulae may serve as characteristics of condition giving the benchmark for valid experiment. This type of work involving exponential distribution was not done previously and provides a deeper insight into the characterisation of its comparative features with respect to normal distribution of fibre length. It can be observed that fibrograms for exponential and normal distributions have close similarity if their mean values are same. However, the survivor or array diagram distinguishes the basic distributions better than the fibrogram. This would be useful in the context that the basic frequency distributions of fibre lengths are somewhat difficult to obtain from direct experiments.

Keywords: Gamma distribution; Jute fibre; Fibre length; Length distribution

Introduction

In jute, the raw fibre as it comes from the plant consists of many single filaments bonded together by adhesion and branching to form a three-dimensional meshy structure. On carding, this meshy structure is broken into individual fibrous elements of different lengths , the longer elements being more complex, since these consists of several single filament still joined together in some way. This complexity will cause the linear density of the longer elements to be greater than that of the single filaments. In addition, the single filaments in the longer elements may be thicker and stronger than those in the shorter ones. On both these counts, the linear density of the longer elements will be greater than that of the shorter ones [1]. When the meshy structure of raw jute reeds is broken down randomly into fibrous elements in a carding machine, the frequency of such elements by number can be expected to decrease exponentially with increase in their length [2]. In this paper a graphical presentation of jute fibre length distribution is made. A study of fibre length-frequency distribution depending on methods of sampling and testing may lead to a general form of Gamma distribution function. It may be mentioned that for Gamma distribution [3] the density function of the random variable x is given by

$$f_x(x) = \left(1/(\Gamma\alpha) \right)\left(\lambda^\alpha x^{\alpha-1} e^{-\lambda x} \right), \alpha \geq 0$$

Whose

mean=α/λ, variance=α/λ^2, standard deviation=$\sqrt{\alpha}/\lambda$ and coefficient of variation=$100/\sqrt{\alpha}$. This distribution has two parameters.

The parameterα is called a shape parameter since, as α increases, the density becomes more peaked. The parameter λ is a scale parameter; that is, the distribution depends on λ only. Variation for the values of α which may serve as behaviour characteristic of condition that appear as parameters in the mathematical formulae giving the benchmark for valid experiment.

Presentation of Fibre Length via Graphics

Figure 1 is the graphical representation of data collected by measuring length of single fibres. The vertical axis is number (quantity)

of fibre, the horizontal axis fibre length increments, and each circle o, represents a single fibre. One could imagine that the o's are end views fibres stacked vertically with each stack consisting only of fibres of same length. To the left hand side are the short fibres, the right hand side the long fibres. Statisticians call Figure 1 a histogram or length frequency distribution diagram. Figure 1 gives all of the information we need to know about fibre length and fibre length distribution. Since it is easier to determine the weight of fibre than to count the number of fibre, histograms and staple diagrams (arrays) are usually constructed on fibre weight basis. For clarity and ease of explanation, fibre number

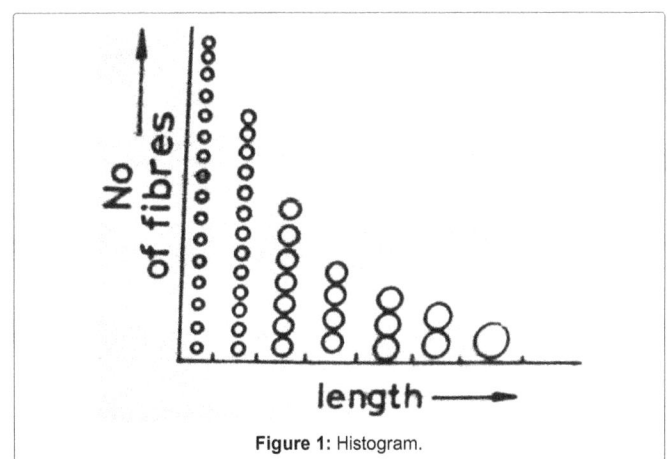

Figure 1: Histogram.

***Corresponding author:** Biswas SK, Textile Physics Section, Department of Jute and Fibre Technology,Institute of Jute Technology, University of Calcutta 35, Ballygunge Circular Road, Calcutta-700 019, India
E-mail: tp_skb@yahoo.com

basis is used in this discussion. Detailed discussion about number and weight basis of testing is made in later section of this paper. Another method of representing data collected by measuring length of fibres is in Figure 2. The vertical axis is number (quantity) of fibre, the horizontal axis is the fibre length and each cylinder represents a fibre. Figure 2a is a picture of fibres arranged in order of length with left hand ends aligned at the vertical axis. Figure 2a is simply a rearrangement of Figures 1 and 2 is the staple diagram and, as in Figure 1 contains all of the information we need to know about end to end lengths. Millions of millions of fibres are involved in fibre and yarn production. Figure 1 and Figure 2a do not contain all the fibres, they contain only a representative number of fibres. More realistic is Figure 2b which is three dimensional having a z-axis. Layer after layer of staple diagrams on to infinity. A bale of fibres could be represented as in Figure 2b. As fibres were rearranged going from Figure 1 to Figure 2b so can the fibres be rearranged as illustrated in Figure 3. To make this rearrangement, fibres of Figure 2b are selected at random along their lengths, the catch points placed on the vertical axis, and fibres thus selected arranged in descending order of the extension distance to the right hand side of the catch point. For illustration, fibre segments marked 1, 2, and 3 in Figure 2b are shown in their appropriate location in Figure 3 accordingly enough fibres are selected to make Figure 3 becomes representative. Figure 3 is a fibrogram and contains all of the information we need to know about fibre length and fibre length distribution. Of the three methods of representative of fibre length, the staple diagram of Figure 2a is easily accepted by our mental processes, the mental picture requires no strain on the imagination and we have a comfortable feeling; thus the staple diagram has been established as the standard of comparison and the bench mark from which other fibre length test methods are compared. However, in yarn spinning and non-oven fabric production, ends of fibres must never be aligned. The fibrogram which is an arrangement of fibre as they

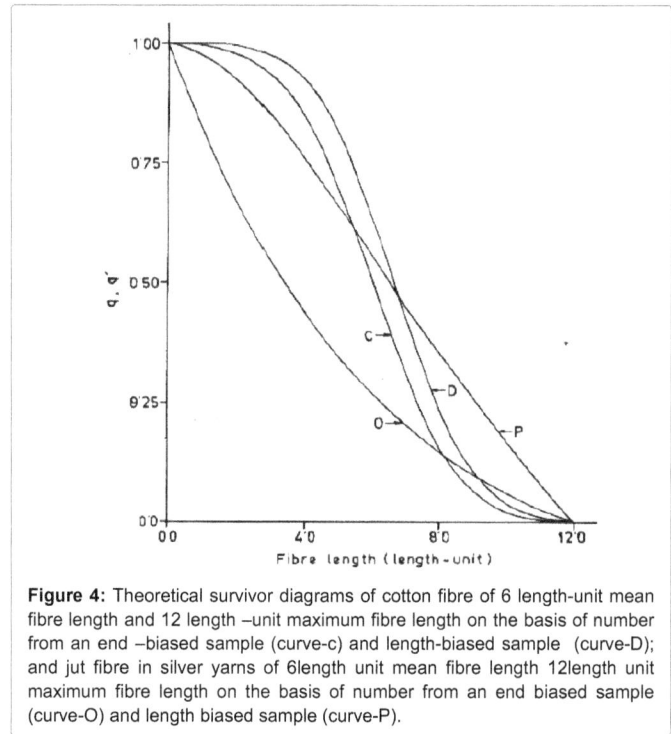

Figure 4: Theoretical survivor diagrams of cotton fibre of 6 length-unit mean fibre length and 12 length –unit maximum fibre length on the basis of number from an end –biased sample (curve-c) and length-biased sample (curve-D); and jut fibre in silver yarns of 6length unit mean fibre length 12length unit maximum fibre length on the basis of number from an end biased sample (curve-O) and length biased sample (curve-P).

are, or as they will be in future processes, becomes the bench mark of comparison concerning fibre performance in yarn spinning and non-oven fabric manufacture [4]. This type of work involving exponential distribution was not done previously and provides a deeper insight into the characterisation of its comparative features with respect to normal distribution of fibre length. The survivor diagrams (curves O and P) and fibrograms (curves Q and R) are also computed and presented in Figure 4 for end-biased and length-biased samples, respectively. It can be observed from Figure 4 that fibrograms for exponential and normal distributions have close similarity if their mean values are same. This is corroborated with the values of the corresponding fibrogram parameters. However the survivor or array diagram which is also still popular in industrial practice distinguishes the basic distributions as evident in Figure 5, better than the fibrogram. This would be useful in the context that the basic frequency distributions of fibre lengths are somewhat difficult to obtain from direct experiments.

Effect of sampling and testing method

In fibre length measurements two kinds of sample are experimentally obtainable through two kinds of sampling methods – end-biased and length biased. An end-biased sample is defined as one in which each fibre of population has equal probability of being included since each fibre has two ends independent of its length or any other characteristics. The end-biased sample is a random sample. A length biased sample is one in which the probability of a fibre being included is directly proportional to its length. In fibre testing, a random selection is restricted due to the essential nature of the fibre that it is much longer than it is thick. Because of this, it is only too easy to take a sample in such a way that it contains far more long fibres than it should. The knowledge of the nature of the bias in favour of the longer fibre would help to avoid the same. In some contexts in which length biased sampling is used it is reasonable to regard the length biased distribution as the object of study [5].

The frequency distributions of fibre length in jute slivers and yarns

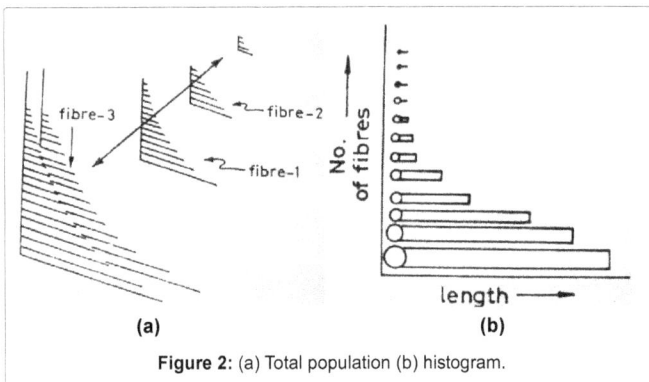

Figure 2: (a) Total population (b) histogram.

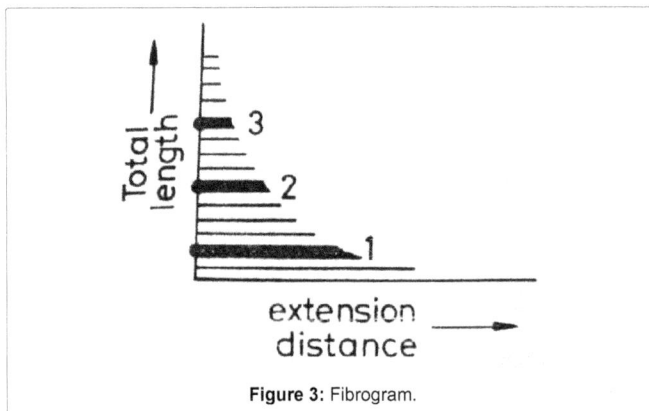

Figure 3: Fibrogram.

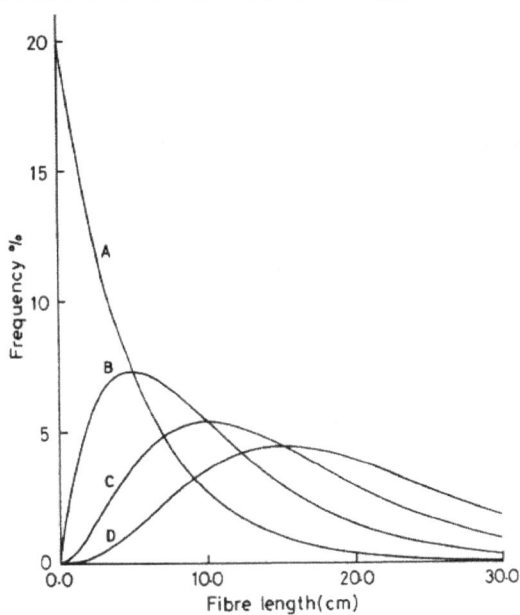

Figure 5: Theoretical Fibrograms of cotton fibre of 6 length-unit mean fibre length and 12 length-unit maximum fibre length on the basis of number from an end-biased sample(curve- E) and length-biased sample (curve-F); and jute fibre in sliver and yarns of 6 length-unit mean fibre length and 12 length-unit maximum fibre length on the basis of number from an end-biased sample(curve- Q) and length-biased sample(curve- R).

have been deduced analytically by Banerjee [2] for different methods of sample preparation. The expected average length of fibres obtained on the basis of number and weight of fibres has been estimated for each of such distribution. The present author [6-8] has shown that the Gamma function provides mathematical short-cut for the analysis of fibre length distribution and average length of jute fibres depending on sampling and testing method to arrive at the results as obtained by Banerjee. The two sampling method i.e., (i) end-biased and (ii) length biased sampling method have been considered. For each of such sampling the expected distribution of fibre length obtained on the basis of (i) number and (ii) weight of fibres has also been estimated. Thus four types of length distributions are obtained such as the frequency of length l tested on the basis of number from end-biased sample (f_{EN} (l)), the frequency of fibre of length l tested on the basis of number from a length-biased sample (f_{LN} (l)), the frequency of fibre of length l tested on the basis of weight from an end-biased sample (f_{EW}(l)) and the frequency of fibre of length l tested on the basis of weight from a length biased sample (f_{LW}(l)) (Figure 6).

The end-biased and number based length distribution of fibre f_{EN} (l), which is expected to be similar to the basic type has been used by Banerjee (1980) to obtain f_{LN} (l) and f_{EW} (l), and in turn f_{LN} (l) has been used to obtain f_{LW} (l). The present author uses the Gamma function to obtain the same result very quickly [6-8] and f_{EW} (l) has been considered [7] to obtain f_{LW}(l). He has also considered all the other possible routes of transformations of one type of length distribution to the other types and presented their rederivation through Gamma function to achieve a considerable amount of simplification [9]. The summary of the results of the distribution is given in Table 1.

Cox [5] describes relations between mean (μ) of the end-biased distribution f(x) and an expectation (mean or average) $E_g(x)$ with respect to length-biased distribution g(x) as

$$Eg\left(X\right) = \mu\left(1+\frac{\sigma 2}{\mu 2}\right)$$

Where μ, σ are the mean and standard deviation of the end-biased distribution f(x).

For Gamma distribution the above relation becomes

$$Eg\left(X\right) = \frac{1}{\lambda}\left(1+\alpha\right)$$

Where α is called a shape parameter of the end-biased distribution f(x).

Computer simulation and modal length

The computer simulations of four frequency distribution function $f_{EN}(l)$, $f_{LN}(l)$, $f_{EW}(l)$ and $f_{LW}(l)$ as given in Table 1 were carried for l=5 cm. The computer programme in F77 is given in Appendix 1. The resulting curves obtained with the help of a Graphic Software are presented in Figure 6. However the proliferation of so many distribution curves depending on the method of sampling and testing may give several points of intersection when these are superimposed. In fact reports of experimental observation of intersection were available in some cases and attempt had been made to attribute suitable technological significance to it [8,10]. The present author has found out analytically all the intersections and showed that intersections depend on the average length of fibre population. Computer simulations are also presented and reported here [8].

He has also found out analytically the modal length of fibre length-frequency distribution depending on methods of sampling and testing [11,12]. The best measure of staple length for general application is given by modal or most frequent length of fibre length-frequency distribution [13].

The modal length l^m can be determined from the condition [14].

at $x = l^m$ (2)

$\frac{d}{dx}f(x)=0$, Using Equations 1 and 2 can be written as

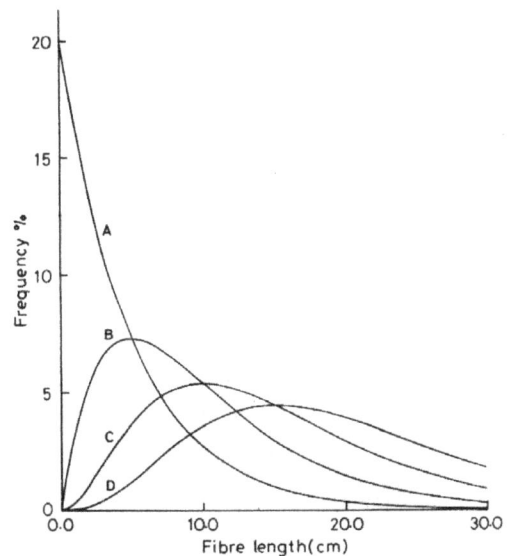

Figure 6: Theoretical length distribution of jute fibres in yarns and slivers. (A) End-biased number based method; (B) Length biased number based method; (C) End-biased weight based method; and (D) Length- biased weight based method.

Method followed	Relationship for frequency distribution	Average length
Basic population number-based	$f(I) = \dfrac{1}{\bar{I}}\exp(-1/\bar{I})$	$\bar{I}$
End-biased sampling number-based	$f_{EN}(I) = \dfrac{1}{\bar{I}}\exp(-1/\bar{I})$	$\bar{I}$
Length-biased sampling number-based	$f_{LN}(I) = \dfrac{1}{\bar{I}^2}\exp(-1/\bar{I})$	$2\bar{I}$
End-biased sampling weight-based	$f_{EW}(I) = \dfrac{1^2}{2\,\bar{I}^3}\exp(-1/\bar{I})$	$3\bar{I}$
Length-biased sampling weight-based	$f_{LW}(I) = \dfrac{1^3}{6\,\bar{I}^4}\exp(-1/\bar{I})$	$4\bar{I}$

Table 1: Summary of results of derivation.

Method followed	Standard deviation	Coefficient of variation (%)
Gamma distribution (shape parameter=α, scale parameter $\lambda = 1/l$)	$\sqrt{\alpha}/\lambda$	$100/\sqrt{\alpha}$
Basic population Number-based Where α=1	$\bar{l}$	100
End-biased sampling Number –based Where α=1	$\bar{l}$	100
Length-biased sampling Number-based Where α=2	$\sqrt{2}\,\bar{l}$	$100/\sqrt{2}=71$
End-biased sampling Weight-based Where α=3	$\sqrt{3}\,\bar{l}$	$100/\sqrt{3}=58$
Length-biased sampling Weight-based Where α=4	$2\,\bar{l}$	$100/2=50$

Table 2: Summary of results of standard deviation and coefficient of variation.

Method followed	Average length from Cox's formula	Modal length from equation (3)
Gamma distribution (shape parameter=α, scale parameter $\lambda=1/l$)	$E_g\left(x\right) = \mu\left(1+\dfrac{\sigma 2}{\mu 2}\right)$	$l^m = (\alpha-1)/\lambda$
Basic population Number-based Where α=1	Not required	0
End-biased sampling Number –based Where α=1	Not required	0
Length-biased sampling Number-based Where α=2	$2\,\bar{l}$	$\bar{l}$
End-biased sampling Weight-based Where α=3	$3\,\bar{l}$	$2\bar{l}$
Length-biased sampling Weight-based Where α=4	$4\,\bar{l}$	$3\bar{l}$

Table 3: Summary of results of average length from cox's formula and modal length from equation (3).

$$\left(1/(\Gamma\alpha)\right)\left(\lambda^{\alpha}\frac{d}{dx}(x^{\alpha-1}e^{-\lambda x})\right)=0$$

$$\Rightarrow \left(\ (\alpha-1)x^{\alpha-2}e^{-\lambda x}-\lambda x^{\alpha-1}e^{-\lambda x}\right)=0$$

$$\Rightarrow l^m = \frac{(\alpha-1)}{\lambda} \tag{3}$$

Motivation and Conclusion

It may be mentioned that for Gamma distribution the density function of the random variable x is given by

$$f(x) = \left(1/(\Gamma\alpha)\right)\left(\lambda^{\alpha}x^{\alpha-1}e^{-\lambda x}\right), \alpha \geq 0$$

Whose

mean=α/λ, variance=α/λ^2, standard deviation=$\sqrt{\alpha}/\lambda$ and coefficient of variation=$100/\sqrt{\alpha}$.

This distribution has two parameters. The parameter α is called a shape parameter since, as α increases, the density becomes more peaked. The parameter λ is a scale parameter; that is, the distribution depends on λ only. A study of fibre length-frequency distribution depending on methods of sampling and testing in Table 1 may lead to a general form of Gamma distribution function if we use $1/l=\lambda$. This remains the motivation of this work. Table 2 gives Standard Deviation and Coefficient of Variation for the respective values of α which may serves as characteristic of condition that appear as parameters in the mathematical formulae giving the benchmark for valid experiment [15,16]. It may be noted that not only average length of fibre but also standard deviation and coefficient variation depend on methods of sampling and testing. Table 3 gives Summary of Results of average length from Cox's formula and modal length from equation (3).

References

1. Banerjee BL (1982) The relation between the length and linear density of jute and flax fibres obtained after carding and drawing operations. J Textile Inst 73: 183-188.

2. Banerjee BL (1980) Effects of sampling and testing methods on the determination of fibre length in jute slivers and yarns. Indian J Textile Res 5: 98-102.

3. Trivedi KS (1994) Probability & Statistics with Reliability, Queuing, and Computer Science Applications, New Delhi, Prentice-Hall of India: 126-210.

4. Biswas SK, Chanda RS (2010) Mathematical Approach to Generating Fibre. Length Distribution via Computer simulation ,In Jute and allied Fibre: Production, Utilisation and Marketing Eds. Palit P, Sinha MK, Meshram JH, Mitra S, Laha SK. Indian Fibre Society, Eastern Region: 83-90.

5. Cox DR (2005) Some sampling problems in technology. In: Symposium on the foundation of survey sampling eds. Johnson NL & Smith H, New York, John, Wiley: 506.

6. Biswas SK (1989) Use of Gamma Function in the Calculation of average fibre length. Indian J Textile Res 14 : 145-146.

7. Biswas SK (1990) Some application of Gamma Function in the Calculation of average fibre length. Jute Development J 10: 26-27.

8. Biswas SK (2010) Application of Gamma Function in the Determination of fibre length in Jute Slivers and Yarns. Modern Textile Journal: 44-47.

9. Biswas SK (1996) Some studies of jute fibre length distribution with Gamma Function. IAPQR Transations 21: 67-72.

10. Sinha NG (1975) Physical aspect of fine Spinning of jute. Indian J Phys 49: 245-248.

11. Biswas SK (2006) Computer and simulation studies of modal length of jute fibre. Textile Trends 49: 33-35.

12. Biswas SK (2010) Computational and Simulation Studies of modal length of jute fibre. J Natural Fibers 7: 111-117.

13. Morton WE, Hearle JWS (1993) Physical Properties Textile Fibres (3rd edn), Manchester The Textile Institute: 94-100.

14. Goon AM, Gupta MK, Dasgupta B (1993) Fundamentals of Statistics Calcutta: The World Press: 264.

15. Biswas SK (2000) Computational and simulation studies of jute fibre length distribution. Indian J Fibre Text Res 25: 221-224.

16. Sinha NG (1975) Physical conditions for finest and best quality yarns from jute. J Text Assoc 36: 51-57.

26

Effect of Slub Yarn Ratio on Single Jersey Knitted Fabric Properties

Fouda A*

Textile Engineering Department, Faculty of Engineering, Mansoura University, Mansoura 35516, Egypt

Abstract

Slub yarns are widely used in Single Jersey knitted fabrics to improve their appearance. This research aims to study the influence of slub yarn ratio on physical and mechanical properties of Single Jersey knitted fabric. Ten single jersey knitted fabrics were produced using four different ratios of slub yarns at two levels of ground yarn counts. Physical properties of knitted fabric were measured such as fabric density, thickness, shrinkage, spirality, color properties and thermal comfort properties. Also, mechanical properties of knitted fabric were measured including bursting strength and abrasion resistance. Results showed that increasing percentage of slub yarns used up to 50% improved fabric density, spirality and shrinkage after repeated washing, while the weight loss percent due to abrasion decreased. Fabric thermal resistance increased and air permeability decreased; while there were no significant changes observed in other measured properties.

Keywords: Slub yarns; Image analysis; Matlab program; Color; Thermal comfort; Bursting; Single Jersey knitted fabric; Abrasion resistance properties

Introduction

Ground slub yarn is a simple fancy yarn whose slub appearance is gained by the variation of yarn linear density during spinning process and no additional yarn or process is required. Slub yarns could be produced by modifying ring spinning frame, such that the intermittent acceleration of the rollers will cause varying degrees of draft to be applied [1].

The usage of slub yarn was spread especially in woven fabric in order to improve its aesthetic appearance with lowest possible cost [2]. At the beginning, it was used as a weft yarn then as warp and weft yarns. Recently, it has been used on circular knitting machines to improve Single Jersey knitted fabrics. Of course, because of the novelty of use, there is scarcity of researches that focus on studying the impact of the use of these yarns on produced fabric properties. This necessitates studying the effect of using such yarn on knitted fabric properties particularly thermal comfort, fabric strength and abrasion resistance after improving its appearance and also optimizing the usage ratio of these yarns to obtain the best fabric quality.

The effect of weft slub yarn parameters including relative thickness, slub length percent and number of slubs per meter on plain woven fabrics geometrical and physical properties were studied. In addition, fabric texture was also investigated using image analysis. Using slub yarns as weft threads in plain woven fabrics resulted in higher fabric bulkiness, higher smoothness, higher tear strength in warp direction and increase in the fabric assistance especially at the higher weft densities. Fabric stiffness increased by the increase in number of slubs per meter and decreased by the increase in slub relative thickness [3].

For Single Jersey knitted fabrics, at constant loop length, as yarn gets finer, yarn diameter decreases and courses spaces increases, hence, courses density decreases theoretically and experimentally. At constant stitch length, as yarn diameter increases, fabric thickness increases, fabric tightness increases and air permeability decreases [4].

Image analysis is applied in detecting fabric defects and extraction of fabric information such as weave, fabric density, yarn count, etc. Two different approaches based on Gabor filters for extracting slubs were studied and compared. The constructed slub extraction technique considers a very important part in the development of a denim fabric recognition system. Denim fabric with slub yarn as warp is quite popular and this fabric recognition system is welcomed by many denim manufacturers [5].

Single Jersey knitted fabric spirality is one of the phenomena which could cause goods to be rejected. In addition to the number of knitting machine feeders, yarn twist liveliness which emerges from increasing twist factor is a key factor that affects this phenomena. Also this phenomena is influenced by fabric tightness, where fabric spirality angle increases by increasing fabric tightness [6].

The dimensional, physical, and visual properties of Single Jersey knitted fabrics manufactured from Chenille yarns were investigated. These parameters were studied, yarn count, pile length, laundering, and dry-cleaning. The surface properties such as softness, smoothness, and luster become much better as the component yarn count becomes finer and the pile length becomes longer. Tumble drying satisfied the end-user's expectations for knitted goods from fine chenille yarns and long pile [7].

Experimental work and tests methodology

The specifications of the used machine are: ALBI circular Single Jersey knitting machine, Gauge 28, Diameter 17 inch and number of feeders are 34. Three different combed waxed yarn counts were spun using 100% Giza 86 Egyptian cotton and 3.6 English twist factor. The conventional yarns are 26 Ne, 32 Ne.

A ground slub yarn is formed of a single structure that has two parts: slub part and base yarn part as shown in Figure 1 [3]. The basic geometrical parameters of slub yarn are slub length (SL), slub distance (SD), base thickness (d) and slub thickness (D). So the slub yarn properties is (32 Ne, Slub relative thickness 3 (D/d) as shown in Figure 1,

---

*Corresponding author: Fouda A, Lecturer, Textile Engineering Department, Faculty of Engineering, Mansoura University, Mansoura 35516, Egypt
E-mail: eabdo3@gmail.com

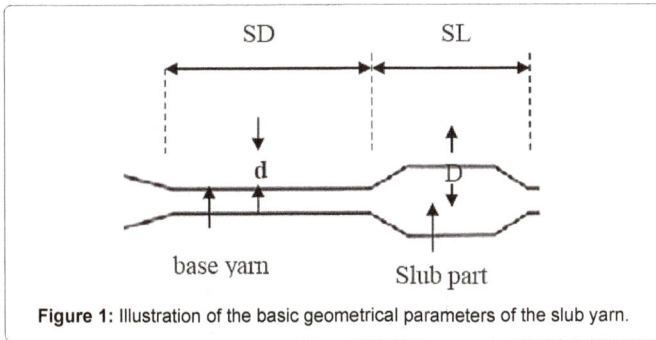

Figure 1: Illustration of the basic geometrical parameters of the slub yarn.

Arrangement (4)		Arrangement (8)		Arrangement (12)		Arrangement (17)	
Feeder No.	Slub Cone	Feeder No.	Slub Cone	Feeder No.	Slub Cone	Feeder No.	Slub Cone
1		1		1		1	
2		2		2	█	2	█
3		3		3		3	
4		4	█	4		4	█
5		5		5	█	5	
6		6		6		6	█
7		7		7		7	
8	█	8		8	█	8	█
9		9	█	9		9	
10		10		10		10	█
11		11		11	█	11	
12		12		12		12	█
13		13	█	13	█	13	
14		14		14		14	█
15		15		15		15	
16		16		16	█	16	█
17	█	17	█	17		17	
18		18		18		18	█
19		19		19	█	19	
20		20		20		20	█
21		21	█	21		21	
22		22		22	█	22	█
23		23		23		23	
24		24		24		24	█
25	█	25	█	25	█	25	
26		26		26		26	█
27		27		27		27	
28		28		28	█	28	█
29		29		29		29	
30		30	█	30		30	█
31		31		31	█	31	
32		32		32		32	█
33		33		33		33	
34	█	34	█	34	█	34	█

Table 1: Slub cone arrangement on knitting machine feeders.

Slub length 5:8 cm and Slub length percent 20%). Thus, the final slub yarn count is 22.65 Ne, total length 16 m, slubs per meter 3.131.

Two controller single jersey knitted fabrics were knitted by conventional 26 Ne for all feeders (this count is approximately equivalent to slub yarn count) and by conventional 32 Ne for all feeders (this count is similar to base count of slub yarn). Then 4 ratios of slub yarns were used (4, 8, 12 and 17 cones) for each conventional yarn count (26 and 32 Ne). Slub yarns were arranged on knitting machine feeders as shown on Table 1.

All fabrics were finished, the 140/3 Luft Rotoplus Thies Jet dyeing machine was used to half bleaching and dyeing of single jersey knitted fabrics. The bleaching solution contained the following ingredients for 200 kg knitted fabrics:

- Sequestering agent for iron (1.5 kg).

- Soap (2 kg).

- Sodium Hydroxide (3 kg).

- Hydrogen peroxide 50% (6 kg).

- Acetic acid (4.5 kg).

- Sequestering agent for water (1.5 kg).

- Leveling agent (1.5 kg)

- Reactive dye S_2G (2.622 kg).

- Softener (8 kg).

Total of 10 samples were knitted keeping the same stitch length and yarn tension. All samples were washed in a home laundering machine for three consecutive washing cycles. The washing process was carried out on (A) program for cotton fabrics at 90°C. Then fabric shrinkage and spirality were tested.

L1956A HP Scanjet scanner was used to investigate only the fabric surface appearance with 4800x9600 dpi, hardware resolution, which works according to light reflection. All samples were captured and analyzed using Matlab software to find the actual light permeability and fabric cover. For this purpose, an EOS450D Canon camera with Lens EF100 mm f/2.8L Macro IS USM, 12.8 Megapixels was used. The camera was fixed at right angle to the Single Jersey knitted samples in order to focus on the fabric sample that is fixed on a lighting box. 40x40 cm fabric samples were weighed 5 times using a digital balance of two decimal digits accuracy.

Fabric cross section was analyzed to find the differences in fabric thickness due to using slub yarns. For this purpose, Fabric was coated with transparent silicon layer, then the sample was cut, and fabric cross-section was captured by the camera. During wales density test, two needles were removed and there was a span of 288 needles. Distance between the two removed needles was measured in the finished fabric and wales density was calculated by equation (1) [4],

$$(WPC) = \frac{288}{distance\ between\ the\ removed\ needles\ (cm)} \qquad (1)$$

Courses density was measured by inserting a different yarn color during knitting and the length of ten repeats was measured. The courses density of finished fabric was then calculated by equation (2) [4]:

$$(CPC) = \frac{№\ of\ feeders \times 10}{length\ of\ ten\ repeats\ (cm)} \qquad (2)$$

Bursting strength was tested on Tinius Olsen Material Testing Machine 500, according to ASTM D3787-2001 applying 50 kgf (N) load range, 95 mm extension range, head speed of 305 mm/min, 90 mm endpoint and 0.1 kgf preload. Abrasion resistance test was performed using Martindale instrument according to ASTM 4966. A color property was measured using Data color 100 spectrophotometer according to ASTM E1164. Thermal comfort characteristics were measured. Alambeta instrument was used to measure thermal conductivity, fabric thickness, dry thermal resistance, and thermal absorptivity values. These parameters were tested according to ISO

EN 31092-1994. Relative water vapor permeability was measured on Permetest instrument based on similar skin model principle as given by ISO 11092. Air permeability was measured on Metefem instrument according to ASTM D737. The working pressure was 100 Pa using 20 cm² fabric samples. Five readings for each fabric sample were recorded. Results were statistically analyzed using SPSS software to test the significance of slub yarn ratio on all Single Jersey knitted fabric tested properties. Table 2 shows the statistical significance results at 95% confidence level, after uses of Univariate Analysis of Variance by SPSS Program (Two Way ANOVA).

Results and Discussion

Fabric Structural analyses

Figure 2a shows fabric knitted from one yarn count 32 Ne while Figure 2b shows fabric knitted from 50% conventional yarns and 50% slub yarns. It is clear from figure and by observing light reflectance, that fabric typical appearance has been improved as shown in Figure 2b compared to Figure 2a.

Figure 3 shows the real fabric images and binary images that were processed in order to find the actual values of covered area for each sample and to investigate the influence of slub yarn usage ratio on fabric light permeability which is an indication for air permeability and water vapor permeability. Figure 3a shows image of controller fabric sample knitted from 32 Ne conventional yarns, while Figure 3b

Property	Slub yarn percent	Ground yarn count
Weight	0.000	0.000
Thickness	0.000	0.000
Shrinkage	0.000	0.000
Spirality	0.000	0.921
Thermal conductivity	0.063	0.002
Thermal absorptivity	0.382	0.000
Thermal resistance	0.000	0.007
Air permeability	0.008	0.000
Vapor permeability	0.341	0.003
Bursting strength	0.238	0.000
Abrasion resistance	0.000	0.027

Table 2: Statistical significance of slub yarn ratio and ground counts on fabric properties.

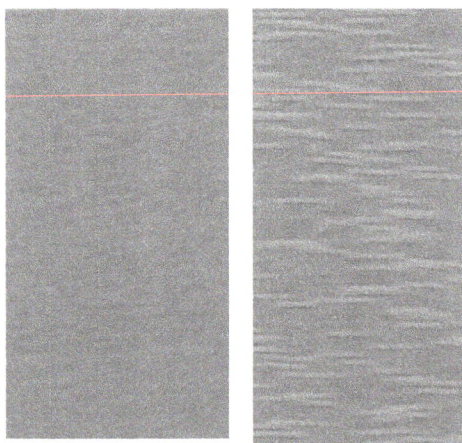

a) Controller sample (all yarns conventional Ne 32)   b) Sample with 50% slub yarns

Figure 2: Images of different Single Jersey knitted samples based on light reflection.

shows this image after processing. It can be seen that light cover ratio is 91.48%. While Figure 3c and 3d also shows the original and processed image for fabric knitted using 50% slub yarns. It can be observed that light cover ratio increased up to 94.94% and this is expected as a result of bigger yarn diameter. All samples were analyzed and processed at same conditions and settings.

Figure 4 shows the effect of both increasing slub yarn ratio and

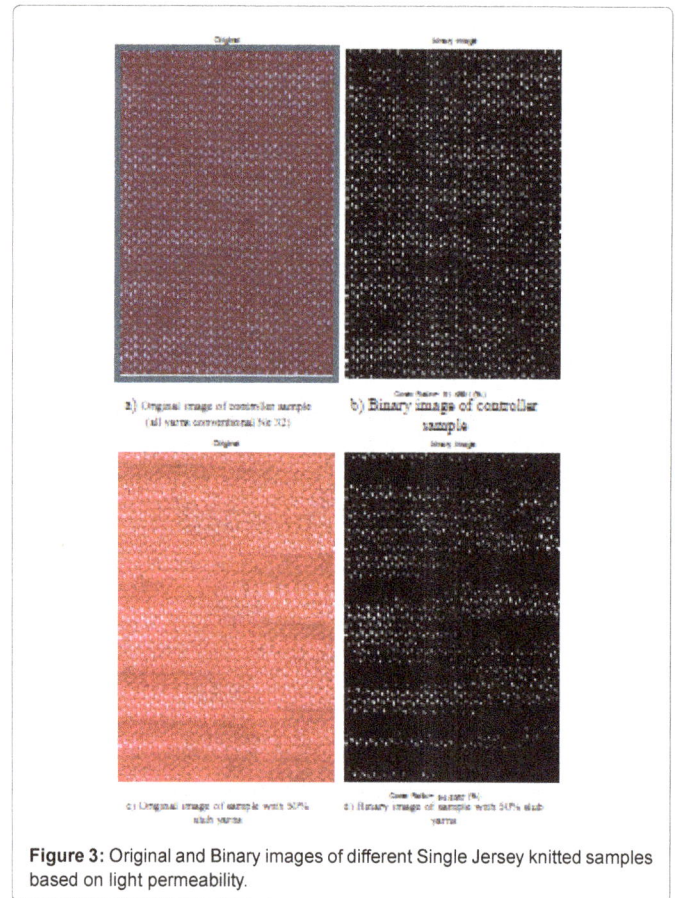

a) Original image of controller sample (all yarns conventional Ne 32)   b) Binary image of controller sample

c) Original image of sample with 50% slub yarns   d) Binary image of sample with 50% slub yarns

Figure 3: Original and Binary images of different Single Jersey knitted samples based on light permeability.

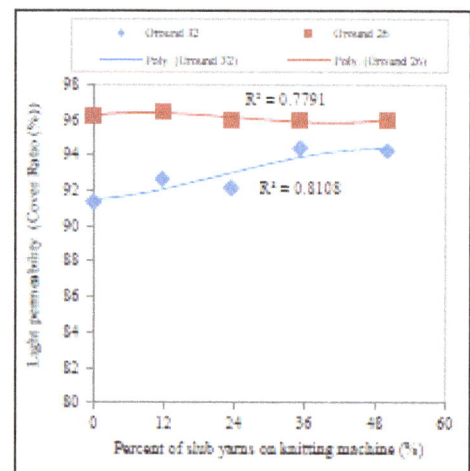

Figure 4: Effect of slub yarn ratio on light permeability of Single Jersey knitted fabric at different ground yarn counts.

ground yarn count on light cover ratio of Single Jersey knitted fabric. By increasing slub yarn ratio from 0 to 50%, values of light cover ratio. Increase by a ratio up to 6% particularly if ground yarn count is equal to slub yarn base count. While the light cover ratio does not change by increasing slub yarn ratio if ground yarn is coarser than base of the slub yarn count. Due to this change in light cover ratio. It is expected that all fabric characteristics are affected especially air permeability and water vapor permeability.

Fabric geometry

Figure 5 explains the influence of slub yarn ratio on fabrics wales and courses density after finishing and repeated washing. Wales density decreases slightly by 3% as slub yarn ratio increases from 0 to 50% either in finished or washed fabrics, however, it is not affected by ground yarn count.

Courses density increases as slub yarn ratio increases and this is due to increasing yarn count with constant stitch length. As yarn diameter

increases courses spaces decrease, consequently density increases. It can be seen from the figure that as yarn count decreases from 32 to 26 Ne, courses spaces decrease and density increases by a ratio up to 6.5% for both finished and washed fabrics. Therefore, as slub yarn ratio changes from 0 to 50%, courses density increases by a ratio up to 10%.

Figure 6 shows the relationship between slub yarn ratio and Single Jersey knitted fabrics weight per square meter. These fabrics are knitted from ground yarns 26 Ne and 32 Ne. As slub yarn ratio increases from 0 to 50%, fabric weight per square meter increases by 28% and this is because average yarn diameter increases generally in produced fabric. Whereas increasing English yarn count results in increasing yarn diameter and fabric weight increases by 38% as shown in Figure 6. Statistical analysis indicates that the effect of ground yarn count and slub yarn ratio is significant.

Fabric shrinkage

Figure 7 shows the relationship between slub yarn ratio and Single Jersey knitted fabric shrinkage in wales and courses direction. As slub yarn ratio increases from 0 to 50%, fabric shrinkage improves in both lengthwise and widthwise approximately by percent 5%. This is because using more yarns with bigger diameter causes an increase in fabric tightness i.e. reduction in fabric pores as shown in Figure 4. Statistical analysis indicates that the effect of slub yarn ratio on fabric shrinkage and ground count is significant.

Fabric spirality

As noticed from Figure 8, when slub yarn ratio increases from 0 to 33%, spirality decreases approximately from 8 to 0.5 degree, and this is due to increase in fabric tightness and light permeability as shown in Figures 3 and 4. Statistical analysis indicates that the effect of slub yarn ratio on fabric spirality is significant while the effect of ground yarn count is non-significant.

Fabric thickness

It is known that fabric thickness is a function of yarn diameter whereas yarn count decreases from 32 to 26 Ne, yarn diameter increases and fabric thickness increases by a ratio up to 10% as shown in Figure 9. It is clear also that increasing the ratio of used slub yarns from 0 to

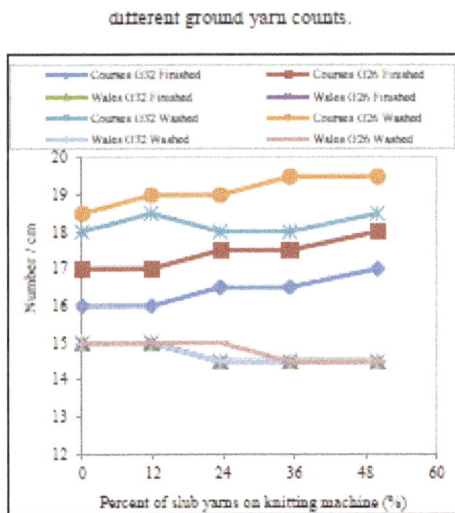

Figure 5: Effect of slub yarn ratio on courses and wales density of finished and washed Single Jersey knitted fabric at different ground yarn counts.

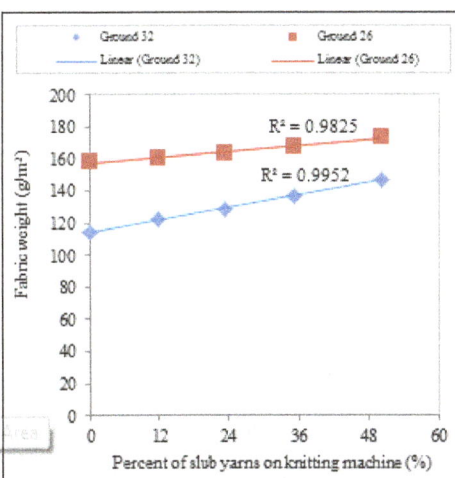

Figure 6: Relationship between slub yarn ratio and weight of Single Jersey knitted fabric at different ground yarn counts.

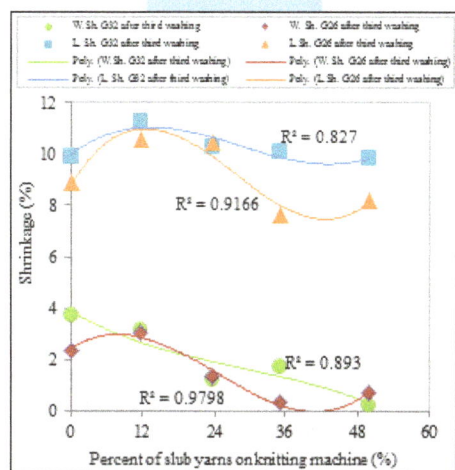

Figure 7: Relationship between slub yarn ratio and shrinkage of Single Jersey knitted fabric at different ground yarn counts.

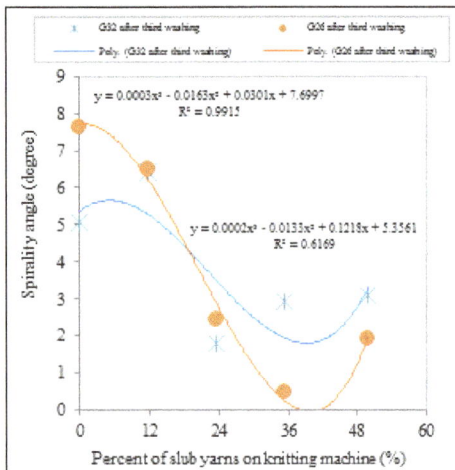

Figure 8: Relationship between slub yarn ratio and spirality of Single Jersey knitted fabric at different ground yarn counts.

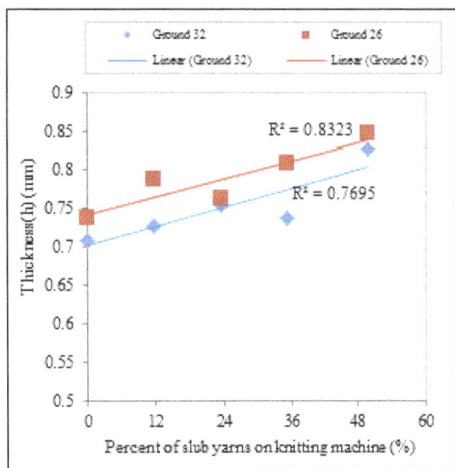

Figure 9: Relationship between slub yarn ratio and thickness of Single Jersey knitted fabric at different ground yarn counts.

Figure 10: Longitudinal cross section of Single Jersey knitted fabrics.

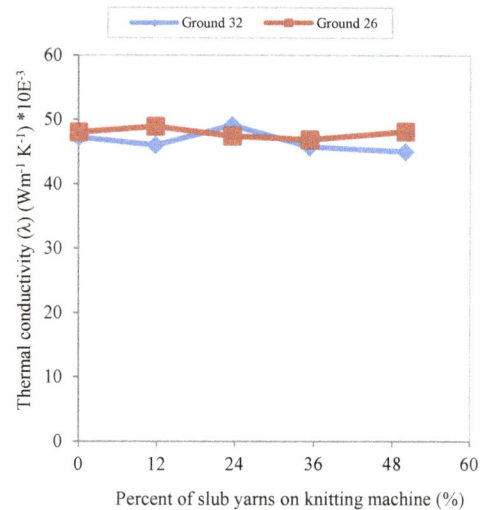

Figure 11: Relationship between slub yarn ratio and thermal conductivity of Single Jersey knitted fabric at different ground yarn counts.

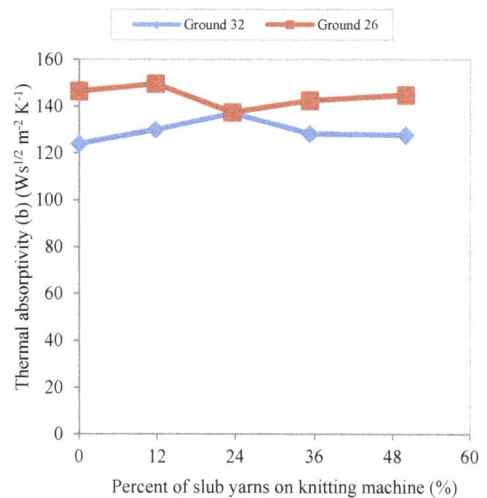

Figure 12: Relationship between slub yarn ratio and thermal absorptivity of Single Jersey knitted fabric at different ground yarn counts.

50%, fabric thickness increases by 17% and this is due to the usage of yarns that have bigger average diameter. This also affects fabric thermal resistance and mechanical properties. Statistical analysis indicates the significance of both variables. To clarify the change in thickness, fabric longitudinal cross section was obtained and captured by high resolution camera and Figure 10 shows the big variation in fabric thickness between the places from ground and slub yarns. Also there is variation in thick places length that is ascribed to the convergence and divergence of thick places between two adjacent slub yarns.

Fabric thermal comfort

Figures 11 and 12 show the relationship between slub yarn ratio, fabric thermal conductivity and fabric thermal absorptivity. It is clear from both figures and from statistical analysis that the effect of slub yarn ratio is insignificant. As shown in Figure 12, fabrics knitted from coarse counts have higher fabric absorptivity values and this is expected because as yarn diameter increases, number of paths in which heat transfers increases causing higher absorptivity.

Regarding fabric thermal resistance, as shown in Figure 13 and by statistical analysis it is evident that slub yarn ratio and ground yarn count affects it significantly, it is clear that when slub yarn ratio increases from 0 to 50%, thermal resistance increases by 20% which is due to greater fabric thickness that is directly proportional to thermal resistance as shown in Figures 9 and 10. While the effect of slub yarn ratio, as shown in Figure 14 and from statistical analysis on relative water vapor permeability is insignificant.

Figure 15 shows the relationship between slub yarn ratio and Single Jersey knitted fabric air permeability. These fabrics are knitted from ground yarns 26 Ne and 32 Ne. As slub yarn ratio increases from 0 to 50%, fabric air permeability increases by a ratio up to 23% and

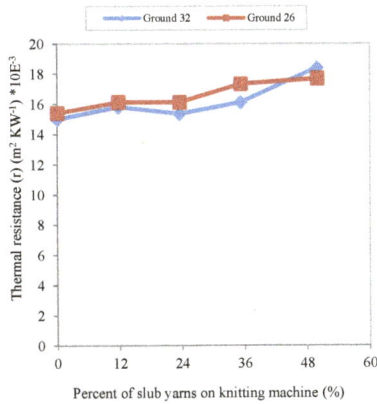

Figure 13: Relationship between slub yarn ratio and thermal resistance of Single Jersey knitted fabric at different ground yarn counts.

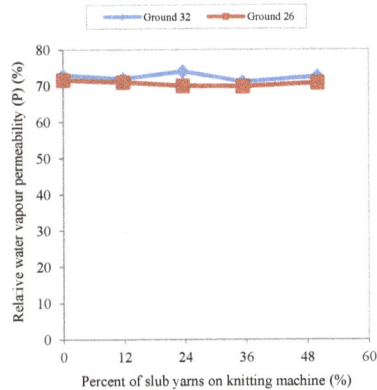

Figure 14: Relationship between slub yarn ratio and relative water vapor permeability of Single Jersey knitted fabric at different ground yarn counts.

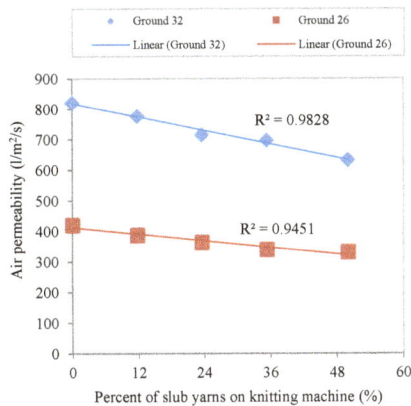

Figure 15: Relationship between slub yarn ratio and air permeability of Single Jersey knitted fabric at different ground yarn counts.

this is because light cover ratio increases by 4% and fabric thickness by 17%. Whereas yarn thickness increases, air permeability decreases. This trend is obvious form Figure 15 where increasing English yarn count from 26 to 32 i.e. reducing yarn diameter, results in increasing air permeability by 95%. And statistical analysis indicates that the effect of both variables on fabric air permeability is significant.

Fabric color properties

It is evident from Figure 16 and by statistical analysis that the effect of slub yarn ratio and ground yarn count on Single Jersey knitted fabrics color properties (LAB) is insignificant.

Fabric strength

Figure 17 shows the relationship between slub yarn ratio and Single Jersey knitted fabric bursting strength, these fabrics are knitted from different ground yarn counts. It is apparent from statistical analysis and the figure that fabric bursting strength is not affected by slub yarn ratio where it increases slightly if ground yarn counts are equal to the slub yarn count base (32 Ne) and decreases slightly if ground yarn counts are similar to the equivalent slub yarn count (26 N_e).

Fabric abrasion resistance

It is clear from Figure 18 that when slub yarn ratio increases from 0 to 50%, abrasion resistance increases by 60% because as slub yarn ratio increases, the probability of existence of apparent thick areas in fabric increases. These areas resist friction instead of the remaining fabric structure which reduces the percentage fabric weight loss. From statistical analysis, the effect of slub yarn ratio is significant unlike the effect of ground yarn count which is non-significant.

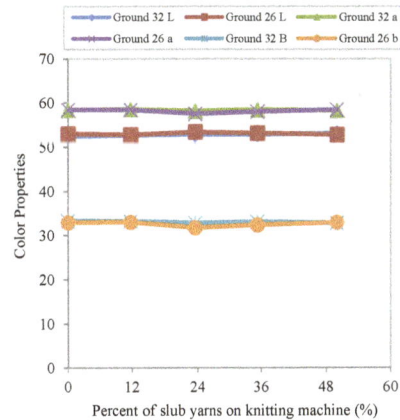

Figure 16: Relationship between slub yarn ratio and color properties of Single Jersey knitted fabric at different ground yarn counts.

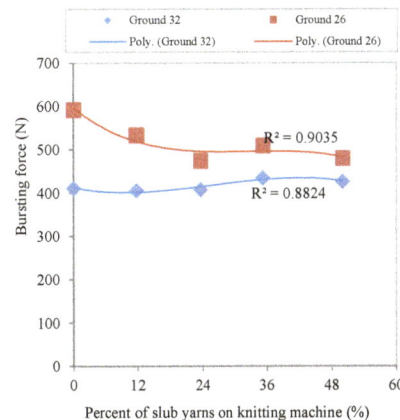

Figure 17: Relationship between slub yarn ratio and bursting strength of Single Jersey knitted fabric at different ground yarn counts.

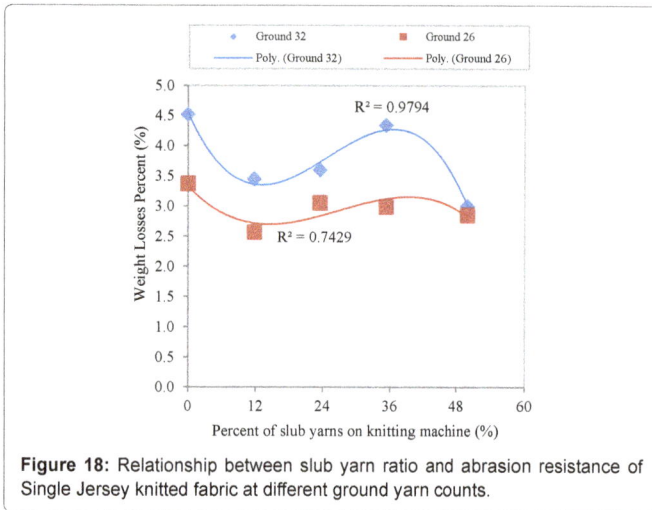

Figure 18: Relationship between slub yarn ratio and abrasion resistance of Single Jersey knitted fabric at different ground yarn counts.

Conclusion

Single Jersey knitted fabric appearance improved after using the slub yarns and its classic texture has changed specially from fabric back side. Fabric weight has increased by 28% and shrinkage ratio has improved by 5%. Also, fabric spirality improved where it reached 0 approximately particularly at 33% slub yarn ratio. Fabric thickness increased by 17% which enhanced thermal resistance by 20% in contrary to fabric air permeability which was reduced by 23%. Finally fabric abrasion resistance improved. While the influence of slub yarn ratio on other tested properties was insignificant.

References

1. Gong RH, Wright RM (2002) Fancy Yarns: Their Manufacture and Application. Woodhead Publishing Limited UK.

2. http://www.Textileworld.com/Issues/2005/January/Features/New_Concepts_For_Fancy_yarns

3. El-Khalek RA (2015) Computer - based system for evaluation and recognition of fancy yarns properties, quality and their applications. Thesis of PhD Degree in Mansoura University.

4. Fouda A, El-Hadidy A, El-Deeb A (2015) Mathematical modeling to predict the geometrical and physical properties of bleached cotton plain single jersey knitted fabrics. Journal of Textiles.

5. Liu X, Wen Z, Su Z, Choi KF (2008) Slub Extraction in Woven Fabric Images Using Gabor Filters. Text Res J 78: 320-325.

6. Fouda AEF (2008) Study of Dimensional Stability of Weft Knitted Fabrics. Thesis of Master Degree in Mansoura University.

7. Nergis BU (2003) Properties of Plain Knitted Fabrics from Chenille Yarns. Text Res J.

27

Evaluating the Effect of Weft Yarn Characteristics on Fabric Texture Using Image Analysis Techniques

Singh JP*

Department of Textile Technology, U.P. Textile Technology Institute Kanpur, India

Abstract

In this research work, an objective evaluation method for assessing aesthetic appearance of apparel based fabric having five different wefts by image processing technique is suggested. Surface texture characteristics such as energy, entropy, inertia have been estimated by different image processing techniques. Fast fourier transform (FFT) power spectrum method is used to estimate the surface texture in terms of fractal dimension (FD). Subjective evaluation of appearance for different fabric samples has been carried out by experts and its correlation with image processing based texture characteristics is determined. Effect of five different weft yarn characteristics on the fabric appearance is investigated. Negative correlation is obtained between fractal dimension & energy while rest of the parameters have resulted good positive correlation. The surface of the fabric with polyester textured filament yarn showed the uniform texture while fabric with cotton slub weft exhibited the rough texture. The results also suggest that all the texture results from image analysis based objective methods are very well correlated with the subjective texture results. A preliminary study was also done to derive the textural parameters from the geometrical parameters of the slub yarn. But the textural parameters do not seem equal to any of the geometrical parameters of the slub yarn. For this, probably a more detailed study is needed with different quality of slub yarn. The research work is in progress to give a mathematical relation between the fabric texture and slub yarn geometrical parameters.

Keywords: Texture; Fractal dimension; Weft yarn; Slub yarn; Textured polyester yarn

Introduction

Apparel based fabrics are popular particularly for their textural properties like fine visual appearance, pleasant to touch, soft hand and comfortable to wear behavior. The appearance of apparel based fabric is normally evaluated by highly experienced people by subjective methods. Although these methods are useful in the overall assessment of textile products, however, they are time consuming and the results lack reproducibility. Many researchers [1-3] have questioned on the reliability and usefulness of these techniques and lead to the development of a more realistic instrumental method to quantify apparel based fabric appearance value. It is thought to be very useful for quality assurance, process control and product development in manufacturing of apparel based fabrics. The textural properties of fabrics are some of the key factors that can be altered by the manufacturers to influence consumers at the point of sale.

Producing different variety of fabric is the main concern on the manufacturers so that they can satisfy the growing demand of the customers. Fabric texture is an important properties that can be altered easily to add variety to the fabrics. Producing fabric by using different types of yarn in the warp has some limitation because it has to pass though the healds and reed. These limitations can be overcome by the weft yarn which provides freedom to use any type of yarn in the fabric. So, the different types of the weft yarn can be used to alter the properties of fabric without facing production and quality difficulties [4]. The exact effect of the changes in yarn characteristics on the fabric texture needs to be understood for better quality of the fabric. In this research work, an attempt has been made to change the fabric texture by changing the weft yarn in the fabric and to correlate these texture changes with the yarn parameters. The paper also aims at verifying the results of image analysis based objective method for texture evaluation with the traditional subjective method.

Theory of Aesthetic Characterization

Texture analysis

Texture describes the surface properties, i.e. smoothness or roughness and the uniformity of the fabric surface. Wavelet transform [4-7] is capable of extracting features that may be used to distinguish between different fabric textures. The spatial grey level co-occurrence measurement method, a statistical approach, can be used for texture measurement. The image size is 512 by 512 pixels The spatial grey level co-occurrence probability function f (i, j, d, a) is a second-order probability density function; it is the relative frequency with which two pixels separated by a distance d in the a direction occur on the image one with i grey level and the other with j grey level. From this definition one can calculate co-occurrence matrices in the directions 0°, 45°, 90° and 135° and from these matrices the following parameters can be calculated.

$$Inertia = \sum_i \sum_j \left[(i-j)^2 \frac{f(i,j,d,a)}{S} \right] \tag{1}$$

$$Energy = \sum_i \sum_j \left[\frac{f(i,j,d,a)}{S} \right]^2 \tag{2}$$

$$Entropy = -\sum_i \sum_j \left[\frac{f(i,j,d,a)}{S} log\left(\frac{f(i,j,d,a)}{S} \right) \right] \tag{3}$$

Where

$$S = \sum_i \sum_j f(i,j,d,a)$$

Energy illustrates the homogeneity of the structure: high energy means more homogeneous texture and vice versa. Similarly entropy

---

*Corresponding author: Singh JP, Department of Textile Technology, U.P. Textile Technology Institute Kanpur, India, E-mail: jpsingh.iitd@gmail.com

and inertia depict structural disorder and local variations respectively; large entropy is indicative of more disordered structure, and high inertia means high local variations. The parameters are based on statistics and information theory.

The estimation of fractal dimension

Bergmann et al. [8] proved that the tactile surface property, such as roughness, can be evaluated by the fractal dimension of the surface. Since Mandelbrot [9] first proposed the term 'fractal', there are many different definitions and different measurements for the fractal dimension. The method we used for fractal dimension estimation is the power spectrum method [10].

The power spectrum estimation is far more accurate than the other methods. As shown in equation (4), the power spectrum P is proportional to the certain power β of the radial frequency ω [11].

$$P(\omega) \propto \omega^{-\beta} \tag{4}$$

Where $\omega \geq 0$. The fractal dimension D of an image is related to the exponent β in above equation.

$$D = \frac{8-\beta}{2} \tag{5}$$

The general algorithm of FD power spectrum estimation is shown in the below figures. The first step is to read the information from a binary image file and then store the information into a matrix. Using the fast Fourier transform (FFT) the information is transformed to the spatial frequency domain [12]. The power spectrum is calculated in order to remove the imaginary part of the FFT result. In the fourth block, the power spectrum is cut into several radial slices and integrated respectively.

The final step consists of determining the exponent β by using linear regression to find the slopes of the fourth step results on a log-log scale, and applying equation (5) to compute D. It is robust to the linear transformation of the image data and the scaling of the image theoretically. The experiment results in the next section also proved its good performance.

Materials and Methods

Materials

Fabric samples were manufactured on the rapier weaving machines (TD - 736A). The details of the sample were given in Table 1, and the images are shown in Figure 1.

Slub yarn: Slub yarn used in this research work is produced in controlled manner on special ring frame. Three important parameters, total slub length percentage, total slub distance percentage and percentage of CV, of the slub yarn were evaluated using conventional method. The results were used as a preliminary study to correlate these parameters with the change in the texture of the fabric Figure 2.

$$Total\ slub\ length\ \% = \frac{\sum_{i=1}^{n} SL_i}{TL} * 100 \tag{6}$$

$$Total\ slub\ length\ \% \ \frac{\sum SD}{TL} * 100 \tag{7}$$

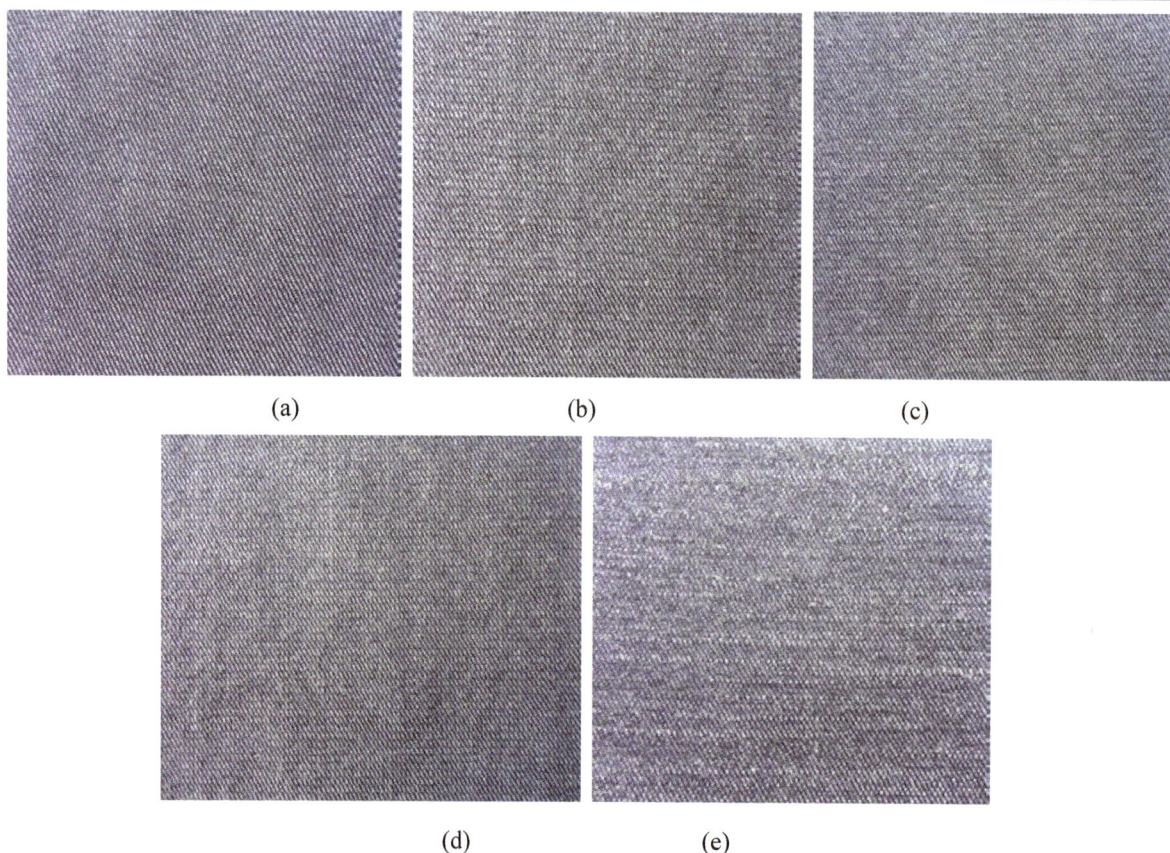

Figure 1: Fabric samples having (a) polyester textured yarn (b) combed cotton yarn (c) P/V yarn (d) cotton lycra yarn (e) cotton slub yarn in weft.

Where SL: Slub length, SD: Slub diameter, and TL: Total length measured.

Methods: Images of 5 fabric samples were taken using an image acquisition box and the method described by researchers [13] and as shown in figure. Energy, Entropy and Inertia were determined by using spatial grey level co-occurrence measurement method as given in equations 1, 2 and 3 respectively. FD was calculated by using FFT. The textural parameters (Energy, Entropy and Inertia) and fractal dimension (FD) were evaluated by using image analysis techniques

according to the algorithm given in Figures 3 and 4 respectively. Energy is multiplied by a factor 10^5 and Entropy is divided by a factor 10 to represent the data graphically.

Based on these flow diagrams, Maltab program have been written to calculate the texture and fractal terms.

Subjective evaluation of surface texture: The selection of judges for subjective evaluation is very crucial as their opinion should match the general opinion of user as well as the manufacturer. The person

S.No	EPI	PPI	Warp count (Ne)	Weft count (Ne)	Weft yarn type	Weave
S1	62	43	30/2 P/V	30/2˙	Polyester textured yarn	2/1 twill
S2	62	43	30/2 P/V	30/2	Combed cotton yarn	2/1 twill
S3	62	43	30/2 P/V	30/2	P/V˙˙ yarn	2/1 twill
S4	62	43	30/2 P/V	30/2	Cotton lycra yarn	2/1 twill
S5	62	43	30/2 P/V	30/2	Cotton slub yarn	2/1 twill

˙Equivalent to 2/30 Ne, ˙˙P/V: Polyester/viscose, EPI: ends per inches, PPI: picks per inches.

Table 1: Sample constructional details.

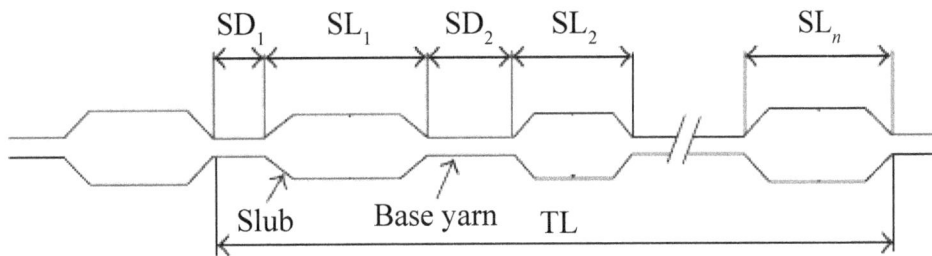

Figure 2: Geometrical parameters of slub yarn (El-khalek, El-Bealy, and El-Deeb, 2014).

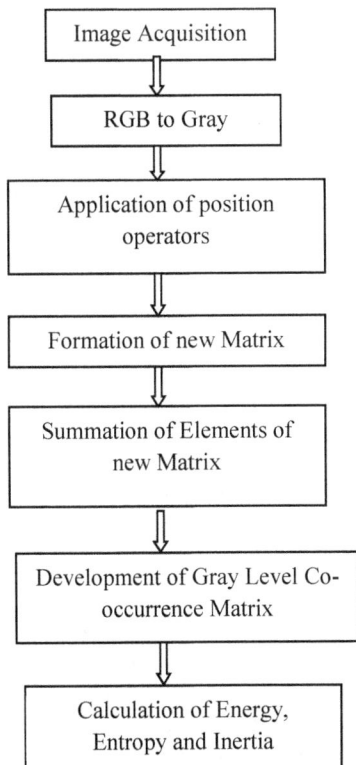

Figure 3: Flow diagram for calculating Energy, Entropy and Inertia.

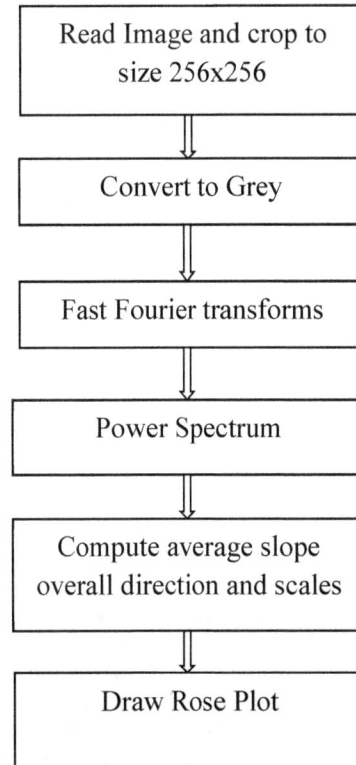

Figure 4: Flow diagram for calculating Fractal dimension.

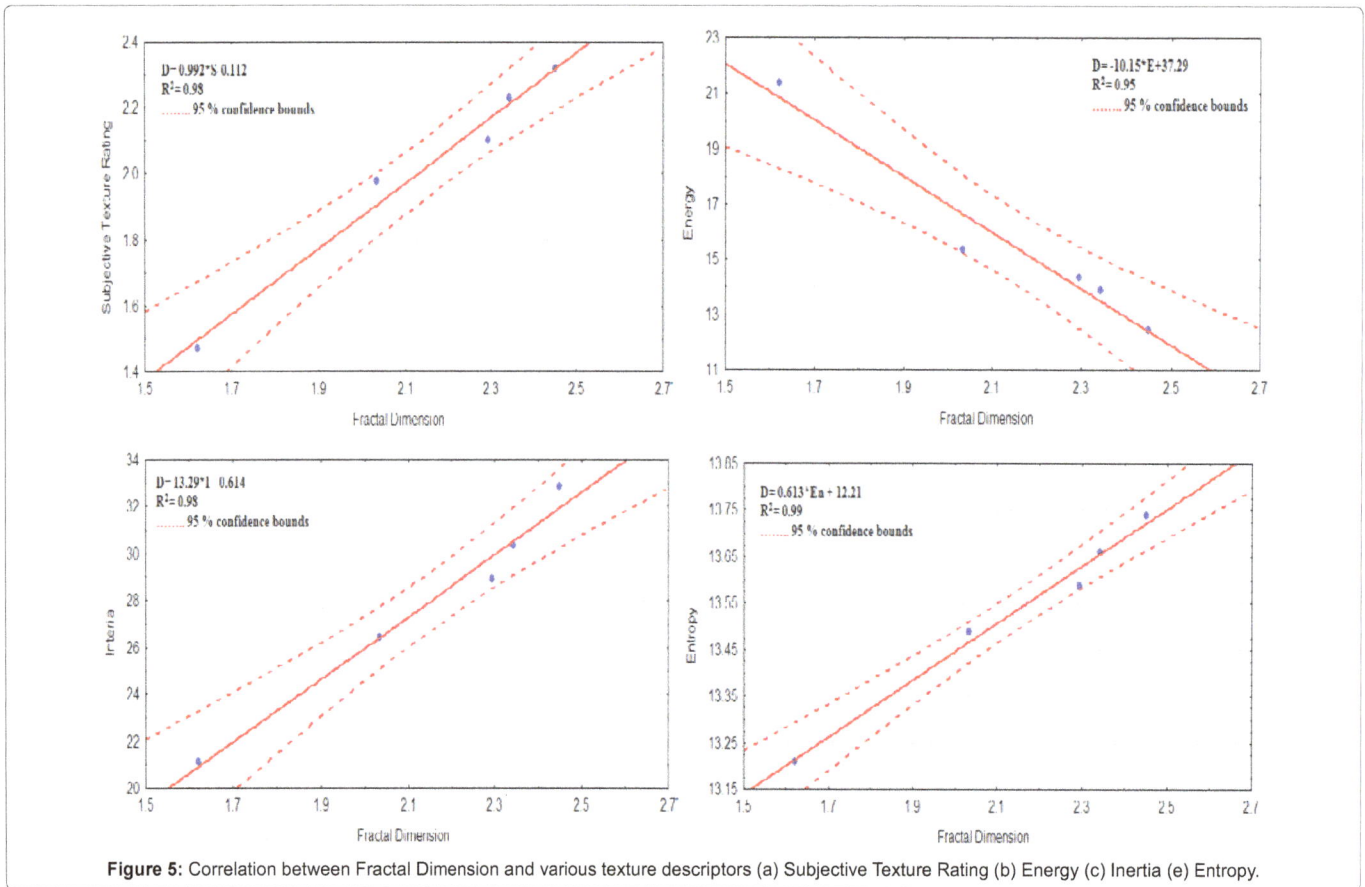

Figure 5: Correlation between Fractal Dimension and various texture descriptors (a) Subjective Texture Rating (b) Energy (c) Inertia (e) Entropy.

working in the sample design and development section of a factory is a link between buyer and manufacturer as he interacts directly with the buyer who place his order based on the survey/opinion of user. The persons working in finishing section and quality assurance section are the key persons involved in manufacturing and maintaining the quality. So, the judges are the experts from these sections of the textile industry. Subjective evaluation of surface texture (uniformity and orientation) was performed by 25 judges and a subjective texture rating (STR) was calculated. Mean value of rating given by all the judges has been taken for further analysis. Ranks were given on the scale of 0 to 5, 0 being the smoothest and 5 being the most rough. Kendall's coefficient of concordance (W) is a measure of the agreement among several (k) quantitative or semiquantitative variables that are assessing a set of n objects of interest. To verify the degree of agreement among judges, the coefficient of concordance has been calculated using the following formula.

$$W = \frac{12 \sum \left\{ SR^2 - n \left(\frac{SR}{n} \right)^2 \right\}}{k^2 \left(n^3 - n \right)} \qquad (8)$$

Where, SR: Rank Sum, k: No. of Judges, n: No. of samples

Energy, entropy and inertia were determined by using spatial grey level co-occurrence measurement method as given in equations (1), (2) and (3) respectively. Fractal dimension was calculated by using fast fourier transform (FFT). The textural parameters (energy, entropy and inertia), and fractal dimension were evaluated by using image analysis techniques according to the algorithms given. Energy is multiplied by a factor 10^5 and Entropy is divided by a factor 10 to represent the data graphically.

Results and Discussion

Fractal dimension and texture descriptor

At the outset, a correlation study was carried out to examine the role of fractal dimension in describing the surface properties of apparel grade fabric. Coefficient of concordance among the judges for subjective surface roughness was found to be 0.79 which can be considered as a good agreement. The results were analysed by using Statistica Software. Image processing based texture descriptors are plotted with fractal dimension and the results are shown in Figures 5a-5d.

From the figures it can be seen that there is a negative correlation between fractal dimension & energy while good positive correlations are observed with rest of the parameters. Higher homogeneous surface gives lower fractal dimension values and higher energy value which explains the negative correlation between fractal dimension and energy. Highly disordered surface gives high entropy value. The correlation coefficient between fractal dimension and the other texture descriptors are in the range of 0.95 to 0.99 and all the points are well within 95% confidence bounds. The results show a similar trend found by the Behera and Singh [1].

Effect of weft variation on the fabric texture

In this section, the results related to the texture change due to weft change are discussed. Since the constructional parameters of fabric, like weave, warp density, weft density etc, were kept constant, the change in the amount of weft yarn exposed per unit area on the fabric surface is also constant for all the fabrics. Figure 6 shows the rose plot of the fractal dimension of various samples. It is clearly visible from Figure

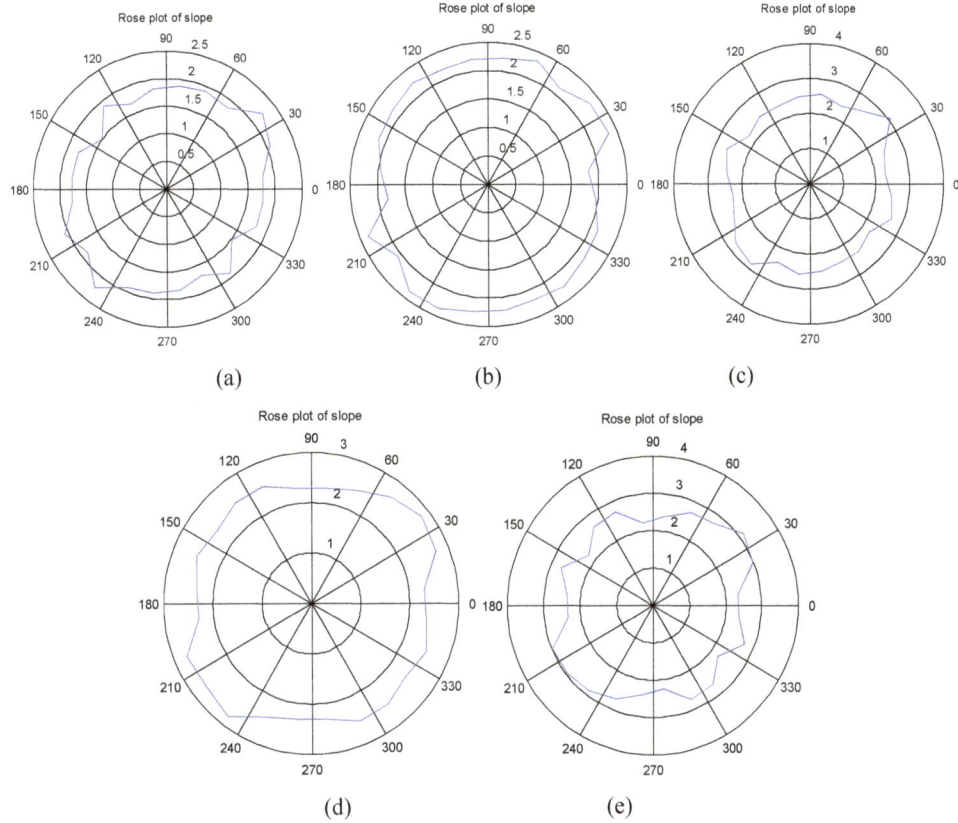

Figure 6: Rose plot of Fractal Dimension for (a) Texturised polyester filament weft (b) Combed cotton weft (c) P/V weft (d) Cotton lycra weft (e) Cotton slub weft.

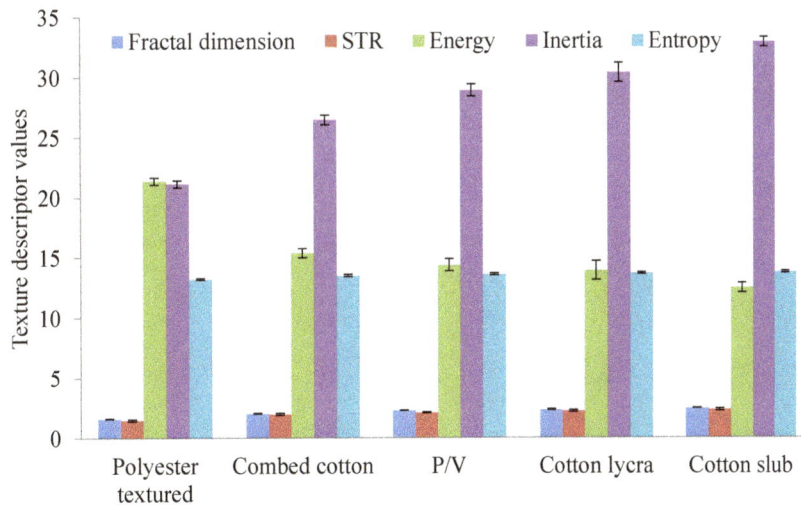

Figure 7: Texture descriptors of the surface of fabrics made from different weft.

6a that minimum fractal dimension, shown by the surface of the fabric woven with textured polyester filament yarn, is minimum among all the samples.

The next higher fractal dimension is shown by the surface of the fabric having combed cotton weft. This increase in fractal dimension is about 26 % from the fabric having polyester textured filament weft. The next higher fractal dimension is shown by the surface of the fabric having P/V weft. This increase in fractal dimension is about 30 % from the fabric having polyester textured filament weft. This increase in the fractal dimension may be attributed to the discontinuous yarn surface made of staple fibre yarn. Lower fractal dimension exhibited by the surface of the fabric having combed cotton weft as compared to the one having P/V weft may be attributed by the smoother yarn surface achieved by the finer tip of the natural cotton fibre.

The next higher fractal dimension is shown by the surface of the fabric having cotton lycra weft. This increase in fractal dimension is about 45 % from the fabric having polyester textured filament weft. The surface of the fabric having cotton slub weft exhibits highest fractal dimension. This increase in fractal dimension is about 57 % from the fabric having polyester textured filament weft [14,15].

Texture descriptors for various fabrics shown in the Figure 7 clearly explain the changes taking place in the fabric surface upon the change of weft yarns. The fabric having polyester textured filament yarn as weft exhibits the homogeneous texture with lower structural disorder and low local variations. These properties continue to deteriorate with the use of combed cotton weft, P/V weft, cotton lycra weft and cotton slub weft sequentially.

The slub yarn were tested for total slub length%, total slub distance % and total cv% and were found 36.9, 63.1 and 46.3 respectively. As compared to the normal cotton yarn, the slub yarn is having these variations. As already mentioned, the other parameters of the fabric are constant for all the samples except the weft yarn; the additional irregularities of the weft yarn should be correlated with the change in the texture of the fabric. The percentage change in the texture descriptors are 25.37, 17.17, 18.95, 24.44, 1.85 in fractal dimension, subjective texture rating, energy, inertia and entropy respectively. The parameters mentioned here are really not equal to any of the parameters of the slub yarn.

Conclusions

Five woven fabric samples have been prepared by changing different yarn in weft and tested for the textural properties. It is found that the fabric with textured polyester filament yarn as weft exhibits the uniform textural properties and fabric having slub yarn as weft exhibits the rough textural properties among the sample tested. The fractal dimension of the surface of the fabric with polyester textured filament weft is lowest and its rose plot is uniform in all the directions. The fractal dimension of the fabric surface keeps on increasing with change in weft from combed cotton to cotton slub weft. The rose plot of all the fabric except cotton slub weft fabric is almost uniform in all the direction which expresses its uniformity in all the direction. The disturbance in the weft yarn smoothness is the main reason behind the variation in the textural features. In this research work, a preliminary study was also done to derive the textural parameters from the geometrical parameters of the slub yarn. But the textural parameters do not seem equal to any of the geometrical parameters of the slub yarn. For this, probably a more detailed study is needed with different quality of slub yarn. The research work is in progress to give a mathematical relation between the fabric texture and slub yarn geometrical parameters. The results also suggest that all the texture results from image analysis based objective methods are very well correlated with the subjective texture results.

References

1. Lewis WJ (1977) Wool Research Organization of New Zealand. Christchurch, New Zealand.

2. Ross DA (1986) Instrumental Objectivity. Textile Horizons 2: 29-30.

3. Wen CY (2001) Defect segmentation of texture images with wavelet transform and a co-occurrence matrix. Textile Research Journal. 71: 743-749.

4. Behera, BK, Singh JP (2014) Objective evaluation of aesthetic characteristics of terry pile structures using image analysis technique. Fibers and Polymers 15: 2633-2643.

5. Durrant PJ, McLaughlin JR (1984) A method for the objective measurement of carpet texture. Wool Research Organization of New Zealand, Communication no C93, Christchurch, New Zealand.

6. Haralick RM, Shanmugam K, Dinstein I (1973) Textural features for image classification. IEEE Transactions on Systems, Man and Cybernetics, SMC 3: 610-620.

7. Behera BK, Mishra R (2006). Objective measurement of fabric appearance using digital image processing. The Journal of The Textile Institute, 97: 147-153.

8. Bergmann M, Herbst I, Wieding VR, Wolter FE (1999) Haptical Rendering of Rough Surfaces using Their Fractal Dimension. In Proc of the First PHANTOM Users Research Symposium, German Cancer Research Centre, Heidelberg, Germany 9-12.

9. Mandelbrot B (1983) The Fractal Geometry of Nature. Freemann, New York.

10. Voss R (1988) Fractal in Nature: From Characterization to Simulation. The Science of Fractal Images, Springer-Verlag, New York 1: 21-70.

11. Zwiggelaar R, Bull CR (1995) Optical Determination of Fractal Dimensions using Fourier Transforms. Optical Engineering 34: 1325-1332.

12. Hosseini SA, Torimuki K (1995) Fourier transforms analysis of plain weave fabric appearance. Textile Research Journal 65: 676-683.

13. Bailey DG, Hodgson RM (1985) Range Filters: Local intensity Sub-range Filters and their Properties. Images Vis. Computer. 3: 99-110.

14. El-khalek R, El-Bealy R, El-Deeb A (2014) A computer based system for evaluating slub yarn charecteristics. Journal of Textiles.

15. Gonzalez RC, Wood RE (1993) Digital Image Processing. Addison-Wesley, New York.

28

Mechanical Property Measurement and Prediction Using Hirsch's Model for Glass Yarn Reinforced Polyethylene Composite Fabric Formwork

Jun Xiong[1] and Mashiur R[2]*

[1]Department of Textile Sciences, Faculty of Human Ecology, University Manitoba, Winnipeg, Canada
[2]Department of Biosystems Engineering, Faculty of Agriculture and Food Sciences, University of Manitoba, Winnipeg, Canada

Abstract

The purpose of this research was to assess the effectiveness of two methods of reinforcing glass yarns, stitching and laminating methods, in a woven (1/1) polyethylene (PE) geotextile fabric used for fabric formwork. Both reinforcement methods improved the tensile properties when the holding load is within the combined breaking load of reinforced glass yarns. However, reinforcing by stitching method provided better tensile properties than the laminating method in respect to creep. The stitched specimens (S-1000 N) and (S-500 N) generated 7.04 mm and 0.53 mm creep respectively compared to the laminated specimens that has produced 7.56 mm (L-1000 N) and 3.11 mm (L-500 N). A Hirsch's model (combined effect of parallel model and series model) was used to predict the modulus of PE + glass yarn composite fabric for stitching method based on the 500 N holding load data. This modulus can be used to calculate the number of glass yarns that should be used in composite for a specific dimension of fabric formwork.

Keywords: Composite; Polyethylene; Glass yarn; Fabric formwork; Hirsch's model; Mechanical property; Lamination and stitching.

Introduction

Fabric formwork is a technology that uses textile fabrics to make molds for concrete casting. Compared to conventional molds that use lumber, plywood and steel, the molds made of fabrics are flexible, easy to use, light weight, permeable and cost less [1]. Other advantages of using fabric formwork include savings in material cost, labour and time [1], rendering blowhole-free smooth surfaces and aesthetic façade impressions [2], resulting in a denser casted surface [3] and extremely beautiful streamline [4-7]. However, due to limited strength and rigidity, fabric formworks are not suitable for large sized structures. Further, when designing the formwork, fabric can only resist tensile forces and cannot retain moisture during concrete curing [8].

Currently, textile woven fabrics (plain: 1/1) are available for fabric formwork and among these fabrics, polyethylene (PE) fabric using various commercial names and patented products is widely used because low cost, lightweight, chemically resistant and durable [8-10]. Further, PE fibre is used in concrete structure to prevent cracking. However, PE is a thermoplastic material that is inherently ductile which causes creep deformation overtime [11]. This creep tendency creates problems in concrete casting. For example, when a concrete column is cast in a tubular PE fabric formwork, due to the hydraulic pressure of the fresh concrete, the bottom section could expand significantly and the column would eventually be out of shape [12]. At the University of Manitoba, the Center for Architectural and Structural Technology (CAST) laboratory is dedicated to research on functional and artistic shapes of architecture object that can be cast from fabric formwork [13]. For fabric formwork, low cost fabric such as PE, polypropylene (PP) and polyester are preferred due to large quantity required in applications. Creep deformation is common in these fabrics and the highest for PE compared to PP and polyester and has been reported as a frequent problem in using a PE fabric for formwork as well as in geotextile applications [14-17].

In concrete casting, the fabric formwork is subject to stress over a period of time. Before the fresh concrete cured to solid form it exerts a constant hydraulic pressure on the fabric formwork [18-20]. Even the pressure, which where resisted by the tensile strength of the fabric, is not high enough to break the fabric it can cause the fabric to creep and expand the dimensions overtime [21-24].

Creep is a very important material property to consider when assessing the effectiveness of textiles as load bearing agents. The conventional stress/strain curve which is commonly used in measuring the strength of a textile fiber is inadequate in measuring creep because when a fabric is used to cast concrete, the instantaneous elastic deformation (IED) and creep together contribute to the overall dimensional change of the fabric [24]. Since IED is almost instantaneous and is inversely related to modulus (stiffness), the change is immediately noticeable and remedies can be made accordingly. Creep, however, is time-dependent and its impact on dimensional change will not show until hours later when the concrete starts to harden. Ideally, both IED and creep of a fabric should be as low as possible to minimize deformation.

When a fabric is used for concrete casts, although the pressure from the fresh concrete is much smaller than its strength, the tendency for the fabric formwork to stretch is still a troublesome phenomenon because, depending on the fabric, it can stretch at a relatively low pressure level. When it does stretch, the dimension of the cast object will deviate from the design parameters.

A potential solution to the problems associated with creep deformation and IED will be to increase the fabric modulus by reinforcing a stiff textile material. The resulting fabric will be low in production cost, stiffer and larger creep resistant, however still be flexible and permeable. A permeable woven specimen would facilitate

***Corresponding author:** Mashiur Rahman, Department of Biosystems Engineering, Faculty of Agriculture and Food Sciences, University of Manitoba, Winnipeg, Canada
E-mail: Mashiur.Rahman@umanitoba.ca

the curing of the concrete by enabling water to bleed out as the concrete hardens [14].

High tenacity technical fibres are used for engineering applications and among all technical fibers, glass fiber is the most commonly used for reinforcement purposes because of its high strength, high modulus, low deformation, minimum IED, near-zero creep, low cost and availability [15]. Traditionally, glass yarns are introduced as the reinforcement by impregnating a sheet of parallel-laid glass yarns in a plastic matrix resulting in a composite material that yields excellent mechanical performance [16]. This type of composite material is widely used to make parts for sports equipment, automobile, heavy-duty machinery, airplanes and mortars [23,24].

The objective of this research is to evaluate and compare two reinforcement methods of glass yarns on PE woven formwork fabric to reduce deformation and increase modulus. The first method was to reinforce a PE fabric by laminating straight laid glass yarn every one-half inch across its 3 inch width. The second method was to affix, by stitching, glass yarns over a PE fabric across its 3 inch width at one-half inch interval. Theoretically, this would impart the necessary rigidity to resist deformation without adding any significant weight.

Materials and Methods

The base fabric for investigation, polyethylene woven fabric (Geotex® 315 ST fabric), was obtained from Propex (Chattanooga, TN, USA). This particular geotextile fabric is used for fabric formwork in CAST lab for concrete casting for more than 10 years. The multifilament C-glass yarn was obtained from Anping Furit Wire Mesh Making Co., (China). The tex of the glass yarn is 320 and contains 200 filament.

Tensile properties of PE fabric and the glass yarns used for reinforcement

The maximum tensile capacity of the original PE fabric was determined by following the procedure set forth in ASTM D5034 [18] with one methodological modification on sample width. Five specimens were subjected to a load at machine speed of 300 mm/min until the specimens broke. To determine breaking force, an Instron 5965 tester was connected to a computer with the Bluehill 2.0 software that recorded data on force, extension and time. The maximum jaw width was 3 inches and the distance between the jaws (effective test length) was 3 inch. The average tensile properties with standard deviation of the original PE fabric are given in Table 1.

The maximum load and extension of the C-glass yarns (320 tex) that were used to reinforce the PE fabric were measured according to the procedure set forth in ASTM D2256-10 [19]. The gauge length specified in ASTM was 250 mm or 500 mm but in this research was reduced to 150 mm because the length of the composite specimens to be tested would be 150 mm in length and have glass yarn reinforcements at the same lengths. Testing the glass yarn's mechanical properties at the same length as their application requirement will better estimate the properties of the glass yarn in the composite phase. Ten specimens

Specimen type	Maximum load (N)	Extension at maximum load (mm)	Maximum strain (%)	IED at zero slope (mm)	Load at IED (N)
PE fabric	3677.1 ± 399.0	26.5 ± 5.8	16.5 ± 3.2	26.5 ± 5.8	3677.1 ± 399.0
Glass yarn	136.8 ± 22.6	4.90 ± 0.46	3.3 ± 0.3	4.9 ± 0.46	136.8 ± 22.6

Table 1: Tensile properties for original (unreinforced) PE fabric and original C-glass yarn.

of 320 tex glass yarns were tested. The average maximum breaking load, extension at break, maximum strain, instantaneous elastic deformation (IED), IED at zero slope and load at IED with their standard deviations are given in Table 1.

Measurement of fabric thickness

The thickness of PE fabric was measured using an electronic low pressure thickness meter supplied by Custom Scientific Instrument (New Jersey, USA).

Incorporating the glass yarns into the PE fabric

Two methods were used to incorporate the glass yarns into the PE fabric–by lamination and by stitching them directly onto it.

Lamination

Glass yarn reinforcement tapes were manufactured for lamination: 6 strands of 320 tex glass yarns were placed evenly on the 2 inch width between two sheets of vinyl heat & bound (Iron-on Clear Cover® manufactured by Kittrich Corp) and pressed at about 143°C for 15 seconds to activate the bounding. Then the composite materials were cooled and cut into two strips measuring 1 inch in width and 10 inches in length. Each strip contains 3 glass yarns. The two strips were then glued (Lepage Grey Pres-TITE multi-purpose stray adhesive) onto to the center section of a PE base fabric which has dimensions of 5 inches width and 10 inches length. During the creep test, the laminated samples were clamped between the jaws of the intron testeras shown in Figure 1. The stiffness of the vinyl sheets is assumed to be insignificant compared to the stiffness of the composite material.

Stitching glass yarns onto PE fabric

A piece of Geotex® 315 ST fabric (PE fabric), 5 inches wide and 10 inches long, was reinforced by stitching 6 strands of 320 tex glass yarns within a 3-inch band at its center. The glass yarns were sewn onto the PE fabric using a single needle, walking foot lockstitch machine (Brother Model number: LS2-B837) at 6 stitches per inch along the length. The stitch is classified as 301 according to ASTM Standard [20], with the glass yarns threaded from the bottom bobbin and an all-purpose polyester thread (manufactured by Coats and Clark) threaded from the top needle. The glass yarns were stitched onto the PE fabric

Figure 1: Laminated glass yarn reinforced PE fabric clamped on Instron testing machine.

one-half inch apart and appeared on the reverse side of the PE fabric. During the creep test, the stitched samples were clamped between the jaws of the intron tester as shown in Figure 2. The stitching process was carried out using an Industrial Sewing machine at K9 Storm in Winnipeg, Canada.

Creep tests

The holding load and holding time for the creep test was established according to the failure load and creep rate of the original base fabric. These two parameters were subsequently used in a series of tests to evaluate the properties of the reinforced specimens. These properties included breaking load, elongation, IED, modulus and time dependent creep of specimens.

Establishing time parameter for creep tests

After the test started, the jaws exerted tension force on the specimens and caused them to elongate. About 1,000 seconds later, the creep rates of the original and reinforced specimens stabilized to a much slower rate and continued to elongate at this rate without any fluctuation. Thus, the maximum elongation time deemed suitable for the three specimens was set at creep behavior for 3600 seconds (one hour) to capture the characteristics of time dependent creep properties.

Establishing holding load parameters for creep tests

The breaking loads, extension, IED at zero slope and Load at IED of the PE specimen and glass yarn were measured to determine the holding loads for the tension-creep test. Test results of these values are given in Table 1.

The breaking load of the PE fabric was used to establish the appropriate holding load which was the constant load of tension applied to the specimens for the time-dependent tension creep test on both the unreinforced specimen and the reinforced specimens. For this purpose, a series of holding loads, which were below the breaking load of PE (Table 1), starting from 2000 N, 1500 N, 1000 N and 500 N were applied for one hour. When subject to 2000 N and 1500 N tensile load, the glass yarns in both reinforced specimens broke quickly in the first 30 seconds indicating that they were not strong enough to sustain the

tensile loads at these two levels. When the load was reduced to 1000 N, the glass yarns in the laminated specimen (Group L-1000) were able to sustain the load for one hour without breaking. The detail sample identifications are provided in Table 2. However, the glass yarns on the stitched specimens (Group S-1000) continued to break during tests within the first 1000 seconds. When the load was reduced to 500 N, the stitched specimens were able to sustain the load for one hour without breaking. Therefore, 1000 N and 500 N loads were selected as parameters for the creep tests. Notice that the breaking strength of the glass yarn is around 137 N, thus theoretically the total strength that 6 glass yarns alone in each reinforced sample can sustain is around 822 N (Table 1).

Statistical analysis

Statistical analysis is conducted using one tailed T test with unequal variance assuming the null hypothesis to be no difference among comparing groups.

Value of t is obtained by:

$$t = \frac{\overline{X_1} - \overline{X_2}}{s_{\overline{X_1} - \overline{X_2}}}$$

$$s_{\overline{X_1} - \overline{X_2}} = \sqrt{\frac{s_1^2}{n_1} + \frac{s_2^2}{n_2}}$$

For which $\overline{X}$ is the average value of the sample group, S is the value of standard deviation and n is the number of tested samples, which in this experiment, is 5 for all groups. Degree of freedom m is calculated by:

$$1/m = C^2/(n_1-1) + (1-C)^2/(n_2-1), \text{ which } C = (S_1^2/n_1)/(S_1^2/n1 + S_2^2/n_2).$$

The hypothesis is rejected if t value is greater than the value of $t_{m, 0.95}$ from the t-distribution table. The values of tm, 0.95 is chosen from the table using the lowest m value calculated among groups to maximizing the value of tm, 0.95 and increase the difficulty to reject null hypothesis.

Results and Discussion

Evaluating the effectiveness of reinforced specimens

To evaluate the effectiveness of the two types of reinforcement, the reduction in total elongation, IED and creep, and modulus of the original PE fabric and the reinforced specimens were measured and compared. Three groups of specimens were tested: Group O represents the original PE fabric; group L represents the laminated specimens and group S represents the stitched specimens (Table 2). The specimens subjected to 1000 N and 500 N holding load were labeled O-1000, O-500, L-1000, L-500, S-1000 and S-500.

Figure 2: Stitched glass yarn reinforced PE fabric clamped on Instron testing machine.

Sample type	Reinforcement method	Applied load (N)	Samples identification
Original PE samples	N/A˙	500	O-500
Original PE samples	N/A˙	1000	O-1000
Glass yarn reinforced samples	Vinyl lamination	500	L-500
Glass yarn reinforced samples	Vinyl lamination	1000	L-1000
Glass yarn reinforced samples	Stitching	500	S-500
Glass yarn reinforced samples	Stitching	1000	S-1000

˙N/A: not applicable

Table 2: Reinforcement methods and sample identification.

Elongation properties

All three groups of specimens with two load parameters were tested to measure elongation, strain (%) and reduction in strain (%). Table 3 shows that under a 1000 N holding load, the laminated specimens showed an elongation of 11.41 mm which was lower than the 12.85 mm elongation for the stitched specimens and the 14.13 mm elongation for the original specimen. Furthermore, the elongation of the laminated specimens reduced by 2.72 mm, which was a 19.25% strain reduction from the original specimen at the same load. The elongation of the stitched specimens was reduced by 1.28 mm, which was a 9.06% strain reduction from the original specimen for the same load. The reduction on the total length of elongation was more in the laminated specimens than in the stitched specimens at 1000 N. However, under statistical analysis ($p \leq 0.05$), the differences between groups under 1000 N loading condition were not significant.

Under holding load of 500 N (Table 3), the reduction in elongation and strain (%) were 3.65 mm and 41.76% respectively for the laminated specimens and 5.90 mm and 67.51% for the stitched specimens. It seems that the impact of reinforcement in elongation reduction was more effective under 500 N than under 1000 N. Furthermore, the stitched specimens showed a larger reduction in elongation and strain (%) than the laminated specimens. A plausible explanation is that the stitched specimens were locked in position at every stitch by the top thread leaving the material very little freedom to slip, enabling the glass yarns to prevent the specimen from stretching. Thus, the stitched specimens at 500 N produced a higher elongation reduction of 5.90 mm compared to the laminated specimens at the same load (3.65 mm). The statistical tests comparing the means of elongations among 3 groups ($p \leq 0.05$) showed that group L-500 had significant lower elongation than group O-500, group S-500 had significantly lower elongation than both group O-500 and L-500.

During the tension creep test under 1000 N, the stitched specimens were not able to sustain the tension and eventually broke. Although they all broke within 500 seconds from the start of the test, some broke suddenly and some broke more slowly. The breaking processes of glass yarns were shown in Figure 3. The step-like segments at the beginning (left side) of the time and elongation curve of the S-1000 specimen were caused by the sudden break of some glass yarns in the stitched specimen. Each sudden break released some elongation instantly when the resistance to stretch provided by the glass yarns was diminished and shifted the curve upwards. This phenomenon did not exist in the unreinforced PE fabric (Sample ID: O-500 and O-1000) and the laminated specimens at 1000 N (Sample ID: L-1000) or in all other specimens subjected to 500 N load (Sample ID: S-500 and L-500).

The laminated specimens at 1000 N showed a smooth curved elongation-time relationship (Figure 3) and the elongation was always lower than that for the original and stitched specimens at the same load. This is because the glass yarns were better able to sustain the 1000 N holding load and limited the fabric from stretching. Compared

to the breaking test results for glass yarns (Table 1), the average total elongation of the laminated specimens at 1000 N (11.41 mm: Table 3) was much higher than the average maximum elongation of the glass yarns themselves (4.9 mm). The maximum load that the 6 glass yarns could sustain was approximately 822 N.

When pulled by the 1000 N load, the PE fabric was much less stiff than the glass yarns, all tension was primarily exerted on the glass yarns and they should have broken at their maximum capacity of 822 N or their breaking elongation around 4.9 mm. However, the glass yarns in the laminated specimens survived the tension exerted over the duration of the test and reached an elongation of 11.41 mm (Table 3), which was more than twice its maximum elongation. This indicates the likelihood that the glass yarns were slipping under the laminated coating to accommodate the stress and thus avoid being broken. Visual inspection of the laminated samples after testing revealed the formation of crimps by the glass yarns under the coating as shown in Figure 4. When the applied stress is higher than the combined strength of reinforced yarns, the laminated glass yarns can slip without breaking. A similar crimp was developed under 500 N for L-500 samples, but the severity was much less.

Thus, the fact that glass yarns in lamination reinforced specimens could slip under the coating to avoid breakage under overloading gives an advantage that overloading are less likely to cause a dramatic failure in fabric formwork when using a glass yarn lamination reinforced fabric. The disadvantage of such a textile is that it has less creep resistance than the stitched glass yarn reinforcement. On the other hand, the advantage of the glass yarn reinforced textile by stitching method is that if the loading condition is within the glass yarn's breaking load, the reinforcement provides much better creep resistance than the laminated textiles. However, its disadvantage was when overloading happens; there is no mechanism to prevent dramatic failure.

The data in Table 3 show that at both holding loads, the original specimens had the largest average elongation after one hour compared to both reinforced specimens under the same load. The total elongation for Group O-1000 and O-500 was 14.13 mm and 8.74 mm respectively (Table 4).

Instantaneous elastic deformation (IED)

The amount of elongation or extension from start to the yielding point is referred as Instantaneous Elastic Deformation (IED). Figure 5 provides the graphic illustration of and Table 5 shows the IED and yielding point at zero, 80%, 60% and 40% slope thresholds.

In cases where a fibre is plastic in nature, for some samples the zero slope moment may not be detected by the BlueHill® (Instron software). However, the undetected zero-slope moment was estimated by observing the yielding points at 80%, 60% and 40% slope thresholds as shown in Table 4. Subsequently, the IED was approximated from those elongations. For example, in the current research, the zero-slope point could not be detected for some test specimens from the Group

Sample group	Elongation (mm)	Total elongation reduction (mm)	Total strain (%)	% of reduction strain	Initial modulus (MPa)	IED at zero slope (mm)	Creep (mm)
O-1000	14.13 ± 3.64	0	9.42	0	586.85 ± 4.27	8.67 ± 0.34	5.46 ± 3.99
O-500	8.74 ± 2.34	0	5.83	0	532.47 ± 4.77	5.98 ± 0.22	2.76 ± 2.11
L-1000	11.4 1 ± 2.94	2.72	7.61	19.25	1263.52 ± 9.18	3.85 ± 0.15	7.56 ± 3.1
L-500	5.09 ± 1.91	3.65	3.39	41.76	1121.04 ± 5.97	1.98 ± 0.11	3.11 ± 2.02
S-1000	12.85 ± 3.31	1.28	8.57	9.06	903.03 ± 6.56	5.81 ± 0.23	7.04 ± 3.55
S-500	2.84 ± 1.23	5.9	1.89	67.51	898.24 ± 3.26	2.31 ± 0.15	0.53 ± 1.05

Table 3: Changes in mechanical properties under 500 N and 1000 N holding loading condition after 1 hour.

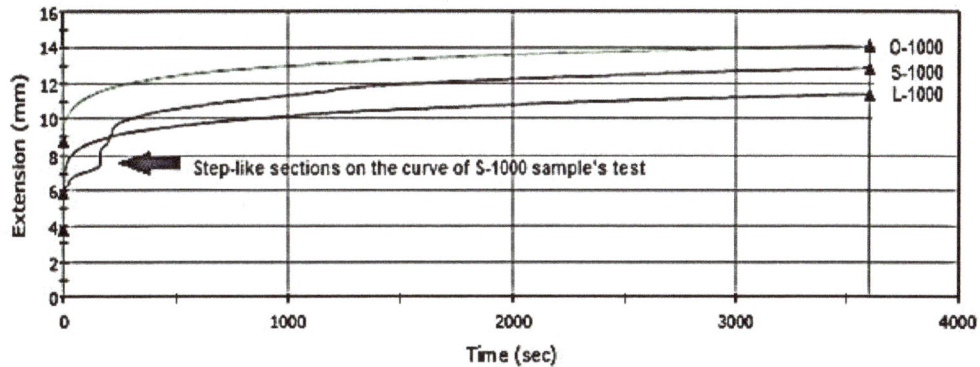

Figure 3: Extension-time curve under 1000 N loading condition tests (for the demonstration purpose curves are selected individual tests which show median results of total elongation in the group to avoid having many curves showing on one graph).

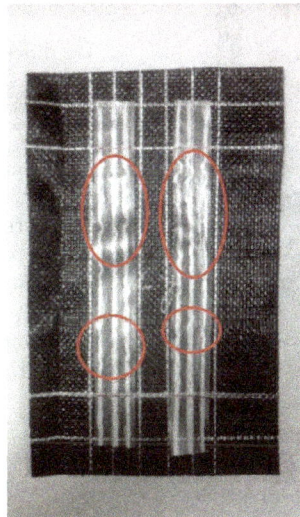

Figure 4: Glass yarns are crimped during the test under 1000 N loading condition (Crimping is marked by the red circles).

Sample ID	IED (mm)	Extension at yield (mm) (zero slope)	Load at yield (N)	Extension at yield (mm) (slope 80%)	Extension at yield (mm) (slope 60%)	Extension at yield (mm) (slope 40%)
O-1000	8.67	8.78	1026.11	8.71	8.71	8.8
L-1000	3.85	3.85	1041.29	3.86	3.96	3.96
S-1000	5.81	5.93	1026.67	2.97	2.97	2.97
O-500	5.98	5.98	499.97	5.97	5.97	6.06
L-500	1.98	Some value not detected	500.77	0.61	1.98	1.98
S-500	2.31	Some value not detected	501.16	0.72	2.31	2.31

Table 4: Yielding elongation at zero slope, 80% maximum slope, 60% maximum slope and 40% maximum slope and load at the yielding points.

S-500. Consequently, the average yielding elongation at zero-slope was calculated by observing the average elongations at 80% (0.72 mm), 60% (2.31 mm) and 40% (2.31 mm) slope thresholds. Since the average elongations at 60% and 40% slope thresholds were identical, the decrease in rate of stretch from 60% to 40% slope threshold occurred at 2.31 mm elongation. Consequently, for the S-500 specimens, the approximate IED is at 2.31 mm elongation at zero-slope (Table 4).

However, Table 4 shows that the load experienced at the yielding elongation is close to the tests' pre-set maximum loads of 1000 N and 500 N. This indicates that the zero-slope was more likely to have been caused by the stopping of the jaws when the pre-set holding loads were

reached rather than the sudden increase of the resistance from the specimens when their initial elasticity was exhausted. In fact, the PE original fabric is a thermal plastic material, thus the boundary between elastic deformation and creep does not exist. It was reported that elastic and plastic deformations happen simultaneously for PE fabrics under stress [21]. Thus, the IEDs obtained in this research from the breaking test (Table 1) and creep-tension tests (Table 3) were not strictly elastic deformation; it may have contained some initial creep. Nevertheless, these values reflect the immediate response of the specimens when tensile force was applied, particularly in the fabric form concrete.

Table 4 also shows that all the original specimens produced the highest IEDs when compared to other groups under the same loads (1000 N: 8.67 mm; 500 N: 5.98 mm) and all the laminated specimens had the lowest IEDs under both loads (1000 N: 3.85 mm and 500 N: 1.98 mm). All the stitched specimens had medium IED values (1000 N: 5.81 mm and 500 N:2.31 mm).

Changes in creep property

The creep results for the three groups of specimens under 1000 N and under 500 N holding loads are given in Table 3. The laminated specimens under both holding loads produced the highest creep values

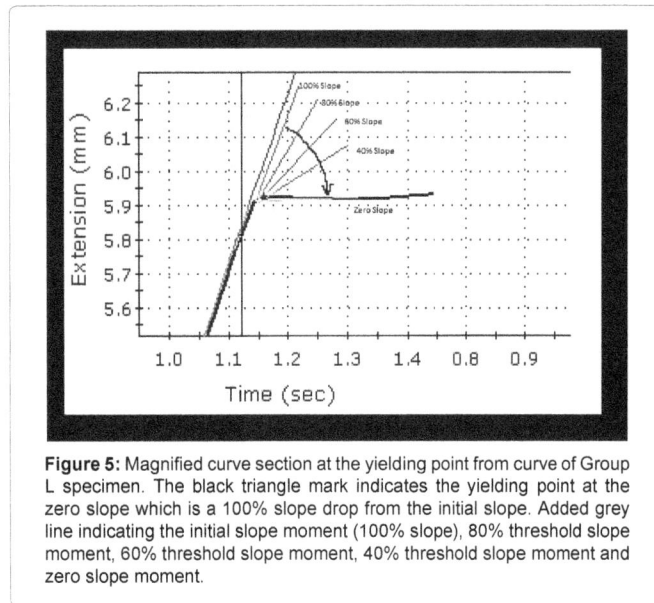

Figure 5: Magnified curve section at the yielding point from curve of Group L specimen. The black triangle mark indicates the yielding point at the zero slope which is a 100% slope drop from the initial slope. Added grey line indicating the initial slope moment (100% slope), 80% threshold slope moment, 60% threshold slope moment, 40% threshold slope moment and zero slope moment.

of 7.7 mm (L-1000 N) and 3.11 mm (L -500 N) respectively. The original specimens under both holding loads produced an average creep of 5.5 mm and 2.76 mm. The stitched specimens under both loads generated 7.1 mm (S-1000 N) and 0.53 mm (S-500 N) respectively.

In the 500 N holding loading condition, the stitched specimens showed more than 80% reduction in creep compared to the original specimen. However, under the 1000 N loading condition the creep of stitched specimens were only 30% higher than that for the original specimens. In contrast, for laminated specimens at 1000 N and 500 N loading conditions, the creep value was increased by 40% and 13% higher than the corresponding original specimens.

Under statistical analysis ($p \leq 0.05$), the differences between groups were not significant for all groups under 1000 N holding loading condition mainly due to yarn slippage in laminated sample and yarn breakage among stitch reinforced sample. There was also not statistical significance among L-500 and O-500 also due to the yarn slippage. However, the significance was found between S-500 and O-500 and S-500 and L-500 specimens.

The differential creep behaviors of glass yarn reinforced PE fabric by stitching method under two holding loading conditions can be explained by the fact that glass yarns broke at 1000 N. When the glass yarns broke, all the stress was transferred to the base fabric and the resistance to stretch provided from the Group S-1000 specimens was no longer better than the Group O-1000 specimens. Additionally, before the glass yarns were broken, the Group S-1000 specimens had lower IED compared to Group O-1000 specimens (IED of Group O-1000: 8.67; IED of Group S-1000 specimens: 5.81; Table 3) and after the glass yarns were broken, the reduced IED that was held by the glass yarns is released to increase the creep. This resulted in the higher creeping rates for S-1000. Under the 500 N holding loading condition, the glass yarns

Glass yarn parameters and equation #		PE fabric parameters and equation #		PE + glass composite and equation #	
Tex	320	Fabric volume width (m)	0.075	Modulus, (*Ec*–upper bound), Eq. # 1	2.089 GPa
Density (kg/m³)	2460	Fabric length (m)	0.15	Modulus, (*Ec*, lower bound), Eq. # 2	560 MPa
Crosssectional area (m2), Eq. # 4	0.00000013	Fabric thickness (m)	0.0002	Modulus (experimental)	898.2 Mpa (From Table 3)
Strain, Eq. # 5	0.033	Fabric volume (m3)	22.5 E-7		
Stress (MPa)	1052	Vt, Eq. #6	0.9506		
Modulus (GPa)	32				
Volume of 6 glass yarn (m³)	0.000000117				
Vg, Eq. # 7	0.0494				

Table 5: Experimental and theoretical (calculated using Equations 1 and 2) modulus using series and parallel models.

Number of yarns per 3 inch width fabric	Volume of yarns (10⁻⁸ m³)	Volume of the base fabric (10⁻⁸ m³)	Relative volume of glass (% Vg)	Relative volume of the base fabric (% Vt)	Predicted resultant modulus (MPa)
5	9.8	225	0.042	0.958	867.2
6	**11.7**	**225**	**0.049**	**0.951**	*898.7 (898.2)
7	13.7	225	0.057	0.943	926.0
8	15.6	225	0.065	0.935	950.1
9	17.6	225	0.072	0.928	971.9
10	19.5	225	0.080	0.920	991.8
11	21.5	225	0.087	0.913	1010.2
12	23.4	225	0.094	0.906	1027.3
13	25.4	225	0.101	0.899	1043.5
14	27.3	225	0.108	0.892	1058.9
15	29.3	225	0.115	0.885	1073.5

*Calculated modulus in the parenthesis

Table 6: Resultant modulus of glass yarn reinforced composite textile using stitching method (holding load: 500 N).

were able to hold the fabric all the way through the end of the test (3600 sec), and thus limit the creep to a very small amount.

Under both holding loading conditions, the laminated specimens produced larger creep rates than the original specimens. As discussed before, the glass yarns in the laminated specimens had some freedom to slip during the test. Since the laminated specimens always had the lowest IEDs among the groups in the same loading conditions, the remaining unstretched portion of IED was released when the glass yarns started to slip. On the other hand, the original

specimens at 1000 N and 500 N always had the largest IEDs that exhausted a large portion of the total elongation and made the specimen more stable during the creep elongation than Group L-1000 and L-500. This observation of reduced creep rates due to larger IEDs can be made useful in some fabric formwork applications where if it is possible to stretch the fabric before using it may help to stabilize the fabric.

Initial modulus

The initial modulus results for all 6 groups of samples are shown in Table 3. Group O-1000 and O-500 samples produced the lowest average modulus of 586.85 MPa and 532.47 MPa respectively; Group L-1000 and L-500 specimens produced the highest modulus at 1263.52 MPa and 1121.04 MPa respectively; and Group S-1000 and S-500 is in the middle at 903.03 MPa for 1000 N and 898.24 MPa for 500 N loading conditions. Under statistical analysis ($p \leq 0.05$), the differences between groups were significant. These results suggest that reinforcement by stitching (Group S-1000 and S-500) and by lamination (Group L-1000 and L-500) increased the modulus of the specimens. Moreover, modulus increases in the laminated specimens were more than double than that of the original PE fabric and the modulus increase in stitch-reinforced specimens was about 50% more than the original (Table 3).

Generally, if a textile has a higher initial modulus, its stiffness is thus higher. When subject to a tensile load, the elongation is expected to be lower. It was not the case in this experiment under the 500 N loading condition (under 1000 N these parameters were not compared as the glass yarns from S-1000 samples were broken during the test). The Group S-500 specimens produced an average initial modulus at 898.24 MPa lower than the L-500 specimens' average initial modulus at 1121.04 MPa but the average total elongation was 2.84 mm lower than the average elongation at 5.09 mm from Group L-500 specimens (Table 4). This indicates that the Group S-500 specimens were less stiffer than the Group L-500 specimens initially during the IED portion of the elongation, then when the elongation proceed to creep, the stiffness of Group L-500 specimens were lowered and produced a higher overall elongation (IED + Creep).

The reason of the differences and changes in the modulus during IED and creep between these two groups of reinforced specimens could be attributed to the configuration of glass yarns in each reinforced specimen. The glass yarns in Group L-500 specimens were pulled straight, laid flat and laminated while in Group S-500, the glass yarns were not laid as straight as in the lamination method because the during the stitching process, the top polyester thread grabs the glass yarn from the bottom bobbin while it was naturally hanging. When glass yarns were sewn onto the PE base fabric there was some crimp introduced to each stitching. Thus at the beginning of the test, the S-500 specimens had more to elongate than group L-500 specimens which resulted in a lower modulus and higher IEDs. During the tests, because the stitched glass yarns in S-500 specimens were more rigidly bound to the PE base fabrics than the glass yarns in L-500 specimens, as soon as the crimps in S-500 specimens were straightened, the specimens became stiffer

than Group L-500 specimens which limited the growth of the creep (which is discussed in the later section) during the rest of the test and eventually resulted in a lower total elongation for S-500 specimens.

The comparison of total elongation, initial modulus, IED and creep for group L-1000 and S-1000 is not discussed because the glass yarns in S-1000 specimens were broken during testing. The elongation property has been improved for both reinforced samples for all loading conditions in the increasing order of S-500 > L-500 > O-500 ($p > 95\%$). As mentioned before even though L-500 had higher modulus and lower IED, the creep development due to yarn slippage underneath the coating caused the total elongation to be higher and thus was out performed by S-500.

Use of Hirsch's model to predict modulus

To reduce the dimension variation in formwork installation, the IED of the formwork can be offset by shortening the dimensions of the fabric used accordingly. The IED can be estimated by predicting the value of initial modulus (Modulus × applied stress = strain, which is IED).

The resultant modulus of a composite material spans a certain range depending on the mechanical interaction between the combined materials. At the upper bound of the range the composite has the highest modulus achieved by perfect bounding which distribute strain evenly within the material. At the lower bound, the composite has the lowest modulus due to complete lack of bounding and strains are developed unevenly under even stress.

The upper bound and lower bound of the theoretical modulus of the composite can be predicted using series and parallel models for two-phase composite materials from the following Equations 1 and 2 [22].

$Ec = Eg \times Vg + Et \times Vt$ → Equation 1 (upper bound assuming perfect bounding, parallel model assuming equal strain), where Ec was the modulus of the composite, Eg is the modulus of the C-glass yarn, Vg is the relative volume of the C-glass yarn in the specimen, Et was the modulus of the textile base, and Vt was the relative volume of the textile base of the specimen.

$Ec = Eg \times Et / (Et \times Vg + Eg \times Vt)$ → Equation 2 (lower bound assuming nonbonding, series model assuming equal stress), where Ec was the modulus of the composite, Eg was the modulus of the C-glass yarn, Et was the modulus of the textile base, Vt was the relative volume of the textile base of the specimen and Vg was the relative volume of the C-glass yarn in the specimen.

Furthermore, the following equations were used to calculate the glass yarn parameters, PE fabric parameter and the composite parameters as shown in Table 5.

The total volume of the composite is: $Vt + Vg = 1$ → Equation 3

Cross Section Area (A) = volume / length → Equation 4

Strain = IED / original length → Equation 5

$Vt = V$ base / V composite → Equation 6 (Vt which is the relative volume of the textile base of the specimen i.e., PE fabric)

$Vg = V$ glass / V composite → Equation 7 (Vg which is the relative volume of the glass)

Thus, the upper and lower bound of modulus can be calculated according to the formula to be 2.09 GPa and 560 MPa which was calculated using the equations 1 and 2, which are given in Table 6. The measured average modulus of the specimens from glass yarn reinforcement by stitching method under 500 N loading conditions was at 898.2 MPa (Table 3) and was between the upper and lower bound of the theoretical modulus.

A Hirsch's model [22] for the modulus of the composite which calculates the modulus from a combined effect of parallel model and series model can be established from the equation 8:

$1/Ec = X (1/(Vg * Eg + Vt * Et)\ Parallel\ model +$

$(1-X)\left(\frac{V_g}{E_g}+\frac{V_l}{E_l}\right)$ series model   ----   Equation 8

Where X represent the portion of the effects of parallel model and was calculated to be 0.515 from the tested Ec at 898.2 MPa (measured modulus of PE + glass yarn composite, Table 3).

Therefore, according to the model, resultant modulus of the composite with additional glass yarns stitched across the width of the PE fabric was computed according to the volumetric ratio. Table 6 lists the computed modulus for the glass yarn reinforced composite using stitching method for 500 N holding load. It is worth mentioning that the predicted and measured modulus for PE + six glass yarn composite is very close, 898.7 MPa and 898.2 MPa respectively (Table 6).

Application implications

From the breaking test of glass yarns, the average breaking strain of the yarns was found to be at 3.3% (average extension at maximum load/original length at 150 mm). If the PE fabric were reinforced with glass yarns at 6 lines of stiches per 150 mm, the composite modulus predicted was 898.7 MPa (Table 6). Thus, the maximum stress that the reinforced fabric can sustain without breaking the glass yarn reinforcements was calculated at 898.7 MPa × 0.033 = 30 MPa. If also assume a safety factor of 2, the fabric is designed to take 15 MPa stress which can be the maximum allowable *Hoop Stress* (HS) at the bottom of the fabric formwork. Using the liquid pressure of fresh concrete and base fabric thickness, one can calculate the maximum height and radius of the fabric formwork from the following Equation 9:

$Hoop\ Stressmax_{max} = \rho gh \times R_{max} / t$ → Equation 9.

Where ρ is the density of the concrete (2400 kg/m³), g is the gravity (9.806 m/s²), h is the height of the column or fabric formwork (m), R is the radius of the fabric formwork (m) and t is the thickness of the base fabric (t = 0.0002 m). If a 3 meters high column (fabric formwork) was designed, the hydraulic pressure from the liquid concrete at the bottom section is 0.071 MPa (ρgh). By solving the Equation 9 we can obtain the allowable column radius R_{max} = 0.042 m. Therefore, the Equation 9 can be used to calculate for different diameter and height of the fabric formwork as shown in Table 7.

If the column dimensions are specified with height and diameter, for example, a column with 2 meter height and 0.15 meter diameter (0.075 meter radius) then the required stiffness of the base fabric can be calculated as [(2400 kg/m³ × 9.806 m/s² × 2 m × 0.075 m / 0.0002 m) × 2(safety factor)]/0.033 (maximum allowable strain) = 1070 MPa. Thus the required amount of glass yarn reinforcement per 3 inch width is 15 yarns as indicated in Table 5 (modulus at 1073.5 MPa).

Conclusions

The elongation property has been improved for both reinforced samples for all holding loading conditions in the increasing order of S-500 > L-500 > L-1000 > S-1000 (p > 95%). All reinforced samples showed an improvement in modulus for both holding loading conditions.

Concrete density (kg/m³)	Fabric thickness (m)	Column height (m)	Column radius (m)
2400	0.0002	3.0	0.042
2400	0.0002	2.0	0.064
2400	0.0002	1.0	0.127

Table 7: Diameter and height of the fabric formwork (Equation 9) (2 glass yarns per inch or 6 for 150 mm).

However, it was observed that the laminated samples produced the stiffer composite due to the straight configuration of glass yarns in the composite fabric. As a result, the modulus of L-500 and L-1000 samples was much higher than the stitch reinforced samples. A significant reduction in IED was noticed for both reinforced samples for all holding loading conditions. This IED value can be considered during the dimension calculation of fabric formwork. The creep has been undesirably increase for both laminated samples and reinforced S-1000 samples due to the slippage and breakage of glass yarns respectively. However, for the S-500 sample, a significant reduction in creep was obtained as the glass yarns' integrity was maintained. It is, therefore recommended that in order to reduce the time dependent creep during fabric formwork, the load should not be exceed the combined breaking load of all reinforced glass yarns. Between these two reinforcement methods, lamination allows more creep but can resist breakage and rupture while the stitching method reduces creep more effectively as long as the loading condition is within the capacity of the glass yarns. Therefore, it can be concluded that stitch method is better than the lamination method. Consequently, for stitch PE + glass yarn composite and 500 N holding load, the Hirsch's model was used to predict the composite modulus from which the dimension of fabric formwork can be calculated. In addition, glass yarn reinforced fabric through stitching method is more cost efficient, has better drape and better permeability than glass yarn reinforced fabric through lamination method.

In general condition, the mechanical properties such as tensile strength, elasticity, shearing resistance and creep rate of a textile woven fabric also depend on the direction of the applied force related to the direction of the weft and warp yarns [25]. The orthotropic property of a conventional textile material is that the strength and the stiffness are at maximum along its weft and warp direction while weaker along off-axes directions [26,27]. Study also showed that maximum stiffness and creep resistance are achieved when a reinforcing material is composited with a woven base fabric according to the orthotropic configuration of the warp and weft because at off-axes direction the low shearing resistance of the woven fabric induces higher plastic deformation [28]. In addition, composite materials have better de-bonding resistance when reinforcement is along the warp or weft direction. In the current study, the reinforcement of glass yarn was conducted in the warp direction, which is justifiable due to the end use of this particular PE + glass composite fabric. The reinforcement may be in the weft direction or in both warp and weft directions depending on the application of the composite fabrics.

Acknowledgement

The first author thanks C.A.S.T. lab at the University of Manitoba for the training, learning and inspiring. Also thanks to K9 Storm Inc., Winnipeg, Manitoba, Canada for the help of making glass yarn stitched PE fabric.

Conflict of Interest

We declare that we have no financial and personal relationships with other people or organizations that can inappropriately influence our work, there is no professional or other personal interest of any nature or kind in any product, service and/or company that could be construed as influencing the position presented in, or the review of, the manuscript entitled "*Mechanical Property Measurement and Prediction Using Hirsch's Model for Glass Yarn Reinforced Polyethylene Composite Fabric Formwork*".

References

1. Abdelgader H, West M, Górski J (2008) State-of-the-art report on fabric formwork. ICBT 8: 93-106.

2. Orr JJ, Darby AP, Ibell TJ, Evernden MC, Otlet M (2011) Concrete structures using fabric formwork. The Structural Engineer 89: 20-26.

3. Lamberton BA (1989) Fabric forms for concrete. Concrete International 11: 59-67.

4. Chandler A (2015) Fabric formwork- prototype or typology. The Journal of Architecture 20: 420-429.

5. Cauberg N, Tysmans T, Adriaenssens S, Wastiels J, Mollaert M, et al. (2012) Shell Elements of Textile Reinforced Concrete Using Fabric Formwork: A Case Study. Advances in Structural Engineering 15: 677-690.

6. Quinn B (2010) Textiles Futures, Fashion, Design and Technology. Berg, New york, USA.

7. Ghaib MA, Gorski J (2001) Mechanical properties of concrete cast in fabric formworks. Cement and Concrete Research 31: 1459-1465.

8. Jha KN (2012) Formwork for Concrete Structures. Tata McGraw Hill.

9. Peurifoy RL, Oberlender GD (2011) Formwork for Concrete Structures (4th edn.). McGraw Hill, New York, USA.

10. Rankilor PR (2000) Textiles in civil engineering. Woodhead Publishing, UK.

11. Hatch KL (2006) olefin fibers. Textile Science Apex: Tailored Text Custom Publishing.

12. Fearn R (2012) Personal communication.

13. West M (2007) Fabric Architecture. CAST laboratory.

14. Mukhopadhyay SK (1993) High performance fibers-glass fiber. Textile Progress 25: 36-45.

15. Barbero EJ (2011) Materials Introduction to composite materials design. Taylor and Francis group, Boca Raton.

16. Hoedt DG (1986) Creep and relaxation of geotextile fabrics. Geotextiles and Geomembranes 4: 83-92.

17. ASTM D 5034-08 (2008) Standard test methods for Breaking Strength and Elongation of Textile Fabrics (Grab Test). Annual Book of ASTM Standards. Section Seven-Textiles 07: 239-246.

18. ASTM D 2256-10 (2010) Standard Test Method for Tensile Properties of Yarns by the Single-Strand Method, Annual Book of ASTM Standards. Section Seven-Textiles 7: 510-523.

19. ASTM D6173 (2008) Standard Practice for Stitches and Seam, Annual Book of ASTM Standard.

20. Cheng JJ, Polak MA, Penlidis A (2011) Influence of micromolecular structure on environmental stress cracking resistance of high density polyethylene. Tunnelling and Underground Space Technology 26: 582-593.

21. Ahmed S, Jones FR (1990) A reviews of particulate reinforcement theories for polymer composites. Journal of Materials Science 25: 4933-4942.

22. Barcikowski M, Krolikowski W (2013) Effect of resin modification on the impact strength of glass-polyester composites. Polimery 58: 450-460.

23. Ghernouti Y, Rabehi B (2012) Strength and Durability of Mortar Made with Plastics Bag Waste (MPBW). International Journal of Concrete Structures and Materials 6: 145-153.

24. Zouari R, Amar SB, Dogui A (2014) Experimental analysis and orthotropic hyperelastic modelling of textile woven fabric. Journal of Engineered Fabrics and Fibers (JEFF) 9: 91-98.

25. Quaglini V, Corazza C, Poggi C (2008) Experimental characterization of orthotropic technical textiles under uniaxial and biaxial loading. Composites: Part A Applied Science and Manufacturing 39: 1331-1342.

26. Zouari R, Amar SB, Dogui A (2010) Experimental and numerical analyses of fabric off-axes tensile test. Journal of the Textile Institute 101: 58-68.

27. Verchery G, Gong XJ (1998) pure tension with off-axis tests for orthotropic laminates, France.

28. Jin F, Chen H, Zhao L, Fan H, Cai C, et al. (2013) Failure mechanisms of sandwich composites with orthotropic integrated woven corrugated cores. Experiments Composite Structures 98: 53-58.

29

A Study on the Effects of Pre-treatment in Dyeing Properties of Cotton Fabric and Impact on the Environment

Asaduzzaman[1,2], Miah MR[1,2], Hossain F[1], Li X[1], Zakaria[1] and Quan H[1,2]*

[1]*School of Chemistry and Chemical Engineering, Wuhan Textile University, Wuhan, Hubei, PR China*
[2]*Wuhan Textile University Graduate (Color Root) Workstation, Songzi, 315600, Wuhan, Hubei, PR China*

Abstract

Pre-treatment of cotton fabric prior to dyeing mainly involves a combined process consisting of scouring and peroxide bleaching. In this study main focus on to find out the major problem facing during dyeing of cotton fabrics. Pretreatment process has a greater impact on whiteness and dyeing properties of fabrics and also on environment. There are two process of pretreatment which is alkaline scouring and bleaching process and another is enzymatic scouring and bleaching process, between this two processes comparison and also observation its effects on whiteness of fabric, dyeing also impact on environment. It is also observe that the 3 gL^{-1} hydrogen peroxide and 2 gL^{-1} sodium hydroxide give the good result on fabric whiteness with low environmental impact. Different pre-treated sample of cotton fabric dyed with reactive dye. The result obtain from dyed samples the combine pre-treatment by enzymatic scouring and bleaching gives good alternative of alkaline scouring bleaching process. Since it produce low BOD, COD and TOC impact on ecological factor. Furthermore it is also added advantages that it produces same whiteness and lower hydrolysation of dyestuff and good exhaustion.

Keywords: Cotton; Pretreatment; Enzymatic scouring; Bleaching; Hydrogen per oxide; Whiteness index; Dyeing

Introduction

Cotton is the oldest and the most important of the textile fibers. It has been used in the East and Middle East for thousands of years and was found in use in America when the continent was discovered. In fact, Cotton is called King Cotton, because of its versatility of its use and certain of its properties. Everyday cotton is the work's predominant textile fiber accounting 50 million tons per year and will maintain the leading position into the next century. We are using more cotton fabrics for our daily life survives in the world. For using of cotton fabric we have design and colored with different processes. Pretreatment is a process for prior fabric dyeing which involves scouring and bleaching. The object of scouring and bleaching is to produce white fabrics by destroying the colorings matter with the help of bleaching agents with minimum degradation of the fiber [1,2]. One of the important steps is scouring , in which the complete or partial non-cellulosic compounds found in cotton are removed, as well as impurities such as machinery and size lubricants [3,4]. The bleaching agents either oxidize or reduce the coloring matter which is washed out and whiteness thus obtained is permanent nature [5]. Now a days, highly alkaline chemicals, such as sodium hydroxide, are usually used for scouring in textile finished fabric production. These chemicals not only remove the impurities, but also attack the cellulose, leading to a reduction in strength, fabric weight loss [6,7]. So, the resulting waste water has a high COD (Chemical oxygen demand), BOD (biological oxygen demand), TDS (total dissolve solid) and higher content of salt. Now a day's textile producers have the possibility of applying a new, effective alternative to chemical scouring, with the enzyme and various other combined processes which are completely meet the requirements of scouring and bleaching [8,9].

The aim of this research work was to investigate the influence of pre-treatment for textile dyeing properties and also effects on the environment. There is a process of pretreatment which includes scouring and bleaching process. We have tried to find out the environment friendly pretreatment process which is based on enzyme. Enzymes have lowest impact on the environment and good results for pre-treatment

process. Reactive dyes form a covalent bond between the dye molecules and the –OH groups of cotton fiber in the dying process. This leads to favorable properties, such as wash fastness. Furthermore, unfix dye reacts with water to form hydrolyzed or oxy-dye that has lost its bonding capacity, and thus cannot be reused [10,11]. The dyes for this research were selected from Novacron series reactive dyes by Huntsman. These dyes had been developed for use in more environmentally friendly dyeing methods due to the fact that the dying process performs by the low salt. The colors of the bleach and dyed samples were determined using CIELAB 1976 color values and color different equations [12].

Experimental

Materials

A 100% cotton interlock knitted fabric 1 x 1(200GSM) was used in this experiment. The interlock knitted fabric that we used as a substrate in this research was produced from combed cotton yarn on a special interlock circular knitting machine. Fabric was supplied by Concept Knitting Ltd, Tongi, Gazipur, Bangladesh.

Chemicals and colorants

Caustic soda flakes (98.5%) and hydrogen peroxide (35%) were collected from ASM chemical Ind. Ltd, Gazipur, Bangladesh, Detergent and wetting agent, Sequestering agent, leveling agent were supplied by Huntsman (BD) Ltd. Glacial acetic acid (99.99%) and Soda ash (Sodium carbonate-99.2%) were supplied by Trade Asia Int'l Pte Ltd.

---

*Corresponding author: Heng Quan, School of Chemistry and Chemical Engineering, Wuhan Textile University, Wuhan, Hubei, PR China
E-mail: quanheng2002@163.com

Novacron Super Black G and Novacron Navy WB were supplied by Taha color Int'l Ltd.

Equipment

Laboratory scale knit fabric exhaust dyeing machine DATACOLOR AHIBA IR Pro were used for pre-treatment and dyeing of fabric. Spectrophotometer DATACOLOR 650 was used for measuring whiteness index of bleached fabric and also K/S value of dyed fabric.

Recipe of processing

The gray fabrics are pre-treated and dyed according to the following recipe which was presented in Tables 1 and 2.

Methods

Pre-treatment

Interlock fabric has a technical face of plain fabric on both sides, but its smooth surface cannot be stretched out to reveal the reverse meshed loop wale's because the wale's on each side are exactly opposite to each other and are locked together [13]. The fabric was pretreated with the liquor ratio of 1:8 using two different procedures for simultaneous scouring and bleaching: modified alkali pre-treatment and enzymatic pre-treatment. The modified alkali pre-treatment and bleaching started at 50 °C in the presence of a 0.4% wetting agent (Invatex CRA) and 0.1% sequestering agent (Invatex CS). The bath was circulated for 5 min, after which 6% sodium hydroxide and a peroxide stabilizer (Clarite CBB) were added. After a further 5 min circulation, 3 gL^{-1} hydrogen peroxide was added, then the bath temperature was raised to 98 °C at a gradient 3 °C/min and maintained for 60 min. The enzymatic pre-treatment and bleaching started at 30°C with an addition of 1 ml/l of a commercial synergistic mixture of enzymes containing pectinases, hemicellulases and cellulases (Baylase EVO), and 1 ml/l of a bleaching stabilizer (Clarite CBB). The bath was then heated to 98 °C at a gradient 3 °C/min. At 75 °C 1 ml/l of the bleaching Stabilizer (Clarite CBB), 1 ml/l of a detergent and wetting agent (Invatex CRA), 2% (owf) sodium hydroxide and 5% hydrogen peroxide were added. At 98 °C the bleaching process continued for 60 min. Both pre-treatment procedures were completed with neutralization and rinse.

Process Chemicals	Function of the chemicals	Amount (gL^{-1})
Invatex CRA	Detergent & wetting agent	0.6
Invatex CS	Sequestering agent	1
Clarite CBB	Peroxide Stabilizer	0.5
Hydrogen peroxide (35%)	Bleaching agent	3
Caustic soda	Scouring agent	2
Invatex PC	Peroxide killer	0.3
Baylase EVO	Enzyme	0.4%

Table 1: Pre-treatment Recipe.

Chemical\Dyes name	Function of the Chemicals/Dyes	Amount (gL^{-1})
Novacron Navy WB	Reactive dye	2%
Glauber salt	Electrolyte	40
Soda Ash	Fixing agent	14
Invadine LUN	Wetting agent	0.5
Invatex CS	Sequestering agent	0.5
Albatex DBC	Leveling agent	0.5
Glacial Acetic acid	Neutralizer	0.8
Eriopon OLS	Soaping agent	1

Table 2: Dyeing Recipe.

Measurement of weight loss %

Scouring of textile fiber loss a remarkable amount of weight (impurities like as oil, fats, wax, salts etc). The scouring effect, thus, can be evaluated based on this weight loss of fiber. Usually, it is calculated from Equation 1 the difference of un-scoured and scoured sample weight, measured in percentage of un-scoured weight of the sample.

$$\textbf{\textit{Weight loss}} \% = \frac{W_1 - W_2}{W_1} \times 100 \tag{1}$$

Where, W_1=Weight of the sample before scouring and, W_2=Weight of the sample after scouring

Absorbency test: The following tests were done to measure the absorbency of cotton fabric samples.

Sinking test: Absorbency may be assed in various ways, the most popular being the sinking time test [14]. Test specimens of 1 cm X 1 cm were cut at random and place on the surface of water. Slowly the fabric samples were wetted and entrapped air was removed. The time taken by the fabric samples to go inside water from floating state and sank in completely was noted down. The shorter the time taken by the specimen to sink in water completely, the greater is its absorbency.

Drop Test: The test was done according to the AATCC Test method 79 that measure a fabric's propensity to take up water, in which water drops are allowed to fall by gravity from a burette placed at a certain height from the fabric surface [14]. A drop of water is allowed to fall from a fixed height onto the taut surface of a test specimen. The time required for the specular reflection of the water drop to disappear is measured and recorded as wetting time.

Whiteness Index: The whiteness of each pretreated and chemically bleached sample was evaluated on the basis of the following CIE equation for illuminant D65 and 1964 10° observer Equation 2:

$$W = Y + 800(x_n - x) + 1700(y_n - y) \tag{2}$$

Where Y is the tristimulus value of the sample; x and y are the chromaticity coordinates of the sample, xn and yn are the chromaticity coordinates for the perfect reflecting diffuser (0.3138 and 0.3310, respectively).

Whiteness of fabric analyzed under spectrophotometer and weight loss were analyzed by taking weights of substrate before and after the bleaching process [15]. The CIE Whiteness Index value (CIE WI) was determined for the bleached fabric using AATCC test method. The whiteness was measured using a DATACOLOR 650, illuminants D-65 [16].

Color Strength measurement

The pre-treated and bleached samples were subsequently dyed with Novacron Super Black G and Novacron Navy WB reactive dyes in the DATACOLOR AHIBA IR laboratory apparatus at a liquor ratio of 1:10. The initial dyeing temperature was 30 °C and the dyebath consisted of 0.5% sequestering agent (Invatex CS), 1% of a leveling agent (Albatex DBC) and electrolyte (40 g/l Glauber Salt). Within 20 mins the dyebath was heated to 60 °C, after which the dye was added (2% o.w.f.). Then the dyebath was further running for 20 mins. Afterwards add to the dye bath of 14 g/l of sodium carbonate (Na_2CO_3). Finally run for the 45 min. After completion of run time rinsing the sample with cold and hot water, as well as neutralization with glacial acetic acid and soaping with 1 g/l of a soaping agent (Eriopon OLS). Then the dyed sample was dried at 70°C and measure K/S value. Color strength K/S was measured on a datacolor -650 Spectrophotometer. These values are calculated using the following Equation 3 "KUBELKA-MUNK" equation:

$$\frac{K}{S} = \frac{(1-R)^2}{2R} \tag{3}$$

Where K is the absorption co-efficient, R is the reflectance of the dyed sample and S is the scattering co-efficient at the wavelength of maximum absorption. The pretreated and dyed samples of the cotton knitted fabric were colorimetrically evaluated using a spectrophotometer (Datacolor-650). The CIE equation was used to calculate the whiteness of the scoured and bleached samples, the CIELAB color values (lightness L*, red/green axis a*, yellow/blue axis b*, chroma C*, and hue h), and the CIELAB color differences (ΔE*) were determined for the dyed samples.

Ecological impact measurement

Ecological studies of the residual baths were performed by analyzing the Biochemical Oxygen Demand (BOD5), according to SIST EN 1899-2, the Chemical Oxygen Demand (COD) according to SIST ISO 6060, the Total Organic Carbon (TOC) according to SIST ISO 8245, the biological degradation as a ratio of BOD5 and COD, and the spectral absorption coefficient SAC according to SIST EN ISO 7887.

Result and Discussion

Effects of hydrogen peroxide on fabric whiteness and weight loss%

In order to reduce the time of bleaching process, increasing concentration of hydrogen peroxide and analyzes the effects on given substrate and compare with standard process. From the Figure 1 it was observe that, an increasing of hydrogen peroxide with temperature reducing the dwell time of the bleaching process. Whiteness increase with the increase of hydrogen peroxide, but at 3 gL⁻¹ we get the maximum whiteness of the fabrics. It is also observe that the whiteness of the alkaline bleach process is more than the enzymatic bleach process. Alkali is indispensable for activating hydrogen peroxide. Hydrogen peroxide in a weak acid or neutral medium has little or knows bleaching action. In an alkaline medium the following equilibrium Equation 4 is set up.

$$H_2O_2 + NaOH \rightarrow Na^+ + HO_2^- + H_2O \tag{4}$$

Hydrogen peroxide is a powerful oxidizing agent that rapidly destroys the natural coloring matters present in cotton without undue oxidative damage to the fibers [7,17]. The effect of alkali, as seen from this equation is to shift the equilibrium to the right to increase the commemoration of per hydroxyl (HO₂-) ion, the bleaching agent and hence the bleaching action is intensified. However, peroxide bleached baths with alkali only are unsTable and they require stabilizers of inorganic or organic nature.

It is also observed that from Figure 2 with the increase of hydrogen peroxide concentration weight loss percentage also increase [18,19]. In general, the time of bleaching is inversely proportional to the temperature of the bleach bath and weight loss of the fabric [17,18].

Effect of processing chemicals and methods on fabric absorbency

Absorbency of a fabric can influence the uniformity and completeness of textile processing by the ability to take in water into fiber, yarn or fabric structure. Scouring imparts consistent and sufficient absorbency apart from enhancing the cleanliness of the material, bleaching further enhances the absorbability and imparts whiteness to the material. From Figure 3 revealed the average time

taken by the samples to sink under their own weight. It is observe that the alkaline bleach fabric and enzymatic bleach fabric absorbency is very closer. For complete sinking of fabric it requires 42 sec for alkaline bleach fabric and 48 sec for enzymatic bleach fabric. Whereas gray fabrics were require more than 13 min. So low sinking time indicates rapid wet ability resulted because of good pretreatment.

The fullness of bleaching as well as the suitability of a fabric for a particular use is dependent upon its ability and propensity to take up water. The absorbency of the samples assessed through drop test Figure 4 showed, there was tremendous improvement in the wettability of enzymatic bleach fabric. It is seen that enzymatic bleach fabric require 8 sec whereas alkaline bleach fabric require 10 sec. But for the gray fabrics it requires more than 10 min. So rapid disappear of water drop indicates the good pretreatment of fabric and it's also good prepare for dyeing.

Effect of caustic soda concentration on the H₂O₂ bleached fabric

The effect of increasing the amount of free caustic soda in bleach liquor containing peroxide stabilizer and H₂O₂ is shown in Figure 5.

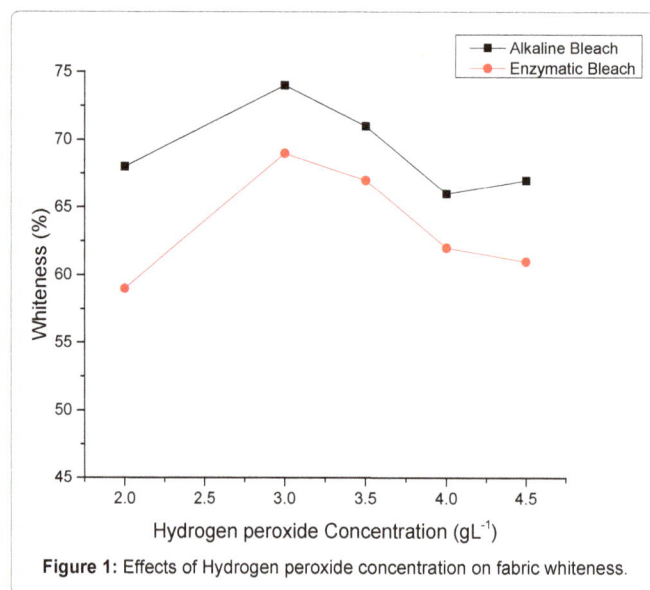

Figure 1: Effects of Hydrogen peroxide concentration on fabric whiteness.

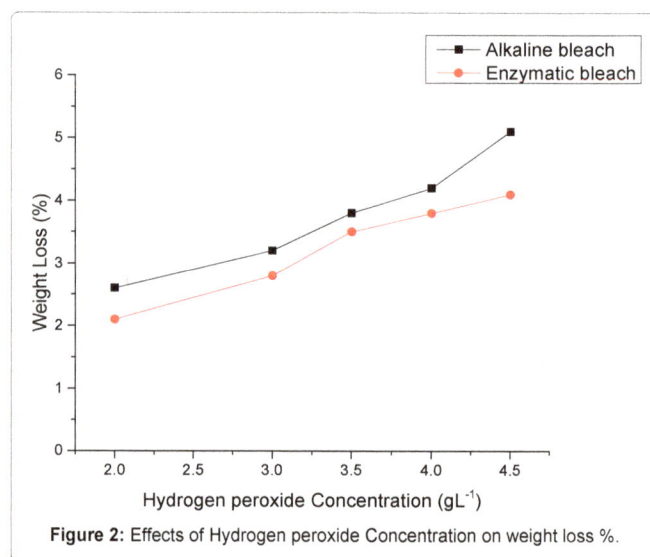

Figure 2: Effects of Hydrogen peroxide Concentration on weight loss %.

It is seen that the whiteness decreases with increasing free caustic concentration. At 2 gL^{-1} caustic soda concentration we got the maximum whiteness of fabric. Alkaline bleach fabrics whiteness is more than that of enzymatic bleach fabric. Peroxide concentration in commercially accepTable bleaching in one stage bleaching is higher than that in the multistage process and peroxide requirement depends on the fabric type. Generally a whiteness of about 76% is obtained with 3 gL^{-1} hydrogen per oxide.

Colorimetrical evaluation

Whiteness (CIE W) and lightness (L*) of gray fabric and different pre-treated bleach fabrics are shown in Table 3 the results show that the lightness L* is higher for bleach fabric than gray fabric. But there is

No significant difference between alkaline bleach and enzymatic bleach sample. However the CIE whiteness obtain from alkaline pre-treatment process (67.4) was higher than the enzymatic pre-treatment process (62.6%), mainly caused by the differences the b* axes.

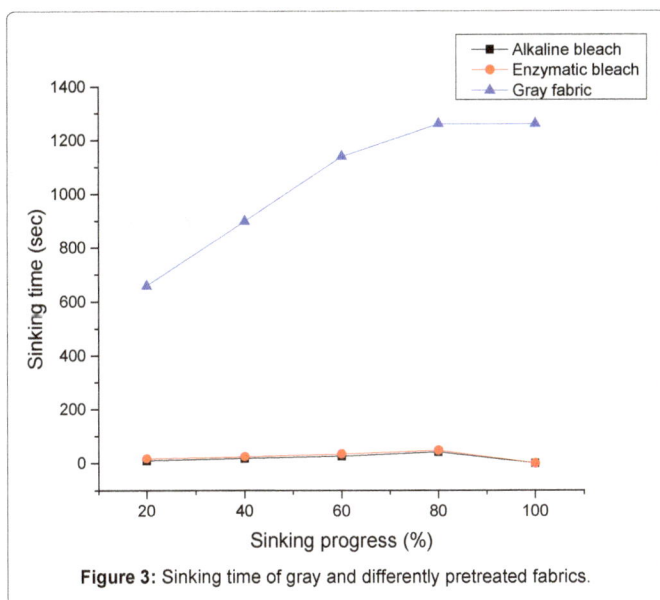

Figure 3: Sinking time of gray and differently pretreated fabrics.

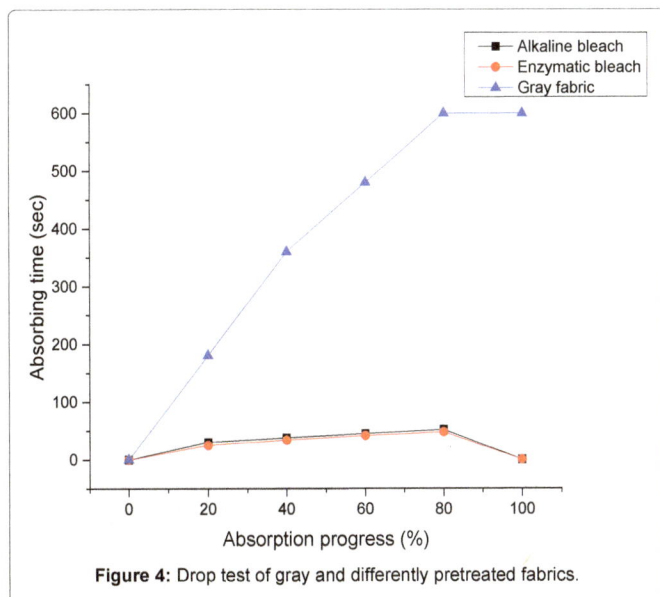

Figure 4: Drop test of gray and differently pretreated fabrics.

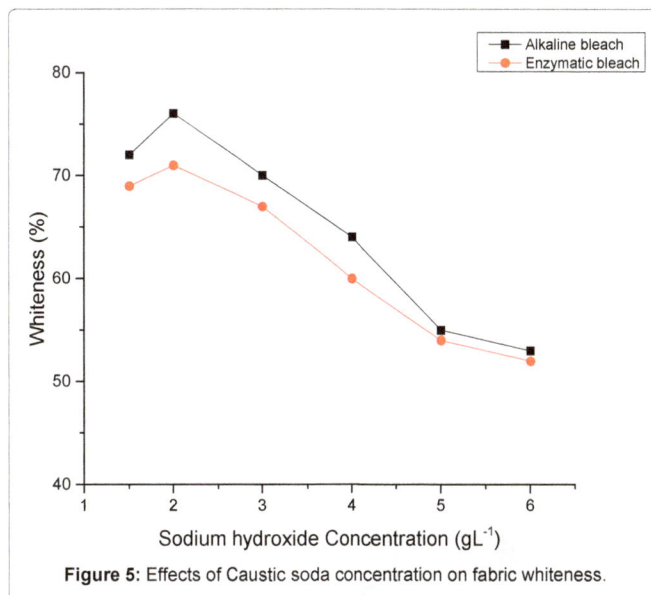

Figure 5: Effects of Caustic soda concentration on fabric whiteness.

Samples	CIE W	L*
Gray Fabric	0	84.3
Alkaline Bleach	67.4	92.5
Enzymatic Bleach	62.6	92.1

Table 3: Whiteness (CIE W) and Lightness (L*) of different samples.

Afterwards, the modified alkali pre-treatment and enzymatic treated cotton fabric were dyed using the two selected reactive dyes (Novacron Super Black G and Novacron Navy WB).

The dyed samples in all concentrations were colorimetrically evaluated by using CIELAB color system. CIELAB color values for 2% added dyestuffs are shown in Table 4. CIELAB color values of the dyed samples confirmed that the pre-treatment process had no important influence on the dyeing properties of the samples dyed using Novacron Super Black G and Novacron Navy WB dyes.

Ecological impact of pretreatment

The results of ecological analysis of the residual pretreatment baths are presented in Figure 6. The ecological analyses of the residual baths clearly showed the difference between the combined pre-treatment and bleaching processes performed. The TOC value of the residual bath after modified alkali pre-treatment was 20% higher than the TOC of the enzymatic pre-treatment waste-water. Since the TOC value gives the amount of total organic compounds present in wastewater, the lowest value of this parameter for the enzymatic pre-treatment is additional proof of its environmentally-friendly character. The COD and BOD5 values give the amount of oxygen needed for the oxidative and biological degradations of organic compounds in wastewater, respectively. The COD for the modified alkali pre-treatment wastewater was 41% and for BOD5 31% higher than that for the enzymatic pre-treatment wastewater. Both wastewaters of the pre-treatment were totally biologically degradable (for modified alkali pre-treatment it was 3.37, for enzymatic pre-treatment 3.43).

The residual baths analyzed were concentrated because the rinsing baths, which certainly lowered these values on an industrial scale, were not included; therefore, the relative comparison between the procedures is even more realistic. It should be noted that both pre-treatment processes have a strong impact on environmental pollution,

Samples	Dyestuffs	L*	a*	b*	C*	H
Alkaline bleach and dyeing	Novacron Navy WB	24.1	-6.2	-15.9	16.9	247.9
	Novacron Super Black G	22.3	-1.8	-3.2	4.1	239.8
Enzymatic bleach and dyeing	Novacron Navy WB	24.5	-6.4	-16.0	16.9	248.0
	Novacron Super Black G	25.2	-2.7	-4.1	5.0	235.1

Table 4: CIELAB Color values of differently pre-treated and dyed samples.

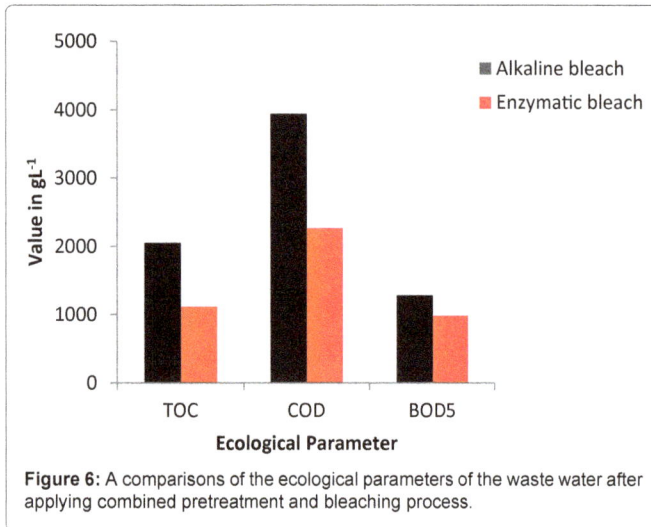

Figure 6: A comparisons of the ecological parameters of the waste water after applying combined pretreatment and bleaching process.

since all the ecological parameters exceeded the limits stated in Slovenian regulations (TOC=60 gL^{-1}, COD=200 gL^{-1}and BOD5=30 gL^{-1}) [20]. However, after combined single-bath scouring and bleaching, the volume of wastewater is significantly reduced. All the ecological parameters analyzed indicated that the degree of pollution of the residual baths was insignificant after enzymatic pre-treatment and bleaching.

Conclusion

It is observe that from results and discussion the combined bath pre-treatment processes performed, comprising scouring and bleaching, have a significant influence on the ecological parameters of the residual baths, but a minor influence on the whiteness color of the dyed cotton fabric using reactive dyes. At hydrogen peroxide conc. 3 gL^{-1} and sodium hydroxide 2 gL^{-1} we have get good results with less environmental impacts. But cotton fabric pretreated with alkaline process and treated with enzymatic process almost similar when fabric dyed with Novacron navy WB, where as a greater difference ($\Delta E^*=4.1$) was obtained with Novacron super black G dyestuff, where combined alkaline process treated fabric produce a darker and less chromatic black color. Now a day's conventional or modified alkaline bleaching process produce more pollutant where as a good alternatives process of pre-treatment for cotton fabric treat with enzymes during bleaching which produce almost same fabric whiteness and depth of shade with lower environmental impact.

Acknowledgements

Author's would like to express sincere thanks to the Concept Knit Composite Ltd. Gazipur, Bangladesh for giving scope of doing this research work on their laboratory and providing the raw materials for this project.. Author's also express sincere appreciation to Md. Mizanur Rahman, General Manager, Concept dyeing Ltd. for his guidance and suggestions throughout the research work. This research was financially supported by the Wuhan Textile University postgraduate innovation team project. Author's would like to express deep appreciation to my supervisor Prof. Heng Quan, Vice Dean, Department of Chemistry and Chemical Engineering, Wuhan Textile University, China for his inspiring guidance, encouraging attitude and valuable suggestions for publishing this research work.

References

1. Shenai V (1991) Technology of Bleaching and Mercerizing. Sevak Publications (2ndedn) New Delhi, India.

2. Trotman E (1975) Dyeing and Chemical Technology of Fibers. London: Charles Griffin and Co. 617-618.

3. Madaras G, Shore J (1993) Batchwise dyeing of woven cellulosic fabrics. Society of Dyers and Colourists.

4. Bae SH, Motomura H, Morita Z (1997) Diffusion/adsorption behaviour of reactive dyes in cellulose. Dyes and pigments 34: 321-340.

5. Shore J (1990) Colorants and auxiliaries.

6. Krässig HA (1993) Cellulose: structure, accessibility and reactivity. Gordon and Breach Science Publ.

7. Islam MR, Farhat FI (2016) Effect of Chemical Concentration on the Pre-treatment Performance of Cotton Woven Fabric. IJCET.

8. Burget L (1994) The Ecological Aspect of Cellulose Fibre Bleaching. Fibres and Textiles in Easten Europe.

9. Buschle-Diller G, Yang XD, Yamamoto Y (2001) Enzymatic bleaching of cotton fabric with glucose oxidase. Textile Research Journal 71: 388-394.

10. Karcher S, Kornmüller A, Jekel M (2002) Anion exchange resins for removal of reactive dyes from textile wastewaters. Water Research 36: 4717-4724.

11. Timofei S (2000) A review of QSAR for dye affinity for cellulose fibres. Dyes and Pigments 47: 5-16.

12. Schanda J (2007) Colorimetry: understanding the CIE system. John Wiley and Sons.

13. Spencer DJ (2001) Knitting technology: a comprehensive handbook and practical guide. CRC press.

14. Choudhury AKR (2006) Textile preparation and dyeing. Science publishers.

15. Jones B (1985) Effect of ozonation and UV irradiation on biorefractory organic solutes in oil shale retort water. Environmental progress 4: 252-258.

16. Xu C (2011) Review of Bleach Activators for environmentally efficient bleaching of textiles. Journal of Fiber Bioengineering and Informatics 4: 209-219.

17. Broadbent AD (2001) Basic principles of textile coloration. Society of Dyers and Colorists West Yorkshire, Canada.

18. Uddin M (2010) Determination of weight loss of knit fabrics in combined scouring-bleaching and enzymatic treatment. J Innov Dev Strategy 4: 18-21.

19. Sheth GN, Musale AA (2005) Single bath bio-scouring and bleaching of cellulosic yarn, Knitted and woven fabrics. Colourage 52: 49-52.

20. Golob V, Vinder A, Simonič M (2005) Efficiency of the coagulation/flocculation method for the treatment of dyebath effluents. Dyes and pigments 67: 93-97.

30

Mechanical and Surface Properties of Thai Cotton Hand-woven Fabric Made from Hand-spun and Machine-spun Yarns

Phoophat P and Sukigara S*

Kyoto Institute of Technology, Kyoto Japan

Abstract

In some areas in Thailand, women weave fabrics from hand-spun cotton as a cottage industry. Cotton hand-spun yarn has an uneven thickness and low twist that can give a unique appearance to the fabric and is not considered a defect. In this study, we measured the characteristic mechanical and surface properties of 16 Thai fabric samples, and divided them into six groups. The effect of using hand-spun yarn as the weft of a fabric was considered. The Kawabata evaluation system was used to compare the characteristic values with reference values to inform the future direction of hand-spun woven textiles in Thailand. The results show the differences between the characteristic values of warp and weft direction are important for tensile and bending properties and for surface roughness. Using hand-spun yarn in the weft direction of the woven fabric affected the surface irregularity, producing more space between yarns, affecting the air resistance value, which was in the reference range for summer suiting materials, although the thickness was larger. There were large differences between the six groups of fabrics varied in yarn count and cover factor, especially in the bending, shear, and compression properties. All characteristic values compared with the reference values show that most of the Thai handloom-woven fabrics surveyed in this study showed stiffness, crispness, and anti-drape values that were not suitable for suiting. However, Thai hand-loom woven fabrics would be suitable for summer jackets and ladies skirt with anti-drape silhouette for hot summer.

Keywords: Hand-spun yarn; Cotton yarn; Handloom woven fabric; Mechanical properties; Surface properties

Introduction

In rural Thailand, in addition to agriculture, women engage in handloom weaving as a cottage industry. In some areas that produce cotton fabric, the process includes cultivating cotton, hand-spinning, dyeing and then hand-weaving. Cotton hand-spun yarn has uneven thickness and twist that can give a unique appearance to the fabric, which conventional fancy yarn cannot reproduce. The advantage of handloom weaving is that many types of hand-spun yarns can be handled to produce a unique woven fabric. However, the handloom woven fabric must be suitable for the end use.

The fabric quality can be considered from three points of view: the suitability of its physical properties for the end use; the aesthetic appeal of the fabric, which attracts consumers; and how kind the product and process is to humans and the environment. We expected handwoven village fabrics to have different properties and uniqueness from regular, machine-woven fabrics.

In fabric manufacturing, yarn imperfections (neps, thick and thin places) affect fabric weavability and thickness variation [1,2]. Therefore, the causes of yarn imperfections from raw materials and yarn processing, and their prevention have been studied in many countries [3-5]. In synthetic fibers, thick and thin yarns, and irregular cross sections have been used to change the fabric hand. The yarn structure obtained from different spinning systems affects the tactile properties, even if the fabric is produced with the same knitted structure [6]. The imperfections in the fabric surface of hand-spun woven fabric have aesthetic value and are not considered defects or an indication of low-quality. A few studies have examined the hand-spun woven textile design process considering the mechanical and surface properties of the fabric. For example, the quality of shawls made from hand-spun and machine-spun yarns were compared to develop the production process for fine pashmina fibers [7,8].

Textile properties are controlled by a combination of yarn structure and fabric structure to produce an end product that is suitable for its intended use. In this study, fabric samples produced by handloom weaving were collected in Thai villages. The mechanical and surface properties were measured with the Kawabata evaluation system (KES) to obtain the characteristic values. The effect of using hand-spun yarn as the weft yarn on the mechanical and surface properties of the fabric was considered. All characteristic values obtained by KES were compared with the reference values reported in the literature, particularly for high-quality hand and for garment appearance, to guide the future direction of hand-spun woven textiles in Thailand.

Materials and Methods

Samples

Sixteen cotton Thai handloom woven fabrics were collected in Thailand. All warp yarns were spun by a spinning machine. For the weft yarns, either hand-spun or machine-spun imitations of hand-spun cotton yarns were used. All fabrics were woven with a plain structure and finished. They were categorized by their yarn count and the cover factor of weft yarns into six groups (Tables 1 and 2). The warp yarn count was in the range of 23.14 to 34.74 tex, and the weft yarn count was much larger than the warp yarn count (Table 2).

Characterization of physical properties

The physical properties of fabrics were measured at 20 °C and 65% relative humidity by using a KES (KES-FB, Kato Tech Co., Ltd.). The mechanical properties, including tensile, shear, bending,

---

***Corresponding author:** Sachiko Sukigara, Kyoto Institute of Technology, Kyoto, Japan, E-mail: sukigara@kit.ac.jp

Range of weft cover factor*	Range of weft yarn count (tex)		
	S (80–110 tex)	M (140-170 tex)	L (210-250 tex)
Loose (11.0–12.5)	1, 2, 3	7, 8	12, 13
Tight (13.5–15.0)	4, 5, 6	9, 10,11	14, 15, 16

*Cover factor: (threads/inch)/√cotton count.

Table 1: Sample numbers of the six categories.

Group	No.	Yarn	Yarn count (tex)		Yarn density (/cm)		Cover factor	Weight (g/m²)	Thickness (mm)	Air Resistance (kPa.s/m)
		Weft	Warp	Weft	Ends	Picks	Weft			
S-loose	1	Hand-spun	14.50 × 2	85.59	16.95	12.28	11.87	170.0	1.024	0.11
	2	Machine-spun	15.54 × 2	82.02	17.45	12.21	11.56	194.5	1.12	0.22
	3	Hand-spun	15.18 × 2	84.36	12.76	13.03	12.51	174.3	1.078	0.13
S-tight	4	Hand-spun	14.91 × 2	109.36	16.55	13.56	14.82	214.0	1.159	0.24
	5	Machine-spun	14.82 × 2	85.59	13.18	14.97	14.47	240.3	1.262	0.41
	6	Machine-spun	15.58 × 2	84.36	16.97	15.38	14.76	215.3	1.103	0.37
M-loose	7	Hand-spun	16.68 × 2	144.03	19.53	8.97	11.25	216.0	1.612	0.11
	8	Hand-spun	11.95 × 2	159.61	19.72	8.62	11.39	205.3	1.221	0.11
M-tight	9	Machine-spun	14.84 × 2	151.42	14.49	10.82	13.92	234.5	1.275	0.30
	10	Hand-spun	11.57 × 2	168.73	19.22	10.27	13.95	235.3	1.243	0.16
	11	Hand-spun	34.53	168.73	14.91	10.30	13.98	287.5	1.685	0.22
L-loose	12	Machine-spun	26.25	210.91	23.29	7.68	11.66	253.5	1.82	0.18
	13	Hand-spun	30.44	246.06	9.91	7.18	11.77	229.0	2.122	0.12
L-tight	14	Hand-spun	16.36 × 2	203.63	18.94	10.01	14.92	290.3	2.011	0.16
	15	Hand-spun	30.44	246.06	8.94	8.81	14.44	267.0	2.155	0.40
	16	Hand-spun	34.74	236.2, 140.6	16.47	9.84	13.87	259.0	1.819	0.32

Table 2: Sample specifications.

compression, surface properties of the fabrics were measured under the standard measuring conditions [9]. The air resistance was measured by using KES-F8API air permeability tester [10]. The characteristic values and KES testers are listed in Table 3. For hand-spun yarns, yarn thickness variation was observed. Yarn diameter and twist angle were measured in 30-cm-long samples by using a 3D microscope (VR-3000, Keyence).

The primary hand value (HV) was rated to evaluate the suitability of the fabric for men's summer suits and women's medium-thick fabrics, calculated by using the Kawabata hand evaluation equations KN101-S and KN-201-MDY [9]. KN101-S was derived based on fabric mechanical data, mainly for wool materials, and KN-201-MDY includes various materials to construct the predictive equation.

Results and Discussion

Mechanical and surface properties

The difference between the characteristic values of warp and weft direction was profound for tensile and bending properties and surface roughness (SMD). Figure 1 shows the fabric extension at 500 N/m (EM) for both warp (EM1) and weft (EM2) directions for each sample. The EM1 values were two to eight times higher than the EM2 values for all fabrics. For wool fabrics for men's suits, the EM2/EM1 values are between 2 and 3 [11]. EM1 for the S-loose and L-loose groups (Table 1) increased with the increase in weft yarn count, whereas it did not for the M-loose group. When the thick hand-spun yarns were used for the weft direction, the yarn crimp ratio of the warp yarn was higher than that of the weft crimp ratio. The higher crimp of the warp yarn and weft cover factor affected EM1. The high EM1 value for cotton fabrics is usually accompanied by low tensile resilience of warp direction (RT1) as shown in Figure 2.

In Figure 3, the bending rigidity along the warp (B1) and weft (B2)

directions are shown. B2 is larger than B1 for all samples, especially for samples with thick weft yarns (sample Nos. 13-15).

SMD of the warp direction (SMD1) is also much larger than that of weft direction (SMD2; Figure 4). When hand-spun yarns were used as the weft yarn with a larger yarn count, the SMD increased more when the sensor moved over thick weft yarns compared with industrial warp yarns.

Fabric surface appearance

When hand-spun yarns are used for fabric, thick and thin parts of the yarn appeared randomly in the fabric surface (Figure 5). Hand-spun yarns inserted in the weft direction were not as even as machine spun warp yarns. This is also why the SMD values for the fabric warp direction were large (Figure 4). For the coefficient of friction (MIU) and mean deviation of MIU (MMD), the difference between the warp and weft directions was small.

Figures 6 and 7 show the distributions of the yarn diameter and twist angle of samples from the six groups, measured by a 3D microscope, respectively. The coefficient of variance(C.V) of these parameters is also shown in figures. Samples No. 13 and 15, which had a higher yarn count, showed more variation in the diameter of thick and thin places than samples No. 1 and 4. However, the distribution of the yarn twist angles did not vary much among samples. The hand spinning process to make these yarns seemed to vary among individuals. Drafting is the part of the spinning process in which a bundle of fiber to be spun is pulled. The more fibers that are drafted, the thicker the yarn will be and vice versa. This is the most important process for controlling the regularity of the yarn thickness. After drafting a certain length, the yarn is held and the other hand turns the wheel to twist the yarn an exact number of times to obtain a consistent twist. Where the yarn is thicker, the twist angle is slightly lower than where the yarn is thinner.

Properties	Parameters symbol	Characteristic value	Unit	KES machines
Tensile	EM	Fabric extension at 500 N/m width	%	KES-FB1
	LT	Linearity of load-extension curve	-	
	WT	Tensile energy	J/m²	
	RT	Tensile resilience	%	
Bending	B	Bending rigidity	μNm	KES-FB2
	2HB	Hysteresis of bending moment	mN	
Shear	G	Shearing stiffness	N/m	KES-FB1
	2HG	Hysteresis of shear force at a shear angle of 0.5°	N/m	
	2HG5	Hysteresis of shear force at at a shear angle of 5°	N/m	
Compression	LC	Lineariry of compression-thickness curve	-	KES-G5
	WC	Compression energy	J/m²	
	RC	Compression resilience	%	
Surface	MIU	Coefficient of friction	-	KES-SE-STP
	MMD	Mean deviation of MIU	-	
	SMD	Geometrical roughness	μm	KES-SE
Air Resistance	R	Air resistance	kPa.s/m	KES-F8-AP1
Construction	T	Fabric thickness at 50 Pa	mm	
	W	Fabric weight per unit area	g/m²	

Table 3: Physical properties of fabrics measured with the KES system

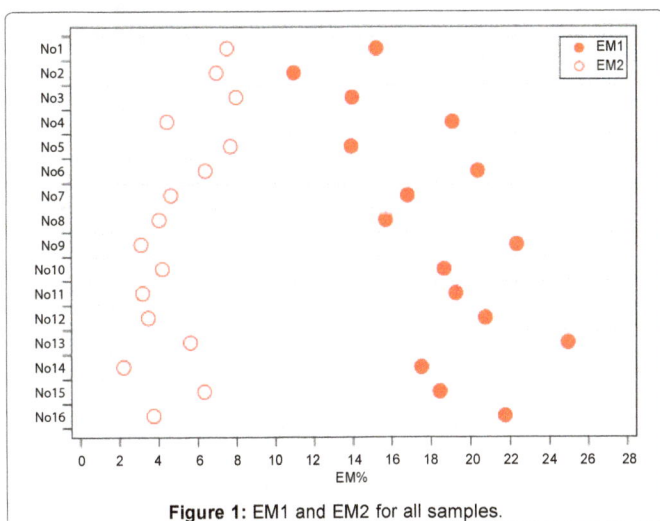

Figure 1: EM1 and EM2 for all samples.

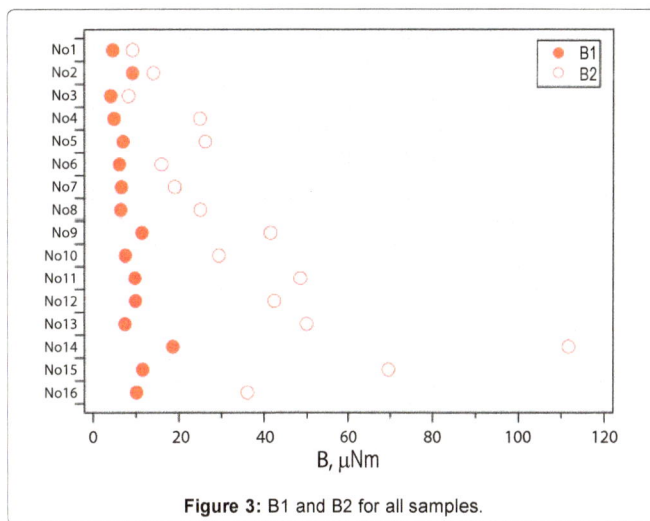

Figure 3: B1 and B2 for all samples.

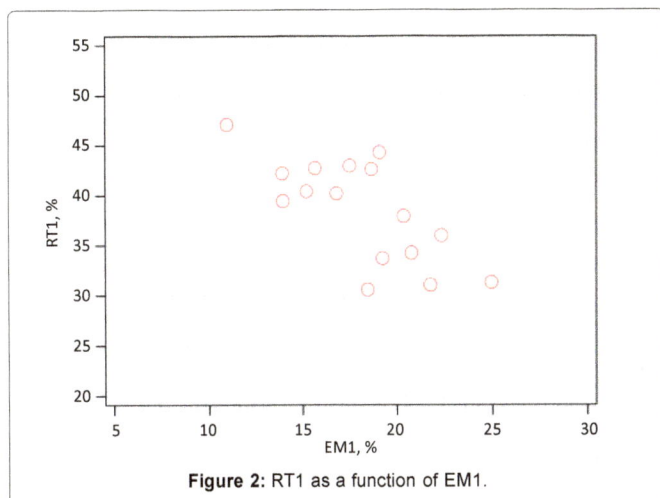

Figure 2: RT1 as a function of EM1.

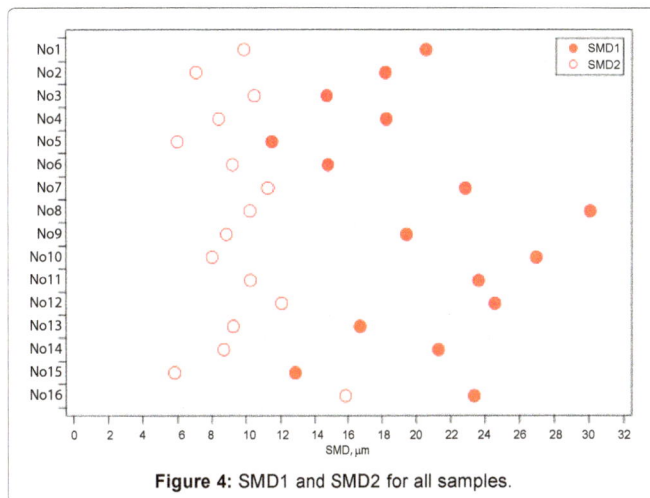

Figure 4: SMD1 and SMD2 for all samples.

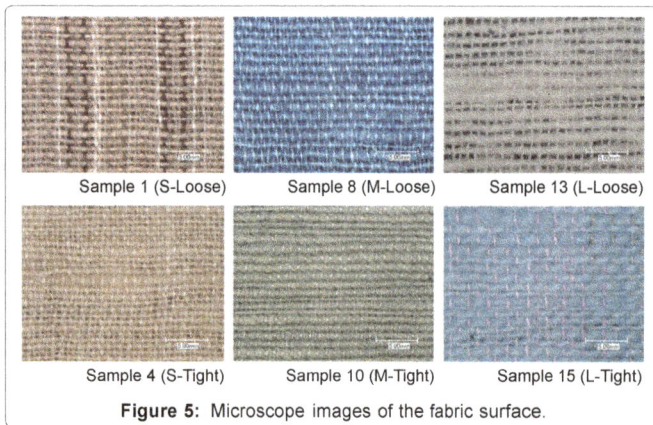
Figure 5: Microscope images of the fabric surface.

Sample 1 (S-Loose)   Sample 8 (M-Loose)   Sample 13 (L-Loose)
Sample 4 (S-Tight)   Sample 10 (M-Tight)   Sample 15 (L-Tight)

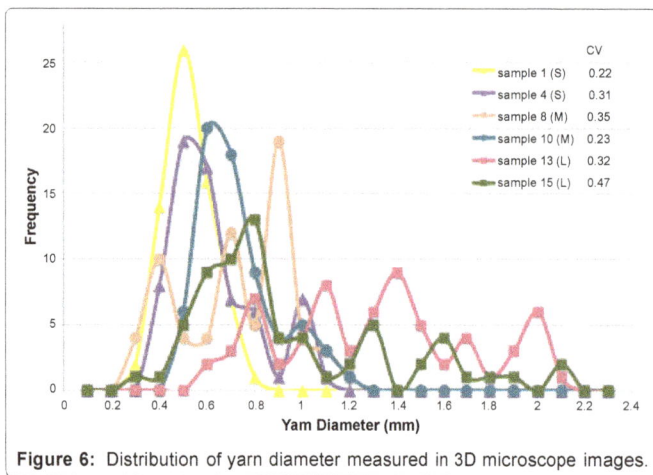
Figure 6: Distribution of yarn diameter measured in 3D microscope images.

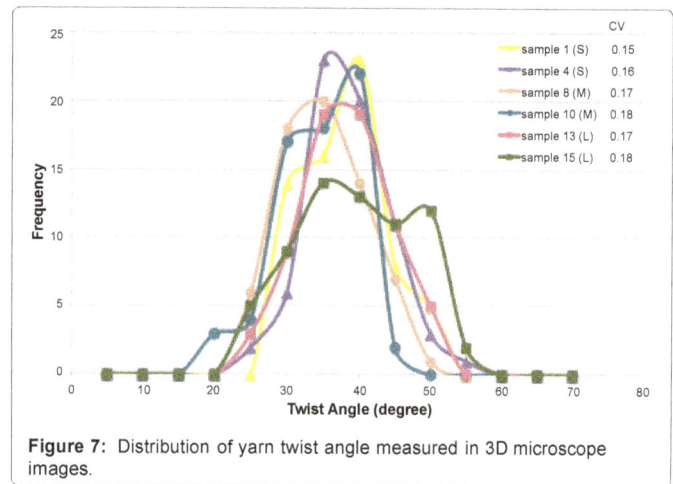
Figure 7: Distribution of yarn twist angle measured in 3D microscope images.

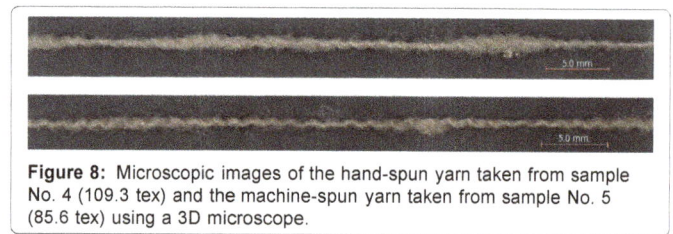
Figure 8: Microscopic images of the hand-spun yarn taken from sample No. 4 (109.3 tex) and the machine-spun yarn taken from sample No. 5 (85.6 tex) using a 3D microscope.

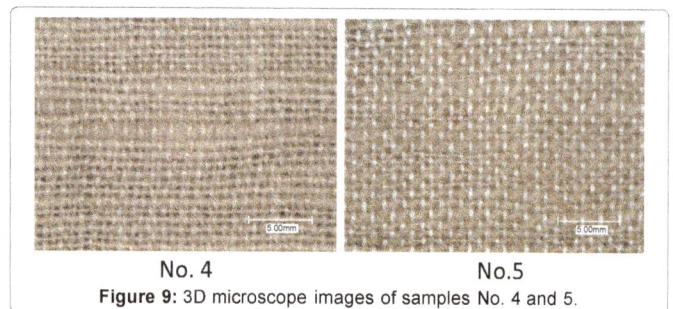
No. 4   No.5
Figure 9: 3D microscope images of samples No. 4 and 5.

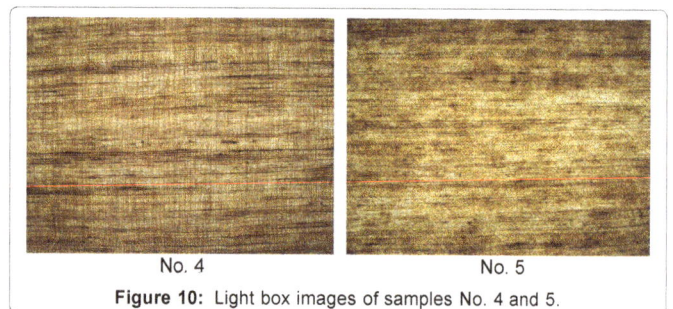
No. 4   No. 5
Figure 10: Light box images of samples No. 4 and 5.

To determine the effect on fabric appearance of the difference between hand-spun yarn and machine-spun imitations of hand-spun yarn, fabric samples No. 4 and 5 were compared. These two fabrics had a similar weave density and weft cover factor (Table 2). Figure 8 shows weft yarns taken from samples No. 4 and 5. The hand-spun yarn from sample No. 4 showed spindle-shaped thick places. The machine-spun yarn taken from sample No. 5 had a more regular crimp. This result is also consistent with EM1 being larger than EM2 (Figure 1).

3D microscope images of the fabric surface of samples No. 4 and 5 are shown in Figure 9 and photographs of the samples on a light box are shown in Figure 10. Sample No. 4 shows space between the yarns in Figure 9 and has clearer horizontal and vertical organization that corresponds to the thick weft yarns in Figure 10. The air resistance (R) of sample No. 5 is larger than sample No. 4 (Table 2).

Future end-use applications of Thai handloom woven fabrics

The mechanical and surface properties were compared with reference values for men's suit fabric. Kawabata and Niwa developed a data chart to indicate targets for developing fabrics [9]. The mechanical and surface properties of one sample from each of the six groups are plotted in a HESC chart for men's summer suiting in Figure 11. The shaded zone in this chart is the excellent zone in which high-quality fabrics fall [9].

For all fabrics, the properties outside of the excellent zone are tensile resilience (RT), bending hysteresis (2HB), shear hysteresis (2HG), compression energy (WC), mean deviation of MIU (MMD),

and thickness. The Thai cotton handloom woven fabrics surveyed in this study were thick and stiff, which are not suitable properties for suit materials. In Figure 12, the thickness of the fabric samples is plotted against air resistance (R). The mean and standard deviation of reference values for men's summer and winter suiting and women's suiting are also shown [12]. The thickness of Thai handloom woven fabrics is large, although the R values are in the same range as summer suiting materials.

There are large differences between the six groups of fabric in terms of yarn count and cover factor, and particularly in bending, shear, and compression properties. Thai woven fabrics with a lower yarn

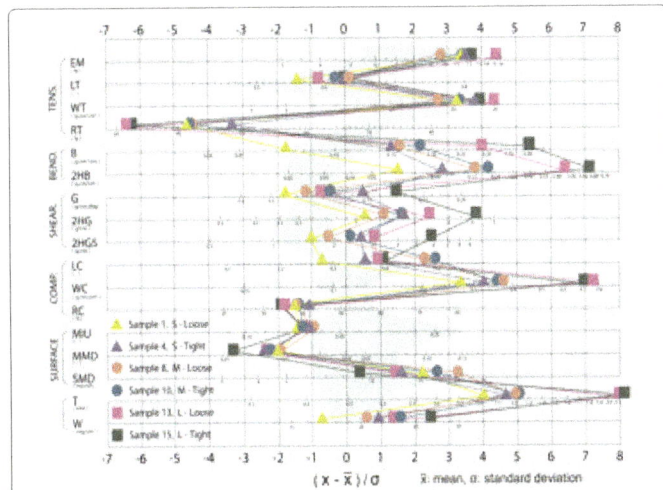

Figure 11: Comparison of mechanical and surface properties of samples from the six groups based on the HESC chart for men's summer suiting. The shaded zone in this chart is the excellent zone in which high-quality fabrics fall.

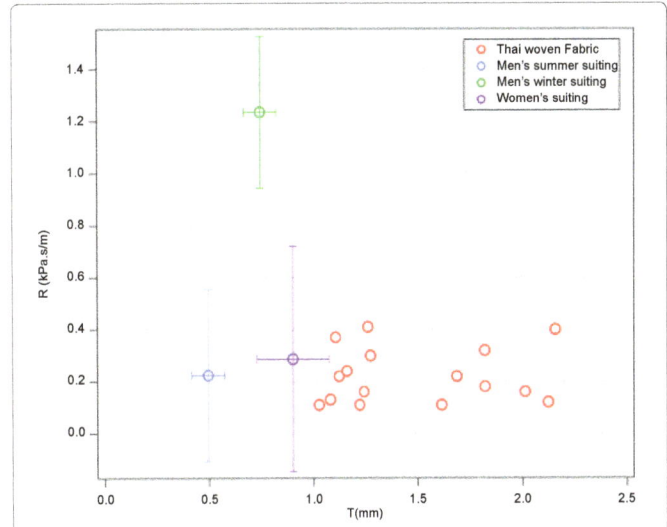

Figure 12: Plot R against fabric thickness.

count and cover factor had a lower bending rigidity (B) and bending hysteresis (2HB), as seen in sample No. 1 (S-loose group) in Figure 11. Thus, woven fabric with a high yarn density and thick hand-spun yarn increased B and 2HB. All fabrics in groups S-loose, M-loose, L-loose (No. 1, 8, and 13) had lower shearing stiffness (G), because the crossing point of the warp and weft moved easily. Compression energy (WC) of all hand-spun woven fabrics had a larger value than that of the high-quality zone of commercial fabrics. The fabrics woven with a larger weft yarn count had higher WC values than fabrics with a smaller yarn count. This was attributed to the thick yarn having a lower twist number, which made the fabric fluffy and soft. However, EM was slightly higher and RT was much lower than the high-quality zone, and thus fiber slippage was possible. This was also because of the properties of cotton fibers. The surface properties MIU and MMD were similar for the six groups.

HV was calculated by using Kawabata hand evaluations KN101-S for men's summer suiting and KN-201-MDY for women's medium-thick fabric. The calculated HVs, such as stiffness (*koshi*), crispness (*shari*), anti-drape (*hari*), fullness (*furukami*), smoothness (*numeri*), and softness (*sofutosa*) are listed in Tables 4 and 5.

HVs of stiffness, crispness, and anti-drape for most of the fabrics were above 5, particularly for the L-tight, L-loose, M-tight fabrics, which mean that the stiffness, crispness, and anti-drape values of these fabrics were too high for men's summer suiting, except for the S-loose fabric. These fabrics had medium stiffness values for women's medium-thick fabric, except L-tight and M-tight fabrics, and L-tight and M-tight fabrics had a low softness, which is not suitable for women's medium-thick fabric.

One of the end-use application for Thai handloom woven fabrics could be casual summer jackets. In this case, weft yarn count must be in the range of 80-170 tex and with low cover factor to satisfy the clothing requirement. The other application is used for lady's dress and skirts. Niwa reported the silhouette design for lady's garments based on the fabric mechanical properties which concerns fabric weight and bending properties [13,14]. We think Thai handloom woven fabrics has the bending properties to generate the silhouette like an anti-drape (hari-type) to forms a skirt with space between fabrics and the body.

Group	Sample No.	Koshi (Stiffness)	Shari (Crispness)	Hari (Anti-drape)	Fukurami (Fullness and softness)
S-loose	1	2.95	5.30	3.52	4.84
	2	5.04	5.99	5.51	5.12
	3	2.17	4.15	2.85	5.81
S-tight	4	6.06	6.63	6.46	4.69
	5	5.65	4.96	5.49	5.12
	6	4.80	5.51	4.93	4.72
M-loose	7	4.55	6.33	5.58	5.00
	8	5.63	7.58	6.52	4.54
M-tight	9	8.30	7.81	8.78	3.76
	10	6.53	7.67	7.37	4.78
	11	8.67	8.48	10.17	4.79
L-loose	12	7.40	7.65	8.69	4.67
	13	7.54	6.55	8.91	4.97
L-tight	14	11.54	9.33	12.80	3.78
	15	9.20	6.27	10.54	5.11
	16	8.12	7.99	8.29	4.75

Table 4: HV calculated by Kawabata hand evaluation KN101-S to characterize the fabrics for men's summer suiting.

In making the garment construction, the warp and weft directions of fabric had better to be considered because of their large different extensibility of fabric.

Conclusion

The mechanical and surface properties of Thai hand-woven fabrics made from hand-spun and machine-spun cotton yarns were investigated. For all fabrics, the warp yarns were machine spun and the weft yarns were hand-spun. The difference between the fabric characteristic values of the warp and weft directions was large for tensile and bending properties and surface roughness (SMD) measured by the KES system.

Using hand-spun yarn in the weft direction of the woven fabric affected the surface irregularity, producing more space between yarns and clear horizontal and vertical lines in the fabrics corresponding to the thick weft yarn compared with machine-spun imitations of hand-spun yarn.

Group	Sample No.	Koshi (Stiffness)	Numeri (Smoothness)	Fukurami (Fullness and softness)	Sofutosa (Softness)
S-loose	1	4.10	5.22	5.32	4.46
	2	5.31	6.00	6.16	4.84
	3	3.82	6.69	6.47	6.21
S-tight	4	5.69	4.87	5.48	3.59
	5	5.88	5.94	6.39	4.56
	6	5.16	5.68	5.98	4.43
M-loose	7	4.70	5.27	6.05	4.71
	8	5.24	4.69	5.19	4.05
M-tight	9	6.49	3.98	5.19	2.71
	10	5.65	4.74	5.25	4.00
	11	6.43	4.74	6.00	3.73
L-loose	12	5.78	5.10	6.26	4.37
	13	5.49	5.87	7.23	5.03
L-tight	14	7.88	3.70	5.87	1.59
	15	6.51	6.30	7.53	4.91
	16	6.43	5.04	6.19	3.12

Table 5: HV calculated by Kawabata hand evaluation KN-201-MDY to characterize the fabrics for women's medium-thick fabric.

The air resistance (R) values were similar to reference values for summer suiting materials even though the handwoven fabrics were thicker. There were large differences between the six groups of fabrics varied in the yarn count and cover factor, particularly in the bending, shear and compression properties. The fabrics with thinner yarn and looser weave had lower bending rigidity (B) and shearing stiffness (G) values. The tensile resilience (RT), bending hysteresis (2HB), shear hysteresis (2HG), compression energy (WC), mean deviation of MIU (MMD), and thickness of all the hand-spun woven fabrics were outside the excellent zone of commercial fabrics. Primary hand values (HV) of stiffness, crispness, and anti-drape were too high for most of the fabrics, and they are not suitable for suits. However, Thai hand-loom woven fabrics would be suitable for summer jackets and ladies skirt with anti-drape silhouette.

Acknowledgement

This work was supported by JSPS KAKENHI Grants, Numbers 24220012 and 15H01764.

References

1. Gupta N (2013) Analysis on the defects in yarn manufacturing process and its prevention in textile industry. International Journal of Engineering Inventions 2: 45-67.

2. Frydrych I, Matusiak M (2002) Predicting the nep number in cotton yarn-determining the critical nep size. Textile Research Journal 72: 917-923.

3. Ochla J, Kisato J, Kinuthia L, Mwasiagi J, Waithaka A (2012) Study on the influence of fiber properties on yarn imperfections in ring spun yarns. Asian Journal of Textile 2: 32-43.

4. Seyam A, Aly EL-Shiekh (1990) Mechanics of woven fabrics part I: Theoretical investigation of weavability limit of yarns with thickness variation. Textile Research Journal 60: 389-404.

5. Zhang W, Iype C, Oxenham W (1998) The analysis of yarn thin places and unevenness with an image-analysis system and program design. The Journal of the Textile Institute 89: 44-58.

6. Suzuki Y, Sukigara S (2013) Mechanical and tactile properties of plain knitted fabrics produced from rayon vortex yarns. Textile Research Journal 83: 740-751.

7. Bumla NA, Wani SA, Shakyawar DB, Sofi AH, Yaqoob I, et al. (2012) Comparative study on quality of shawls made from hand- and machine-spun pashmina yarns. Indian Journal of Fibre and Textile Research 37: 224-230.

8. Shakyawar DB, Raja ASM, Wani S A, Kadam VV, Pareek P K (2015) Low-stress mechanical properties of pashmina shawls prepared from pure hand spun, machine spun and pashmina-wool blend yarn. The Journal of the Textile Institute 106: 327-333.

9. Kawabata S, Niwa M (1995) Objective Measurement of Fabric Hand: InRaheel M Modern textile characterization methods: (1st edtn) Marcel Dekker Inc., New York, USA, 329-354.

10. Kawabata S (1987) Development of an automatic air-permeability tester. Journal of the Textile Machinery Society of Japan (predecessor journal of Journal of Textile Engineering) 40: T59-T67.

11. Ito K,Kawabata S (1985) Conception of the automated tailoring controlled by fabric objective- measurement data. InKawabata S, Postle R, Niwa M, Objective measurement: applications to product design and process control: The Textile Machinery Society of Japan, Osaka, Japan, 175-181.

12. Nakanishi M, Niwa M (1989) Studies on the air permeability of clothing material (Part 1) Air resistances of clothing materials of the different end-uses. Journal of Home Economics of Japan 40: 797-804.

13. Ayada M, Niwa M (1990) the relations between silhouette of gathered skirt and mechanical properties of fabrics (part1). Journal of Home Economics of Japan 41: 313-320.

14. Niwa M, Nakanishi M, Ayada M, Kawabata S (1998) Optimum silhouette design for ladies' garments based on the mechanical properties of a fabric. Textile Research Journal 68: 578-588.

31

The Influence of Drying Regimes in Moisture of Raw Cotton and its Components

Kayumov Abdul-Malik Hamidovich*

Tashkent Institute of Textile and Light Industry, Tashkent, Uzbekistan

Abstract

Choosing the drying mode of cotton, from the point of view dehydration its components up to the normalized humidity is an actual problem. In this article the pattern of changing of the humidity of a fibre, seeds in various modes, and frequency rates of drying by which it will be possible to recommend optimum parameters of drying for uniform dehydration cotton and its components are determined.

Keywords: Cotton seed; Fibers; Drying; Regime; Dryer; Dry agent; Moisture

Introduction

In the drying and cleaning shop (DCS) of storage units, raw cotton with moisture content up to 19% is exposed to a single dried, with a humidity of 19-29% double dried, and more than 29% triple dried to reduce their moisture content to 14% for low and to 11% for the first commercial varieties [1].

For cleaning shops (CS) of cotton plants, recommended to bring the single drying moisture of raw cotton, to 7-8% for the first and 8-9% for low sorts respectively. Moreover, the initial moisture content of raw cotton, entering the CS, must not exceed the 14%.

These recommendations include cotton ginneries drying in a continuous chain to a moisture content of 7-10% depending on the sorts and cotton varieties. However, the recommendations are rigidly installed efficiency of raw cotton passing through the drying units, compliance with which is in practice not always possible. Cotton factories often operate with a capacity of less than recommended, and the temperature of the drying agent remains constant or decreases unnecessarily. It should be noted that the cotton-drying mode does not guarantee the fiber moisture within the norms [1], as fiber moisture has a significant impact on the quality of yarn. These facts have been confirmed in several studies [2-4].

Currently, the DCS and CS of cotton factories used two – 2SB-10 dryer or SBO (SBT) for the preparation of the cotton to the storage and processing. High quality cotton fiber and seeds can be obtained by rationally using and applying the optimum drying regimes in these dryers. Studies conducted on cotton factories showed that the connectivity option dryers significantly affect the quality of the fiber [5].

The results showed that for each drying embodiment same with the initial moisture content, depending on the embodiment dryer's connection achieved varying degrees of defects and littered content of fiber. This shows that the rationale choice version of drying with optimal values of temperature of drying agents depending on the initial moisture content of cotton can significantly improve the performance of the producing cotton fiber.

Moreover, the moisture content of fiber and seed for each embodiment of drying with the same of initial moisture content and moisture of raw cotton after drying, depending on the embodiment of dryers connection obtain different results. This shows that the connection option of dryers significantly affects the uniformity of drying. It is obvious that, in the different ways of drying, the temp of

heating of raw cotton is different as well as the intensity of the drying process depends on the heating temperature of raw cotton components.

In practice, single, double and parallel drying are widespread. However, even these options do not meet the requirements due to insufficient substantiation. It concerns, primarily using of different ways of drying at each initial moisture content, depending on the temperature of the drying agent and efficiency of skipping dried cotton to provide producing high quality fiber and seeds. It follows that, the important task is to choose the optimal variant of drying of modern cotton varieties depending on their initial moisture content.

Objects and Methods

The drying process, which depends on a number of factors, should be research under the production conditions using modern methods of planning and carrying out the experiment, which would allow finding solutions close to the optimal and minimal costs.

One of the closest methods to the target is a method of mathematical planning of the experiment [6]. For our research method, a full factorial experiment like FFE 2^3 is selected. Researches were conducted on 2SB-10 dryer when drying agent temperature T=100 ($-x_3$) and 200°C ($+x_3$), performance 3.5 (x_2) and 10 t/h ($+x_2$) for wet raw cotton. The object of the study was raw cotton varieties of C-6524, II of commercial sort, with initial moisture content W=10.5 ($-x_1$) and 22.3% ($+x_1$).

The experiments were at single, double, and triple drying.

Results and Discussion

The analysis of the regression equations shows that all the factors taken substantially affect the output parameters, either alone or in collaboration.

Mathematical processing of experimental results let gets the separate regression equations for each multiplicity of drying.

*Corresponding author: Kayumov Abdul-Malik Hamidovich, Tashkent Institute of Textile and Light Industry, Tashkent, Uzbekistan
E-mail: abdul-malik2017@mail.ru

We analyzed the regression equations to determine the effect of drying agent temperature and drying performance on moisture of raw cotton and its components can be written as follow:

for a single drying: for moisture of raw cotton: $Y_1=12,8+4,75x_1+0,97x_2-1,10x_3-0,47x_1x_3$; for moisture of fibre: $Y_2=7,98+3,07x_1+1,3x_2-1,66x_3+0,3x_1x_2-0,69x_1x_3$; for moisture of seed: $Y_3=14,76+5,46x_1+0,72x_2-0,74x_3-0,11x_1x_2-0,39x_1x_3$;

-for double drying: moisture of raw cotton: $Y_1=9,81+3,85x_1+1,54x_2-1,68x_3-0,66x_1x_3$; moisture of fiber: $Y_2=5,64+2,08x_1+1,25x_2-1,63x_3-0,58x_1x_3$; moisture of seeds: $Y_3=10,87+4,18x_1+1,53x_2-1,58x_3-0,62x_1x_3$;

-for a triple drying: for moisture of raw cotton: $Y_1=7,49+3,13x_1+2,1x_2-1,94x_3-0,75x_1x_3$; moisture of fiber: $Y_2=4,45+1,78x_1+1,51x_2-1,74x_3-0,67x_1x_3-0,34x_2x_3$; moisture of seeds: $Y_3=7,99+3,33x_1+2,14x_2-1,91x_3-0,77x_1x_3$.

To analyze these dependencies organized numerical calculation of output parameters for different values of the main factors. The results of numerical calculation after mathematical processing are presented in the form of graphs in Figures 1-7.

Analysis of the graphs shows that when single drying (Figures 1 and 2) with increasing the temperature of the drying agent and decreasing of performance on wet cotton, the moisture separation increases. The fiber intensively dried than seeds, so the fiber is in direct contact with the drying agent and heats up quickly.

After first multiplicity of drying of raw cotton and it is feed to the second multiplicity by pneumatic transport. Then the fibers are cooled, but the seed of coated fiber is not cooled. Therefore, generally seeds dried in the beginning of the second drum, the graphs show that the fibers are dried slowly in a low productivity than the raw cotton (Figures 3 and 4).

It was previously established that, with a lower productivity, the

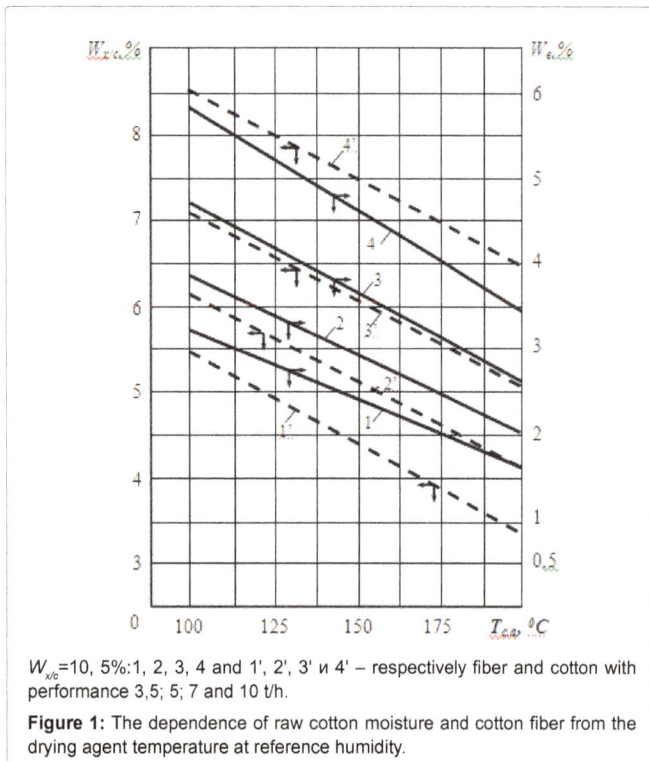

$W_{x/c}=10,5\%$:1, 2, 3, 4 and 1', 2', 3' и 4' – respectively fiber and cotton with performance 3,5; 5; 7 and 10 t/h.

Figure 1: The dependence of raw cotton moisture and cotton fiber from the drying agent temperature at reference humidity.

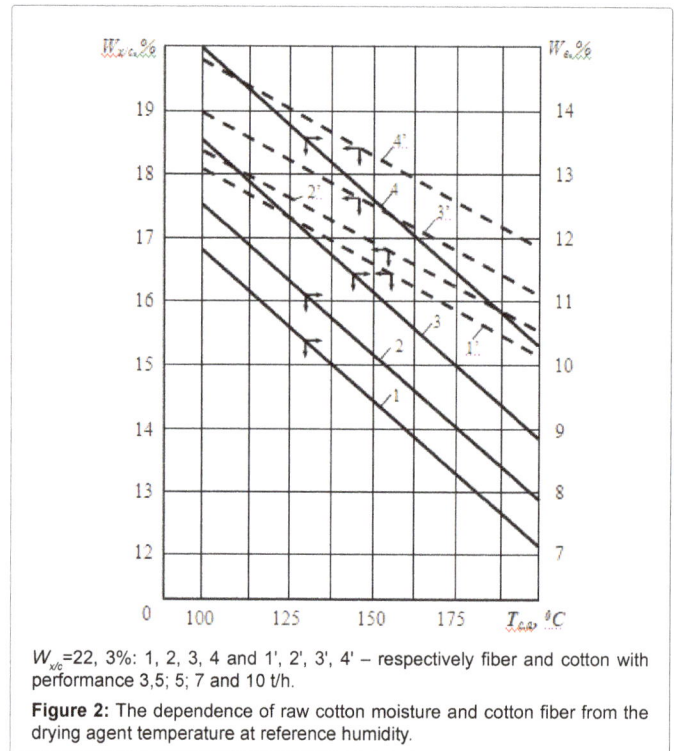

$W_{x/c}=22,3\%$: 1, 2, 3, 4 and 1', 2', 3', 4' – respectively fiber and cotton with performance 3,5; 5; 7 and 10 t/h.

Figure 2: The dependence of raw cotton moisture and cotton fiber from the drying agent temperature at reference humidity.

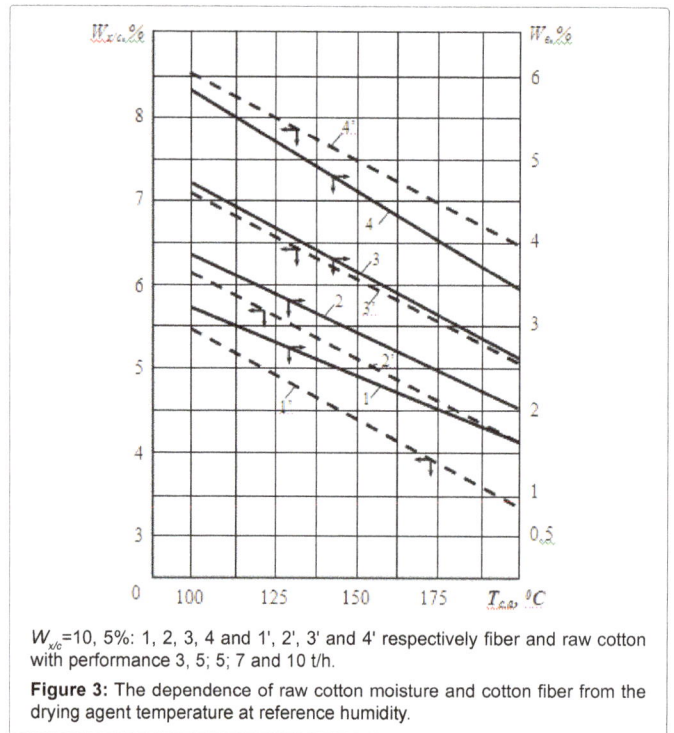

$W_{x/c}=10,5\%$: 1, 2, 3, 4 and 1', 2', 3' and 4' respectively fiber and raw cotton with performance 3, 5; 5; 7 and 10 t/h.

Figure 3: The dependence of raw cotton moisture and cotton fiber from the drying agent temperature at reference humidity.

heating of seeds is greater than at high fiber productivity, while cooling in pneumatic transportation; it is obvious that they are moistened and in the second drum are dried more slowly. In addition, with less moisture, the fibers do not have free moisture and, therefore, with a lower moisture content of the fiber, more heat is required. With high humidity, the fibers have more free moisture, therefore at a productivity of P=10 ton/h (Figures 3 and 4) are dried faster than raw cotton. A similar picture is observed when the cotton is triple dried (Figure 5).

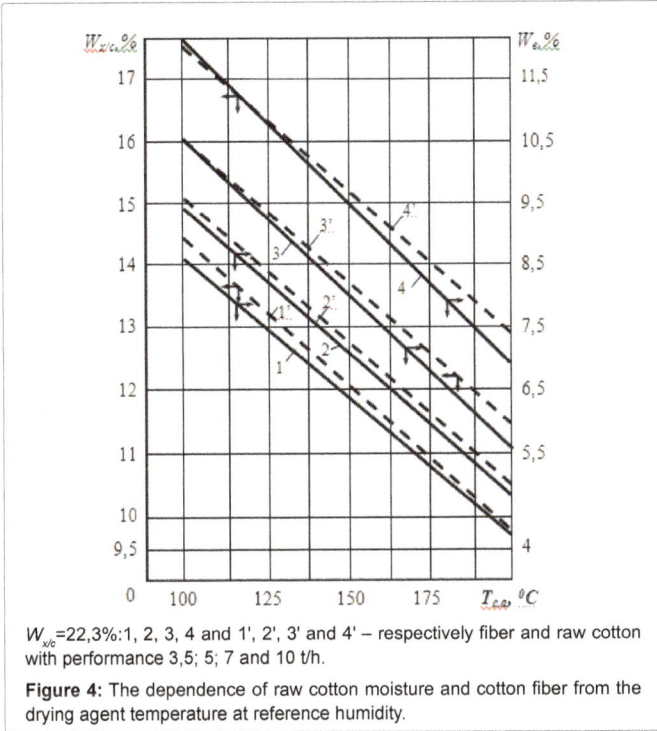

$W_{x/c}$=22,3%:1, 2, 3, 4 and 1', 2', 3' and 4' – respectively fiber and raw cotton with performance 3,5; 5; 7 and 10 t/h.

Figure 4: The dependence of raw cotton moisture and cotton fiber from the drying agent temperature at reference humidity.

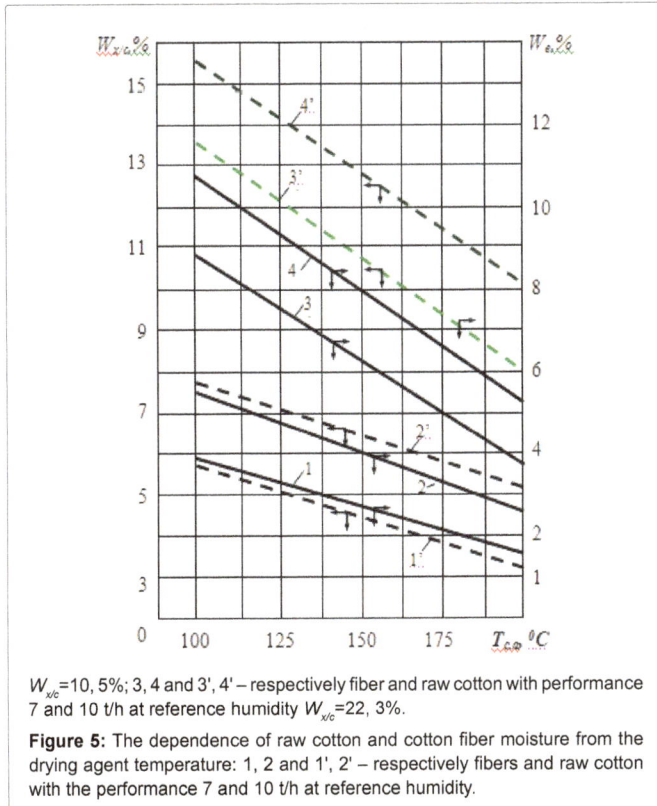

$W_{x/c}$=10, 5%: 1, 2, 3, 4 and 1', 2', 3', 4'- respectively with performance 10, 7, 5 and 3,5 t/h with single and double drying; 1", 2" – with triple drying, respectively with performance 10 and 7 t/h.

Figure 6: Dependence of seed moisture from the drying agent temperature at reference humidity.

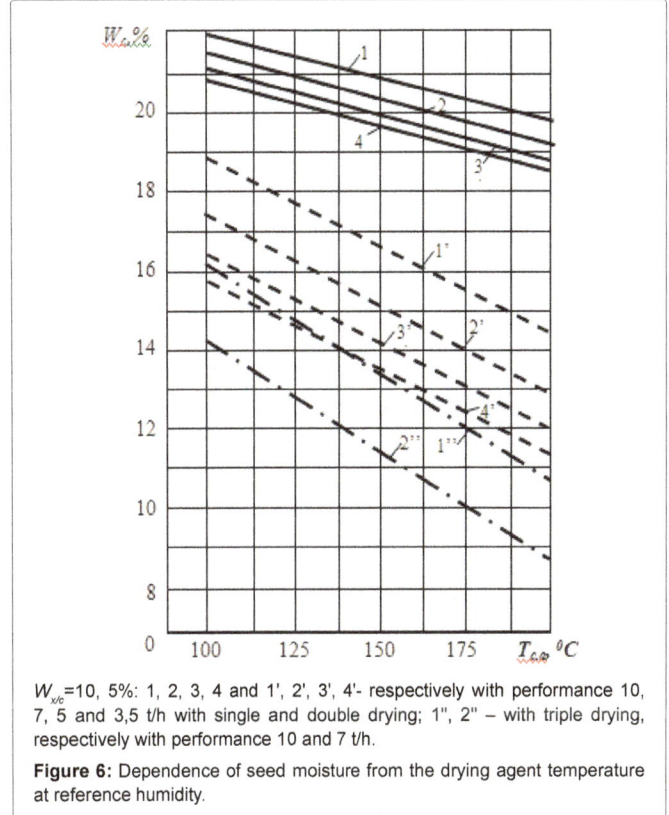

$W_{x/c}$=10, 5%; 3, 4 and 3', 4' – respectively fiber and raw cotton with performance 7 and 10 t/h at reference humidity $W_{x/c}$=22, 3%.

Figure 5: The dependence of raw cotton and cotton fiber moisture from the drying agent temperature: 1, 2 and 1', 2' – respectively fibers and raw cotton with the performance 7 and 10 t/h at reference humidity.

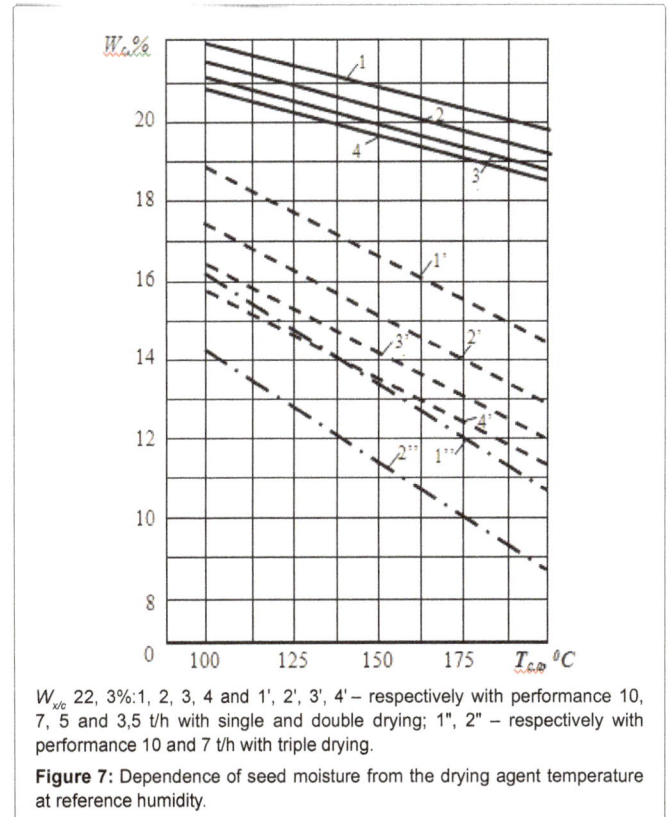

$W_{x/c}$ 22, 3%:1, 2, 3, 4 and 1', 2', 3', 4' – respectively with performance 10, 7, 5 and 3,5 t/h with single and double drying; 1", 2" – respectively with performance 10 and 7 t/h with triple drying.

Figure 7: Dependence of seed moisture from the drying agent temperature at reference humidity.

It can be seen from the graphs (Figures 6 and 7) that, with increasing frequency of drying, moisture removal from the seeds increases. This is explained by the fact that in the first fold of the drying the seeds are mainly heated, and the moisture extraction from the seeds is slower, i.e., the heating process is in progress. In subsequent drying processes (2-3-fold) there is an intensive drain of moisture from the seeds. It is known [7] that, for the intensification of the drying process it is necessary to heat the seeds, followed by drying the raw cotton in the drum dryers.

Conclusion

The regularities of the distribution of moisture between the components of raw cotton during sequential and parallel work of drum dryers are determined.

Moisture of fiber and seeds with each drying option at the same initial humidity and humidity of raw cotton after drying, depending on the option of connecting the dryers, different indicators are obtained.

The moisture content of raw cotton and its components after drying is affected to a greater extent by the initial moisture content of raw cotton, the higher the initial humidity, the higher the moisture content of raw cotton and its components after drying. With the increasing of drying rate, the influence of the initial moisture content of raw cotton on the parameters Y_1-Y_3 decreases.

For the moisture content of raw cotton, with a single drying, the regression coefficient is b_1=4.75; at a double drying 3.85, and with a triple drying 3.13. For the moisture content of the fiber (Y_2) with a single drying b_1=3.07; with a double drying 2.08, with a triple 1.78. For the moisture content of the seeds (V_3), with a single drying b_1=5.46, with a double drying 4.18, with a triple 3.33. The decrease in the regression coefficient b1 with the increase in the drying rate is explained by the fact that the residual moisture content of raw cotton and its components, which is the initial moisture for the subsequent multiplicity, gradually decreases, as cotton enters the subsequent dryers.

Also, humidity of raw cotton and its components after drying significantly influences the temperature of the drying agent. The higher temperature of the drying agent, the lower moisture content of raw cotton and its components after drying. However, the effect of the temperature of the drying agent on the moisture content of raw cotton and its components is not the same.

For example, in a single drying for fiber moisture, the regression coefficient is b3=1.66, for raw cotton moisture 1.1, and for seed moisture 0.74. This means that increasing the temperature of the drying agent drastically reduces the moisture content of the fiber, while the rate of moisture reduction of raw cotton and seeds is much lower.

With a double and triple drying of raw cotton, the regression coefficient b3 of the temperature of the drying agent is, respectively, 1.68 and 1.94 for the moisture content of the raw cotton; 1.58 and 1.91 for the moisture content of the seeds and 1.76 and 1.74 for the moisture content of the fiber. This shows that an increase in the rate of drying contributes to a more uniform drying of raw cotton components.

The efficiency of the drying drum for wet cotton also significantly affects the moisture content of raw cotton and its components after drying. From the regression equations (Y_1, Y_2 and Y_3), it can be seen that when the raw cotton in a single drying, the productivity of the drying drum on moisture of raw cotton significantly affects the moisture content of the fiber (b_2=1.3) than the moisture content of the raw cotton (b_2=0.97), and seeds (b_2=0.72). This means that the more moisture of the raw cotton in the drum, the greater the moisture content of the fiber. It follows that in order to avoid fiber drying, it is necessary to feed wet cotton raw material to drums with higher capacity. It is evident from the regression equations that when double drying b_2=1.54 for moisture content of the raw cotton, b_2=1.25 for the moisture content of the fiber.

The results can be used to develop optimal drying modes, depending on the initial moisture content of raw cotton, the performance of the dryer on wet cotton.

References

1. Sharma (2014) New trends in cotton ginning & cotton seed processing, Dhaka, Bangladesh.

2. Dimitrovsk K, Zupin Z (1981) Mechanical Properties of Fabrics from Cotton and Biodegradable Yarns Bamboo, SPF, PLA in Weft. Woven Fabric Engineering: 28-46.

3. Ochola J, Kisato J, Kinuthia L, Mwasiagi J, Waithaka A (2012) Study on the Influence of Fiber Properties on Yarn Imperfections in Ring Spun Yarns. Asian Journal of Textile 2: 32-43.

4. Ippolitov YY (1960) Influence of air and moisture parameters of raw cotton on the spinning process. Moscow.

5. Kakade RH, Das H, Ali S (2011) Performance evaluation of a double drum dryer for potato flake production. J Food Sci Technol 48: 432–439.

6. Sevostyanov AG (2007) Methods and means of researching the processes of the textile industry. MSTU after named AN Kosigin, Moscow.

7. Saidov S (1989) Increasing the efficiency of drum dryers using two-stage drying of raw cotton: PhD thesis. Tashkent.

32

UV Protection Properties of Cotton, Wool, Silk and Nylon Fabrics Dyed with Red Onion Peel, Madder and Chamomile Extracts

Gawish SM, Helmy HM*, Ramadan AN, Farouk R and Mashaly HM

Textile Research Division, National Research Centre, El-Buhouth st., Dokki, Cairo, Egypt

Abstract

Red onion peel, madder, chamomile and red onion/chamomile mixture (40 g/l, 50%wt) are extracted using aqueous solution at different temperatures (80-100°C). Cotton, wool, silk and nylon fabrics are dyed using a traditional method at pH 4 (except cotton at pH 8) using 25-40 g/l colorant extract, at specified temperature and time, without or with potash alum or $FeSO_4$ (1.5 g/l). Colorimetric data (L*, a*, b*), Color strength (K/S) and ultraviolet protection factor (UPF) are measured. Fastness properties of control and mordanted fabrics are assessed. UPF measurements reveal excellent UV protective properties for different fabrics colored with red onion, madder, chamomile and red onion peel/chamomile mixture (50% wt, 40 g/l) with or without mordants. Dyeing of fabrics without or with mordants gives excellent UPF factors (50+), except nylon which has an insufficient UPF with madder. These natural colored fabrics are suggested to prevent skin cancer and their colorants are recommended for textile coloration industry.

Keywords: Red onion peel; Madder; Chamomile; Dyeing; Fabrics; UV protection

Introduction

Recently, a great attention for application of natural colorants is survived for agriculture availability, ease and safe production. Synthetic colorants produce different shades which are available in low price, but cause environmental pollution, so natural dyes are a good alternative for textile coloration. Many researchers studied the functional finishing of textiles using natural dyes [1-4]. Also, extraction of colouring matter from Flos Sophorae and UPF values of silk were investigated [5].

Many researches have been conducted to determine the influence of the UV rays on different living organisms, particularly humans [6-14] as there is a strong correlation between skin cancer and UV dose. Hurtful UV radiation can be absorbed by Skin cells through sunlight, then the body gets red of these radiation by excretion. However, an extra dosage of UV radiation has some repercussions as it can cause skin cancer, damage of cells and inflammation of human skin which are considered the clear outcomes of erythema or sunburn [15].

Related work about UPF of natural colored yarns with mordants and its biological efficiency was reported [16]. Similar work about natural colorants and their application on selected fabrics, which provided antibacterial and ultraviolet protective level [17,18]. A study for colorimetric determination and antimicrobial analysis of colored fabrics using "Peony, Pomegranate, Clove, Coptis Chinenis and Gallnut extracts" was performed against different micro-organisms [19,20]. Fastness and UPF properties of colored fabrics using EucaLyptus were studied [21].

The following are some natural dyes which are studied in this article and their chemical structure (Tables 1-3)

Red onion peel: is a natural by-product waste from food industry. It gives bright reddish brown color in textile coloration. It is grown all over the world. The outer layer was known as a red onion peel which is a by-product waste of the food industry. The color of red onion peel extract is reddish brown. Previous studies on red onion peel [22,23] investigated the flavonoids which contain large amounts of flavonols, flavones, and anthocyanidins (flavylium cations dyes). The most known of the last compound is cyanidin dyes derivatives (Table 3).

Figure 1 shows the relationship of absorbance/wavelength of colorant at different pH was studied [24]. The absorbance of red onion colorant extract has two peaks; at 290 and 363 nm in UV region. The colorant shape and stability is similar in acidic medium and maximum absorption is achieved at wavelength 363 nm. At pH 8 the curve has different shapes and maximum absorption is achieved at wavelength 376 nm. High UV absorption of the colorant reveals that it could be extracted at the boil in acidic conditions. Figure 1 shows the colorant absorbance radiation in the UV-B region (290-320 nm) and UV-A region (320-400 nm) (Table 4).

Madder: It is an important natural colorant, as it is recognized in Egyptian pharaonic textile. Its constituents are anthraquinone compounds containing hydroxyl auxo-chromic groups [25]. Madder roots were reported to contain about 35 anthraquinone compounds (Table 4). The major components of Indian madder (Rubia cordifolia) are purpurin 66%, 1% munjistin and 10% nordamnacanthal. The major component of European madder (Rubia tinctoriun) is alizarin. The auxochromic groups (OH, COOH) in madder colorants are able to form complex compounds, so they are called mordant dyes depending on the appropriate groups which are capable of forming complex with the metal ion (Table 5).

Chamomile: It is an Egyptian crop [26,27]. The duration of cultivation is 1-2 years and the structure of chamomile is illustrated in Table 5. Terpenoids and flavonoids are found in the dried flowers of chamomile, which have a healing effect. Many human diseases can be cured by the chamomile preparations such as; insomnia, gastrointestinal disorder, inflammation, menstrual disorder, rheumatic pain, ulcers, muscle, spasms, hemorrhoids, wounds, and hay fever.

*Corresponding author: Hany M. Helmy, Textile Research Division, National Research Centre, El-Buhouth st., Dokki, Cairo, Egypt
E-mail: hany_helmy2001@yahoo.com

Plant	Botanic name	Colour index CI	Components of colors	Family
Red onion peel	Allium cepa L	Red, yellow or, orange	Red, yellow or orange colours	Flavonoid (flavonols flavones) and cyanidin
Madder genus (European)	Rubia tinctoriumL	Natural Red 8	Natural Red	Alizarin
Madder genus (Idian)	Rubia Cordifolia L	Natural Red 16	Natural Red	-Rubladin -Purpurin -Nardamna canthal - Lucidin
Chamomile	Composite	Yellow	Flavonoid	Anthemins tinctorial

Table 1: Characteristics of red onion peel, madder, and chamomile.

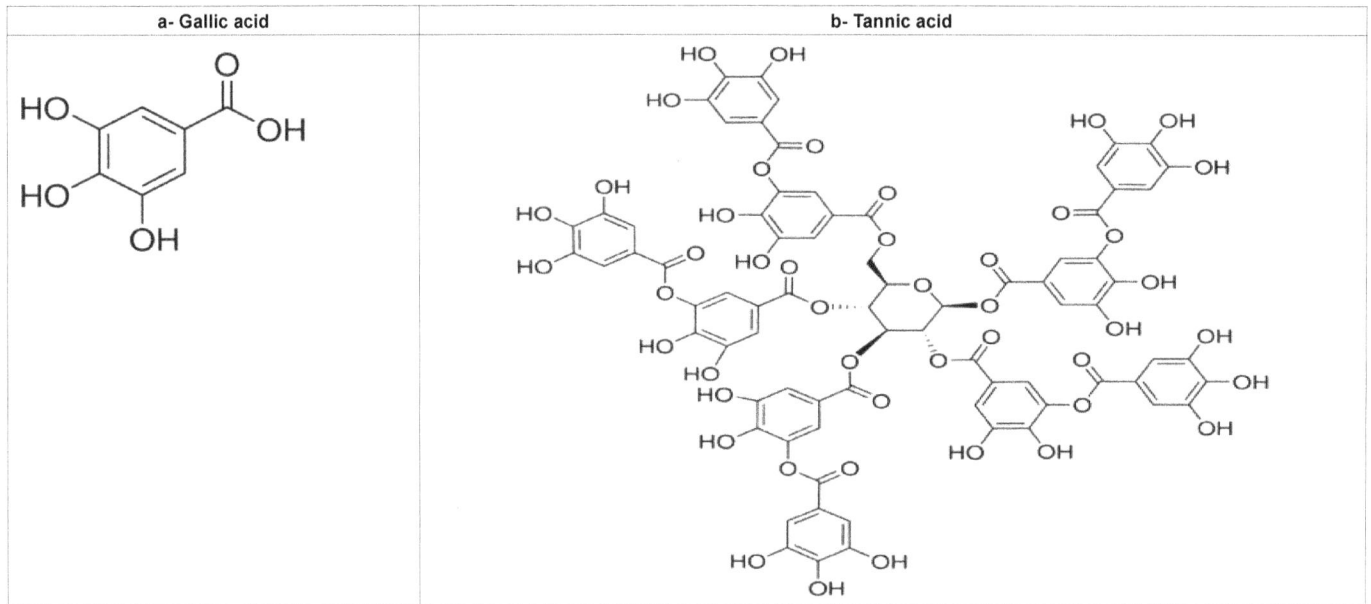

a- Gallic acid	b- Tannic acid

Table 2: Chemical structure of Tannic acid and Gallic acid (found in red onion peel extract, madder and chamomile).

Chemical structure of phenolic compounds of red onion peel				
Compounds	Structure		R_1	R_2
Flavonols		Quercetin	OH	H
		Kaempferol	H	H
		Myricetin	OH	OH
Flavones		Apigenin	H	
		Luteolin	OH	
Anthocyanidines		Cyanidin	H	OH
		Delphinidin	OH	OH
		Pelargonidin	H	H
		Peonidin	OH	OMe

Table 3: Chemical composition of red onion peel extract.

The oil chamomile is paramount as it can be used in a wide scale in cosmetics and aromatherapy [28].

In the present investigation, Red onion peel, madder, chamomile (25-40 g/l) and chamomile/red onion peel mixture (50% wt.) are extracted using aqueous solution at different temperatures (80-100°C). Then these extracts are used for coloration of different fabrics. UV protective properties of the colored fabrics are studied. The coloration of fabrics using natural colorants is performed without or with mordants (1.5 g/l) to minimize pollution. Color measurements, fastness properties, and UPF properties are also determined.

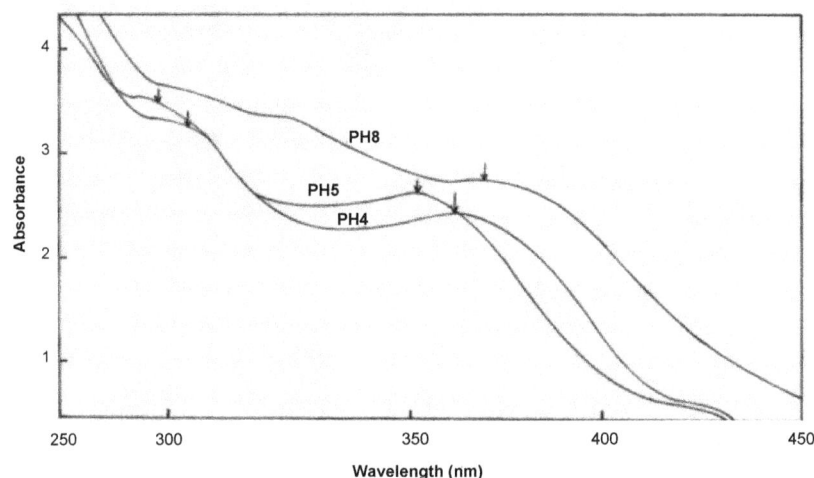

Figure 1: UV-VIS Spectrum of 12.5 gm/L of red onion peels extracted at 95°C and pH 4-8.

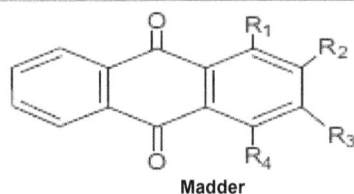

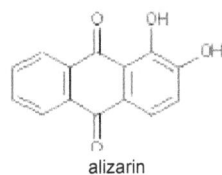

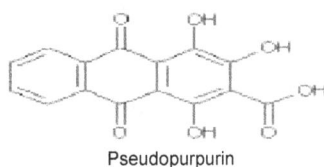

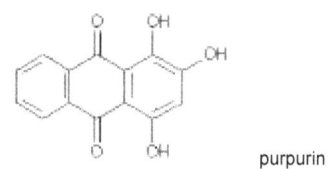

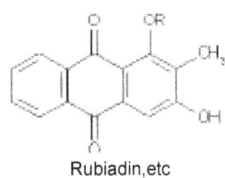

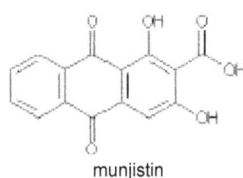

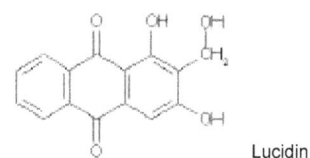

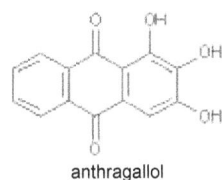

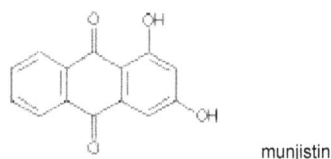

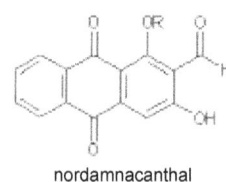

Table 4: Chemical structures of madder (anthraquinone components).

Materials and Methods

Materials

Natural dyes such as red onion peel extract were used, a vegetable by-product waste from the onion food industry, in addition to the imported madder and locally cultivated chamomile crop.

Chemical reagents: Chemical reagents used for adjusting the pH are glacial acetic acid and anhydrous sodium carbonates. The mordants used are potash alum ($Al_2 K_2(SO_4)_4.24H_2O$), ferrous sulphate ($Fe SO_4 7H_2O$) and Tannic acid for cotton fabric. Non-ionic detergent (Triton X-100) and sodium sulphate are used in the washing process.

Fabrics: Different natural fabrics are used such as cotton (weight 133.78 g/m², thickness 0.37 mm), wool (308.8 g/m², 0.55 mm), silk (58.08 g/m², 0.18 mm) and nylon fabric (89.14 g/m², 0.31 mm). All the fabrics were produced by The Misr Spinning and Weaving Company, El-Mahalla El-Kubra with the exception of silk.

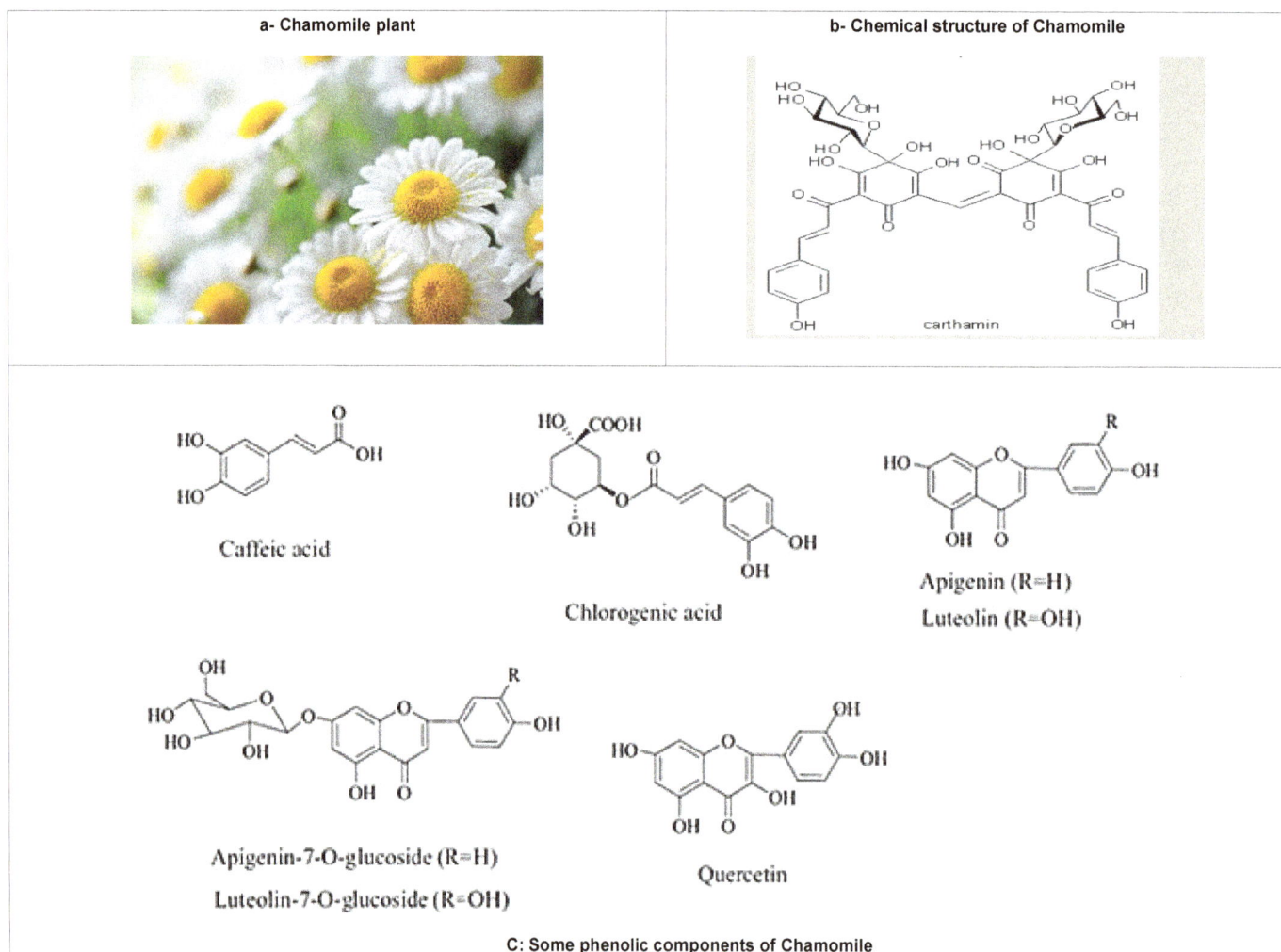

| a- Chamomile plant | b- Chemical structure of Chamomile |

carthamin

C: Some phenolic components of Chamomile

Caffeic acid

Chlorogenic acid

Apigenin (R=H)
Luteolin (R=OH)

Apigenin-7-O-glucoside (R=H)
Luteolin-7-O-glucoside (R=OH)

Quercetin

Table 5: Chamomile and some of its phenolic components.

Methods

Extraction Method: Red onion peel (40 g) was extracted in 1 L of boiled water for 1 h, and then the extract was cooled down to room temperature and filtered to get rid of insoluble materials. The percentage of extracted colorants was calculated and the absorbance of the colorant extract was measured to determine the λ_{max} of the colorant [29]. Also madder and chamomile were extracted in hot water at 80°C using different concentrations (25-40 g/l). Extraction percentage (%) of different colorants was as follow; red onion peel 15, madder 29, chamomile 37.

Scouring of fabrics:

- Silk fabric was added to a bath containing 2 g/l nonionic detergent (Triton X-100) and water at liquor ratio 1:50. Then the temperature of the entire bath was raised to 60°C and was maintained for 1 hr. Finally, Silk was rinsed with warm water and air dried.

- Wool and nylon fabrics were added to a bath containing 2 g/l nonionic detergent and 5 g/l sodium sulfate at liquor ratio 1:50. Then the temperature of the entire bath was raised to 45°C and the temperature remained constant for 30 min. Finally, wool and nylon were rinsed with warm water and air dried.

- Cotton fabric was added to a bath containing 2 g/L Na_2CO_3 and 5 g/l nonionic detergent at liquor ratio 1:50. Then the temperature of the entire bath was raised to boiling and was maintained for 1 h, then rinsed and air dried.

Pre-mordanting method: In the pre-mordanting method, fabrics were treated before dyeing process. Wool, silk, and nylon fabrics were added to a bath containing alum or $FeSO_4$ (1.5 g/l) at liquor ratio 1:50. Then the temperature of the entire bath was raised to 45°C and continues stirring at this temperature for 30 min. The mordanted fabrics were rinsed with warm water to remove the excess mordants.

Cotton fabrics were treated with 8% tannic or gallic acid (owf) at liquor ratio 1:30 then the temperature of the entire bath was raised to 90°C and hold for 30 min. Then cotton fabrics were removed from this bath and put in another bath containing alum or ferrous sulphate (1.5 g/l) at liquor ratio 1:50. Then the temperature of the entire bath was raised to 90°C and maintained for 30 min, rinsed and air dried (Table 1).

Dyeing method: Dyeing of cotton fabrics was carried out using exhaustion method utilizing infrared dyeing machine (Roaches Co., England) with natural colorant extract (25/40 g/l). Cotton fabric was pre-mordanted with 8% tannic acid (owf) and then treated with alum (1.5 g/l) at 100°C for 60 min. After extraction of natural colorant, it was added to the dyeing bath at pH 4 for all fabrics except cotton at pH 8, using liquor ratio 1:40 at a specified temperature (80°C for all fabrics and 100°C for cotton fabrics). The temperature of the entire bath was raised to 45°C and the maintained constant for 30 min. The dyeing procedure was started at 40°C, and then the temperature of the dyeing bath was raised to 100°C and was maintained for 1 h. Then, the fabric was thoroughly rinsed with warm water and soaped using 2 g/l nonionic detergent (Triton X-100) at 50°C for 30 min then rinsed with water and air dried.

Colorfastness to washing: The colorfastness to washing test was implemented according to ISO 105-CO2. It was carried out in a Launderometer (ATLS-Germany); the test specimen was compressed with the two adjacent fabrics (cotton and wool) in contact with the main sample. The test composite specimen was added to the bath containing 5 g/l nonionic detergent using liquor ratio 1:50 and the temperature of the bath was raised to 50°C and remained constant for 45 min. Then, the test composite specimen was removed from the bath, rinsed with water, squeezed, opened and air dried. The grayscale was used to assess the color change of the colored sample and the staining on the two adjacent undyed fabrics after washing [30].

Colorfastness to light: Colorfastness to light test was performed in a Ci3000+Xenon-Ometer® manufactured by Atlas Materials Testing Solutions. Two borosilicate filters were used to simulate outdoor conditions. Fabrics were exposed to UV for 35 hours according to ISO 105-B02 [31].

Colorfastness to rubbing: Colorfastness to rubbing test was performed according to test method ISO 105-X12, using a crockmeter for conducting the dry and wet fastness test. The gray scale was used to assess the staining of colored sample on dry and wet fabrics [32].

Color measurement

Color strength (K/S): The color strength K/S of dyed fabric was measured using a Datacolor Spectra Flash SF600X (Datacolor) and was assessed using Kubelka-Munk equation [33,34]:

$$K/S = \frac{(1-R)}{2R} - \frac{(1-RO)}{2RO}$$

Where:

R=Decimal fraction of the reflectance of the dyed fabric.

R0=Decimal fraction of the reflectance of the undyed fabric.

K=Absorption coefficient.

S=Scattering coefficient.

Color data CIE LAB space: The total difference CIE (L*, a*, b*) was measured using the Hunter-Lab spectrophotometer (model: Hunter Lab DP-9000).

CIE (L*, a*, b*) between two colours each given in terms of L*, a*, b* is calculated from:

L* value: indicates lightness, (+) if sample is lighter than standard, (-) if darker.

a* and b* values: indicate the relative positions in CIE Lab space of the sample and the standard, from which some indication of the nature of the difference can be seen.

Measurement of UPF factor: The ability of the dyed fabric to block UV light is given by the ultraviolet protection factor (UPF) value. The measurement of UPF was performed in UV/Visible Spectrophotometer 3101, using an integrating sphere loaded with the

fabric sample from 290 nm at an interval of 10 nm. The measurements of the UV-penetration characteristics of the compressed fabrics were carried out in the range of 290-400 nm using the UV penetration and protection measurements system. Before measurement, the fabric was conditioned at normal temperature and pressure (NTP) for 24 h. During the measurement, four scans were performed by rotating the sample 90° each time and the spectral data were recorded as the average of these four scans.

The equation used by the software to calculate the UPF value for a flat [35], tensionless dry fabric:

$$UPF = \frac{\sum_{290}^{400} E_ë \; S_ë \; \Delta_ë}{\sum_{290}^{400} E_ë S_ë \; T_ë \; \Delta_ë}$$

Equation for calculating UPF value   (1)

Where: (UPF)-ultraviolet protection factor value through fabrics, E_λ relative erythemal spectral effectiveness (W/m²nm⁻¹), S_λ solar spectral irradiance (Melbourne), $\Delta\lambda$-measured wavelength interval (nm), and T_λ spectral transmittance of the sample (%). The percentage blocking of UVA range (315-40 nm) and UVB range (315-290 nm) was calculated from the transmittance data.

Generally, in all the cases in which UPF reached at least the good protection level and UVA transmittance was also below 5%, which the European standard for Sun Protective Clothing [36] and also the Chinese National standard GB/T18830-2002 [5] consider the threshold above which photosensitive skin disorders, like chronic actinic dermatitis and solar urticaria, can be aggravated (Table 6) [37].

Results and Discussions

Effect of mordants on K/S and colour data of pre-mordanted dyed cotton, wool, silk and nylon fabrics with red onion peel, madder, chamomile, and mixture of red onion peel/chamomile extracts [25,29]

Effect of mordants on K/S and colour data of pre-mordanted dyed cotton, wool, silk and nylon fabrics with red onion peel extract: Table 7 shows that all mordanted dyed fabrics displayed higher color strength values as compared to non-mordanted ones with this order; FeSO₄>alum>control fabrics. This may be linked to the mordanting process that increases interaction between colorant and fabrics through coordination complex formation, which, after all, boosts the colorant uptake.

Table 7 indicates that the darkest color hue is achieved with FeSO₄ compared to mordanted fabric with alum and control fabric, which is clear from the decrease of L˙ values of dyed mordanted fabrics with FeSO₄. From all colorimetric data, the difference of color is due to the type of mordant used. In case of cotton, tannic acid was combined with alum/FeSO₄ for metal complexes formation, which produced a dark fabric shade.

Table 7 reveals the colorimetric data of the dyed fabrics, The largest

UPF range	Protection category	UVBE$_{eryt}$ transmittance (%)
<15	Insufficient protection	> 6.7
15-24	Good protection	6.7-4.2
25-39	Very good protection	4.1-2.6
40-50, 50+	Excellent protection	≤ 2.5

Table 6: UPF categories with relative transmittance and protection level.

Fabric	Mordant	Colour strength (K/S) (λ_{max}: 380)	L*	a*	b*	UPF range	UPF assessment	Transmittance (%)	
								UV-A 315-400 nm	UV-B 290-315 nm
Cotton	Dyed control + Tannic acid	3.21	66.41	7.41	17.14	30	(30+) Very good	3.8	3.18
	Tannic acid + Alum	3.81	63.43	7.26	24.69	34	(30+) Very good	3.47	2.8
	Tannic acid + FeSO$_4$	4.72	48.77	3.77	11.96	56	(50+) Excellent	2.09	1.75
Wool	Dyed control	28.62	35.60	22.41	25.62	2776	(50+) Excellent	0.04	0.04
	Alum	30.75	32.56	14.81	30.08				
	Ferrous sulphate	32.95	20.24	5.39	7.86	8338	(50+) Excellent	0.01	0.01
Nylon	Dyed control	25.35	48.90	12.55	43.19				
	Alum	25.80	49.82	11.30	37.98				
	Ferrous sulphate	26.34	44.19	11.49	27.77				
Silk	Dyed control	14.33	47.10	15.95	23.92	156	(50+) Excellent	0.75	0.59
	Alum	15.07	47.20	13.92	28.56				
	Ferrous sulphate	15.51	32.42	5.88	12.08	385	(50+) Excellent	0.29	0.25

Table 7: Effect of mordants on K/S and color data of cotton, wool, silk and nylon fabrics colored with colorants extracted from red onion extract (40 g/l).

Fabric	Mordant	Colour strength (K/S)	L*	a*	b*	UPF	UPF assessment	Transmittance (%)	
								UV-A 315-400 nm	UV-B 290-315 nm
a- madder (λ_{max}: 370)									
Wool	Dyed control	9.32	45.67	28.34	25.72	780	(50+) Excellent	0.25	0.11
	Alum	11.99	40.35	35.35	27.48	1422	(50+) Excellent	0.12	0.06
	FeSO$_4$	22.32	23.01	9.44	8.01	4395	(50+) Excellent	0.03	0.02
Silk	Dyed control	2.08	64.91	23.69	17.27	34	(30+) Very good	5.64	2.23
	Alum	3.93	54.34	34.21	24.18	77	(50+) Excellent	2.95	1
	FeSO$_4$	6.55	42.65	7.69	7.20	-		-	-
Nylon	Dyed control	9.18	50.38	42.57	26.55	13	(10+) Insufficient	9.56	7.23
	Alum	7.17	47.89	44.97	26.57	14	(10+) Insufficient	7.84	6.67
	FeSO$_4$	24.35	43.93	31.98	24.95	-		-	-
b- chamomile (λ_{max}: 365)									
Wool	Dyed control	15.74	65.39	3.44	26.96	294	(50+) Excellent	0.77	0.27
	Alum	17.45	59.68	6.3	39.44	-			
	FeSO$_4$	22.28	38.4	4.55	19.75	590	(50+) Excellent	0.26	0.15
Silk	Dyed control	3.16	79.51	0.47	19.52	-			
	Alum	3.75	75.9	2.77	80.16	-			
	FeSO$_4$	8.65	54.08	4.69	23.27	-			
Nylon	Dyed control	9.88	80.80	-1.45	15.47	-			
	Alum	10.21	72.92	-1.41	90.12	-			
	FeSO$_4$	12.32	65.14	8.38	27.78	-			

Table 8: Effect of mordants (1.5 g/l) on color strength and color data of wool, silk and nylon fabrics colored with colorants extracted from madder (25 g/L) and chamomile (40 g/L).

decrease of a* and b* values has occurred with FeSO$_4$ more than alum and control samples, that means a little shifted towards green co-ordinate in red yellow zone of CIE Lab color space.

Effect of mordants on K/S and colour data of pre-mordanted dyed cotton, wool, silk and nylon fabrics with madder extract: Table 8a illustrates the effect of mordants on dyed wool, silk and nylon fabrics using madder extract on K/S and color data.

There is a significant enhacment of K/S of pre-mordanted colored fabrics using alum and FeSO$_4$ as a result of metal complex formation among dye, fabric, and mordant.

From Table 8a, it is clear that pre-mordanted colored fabric with $FeSO_4$ gives the highest K/S among pre-mordanted colored fabrics with alum and control fabrics. This is attributed to the change of ferrous to ferric form through oxygen of the air. Ferrous and ferric forms are on the surface of the fabric and their spectra overlapped, resulting in change of K/S and consequently dark hue was obtained [24]. The color of fabrics is dependent on the mordant used.

Table 8a shows that adding mordants leads to a decrease of L^* values, that means darker shade is obtained. Pre-mordanted colored fabric with $FeSO_4$ gives the darkest shade among pre-mordanted colored fabrics with alum and control colored fabrics.

From Table 8a, it is clear that a^* and b^* values of pre-mordanted colored fabrics with alum increase compared to the control fabrics, that means a little shifted towards red co-ordinate in red yellow zone of CIE Lab color space. Conversely, a^* and b^* values of pre-mordanted colored fabrics with $FeSo_4$ decrease in comparison to control colored fabric, that means a little shifted towards green co-ordinate in red yellow zone of CIE Lab color space. The colored nylon fabric is not affected by the alum treatment in comparison to the control colored fabric.

Effect of mordants on K/S and colour data of pre-mordanted dyed cotton, wool, silk and nylon fabrics with Chamomile extract: Table 8b shows the results of mordanted dyed wool, silk and nylon fabrics with chamomile extract in comparison to non-mordanted dyed one. The effect of mordants on K/S and colorimetric data is demonstrated.

For wool fabric, K/S increases with $FeSO_4$ to 22.28 and for alum to 17.45 in comparison to control (15.74). For silk fabric, K/S increases with $FeSO_4$ to 8.65 and for alum to 3.75 in comparison to control (3.16). For nylon fabrics, K/S increases with $FeSO_4$ to 12.32 and to 10.21 using alum in comparison to control (9.88). Pre-mordanted colored fabrics with $FeSO_4$ gives the highest K/S, then Pre mordanted colored fabrics with alum and everyone one of them is higher than the colored control fabrics.

In general, the dyeing affinity of fabrics is dependent on the content of functional polar groups in fabrics. It is well known that the number of functional group in wool is larger than that of silk, and polarity of protein fibres is higher than that of cellulose fibre. The K/S value was in the order of wool>nylon>silk>cotton fabric for all natural colorants. It was found that this order of dyeing affinity matched the order of polarity/functional group content of fabrics very well.

For L^* values (lightness), the fabrics shade becomes darker by using

mordants and fabrics mordanted with $FeSO_4$ is darker than fabrics mordanted with alum and everyone of them is darker than the colored control fabrics.

Also, the increase of a^* and b^* values of mordanted colored fabric with alum, means a little shift towards red co-ordinate in red yellow zone of CIE Lab color space. in the opposite, the decrease of a^* and b^* values of mordanted colored fabric with $FeSO_4$, means a little shift towards green co-ordinate in red yellow zone of CIE Lab color space, except mordanted colored nylon fabric with $FeSO_4$, which shifted towards red co-ordinate.

Effect of mordants on K/S and colour data of pre-mordanted dyed cotton, wool, silk and nylon fabrics with mixture of red onion peel/chamomile extract: Table 9 shows the effect of mordants on color strength and color data of wool, silk and nylon fabrics colored with natural colorants extracted from a mixture of red onion peel/chamomile (50% wt).

Mordants have a slight effect on K/S of wool and silk fabrics. For nylon, the mordants don't improve the K/S compared to the colored control fabric. It is clear from Table 8 that the color of fabrics depends on the type of mordant used.

For L^* values (lightness), the shade of pre-mordanted wool and silk fabrics with $FeSO_4$ turns into darker shade than with alum. For nylon, a little change is observed by using $FeSO_4$ and alum.

For a^* and b^* values, a little shift towards red co-ordinate in red yellow zone of CIE Lab color space in case of pre-mordanted colored fabric with alum in comparison to colored control fabrics in case of wool and silk fabrics. While using $FeSO_4$ as a mordant leads to a little shift towards green co-ordinate in red yellow zone of CIE Lab color space in comparison to the colored control fabrics in case of wool and silk fabrics. However, using $FeSO_4$ and alum as mordants with colored nylon fabrics have a slight effect on the color compared to the control one (there is no remarkable shift of color as for wool and silk fabrics).

Colorfastness properties

Table 10 reveals that pretreatment with tannic acid/$FeSO_4$ and $FeSO_4$, improves colorfastness to washing of cotton. Mordants don't affect colorfastness to rubbing.

Also, Table 10 illustrates all colorfastness properties of colored control and pre-mordanted colored fabrics with red onion peel. From these results, it is obvious that colorfastness to washing and rubbing are fair to very good. While, colorfastness to light is very good to excellent.

Fabric	Mordant	Colour strength K/S (λ_{max}: 385)	L^*	a^*	b^*	UPF	UPF assessment	Transmittance (%)	
								UV-A 315-400 nm	UV-B 290-315 nm
Wool	Dyed control	26.23	43.19	20.03	26.55	2848	(50+) Excellent	0.04	0.03
	Alum	29.0	41.69	14.22	34.85	-	-	-	-
	$FeSO_4$	28.65	25.06	15.98	12.40	4294	(50+) Excellent	0.02	0.02
Silk	Dyed control	9.71	56.65	16.95	21.53	-	-	-	-
	Alum	9.35	55.13	10.24	34.48	-	-	-	-
	$FeSO_4$	13.59	39.03	4.17	16.62	196	(50+) Excellent	0.58	0.49
Nylon	Dyed control	25.12	51.22	10.82	31.08	-	-	-	-
	Alum	20.65	54.39	17.66	35.12	-	-	-	-
	$FeSO_4$	24.39	45.78	8.72	28.00	-	-	-	-

Table 9: Effect of mordants (1.5 g/l) on color strength and color data of wool, silk and nylon fabrics colored with colorants extracted from a mixture of red onion peel/chamomile (40 g/L, 50% wt).

Fabric	Mordant	Rubbing fastness		Washing fastness			Light fastness
		Dry	Wet	St*	St**	Alt.	
Cotton	Dyed control	3-4	4	3-4	4	4	5-6
	Tannic acid /alum	3	4	3-4	4	4	5-6
	Tannic acid/FeSO$_4$	3	4	3-4	4-5	4-5	5-6
	Alum	4-5	3	3	3	3-4	6
	FeSO$_4$	3	4	3-4	4-5	4-5	5-6
Wool	Dyed control	3-4	3-4	2-3	2-3	3-4	5-6
	Alum	2	1-2	3	3	4	6
	FeSO$_4$	1-2	2	3	3	3-4	6
Silk	Dyed control	3	3-4	3-4	3	4	6
	Alum	2-3	3	3-4	2-3	4	6
	FeSO$_4$	2	3	3-4	3	4	6
Nylon	Dyed control	4-5	3-4	2-3	2-3	3-4	6
	Alum	4-5	3	3	3	3-4	6
	FeSO$_4$	3	2-3	2-3	3	3-4	6

St.* staining on cotton
St. ** staining on wool
Alt. Alteration

Table 10: Colorfastness of cotton, wool, silk and nylon fabrics colored with colorants extracted from red onion peel.

Table 11a clarified the colorfastness properties of colored control and pre-mordanted colored fabrics with madder. Comparing colorfastness properties, it is clear that colorfastness to washing is fair to very good. Also, colorfastness to rubbing is fair to good. While, colorfastness to light is very good to excellent. It is clear that pre-treatment of fabrics using mordants enhances colorfastness to washing, except for FeSO$_4$/wool and FeSO$_4$/nylon. Colorfastness to Light and rubbing doesn't improve by mordants.

Table 11b illustrates colorfastness properties of colored control and pre-mordanted colored fabrics with chamomile. By comparing colorfastness properties, it is appear that, colorfastness to washing and to light is very good to excellent. Also, it is clear that pre-treatment with mordants improves colorfastness to light and to washing of all fabrics, but colorfastness to rubbing doesn't improve by using mordants.

Table 12 illustrates colorfastness properties of colored control and pre-mordanted colored fabrics with red onion/chamomile mixture (40 g/l, 50% wt). These results show that colorfastness to washing is good to very good and colorfastness to rubbing is fair to good. While, colorfastness to light is very good to excellent. Pre-treatment with mordants enhances colorfastness to light and washing, but colorfastness to rubbing is not affected by mordants.

UPF properties of colored fabrics

In this study the measurement of UPF based on UV protective properties for colorants namely; red onion peel, chamomile, madder and red onion peel/chamomile mixture (50% weight) of colored fabrics is investigated [5,35,37-40]. Coloration is performed without mordants (control) or using tannic acid (2.6 g/l for cotton) or with a low concentration of FeSO$_4$ or alum (1.5 g/L).

Figures 2-5 of UV transmittance Percentage of different colorants and Figure 6 of UPF and UV-A of all colored wool with different colorants show that wool fabrics colored with 4 colorants used, acquire excellent UPF values (50+) and the order of the UPF values are as follows, red onion peel>mixture of red onion peel/chamomile>madder>chamomile. Also, the addition of FeSO$_4$ as a mordant increases the values of UPF and show excellent UPF values (50+) with this order, red onion peel+FeSO$_4$>madder+FeSO$_4$>mixture

of red onion peel/chamomile+FeSO$_4$>chamomile+FeSO$_4$. Also, the values of T (UV-A) in both cases (mordanted and non-mordanted fabrics) are lower than 1%. Results of Tables 7-9 reveal that increasing K/S of colored wool fabrics with/without mordants accompanied by increasing UPF of these fabrics.

Accordingly, it can be concluded that wool fabrics are ranked as "excellent UV protection" after dyeing with or without a mordant because wool fabric has some properties such as; low porosity and high weight and thickness. Hence, wool fabric gives a high UPF by allowing the penetration of less UV. All wool fabrics colored with 4 colorants extracts without/with mordant (alum-FeSO$_4$) have much higher UPF values and lower T(UVA) and T(UVB).

Figures 2-5 of UV transmittance Percentage of different colorants and Figure 7 of UPF and UV-A of mordanted colored silk fabrics with different colorants using FeSO$_4$ and alum, acquire excellent UPF values

Fabric	Mordant	Rubbing fastness		Washing fastness			Light fastness
		Dry	Wet	Alt	St*	St**	
a- Madder							
Wool	Dyed control	3	2-3	3-4	3	2	6
	Alum	2	2	3-4	3-4	3-4	6
	FeSO$_4$	2	1-2	3-4	2-3	2	5-6
Silk	Dyed control	3-4	4-5	2-3	3-4	3	5
	Alum	2-3	3	3-4	4	3-4	5
	FeSO$_4$	2	3-4	3-4	4	4	5
Nylon	Dyed control	4-5	4	2-3	4	3	4-5
	Alum	2	2	3	4	3-4	4
	FeSO$_4$	2	2	3-4	3	3-4	4-5
b- Chamomile							
Wool	Dyed control	4-5	3-4	4	4-5	4	4-5
	Alum	3	2	5	5	5	6
	FeSO$_4$	2	2	5	4-5	5	6
Silk	Dyed control	4-5	5	4-5	4-5	4-5	5-6
	Alum	4	4-5	5	5	5	5-6
	FeSO$_4$	2-3	2	5	5	5	6
Nylon	Dyed control	5	4-5	4-5	4-5	4-5	5-6
	Alum	3-4	3	5	5	5	5
	FeSO$_4$	2	2	4-5	4-5	5	5

St.* staining on cotton St. ** staining on wool Alt. Alteration

Table 11: Colorfastness of wool, silk and nylon fabrics colored with colorants extracted from madder and chamomile.

Fabric	Mordant	Rubbing fastness		Washing fastness			Light fastness
		Dry	Wet	Alt	St*	St**	
Wool	Dyed control	3-4	3	4-5	3	3	6
	Alum	3	2-3	5	3-4	3	5-6
	FeSO$_4$	2	2	5	4	3-4	6
Silk	Dyed control	3-4	4	4	4	3-4	5
	Alum	3	3	5	5	3-4	5
	FeSO$_4$	2-3	3	5	5	4-5	6
nylon	Dyed control	4-5	4	4	3-4	3	4-5
	Alum	3-4	2-3	4-5	3-4	3-4	5
	FeSO$_4$	3	2-3	4	3-4	3-4	5-6

St.* staining on cotton

St. ** staining on wool

Alt. Alteration

Table 12: Colorfastness of wool, silk and nylon fabrics colored with colorants extracted from mixture of red onion/ chamomile (40g/l, 50% wt).

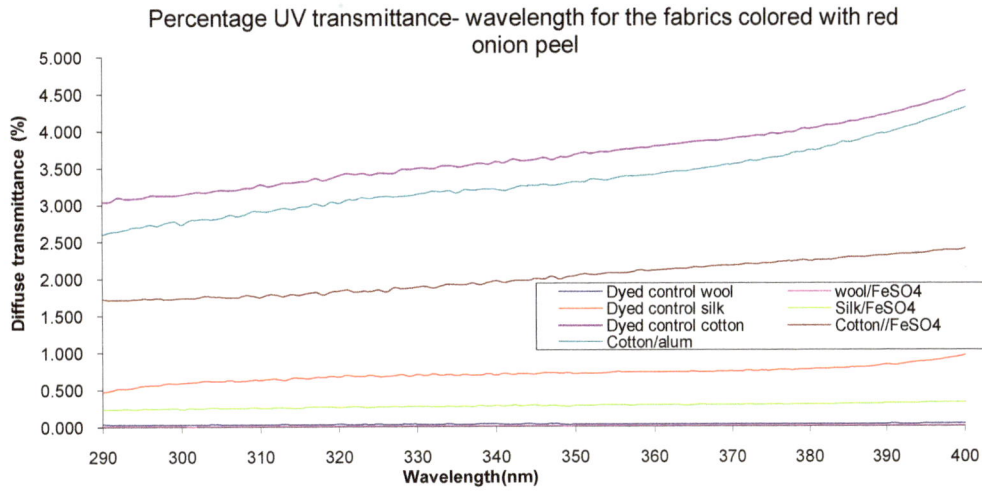

Figure 2: Percentage UV transmittance- wavelength for the fabrics colored with red onion peel.

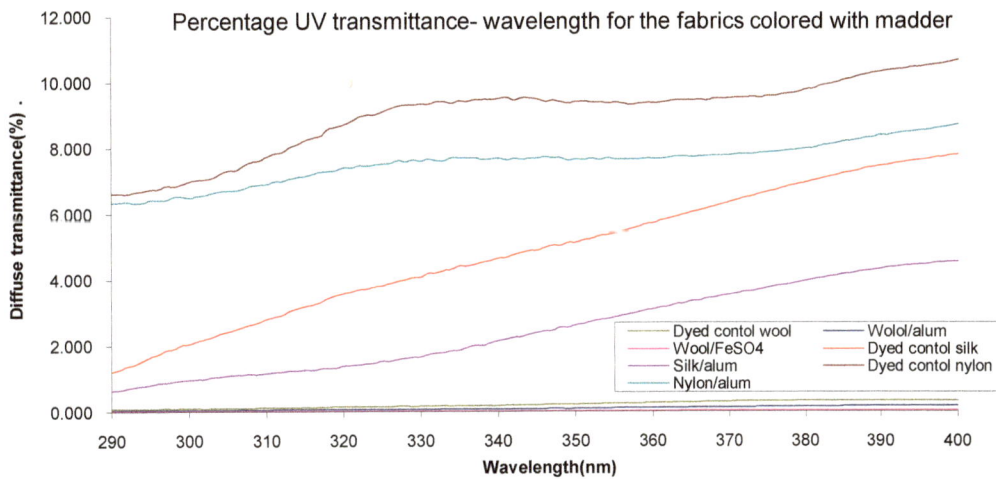

Figure 3: Percentage UV transmittance- wavelength for the fabrics colored with madder.

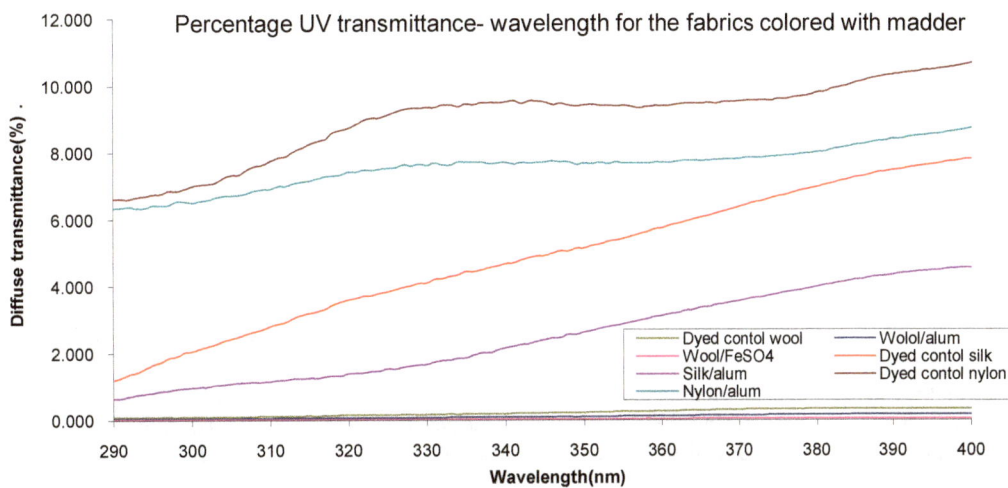

Figure 4: Percentage UV transmittance- wavelength for the fabrics colored with chamomile.

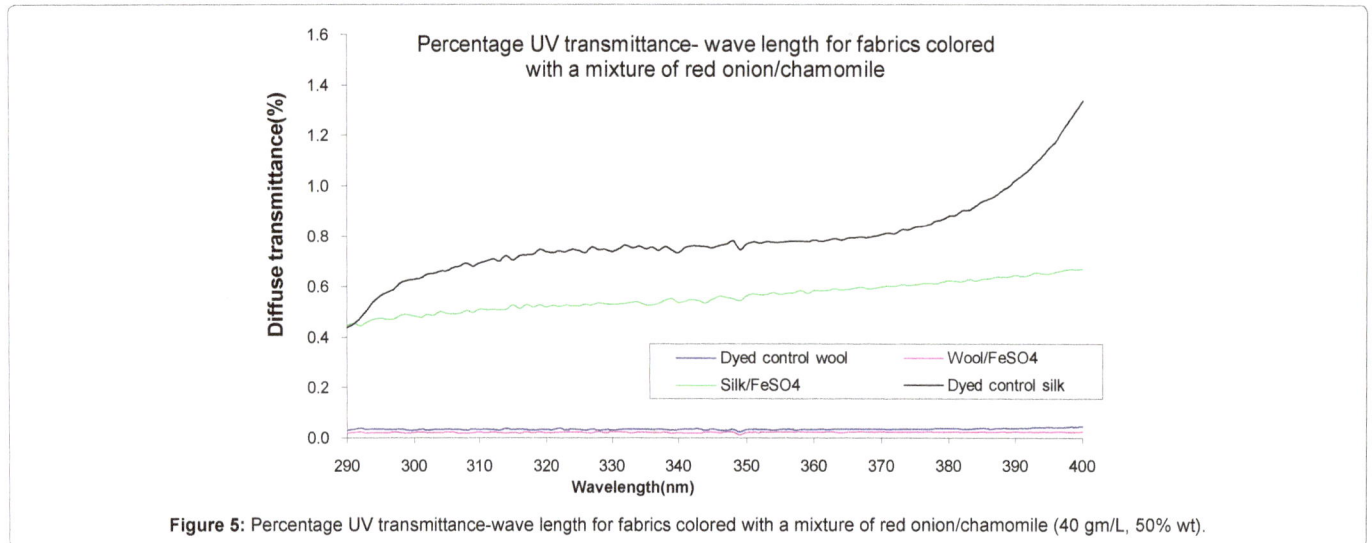

Figure 5: Percentage UV transmittance-wave length for fabrics colored with a mixture of red onion/chamomile (40 gm/L, 50% wt).

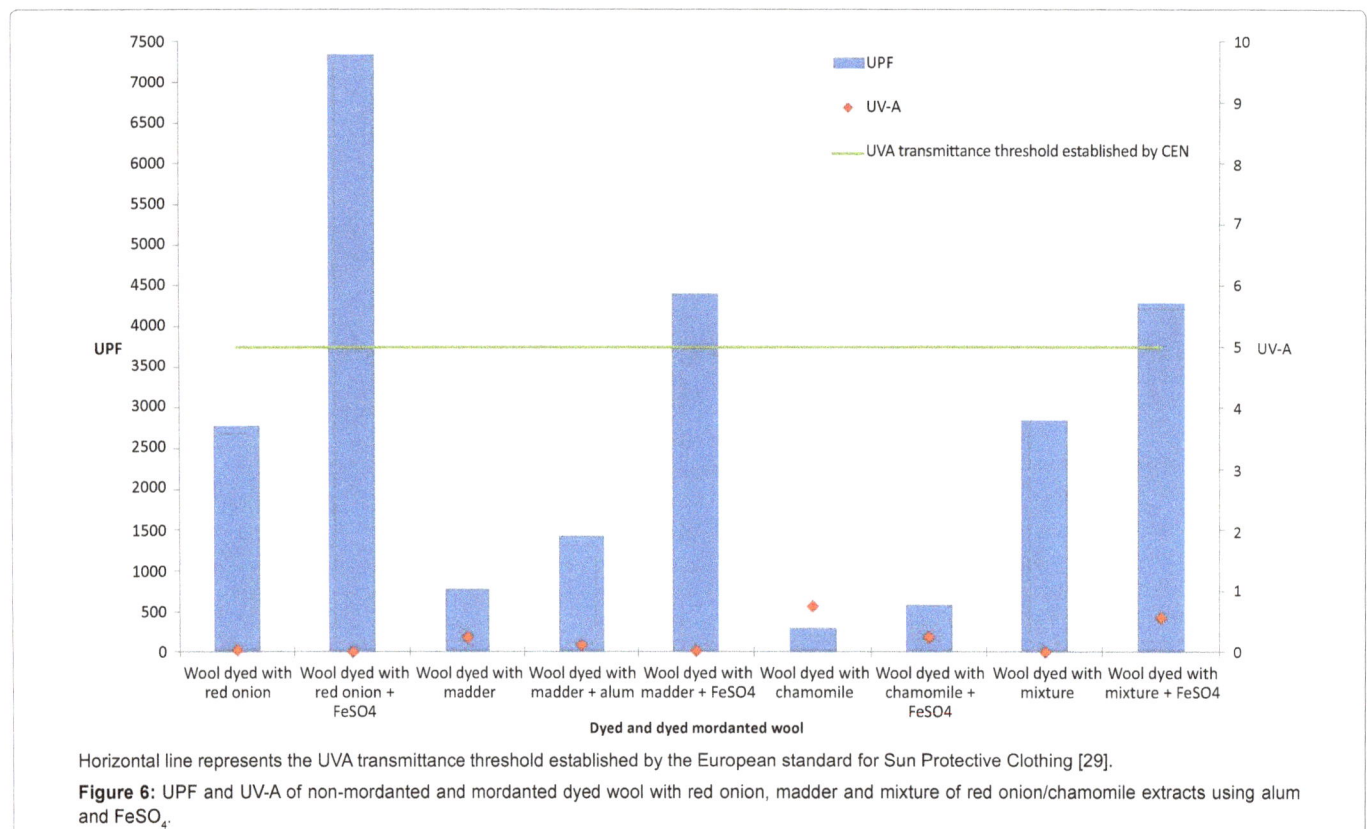

Horizontal line represents the UVA transmittance threshold established by the European standard for Sun Protective Clothing [29].

Figure 6: UPF and UV-A of non-mordanted and mordanted dyed wool with red onion, madder and mixture of red onion/chamomile extracts using alum and $FeSO_4$.

(50+) according to this order; red onion peel+$FeSO_4$>mixture of red onion peel/chamomile+$FeSO_4$>madder+alum and the values of T (UV-A) is lower than 5%.

From Figure 7, it is clear that different mordants have different impacts on the spectral transmittance of silk fabrics colored with 4 used colorants. Comparing with the colored silk fabric without mordant, the values of the spectral transmittance are reduced, but the UPF values are augmented by using mordants such as $FeSO_4$ and alum and show excellent UPF values.

In this experiment, the UPF of the fabrics colored with these natural

dyes was higher than 50 according to Figure 7, and the value of the T (UV-A) was lower than 5%, except with madder (without mordant).

Actually, when the UPF value of the colored silk fabric was higher than 50 and the value of the T (UV-A) was lower than 5%, the colored silk fabric was supposed to be a solar ultraviolet protector [40]. This means that the fabrics colored with natural dyes can greatly prevent the penetration of UVR. Consequently, these colored silk fabrics could effectively shield skin from sunlight UV radiation.

From Figures 2-5 of UV transmittance Percentage of different colorants and Figure 8 of UPF and UV-A of mordanted colored cotton

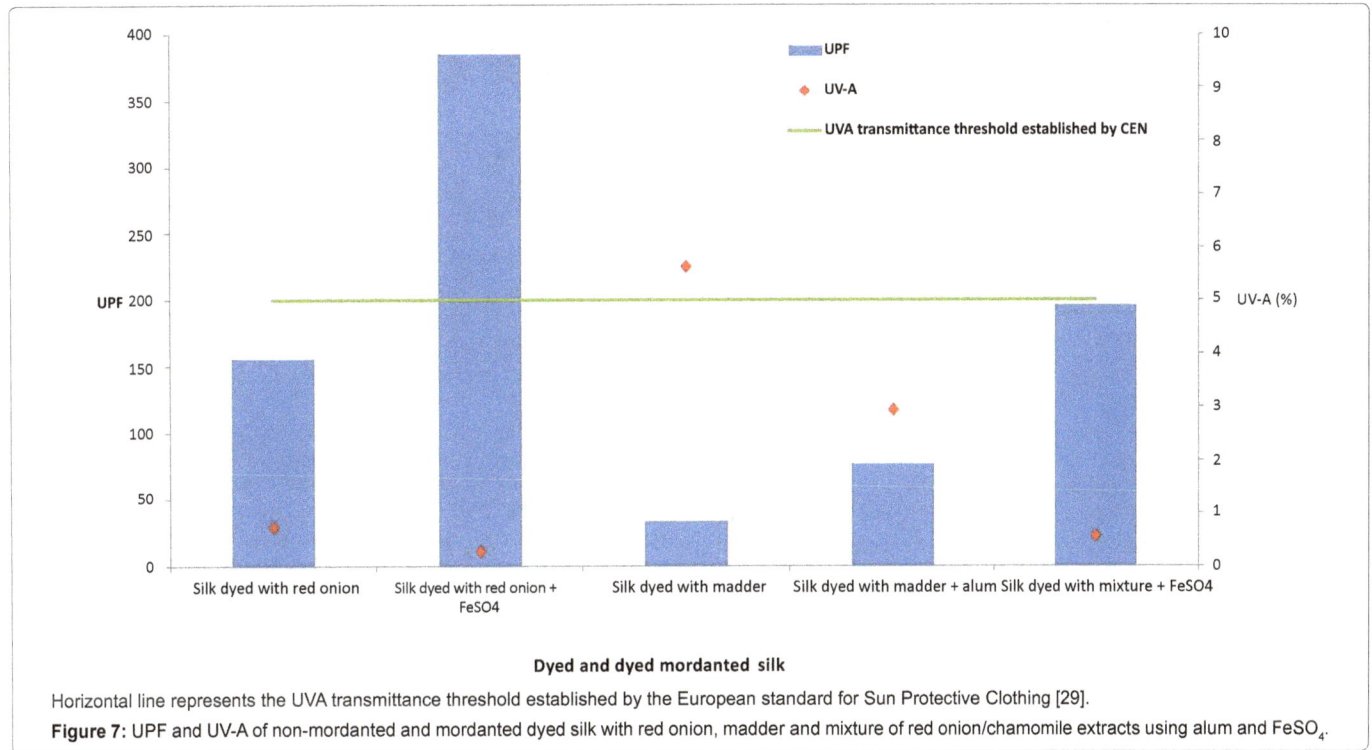

Horizontal line represents the UVA transmittance threshold established by the European standard for Sun Protective Clothing [29].

Figure 7: UPF and UV-A of non-mordanted and mordanted dyed silk with red onion, madder and mixture of red onion/chamomile extracts using alum and FeSO$_4$.

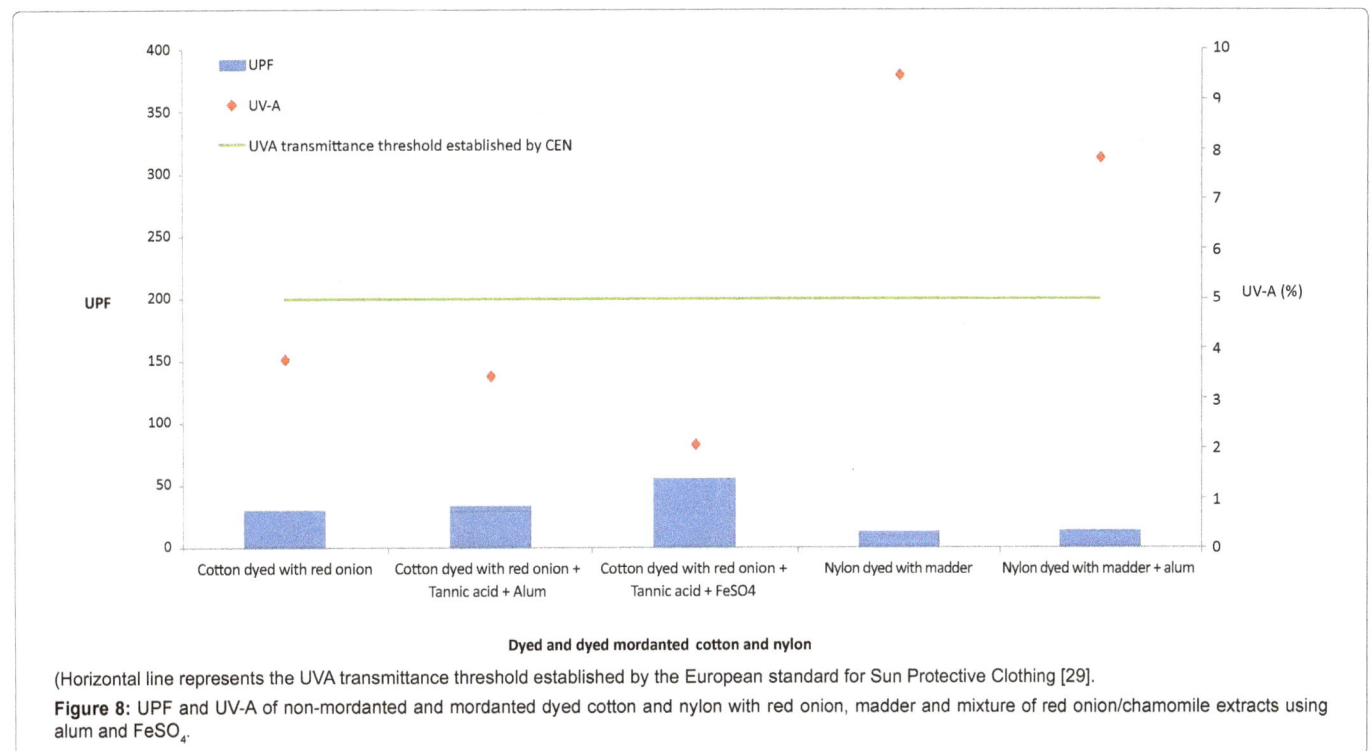

(Horizontal line represents the UVA transmittance threshold established by the European standard for Sun Protective Clothing [29].

Figure 8: UPF and UV-A of non-mordanted and mordanted dyed cotton and nylon with red onion, madder and mixture of red onion/chamomile extracts using alum and FeSO$_4$.

and nylon fabrics with different colorants using alum and FeSO$_4$, it is obvious that non-mordanted/mordanted colored cotton fabric acquire very good UPF values (50+) according to this order; red onion peel+tannic acid+FeSO$_4$>red onion peel+tannic acid+alum>red onion peel+tannic acid (without mordant) and the values of T (UV-A) is lower than 5%.

The tannin molecules contain functional groups which absorb UV radiation, in this way, the colored cotton fabric treated with tannin-demonstrates higher UV protection than fabric without tannic acid [40], as appeared by the higher UPF value and a lower T (UV-A) rate. Cotton fabrics indicated an additional enhancement of UV protection which could be accounted for high uptake of tannin because of formation of the complex with metallic salt (mordant). Tannin molecules which turned into the ligands in the metal tannate complex hold their capability to absorb UV and, alongside the metal ions from the mordant, the overall UV protection effectiveness of the cotton fabrics is assigned to them.

Conversely, it is clear from Table 8, Figures 2-5 and 8 that the colored nylon fabric with madder has insufficient UPF value (13) as well as with alum (14) and the values of T (UV-A) is higher than 5%, which mean that this fabric hasn't any protection against UV radiation (Table 8a and Figure 8).

This study is important to increase natural dyes cultivation in our country for domestic market in addition to produce UV protective apparel clothing and carpets (wool and silk), curtains, and blankets with beautiful colors.

Conclusions

The coloration of fabrics with colorants of red onion peel, madder, chamomile and red onion/chamomile mixture (50% wt) extracts without or with mordants have revealed excellent protective level UPF (50+) for fabrics with exception of nylon, or that mordanted with alum. Colored madder gives insufficient protection with nylon fabric. Control cotton colored with red onion peel and that mordanted with alum have very good protection .While cotton mordanted with $FeSO_4$ has an excellent one. Pretreatment of fabrics with mordants enhances colorfastness to washing and light. Mordanting of fabrics doesn't enhance wet and dry rubbing fastness. All colored fabrics acquire very good to excellent colorfastness to light, but colorfastness to washing is fair to very good. UPF calculations have shown excellent values for all colored control fabrics or mordanted fabrics according to this arrangement; wool fabrics>silk fabrics>cotton fabrics>nylon fabrics, which confirmed UV protection of fabrics colored with the 4 mentioned colorants. When the UPF value of the colored fabrics was higher than 50 and the value of the T (UV-A) was lower than 5%, then the colored fabrics with colorants such as red onion, madder, chamomile, red onion peel/chamomile mixture were supposed to be a solar ultraviolet protector. This further explained that the fabrics dyed with natural dyes strongly blocked UV radiation. Thus, these colored fabrics could efficiently protect skin from solar UV radiation. The above colorants are friendly environment and their colored fabrics are suggested to use in preventing skin cancer of humanity in the world.

Acknowledgment

The authors gratefully acknowledge the support of project no 9163 (2014), awarded by Scientific Technology and Development funds (STDF) - "Academy of Scientific Research and Technology", Egypt.

References

1. Vankar PS, hanker RS, Wijayapala S (2009) Dyeing of cotton, wool and silk with extract of Allium cepa. Pigm Resin Technol 38: 242-247.

2. Dumitrescu I, Visileanu E, iculescu MN (2004) Natural dyes obtained from plants and vegetable wastes. Colourage 51: 121-129.

3. Green CL (1995) Natural colorants and dyestuffs, Non-Wood Products and Energy Branch. Food and agriculture organization of the United Nations.

4. Gulrajani ML (2001) Present status of natural dyes. Indian J Fibre Text 26: 191-201.

5. Wang L, Wang N, Jia S, Zhou Q (2009) Research on Dyeing and Ultraviolet Protection of Silk Fabric Using Vegetable Dyes Extracted from Flos Sophorae. Textile Research Journal 79: 1402-1409.

6. Reinert G. Fuso F. Hilfiker R, Schmidt E (1997) UV Protecting Properties of Textile Fabrics and Their Improvement. AATCC Review 29: 31-43.

7. Dayal A, Aggarwal AK (1998) Textiles and UV Protection. Asian Textile Journal 9: 62-68.

8. Rupp J, Bohringer A, Yonenaga A, Hilden J (2001) Textiles for Protection Against Harmful Ultraviolet Radiation. International Textile Bulletin 6: 8-20.

9. Holme I (2003) UV Absorbers for Protection and Performance. International Dyer 4: 10-13.

10. Menter JM, Hatch KL (2003) Clothing as Solar Radiation Protection. Curr Probl Dermatol 31: 50-63.

11. Chriskis JI (1995) Sun Protection of Apparel Textiles. Paper presented at the Proceedings of 3rd Asian Textile Conference.

12. Bajaj P, Kothari VK, Ghosh SB (2000) Some Innovations in UV Protective Clothing. Indian J of Fibres and Textile Research 35: 315-329.

13. Gerber B, Mathys P, Moser M, Bressoud D, Fahrlander CB (2002) Ultraviolet Emission Spectra of sunbeds. Photochemistry and Photobiology 76: 664-664.

14. Desai AA (2003) Clothing that offer Protection against Ultraviolet Radiation. Textile Magazine 1: 77-79.

15. Baliarsingh S, Behera PC, Jena J, Das T, Das NB (2015) UV reflectance attributed direct correlation to colour strength and absorbance of natural dyed yarn with respect to mordant use and their potential antimicrobial efficacy. J Clean Prod 102: 485-492.

16. Gupta D, Jain A, Panwar S (2005) Anti-UV and anti-microbial properties of some natural dyes on cotton. Indian J Fibre Text 30: 190-195.

17. Deepti G, Ruchi (2005) UPFcharacteristics of natural dyes and textiles dyed with them. Colourage 54: 75-80.

18. Feng XX, Zhang LL, Chen JY, Zhang JC (2007) New insights into solar UV-protective properties of natural dye. J Clean Prod 15: 366-372.

19. Mongkholrattanasit R. Kryštůfek J, Wiener JMV (2011) Dyeing, Fastness, and UV Protection Properties of Silk and Wool Fabrics Dyed with Eucalyptus Leaf Extract by the Exhaustion Process. Fibres Text East Eur 19: 94-99.

20. Kanchana R, Fernandes A, Bhat B, Budkule S (2013) Dyeing of textiles with natural dyes - an eco-friendly approach. International Journal of ChemTech Research 5: 2102-2109.

21. Hussein A, Elhassaneen Y (2014) Natural dye from red onion skins and applied in dyeing cotton fabrics for the production of women's headwear resistance to ultraviolet radiation (UVR). J American Sci 10: 129-139.

22. Oancea S, Draghici O (2013) pH and thermal stability of anthocyanin-based optimised extracts of romanian red onion cultivars. Czech J Food Sci 31: 283-291.

23. Hwang EK, Lee YH, Kim HD (2008) Dyeing, fastness, and deodorizing properties of cotton, silk, and wool fabrics dyed with gardenia, coffee sludge, Cassia tora. L, and pomegranate extracts. Fibers Polym 9: 334-340.

24. Balazsy AT, Eastop D (1998) Chemical Principles of Textile Conservation. Routledge Series in Conservation and Museology, Routledge edn. Butter Worth and Heinemann Publishers NY, USA.

25. Vankar PS (2007) Handbook on natural dyes for industrial applications. National Institute of Industrial Research, New Delhi, India.

26. Srivastava M, Misra N, Singh O, Khanam Z (2011) Chamomile (Matricaria chamomilla L.): An overview.

27. Srivastava JK, Shankar E, Gupta S (2010) Chamomile: A herbal medicine of the past with bright future. Mol Med Report 3: 895-901.

28. Bechtold T, Mussak R (2009) Handbook of natural colorants. Wiley, Chichester, UK.

29. Tests for colour fastness: Colour Fastness to washing Test (1989).

30. Tests for colour fastness: Colour fastness to artificial light: Xenon arc fading lamp test (1988). Textiles.

31. Tests for colour fastness: Colour fastness to rubbing (1987). Textiles.

32. Alvarez J, Lipp-Symonowicz B (2003) Examination of the absorption properties of various fibres in relation to UV radiation. Autex Research Journal 3: 72-77.

33. Amirshahi SH, Pailthorpe MT (1994) Applying the Kubelka-Munk equation to explain the color of blends prepared from precolored fibers. Textile Research Journal 64: 357-364.

34. Gawish SM, Ramadan AM (2013) Effect of nano zinc oxide onto polypropylene properties yarns I-UV protective polypropylene/nano zinc oxide yarns using compounding melt extrusion and spinning technique. Int J Text Sc 2: 121-125.

35. Gambichler T, Laperre J, Hoffmann K (2006) The European standard for sun-protective clothing: EN 13758. Journal of the European Academy of Dermatology and Venereology 20: 125-130.

36. Krutmann J (2000) Ultraviolet A radiation-induced biological effects in human skin: relevance for photoaging and photodermatosis. Journal of Dermatological Science 23: S22-S26.

37. Grifoni D, Bacci L, Di Lonardo S, Pinelli P, Scardigli A, et al. (2014) UV protective properties of cotton and flax fabrics dyed with multifunctional plant extracts. Dyes Pigments 105: 89-96.

38. Hou X, Chen X, Cheng Y, Xu H, Chen L, et al. (2013) Dyeing and UV-protection properties of water extracts from orange peel. J Clean Prod 52: 410-419.

39. Grifoni D, Bacci L, Zipoli G, Albanese L, Sabatini F (2011) The role of natural dyes in the UV protection of fabrics made of vegetable fibres. Dyes and Pigments 91: 279-285.

40. Hoffmann K, Kesners P, Bader A, Avermaete A, Altmeyer P, et al. (2001) Repeatability of in vitro measurements of the ultraviolet protection factor (UPF) by spectrophotometry with automatic sampling. Skin Research and Technology 7: 223-226.

Permissions

The contributors of this book come from diverse backgrounds, making this book a truly international effort. This book will bring forth new frontiers with its revolutionizing research information and detailed analysis of the nascent developments around the world.

We would like to thank all the contributing authors for lending their expertise to make the book truly unique. They have played a crucial role in the development of this book. Without their invaluable contributions this book wouldn't have been possible. They have made vital efforts to compile up to date information on the varied aspects of this subject to make this book a valuable addition to the collection of many professionals and students.

This book was conceptualized with the vision of imparting up-to-date information and advanced data in this field. To ensure the same, a matchless editorial board was set up. Every individual on the board went through rigorous rounds of assessment to prove their worth. After which they invested a large part of their time researching and compiling the most relevant data for our readers.

The editorial board has been involved in producing this book since its inception. They have spent rigorous hours researching and exploring the diverse topics which have resulted in the successful publishing of this book. They have passed on their knowledge of decades through this book. To expedite this challenging task, the publisher supported the team at every step. A small team of assistant editors was also appointed to further simplify the editing procedure and attain best results for the readers.

Apart from the editorial board, the designing team has also invested a significant amount of their time in understanding the subject and creating the most relevant covers. They scrutinized every image to scout for the most suitable representation of the subject and create an appropriate cover for the book.

The publishing team has been an ardent support to the editorial, designing and production team. Their endless efforts to recruit the best for this project, has resulted in the accomplishment of this book. They are a veteran in the field of academics and their pool of knowledge is as vast as their experience in printing. Their expertise and guidance has proved useful at every step. Their uncompromising quality standards have made this book an exceptional effort. Their encouragement from time to time has been an inspiration for everyone.

The publisher and the editorial board hope that this book will prove to be a valuable piece of knowledge for researchers, students, practitioners and scholars across the globe.

List of Contributors

Kamel MM, Helmy HM, Meshaly HM and Abou-Okeil A
Textile Research Division, National Research Centre, Dokki, Cairo, Egypt

Rahman F
Department of Petroleum and Chemical Engineering, Faculty of Engineering, Institute Technology Brunei, Brunei

Md. Touhiduzzaman
Department of Fabric Manufacturing Engineering, Bangladesh University of Textiles, Dhaka, Bangladesh

Rashid KMM
Department of Yarn Manufacturing Engineering, Bangladesh University of Textiles, Dhaka, Bangladesh

Md. Syduzzaman
Department of Textile Engineering Management, Bangladesh University of Textiles, Dhaka, Bangladesh

Min Li, Yufei Chen, David Hinks and Nelson R Vinueza
Department of Textile Engineering, Chemistry and Science, North Carolina State University, Raleigh, NC 27695, USA

Bagwan AS and Pawar S
Mukesh Patel School Of Technology, Management and Engineering, Shirpur, Dhule, India

Policepatil R
Spentex Pvt. Ltd. Baramati, Pune, Maharashtra. India

Ghada Ali Abou-Nassif
Fashion Design Department, Design and Art Faculty, King Abdul Aziz University, Jeddah, Saudi Arabia

Cruz J, Sampaio S and Fangueiro R
Centre for Textile Science and Technology, University of Minho, Portugal

Asagekar SD
Textile and Engineering Institute, Ichalkaranji, Maharashtra, India

Chattopadhyay DP and Patel BH
Department of Textile Chemistry, Faculty of Technology and Engineering, The Maharaja Sayajirao University of Baroda, Vadodara, India

Alsalameh KA, Karnoub A, Najjar F, Alsaleh F and Boshi A
Department of Textile and Spinning, Faculty of Mechanical Engineering, University of Aleppo, Syria

Kumar S and Chatterjee K
Technological Institute of Textiles and Sciences Bhiwani, India 127021

Padhye R and Nayak R
School of Fashion and Textiles, RMIT University, Brunswick, Australia 3056

Parthiban M
Department of Fashion Technology, PSG College of Technology, Coimbatore, India

Thilagavathi G and Viju S
Department of Textile Technology, PSG College of Technology, Coimbatore, India

Perumalraj R
Bannari Amman Institute of Technology, Sathyamangalam, Erode, India

Douissa NB, Dridi-Dhaouadi S and Mhenni MF
Research Unity of Applied Chemistry and Environment, Department of Chemistry, Faculty of Sciences, University of Monastir, 5019 Tunisia

Carvalho C and Santos G
CIAUD, University of Lisbon, Portugal

Jablonski MR
Department of Civil Engineering and Mechanics, University of Wisconsin, Milwaukee, 3200 N. Cramer St. Milwaukee, WI 53211, USA

Ranicke HB and Reisel JR
Department of Mechanical Engineering, University of Wisconsin, Milwaukee, 3200 N. Cramer St., Milwaukee, WI 53211, USA

Qureshi A and Purohit H
Environmental Genomics Division (EGD), National Environmental Engineering Research Institute (NEERI), Nagpur India, Nehru Marg, Nagpur, 440020, India

Satyanarayana KG
Honorary Professor, Poornaprajna Institute of Scientific Research (PPISR), Sy. No. 167, Poornaprajnapura, Bidalur Post, Devanahalli, Bangalore-562 110, Karnataka, India

Saggiomo M, Gloy YS and Gries T
Institut für Textiltechnik (ITA) der RWTH Aachen University, Aachen, Germany

Sarwar Z, Azeem A, Munir U and Abid S
Department of Postgraduate Studies, National Textile University, Faisalabad, Pakistan

Patel S and Sharan M
Faculty of Family and Community Sciences, Department of Clothing and Textiles,The Maharaja Sayajirao University of Baroda, Vadodara, India

Chattopadhyay DP
Faculty of Technology and Engineering, Department of Textile Chemistry, The Maharaja Sayajirao University of Baroda, Vadodara, India

El Messiry M and EL Deeb R
Textile Engineering Department, Faculty of Engineering, Alexandria University, Egypt

Sumithra M
Department of Textile and Apparel Design, Bharathiar University, Coimbatore, India

Amutha R
Department of Costume Design and Fashion, PSG college of Arts and science, Coimbatore, India

Periyasamy AP
Department of Material Engineering, Technical University of Liberec, Studentska, 46117, Liberec, Czech Republic

Kandagor V, Prather D and Fogle J
The University of Tennessee, Knoxville, Tennessee, USA

Bhave R
University of Georgia, Athens, Georgia, USA

Bhat G
Oak Ridge National Laboratory, Oak Ridge, Tennessee, USA

da Silva de Santis SH and Franco B
University of Campinas, Campinas, Brazil

Marcicano CJPP
Univesity of São Paulo, Brazil

Biswas SK
Textile Physics Section, Department of Jute and Fibre Technology,Institute of Jute Technology, University of Calcutta 35, Ballygunge Circular Road, Calcutta, India

Fouda A
Textile Engineering Department, Faculty of Engineering, Mansoura University, Mansoura 35516, Egypt

Singh JP
Department of Textile Technology, U.P. Textile Technology Institute Kanpur, India

Jun Xiong
Department of Textile Sciences, Faculty of Human Ecology, University Manitoba, Winnipeg, Canada

Mashiur R
Department of Biosystems Engineering, Faculty of Agriculture and Food Sciences, University of Manitoba, Winnipeg, Canada

Hossain F, Li X and Zakaria
School of Chemistry and Chemical Engineering, Wuhan Textile University,Wuhan, Hubei, PR China

Asaduzzaman, Quan H and Miah MR
School of Chemistry and Chemical Engineering, Wuhan Textile University,Wuhan, Hubei, PR China Wuhan Textile University Graduate (Color Root) Workstation, Songzi, 315600, Wuhan, Hubei, PR China

Phoophat P and Sukigara S
Kyoto Institute of Technology, Kyoto Japan

Kayumov Abdul-Malik Hamidovich
Tashkent Institute of Textile and Light Industry, Tashkent, Uzbekistan

Gawish SM, Helmy HM, Ramadan AN, Farouk R and Mashaly HM
Textile Research Division, National Research Centre, El-Buhouth st., Dokki, Cairo, Egypt

Index